M000248152

# NEW YORK SUBWAYS

6909
5-23-22
1097

# NEW YORK SUBWAYS

**AN ILLUSTRATED HISTORY OF NEW YORK CITY'S TRANSIT CARS**

*Centennial Edition*

**Gene Sansone**

The Johns Hopkins University Press
*in association with the New York Transit Museum*
BALTIMORE AND LONDON

© 1997 MTA New York City Transit
All rights reserved.

Originally published by New York Transit Museum Press, 1997
Johns Hopkins edition, *Evolution of New York City Subways: An Illustrated History of New York City's Transit Cars, 1867–1997,* 2002; centennial edition, *New York Subways: An Illustrated History of New York City's Transit Cars,* 2004
Printed in the United States of America on acid-free paper

9 8 7 6 5 4 3 2 1

The Johns Hopkins University Press
2715 North Charles Street
Baltimore, Maryland 21218-4363
www.press.jhu.edu

Library of Congress Cataloging-in-Publication Data
Sansone, Gene.
New York subways : an illustrated history of New York City's transit cars / Gene Sansone—Centennial ed.
p. cm.
Rev. ed. of: Evolution of New York City subways. 2002.
ISBN 0-8018-7922-1
1. Subways—New York (State)—New York—History. 2. Local transit—New York (State)—New York—History. I. Sansone, Gene. Evolution of New York City subways. 2002. II. Title.
HE4491.N65S26 2004
388.4'2'097471—dc22
2003024796

A catalog record for this book is available from the British Library.

Photo credit for p. 175: Staten Island Car (TA interior).
Courtesy of Railroad Avenue Enterprises.

*To all the employees, contractors, suppliers, and consultants whose work, products, and services have made the New York City Subway the biggest and best in the world. Without the daily "miracle" of our subways moving more than 4.6 million passengers, New York City would not today be the capital of the entire world. This volume also honors the memory of Frank Scimone, one of the driving forces that helped to ensure successful completion of this work.*

# Contents

Foreword ix

List of Abbreviations xi

**Introduction 1**

**The Locomotive Era 15**

**Historic IRT Cars 27**

**Vintage BMT Cars 101**

**R Type New York City Cars 179**

Appendixes

- A. Subway Car Fundamentals 285
- B. Historic Subway Cars of the New York Transit Museum 289
- C. Specials 291
- D. The State-of-the-Art Car of 1974 299
- E. Redbird Reefings 303

Glossary 307

Index to Cars 317

A Few Words about This Edition 321

Metropolitan Transportation Authority 323

About the Author 325

**Technical Data**

***IRT Technical Data 329***

***BMT Technical Data 391***

***R Type Technical Data 431***

COLOR GALLERY FOLLOWS PAGE 168

# Foreword

*Top:*
IRT subway groundbreaking ceremonies (March 24, 1900).

*Bottom:*
IRT opening day (October 27, 1904).

If a visitor were to ask a New Yorker, "What's the fastest way to get to Lincoln Center?" or "How do I get to Wall Street to see the Stock Exchange?" nine times out of ten that New Yorker will tell the visitor to take the subway. From the historic buildings of Brooklyn Heights to the Theatre District of Manhattan to Van Cortlandt Park in The Bronx, New York City has evolved over the years and still continues to make history. One aspect of this city that has helped make it unique and like no other city in the world is its subway system.

New York City's forefathers saw the arrival of the first rapid transit railcar in the late 1800s. Made of wood and steel, the subway cars rattled into public service in 1904 carrying thousands of people. As the city grew, so too did the number of subway cars that carry millions of passengers on a daily basis.

Today's subway car is more durable and has a much longer life than its predecessors. Over the past one hundred years, leather straps have been replaced by stainless steel ergonomic bars, rattan-covered seats have changed into fiberglass, and metal fans have evolved into effective air-conditioning systems.

The first of the New Millennium Trains took center stage in the year 2000 and proved once again how unique New York's subway system is in its dedication to serving its passengers. The new trains showcased state-of-the-art technological engineering and worldwide award-winning design. Virtually every aspect of the subway car has been improved to serve a new century of subway riders.

This centennial edition, brought to us by Gene Sansone and the New York Transit Museum, illustrates an important era in New York's subway car history in terms of design, engineering, and technological achievement. It is more than a comprehensive record of the miracle of evolution. It illustrates that the subway car is the cornerstone of New York City's exceptional transit system.

I invite you as a reader to explore the subway cars of the past with respect, enjoy the cars of today for what they have to offer you as a passenger, and look to the future and imagine what the cars of tomorrow will be like.

PETER S. KALIKOW
Chairman, Metropolitan Transportation Authority

# Abbreviations

| | |
|---|---|
| A | ampere |
| A/C (AC) | air conditioning |
| AC | alternating current |
| ACF | American Car & Foundry Company |
| AH | ampere hour |
| AML | automatic motor type "L" (brake system) |
| AMUE | automatic motor universal electric |
| AP | all purpose (journal bearing) |
| ARRG'T | arrangement |
| ASC | American Steel Casting |
| ASF | American Steel Foundry |
| ATC | automatic train control |
| ATL | automatic train line |
| ATO | automatic train operation |
| ATP | automatic train protection |
| ATS | automatic train supervision |
| B&O | Baltimore and Ohio Railroad |
| BMT | Brooklyn–Manhattan Transit Corporation |
| BOWERS, DURE | Bowers, Dure & Company |
| BRADLEY | Osgood Bradley |
| BRILL | J. G. Brill Company |
| BRT | Brooklyn Rapid Transit Company |
| BU | Brooklyn Union |
| BUDD | The Budd Company |
| CBTC | communications-based train control |
| CEV | conductor emergency valve |
| CFM | cubic feet per minute |
| CHRIS | Christensen Engineering Company |

| | |
|---|---|
| CINCINNATI | Cincinnati Car Company |
| CONSOLIDATED | Consolidated Car Heating Company |
| CSF | cast steel frame |
| CYL | cylinder |
| DC | direct current |
| ELEC | electric |
| FAI | first article inspection |
| FIXT | fixture |
| GE | General Electric Company |
| GILBERT | Gilbert & Bush |
| GOH | general overhaul |
| GSC | General Steel Castings Corporation |
| GSI | General Steel Industries |
| HARLAN | Harlan & Hollingsworth |
| HEY | Heywood-Wakefield |
| HP | horsepower |
| HV | high voltage |
| HVAC | heating, ventilating, and air conditioning |
| IND | Independent Subway |
| IRT | Interborough Rapid Transit Company |
| JEWETT | Jewett Car Company |
| J. STEPHENSON/ STEPHENSON | John Stephenson Company |
| KHI | Kawasaki Heavy Industries |
| KMM | Kawasaki Motors Manufacturing |
| KRC | Kawasaki Rail Car |
| KW | kilowatt |
| LACONIA | Laconia Car Company |
| lbs | pounds |
| LFM | Locomotive Finished Material Company |
| LV | low voltage |
| MCM | motor cam magnetic |
| MDBF | mean distance between failures |
| MG | motor generator |
| MK | Morrison Knudsen |
| mph | miles per hour |
| mphps | miles per hour per second |
| MR | Midland Ross |
| MTA | Metropolitan Transportation Authority |
| MU | multiple-unit |
| MUDC | multiple-unit door control |

| | |
|---|---|
| NTTT | New Technology Test Train |
| NYAB | New York Air Brake Company |
| NYCT | New York City Transit |
| NYCTA | New York City Transit Authority |
| NYERR | New York Elevated Railroad |
| NYRT | New York Rapid Transit Corporation |
| PA | public address |
| PCC | Presidents' Conference Committee |
| PCM | pneumatic cam magnetic |
| PRESSED STEEL | Pressed Steel Car Company |
| R (plus contract no.) | rolling stock contract |
| RTO | rapid transit operation |
| SCM | simplified cam magnetic |
| SIR / SIRT | Staten Island Rapid Transit Railway Company |
| SIRTOA | Staten Island Rapid Transit Operating Authority |
| SMEE | straight air "ME" type valve self-lapping electric overlay |
| SMS | scheduled maintenance system |
| SOAC | State-of-the-Art Car |
| S/S | Stone Air (formerly Stone Safety Corporation) |
| STANDARD STEEL | Standard Steel Car Company |
| V | volt |
| VAC | volts, alternating current |
| VDC | volts, direct current |
| WABCO | Westinghouse Air Brake Company |
| WASON | Wason Manufacturing Company |
| WEST-AMRAIL | Westinghouse Amrail |
| WH | Westinghouse Electric Company/Corporation |
| WTB | Waugh |
| # | Location given as a reference number |
| [ ] | Quantity indicated within brackets |
| N/A | Not available |

# NEW YORK SUBWAYS

SHOP
BLEECKER ST. & BROADWAY
97

STORAGE.
39
E.E.RICHARDS
37

# INTRODUCTION

This book describes and illustrates all the passenger rolling stock that has ever operated in the New York City Transit rail system, known around the world as the New York City Subway. Although there are many interesting and well-written works on this subject, perhaps only this volume covers every piece of equipment involved in successfully moving the billions of passengers who have ridden the subway and elevated lines since the late nineteenth century.

The story of the subway clearly illuminates the social and economic history of the city itself. In the 1800s, as New York attempted to grow, it practically choked on its own population. Horse-drawn omnibuses competing on the dirt streets of the city for transport space could barely eke out an average speed of one mph. Living space surrounded the city's center, but the commuting time was impossible. Without elevators, the city could not build up—skyscrapers remained in the future.

The first street railway in the world was built in 1832 in New York City, and steel wheel on steel rail began to replace wagon wheel on muddy street to move people. But these rail systems also fell victim to street congestion: an 1852 report counted one horse car every fifteen seconds passing through the intersection of Chambers Street and Broadway. The city's population had nearly tripled from 200,000 people in 1825 to over half a million by 1850 and continued to grow. There was only one thing left to do—move transit off the streets.

One of the first efforts was Charles Harvey's elevated system, a single cable car operated by stationary steam engines. The system was demonstrated on December 7, 1867, on a structure built over Greenwich Street. By April 1871, it had reached Thirtieth Street and Ninth Avenue, and was operated with steam locomotives instead of cable.

*Top*:
Metropolitan Street Railway–Bleecker Street Horsecar Line (1917).

*Bottom:*
Charles T. Harvey testing the first elevated (1867).

Alfred Ely Beach, publisher of *Scientific American,* took transit underground. After being rebuffed by "Boss" William M. Tweed, who controlled most construction, Beach secretly built a twenty-two-passenger, air-propelled vehicle under Broadway in 1870. The luxurious system, complete with a fountain stocked with goldfish and a grand piano in the station, was built on the same principle as a peashooter, that is, it used air as a means of propulsion. Although successful for a short distance (reportedly 400,000 passengers took the 300-foot trip), the pneumatic system ran into both political and practical problems and never progressed past the demonstration stage.

### New York's First Subway Car

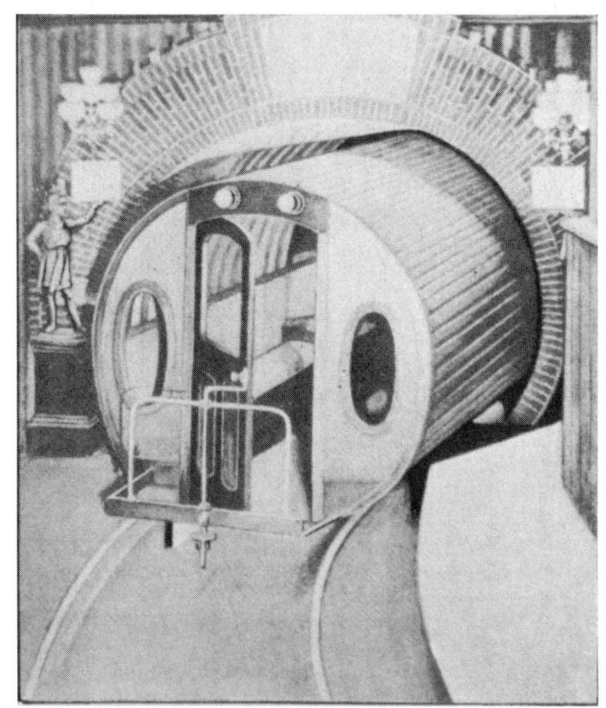

**The construction of New York's first elevated railroad (El) nearly coincided with the building of the first subway in the city. The Beach Pneumatic Transit, although an experimental project, demonstrated that people would accept the idea of riding under the streets of New York City. Construction began in 1869, and the tunnel opened to the public on February 26, 1870. Using air as a means of propulsion, it extended approximately a city block from Warren to Murray Streets under Broadway. At first, people were permitted to walk through the tunnel for a charge of twenty-five cents (for charity).**

**There was only one car built and it was placed in the tunnel in mid-1870. The car occupied the tube and shuttled from the one station to the end of the shaft and back. It had a wooden body with a metal underframe and had a door and two circular windows at each end. The dimensions of the cylindrical car were 18 feet long with about an 8-foot outside diameter. The seating was all longitudinal with places for eighteen to twenty-two passengers. Zircon lamps were used to light the car. This image illustrates how it appeared during operation.**

The first Sixth Avenue elevated train (1878).

Even though London had started its Underground in 1863, the image of riding through tunnels behind steam locomotives was not popular in New York, so the above-ground "elevated" system continued to grow. Structures were placed in the streets, and railcars began to move above the heads of New Yorkers in increasing numbers. With their own right of way, these steam-powered trains could reach the speeds necessary to make travel to the "suburbs" of Harlem practical. In 1883 a cable car line linked Brooklyn and Manhattan over the new Brooklyn Bridge and in 1885 the Brooklyn elevated lines began operation.

Meanwhile, some major milestones in the history of railroading were reached. George Westinghouse developed the automatic air brake in 1872. Next, in 1888, the first of Frank Julian Sprague's electric propulsion breakthroughs occurred when he demonstrated the first practical electrical street railway in Richmond, Virginia. Then, near the turn of the century, Sprague devised the multiple-unit control.

Two trains passing each other at Coenties Slip (ca. 1884).

With this invention, individually powered electric cars could be synchronized so they all accelerated and decelerated together, thus reducing station-to-station running times.

The modern electric rapid transit train was emerging, and, by mid-1903, all steam-powered elevated passenger trains in Manhattan (commonly called Els) had been converted to electric operation. As the Els took the "long-distance" commuters to and from the city's center, streetcars and commercial traffic found negotiating the streets easier and faster. At the turn of the century, New York had a viable system, but it was clear that more was needed. Financier August Belmont took up the challenge, and once again transit went under the streets. The original Interborough Rapid Transit (IRT) subway line, which opened October 27, 1904, started at City Hall, ran up the east side under Fourth Avenue and Park Avenue South, then turned west just south of Grand Central. After running a few blocks under Forty-second Street, it turned north again under Broadway, fi-

The South Ferry terminal of the Manhattan Elevated Lines (ca. 1910).

The first IRT subway inspection trip (1904).

nally terminating at 145th Street. From the Brooklyn Bridge station to Ninety-sixth Street there were four tracks with the inner two introducing true express service. While local trains slogged through nineteen stations between those two stops (including three now closed stations), express trains stopped only at three—Union Square, Grand Central, and Seventy-second Street.

Emboldened by the instant success of the original subway, the city entered into the "Dual Contracts" with the IRT and the Brooklyn Rapid Transit (BRT) (later Brooklyn–Manhattan Transit, or BMT). The IRT preferred cars 51 feet long and 8 feet 9 inches wide, because that was the size of the original subway tunnels, but all of the BRT and the portions of the IRT built under the Dual Contracts could handle the BRT cars, which were 67 feet long and 10 feet wide. The IRT simply extended the edges of its stations to reach the narrower cars. The Astoria Line was actually converted from IRT-size operation to BMT size by cutting back the platform edges after World War II, and, had the proposed Second Avenue Subway not been a victim of the city's financial crisis in the 1970s, the No. 2 and No. 5 Lines above 149th Street would have been converted as well.

As the twentieth century got under way, and for a considerable time afterward, the heavy rail fleet was an amazing amalgam of electric cars and converted steam coaches. The subway lines, however, started to standardize car appearance even though the propulsion, door, and brake systems varied considerably between car types.

The IRT was fairly staid, staying very close to its original 1904 design with little change until the 1938 World's Fair cars, which were single-ended with ogee roofs. The BRT began with a fairly revolutionary car as their Standard. Later known as B types, the 67-foot-long cars were all glass—even the door pockets and off-side front bulkheads had windows. These cars also had drop seats to add seating capacity in the non-rush hours at the expense of doorways, air-operated storm doors, and interior conductor controls. Roll signs and reverser-controlled running lights completed the advanced design. These cars were followed by the articulated D types and Multi-Sections, with a final experimental design called the Bluebirds.

The first IRT ten-car train (1911).

New 67-foot BRT cars on the Sea Beach Line (1915).

In the early days of subway service, the all-steel subway cars were joined underground by Composite (wood and steel construction) cars. The IRT had ordered Composite cars for the original service because the car building industry was not yet ready to produce steel cars in quantity. The IRT had removed their Composite cars from subway service, and after the horrific Malbone Street wreck of November 1, 1918, in which wood elevated cars operating in a tunnel were literally torn apart after they hit the tunnel walls, wood cars rarely appeared in subway service. This accident also resulted in the bankruptcy of the BRT, which was reorganized as the BMT Corporation.

The new Independent Subway (IND) Standard (known as the R1 or R1–R9) came on the scene in the 1930s, and was a shorter, conservative version of the BMT Standard. It improved passenger loading with four pairs of undivided double-leaf door openings (compared to three on the BMT Standard with center posts) and continued other BMT efforts such as end destination and route signs. But the IND R1 reflected a retreat in other areas, such as dropping the reverser-controlled running lights.

The first Triplex (1925).

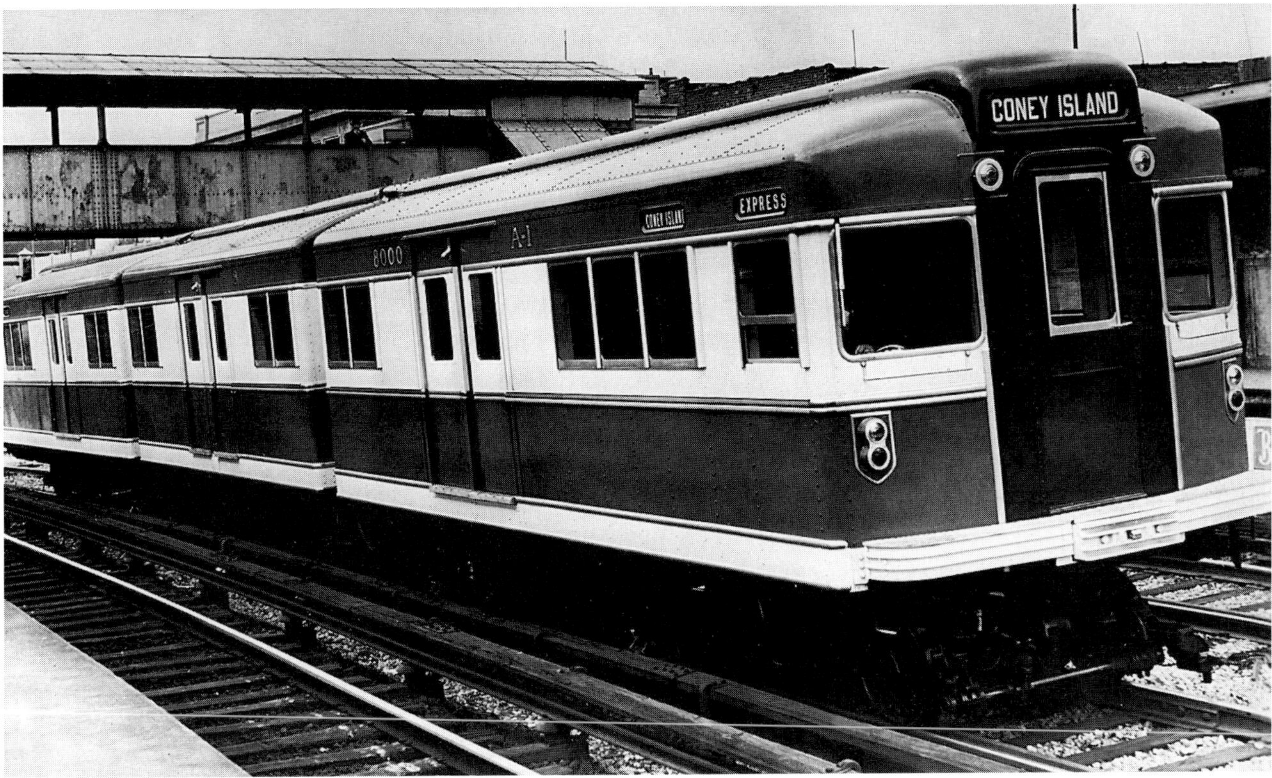


The new Bluebird on the Sea Beach Line (1939).

Whether imaginative or dull and predictable, the IRT and BMT were basically operating companies, running a rail system whose construction costs were paid for by the city. The two companies supplied tracks, signals, rolling stock, and power. The city undertook more construction after the Dual Contracts ended, but when the first of these new lines was ready, the country was in the grip of the Great Depression. Squeezed by rising operating costs and a political situation that made it impossible to hike fares, the IRT and BMT refused to accept stewardship of the new line. With no other entity stepping forward, New York City found itself in the subway business and established the IND Lines. But the IND system was itself a victim of the Great Depression. In the grand master scheme, no city resident would be more than a half mile from a subway, but funding ran out very early. Nonetheless, if one knows where to look, the remnants of construction provisions for this elaborate network are abundantly evident, even today.

After World War II, abandonment of the elevated lines increased in pace and the wooden elevated cars disappeared. A new type of subway car debuted. Starting with the R10 cars, fluorescent interior lighting and new braking systems with "dynamic" brakes were introduced. The cars went through a number of evolutionary changes in appearance and systems as experience was gained. (The exception was the "revolutionary" ten-car R11 fleet—the prototype for a never-ordered 400-car fleet—which had significantly different and advanced concept systems.) R10/R12/R14 cars had ogee roofs, but starting with the R15 cars, the roof became an arc (and remains so even today). Despite differences in window shape, construction material, and interior seating, R15–R38 cars were similar in appearance.

As R cars continued to be produced, significant changes were made. Door controls were moved from outside the cars to the interior. Public address (PA) systems were added and propulsion systems became more sophisticated. Cars went from single, independent cars to married pairs, to the A-B style of the R44 and R46 classes. With the R40, car shape changed significantly once again; car length of the R44s increased to 75 feet, which should have resulted in decreased maintenance costs since eight cars did the job of ten. The R44 and R46 cars were ordered to run on a line that never made it into existence: the Second Avenue Subway and high-speed (up to 70 mph) lines connecting Manhattan and Queens. In the late 1960s, New York City prepared to build the long-delayed replacement for the Second Avenue El. Work started, but the city's fiscal crisis in the 1970s killed it once more. The R44 and R46 cars had been equipped with cab sig-

R1s being delivered at 207th Street Yard (1931).

R10 at 207th Street Yard (1948).

R46 (1975).

R62s being tested on the Flushing Line at Thirty-third (Rawson) Street (1983).

naling and automatic train operation (ATO) for operation on the new subway. With the cancellation of construction, these features were never used and they were later made obsolete by advances in the state-of-the-art. Thus, the last vestiges of this capability were removed during the retrofit and overhaul of these car classes.

Improvements were not limited to new car orders. Existing cars were extensively overhauled, and propulsion systems, air-conditioning systems, brakes, and the like were upgraded or replaced entirely. Fleets were refitted with improved door safety systems and new passenger amenities during in-house and outside overhaul projects.

In this third century of high-tech rail service in New York City, the pace of change could be as radical as the change from steam and cable to electric at the turn of the last century. Mindful of the groundwork that needs to be laid, the New Technology Test Trains (NTTTs—the R110s) came on the scene in 1992 to test new technology before incorporating it into fleets of new cars, such as the R160s—a contract in progress as of this writing.

R110As and R110Bs on the Smith and Ninth Streets curve (1993).

IRT Steinway cars at Grand Central on the Flushing Line (ca. 1940s).

What we attempt to do in this book is take a car's-eye view of the voyage from steam and cable to modern electrically propelled cars. We fill in the information gaps where possible, but some voids and conflicting information remain because the operators of the time seem to have been more interested in making history than in recording it. The railscape of the city has changed drastically from that first cable car above Greenwich Street 136 years ago, but the purpose of the New York City transit system has remained unchanged: to move large numbers of people to and from their places of work and centers of entertainment and leisure both quickly and safely. When we do our job right, tens of thousands of employees make sure that more than 6,000 railcars—weighing in at a half billion pounds and making over 8,000 trips a day on 650-plus miles of main line track and stopping at over 450 stations—blend right into the cityscape, as unnoticed and unremarkable as streetlight poles. What better compliment could we have from one and a half billion passengers a year?

# THE LOCOMOTIVE ERA

Prior to electric operation of the railroad, the coal-burning locomotive was the primary motive power for the surface and elevated lines in the New York City area. (There were a few experiments using other fuels.) In a steam locomotive, the fuel heats water in a boiler. The resulting steam is directed into cylinders whose pistons drive the wheels on the locomotive. New York City had banned the use of steam locomotives within the city limits following an 1839 incident in which a locomotive blew up, killing the crew and injuring several others. Over the years, encroachments on the rule were made. By the 1860s, steam-powered fire engines in the streets and steam boilers in schools made the long-standing opposition give way. Eventually, an elevated line petitioned for permission to change from cable cars to cars hauled by steam locomotives.

Pioneer.

Steam locomotives were first employed on the Greenwich Elevated (Ninth Avenue) Line in Manhattan on April 20, 1871, and then by other lines in the city. Handren & Ripley built the first type of locomotive, the Pioneer, in 1871. The Pioneer had a 6-inch by 8-inch cylinder and four 30-inch spoke drive wheels. It was powered by a two-cylinder vertical steam engine geared to one axle with side rods connecting the second axle. The Pioneer was known as a "dummy" because the steam engine was completely enclosed to disguise its nature. The classification of the locomotive was 0-4-0 (no pilot wheels, four driving wheels [two axles], and no trailing truck).

THE NEW YORK
SPUYTEN DUYVIEL

17
6TH AVE. SOUTH FERRY

Over the next few years, various classes of locomotives were introduced with improved accommodations for the engineers, larger fuel capacity for the longer service, increased efficiency, and greater horsepower and speed to compensate for weight limitations on elevated structures. There was a great variety of locomotives. Weights ranged from 33,000 to 47,000 pounds and drive wheels were between 38 and 42 inches. Locomotives were between 23 feet and 36 feet long, and heights ranged from about 10 feet 6 inches to 12 feet 8 inches. The horsepower varied from 85 to 125. Most locomotives had kerosene-fueled headlights.

In 1877, Matthias Nace Forney proposed a different type of 0-4-4 locomotive that could easily run in either direction. The Forneys negotiated curves as sharp as 56 feet and had an average speed of 15 mph. Maximum speeds were 45 mph on the express lines and 25 mph on the local lines. In later years, the Forneys operated on the Second, Third, Sixth, and Ninth Avenue Lines in Manhattan and the West End, Fifth Avenue, Myrtle Avenue, Lexington Avenue, Broadway, Fulton Street, and Brighton Lines in Brooklyn. At their peak, hundreds of Forneys were operating in Manhattan and Brooklyn.

With the introduction of electricity, steam locomotives were gradually replaced by motorized cars. The first line to be affected was Second Avenue in December 1901, and the last was Ninth Avenue on February 18, 1903, both in Manhattan. The last Manhattan Forneys were locomotive Nos. 137 and 295, which were scrapped in February 1943. These locomotives were probably used as stationary boilers before being scrapped. In Brooklyn, the changeover started with the New York and Sea Beach Railroad in 1898, and the last steam locomotive ran on the Canarsie Line (Canarsie and Rockaway Beach Railroad) in 1906.

3055

46

BROADWAY
FERRY
16

16

92

B.E.R.R.

92

| *Year Ordered and/or Built* | *Car Numbers* | *No. of Cars* | *Elevated Name/Designation* | *Description (page)* | *Technical Data (page)* |
|---|---|---|---|---|---|
| 1868 | 1–3 | 3 | Manhattan Elevated, Trailer | 30–31 | 332 |
| 1872–73, 1875 | 1–16 | 16 | Manhattan Elevated, Trailer | 32 | 333 |
| 1875–77 | 17–39 | 23 | Manhattan Elevated, Trailer | 33 | 334 |
| 1877 | 40–41 | 2 | Manhattan Elevated, Trailer | 34 | 335 |
| 1878–79 | 40–119, 150–205 | 136 | Manhattan Elevated, Trailer | 35–37 | 336 |
| 1878–79 | 120–149, 206–242 | 67 | Manhattan Elevated, Trailer | 35–37 | 337 |
| 1885 | 1–39 | 39 | Manhattan Elevated, Trailer | 41–44 | 338 |
| 1907 | 11 | 1 | Manhattan Elevated, Trailer | 44–47 | 339 |
| 1901–2 | 40–241 | 202 | Manhattan Elevated, Motor | 44–47 | 340 |
| 1907 | 242 | 1 | Manhattan Elevated, Trailer | 44–47 | 341 |
| 1879 | 243–292 | 50 | Manhattan Elevated, Trailer | 35–37 | 342 |
| 1882, 1885 | 293–369 | 77 | Manhattan Elevated, Trailer | 41–44 | 343 |
| 1886–87 | 370–500 | 131 | Manhattan Elevated, Motor | 41–44 | 344 |
| 1878 | 501–540 | 40 | Manhattan Elevated, Trailer | 38–41 | 345 |
| 1878 | 541–580 | 40 | Manhattan Elevated, Trailer | 38–41 | 346 |
| 1879 | 581–600 | 20 | Manhattan Elevated, Trailer | 38–41 | 347 |
| 1879–80 | 601–699 | 99 | Manhattan Elevated, Trailer | 38–41 | 348 |
| 1880 | 700–728 | 29 | Manhattan Elevated, Motor | 41–44 | 349 |
| 1880–81 | 729–790 | 62 | Manhattan Elevated, Trailer | 41–44 | 350 |
| 1881, 1885–87, 1889–91, 1893 | 791–1120 | 330 | Manhattan Elevated, Motor | 41–44 | 351 |
| 1902–3 | 1121–1218 | 98 | Manhattan Elevated, Motor | 44–47 | 352 |
| 1902 | 1219–1254 | 36 | Manhattan Elevated, Trailer | 48 | 353 |
| 1903 | 1255–1314 | 60 | Manhattan Elevated, Motor | 44–47 | 354 |
| 1904 | 1315–1414 | 100 | Manhattan Elevated, Trailer | 44–47 | 355 |
| 1907–8 | 1415–1528 | 114 | Manhattan Elevated, Trailer | 44–47 | 356 |
| 1907 | 1529–1612 | 84 | Manhattan Elevated, Motor | 44–47 | 357 |

| *Year Ordered and/or Built* | *Car Numbers* | *No. of Cars* | *Elevated/Subway Name/Designation* | *Description (page)* | *Technical Data (page)* |
|---|---|---|---|---|---|
| 1910 | 1613–1672 | 60 | Manhattan Elevated, Motor | 44–47 | 358 |
| 1909–11 | 1673–1752 | 80 | Manhattan Elevated, Trailer | 44–47 | 359 |
| 1911 | 1753–1812 | 60 | Manhattan Elevated, Motor | 44–47 | 360 |
| 1904–8 | 1600–1629 A, B, C | 90 | Q (Modified 1950) | 53–54 | 361 |
| 1903–4 | 2000–2159, 3000–3339 | 500 | Composite, LV Motor | 56–60 | 362 |
| 1902 | 3340, 3341 | 2 | Composite samples, Motor | 55–56 | 363 |
| 1903 | 3342 | 1 | Steel car sample, Motor | 61 | 364 |
| 1904 | 3344 | 1 | Mineola (Director's Car), Motor | 62–63 | 365 |
| 1904 | 3350–3513, 3515, 3516 | 166 | Modified Gibbs, HV Motor | 63–67 | 366 |
| 1904 | 3407, 3419, 3421, 3425, 3427, 3429, 3435, 3445 | 8 | Modified Gibbs, HV Motor | 67–68 | 367 |
| 1904–5 | 3514, 3517–3649 | 134 | MUDC Gibbs, HV Motor | 63–67 | 368 |
| 1907–8 | 3650–3699 | 50 | Modified Deck Roof, HV Motor | 68–70 | 369 |
| 1909 | 3700–3756, 3815, 3915 | 59 | Modified Hedley, HV Motor | 71–72 | 370 |
| 1909 | 3757–3814, 3816–3914, 3916–4024 | 266 | MUDC Hedley, HV Motor | 71–72 | 371 |
| 1915 | 4025–4036 | 12 | Steinway, LV Motor | 73–74 | 372 |
| 1915 | 4037–4160 | 124 | Flivver, LV Motor | 74–77 | 373 |
| 1915 | 4161–4214 | 54 | Flivver, LV Trailer | 74–77 | 374 |
| 1915 | 4215–4222 | 8 | Steinway, LV Motor | 78–80 | 375 |
| 1915 | 4223–4514 | 292 | HV Trailer | 80–81 | 376 |
| 1916 | 4515–4554 | 40 | LV Trailer | 82–84 | 377 |
| 1916 | 4555–4576 | 22 | Steinway, LV Motor | 78–80 | 378 |
| 1916 | 4577–4699, 4719 | 124 | LV Motor | 82–84 | 379 |
| 1916 | 4700–4718, 4720–4771 | 71 | Steinway, LV Motor | 84–85 | 380 |
| 1916 | 4772–4810 | 39 | LV Motor | 82–84 | 381 |
| 1916 | 4811–4825 | 15 | LV Trailer | 82–84 | 382 |
| 1917 | 4826–4965 | 140 | LV Trailer | 86–88 | 383 |
| 1917 | 4966–5302 | 337 | LV Motor | 86–88 | 384 |
| 1922 | 5303–5402 | 100 | LV Trailer | 89–90 | 385 |
| 1924 | 5403–5502 | 100 | LV Motor | 90–92 | 386 |
| 1925 | 5503–5627 | 125 | LV Motor | 92–94 | 387 |
| 1925 | 5628–5652 | 25 | Steinway, LV Motor | 95–96 | 388 |
| 1938 | 5653–5702 | 50 | World's Fair Steinway, LV Motor | 97–99 | 389 |

# HISTORIC IRT CARS

### The Manhattan Railway Company and the Interborough Rapid Transit Company

The elevated system on Manhattan Island began inauspiciously with Charles T. Harvey's original one-track line on Greenwich Street. From this one company and a succession of other companies evolved what would become the Manhattan Railway Company, eventually to be joined with the IRT. The West Side and Yonkers Patent Railway Company was first organized in 1867. In 1868 it was reorganized as the West Side Patented Elevated Railway, which eventually became the New York Elevated Railroad in 1871. The Gilbert Elevated Railway was formed in 1872 and renamed Metropolitan Elevated Railway in 1878. New York Elevated controlled the Ninth Avenue and Third Avenue Lines, while Metropolitan Elevated was responsible for the Sixth Avenue and Second Avenue Lines. Both the New York Elevated and the Metropolitan existed separately but in the shadow of the Manhattan Railway Company, which was created in 1875 to ensure the survival of the rapid transit system in the event that either or both companies failed. In 1879, by joint agreement, the Manhattan Railway became the sole operator and owner of all the elevated lines on Manhattan Island. The Suburban Rapid Transit Company was organized in 1880 and started operation in 1886. In 1891 this company was taken over and became the Suburban Division of the Manhattan Railway Company. This situation continued until 1902, when the IRT leased the Manhattan Railway Company for 999 years retroactive to 1875. This allowed the IRT to use the Manhattan system to break in the first subway cars (the Composites) and gain operating experience prior to the opening of the subway on October 27, 1904.

Operation of the Manhattan Railway system during the Age of Steam was straightforward. There were four main elevated lines in Manhattan (the Second, Third, Sixth, and Ninth Avenue Lines) as well as the Suburban branch in The Bronx. The Suburban Line made connections with the Second and Third Avenue Lines at 129th Street and Second Avenue in Manhattan. After through trains ran to lower Manhattan from The Bronx in 1896, the Suburban Line became an extension of the Third Avenue Line. The steam engines could operate in both forward and reverse with equal ability. Those on Second and Third faced downtown, although in at least one instance, those on the Second Avenue Line faced uptown. On Sixth and Ninth, the engines faced uptown. With coaling facilities at the Northern terminals and close to the Southern terminals, the engines could be kept watered and coaled with relative ease. Maximum train length was five cars. The cable cars, which were converted into trailers to be hauled by the steam dummies, ran only on Ninth Avenue. The Shad Belly cars were probably also used only on the Ninth Avenue Line. The Side Door cars may have been used on the Third Avenue Line in addition to the Ninth Avenue Line, but we cannot be certain about this. After the Second Avenue Line was completed and opened in March 1880, there was a gradual mixing of cars in the same trains and on the different lines. The New York Elevated Railroad fourteen-window cars were kept together and eventually assigned to the Second Avenue Line exclusively. All of the other cars were spread out over the system without regard to type.

The first new cars for electric service were ordered in 1901 and were built by American Car & Foundry Company (ACF) (fifty cars) and Wason Manufacturing Company (initial order of fifty plus an add-on of 200 cars delivered in 1902–3). Wason delivered an additional order of sixty cars in 1903. The newly designed car, with six double-paired windows on each side, became known as the Manhattan Standard. (This nomenclature had previously been applied to the standard steam coaches.) After the multiple-unit door control (MUDC) fleet was created in 1923–24, all gate cars in service, regardless of type, became Manhattan Standards. General Electric Company (GE) received the contract for the motors, controllers, and other electrical equipment for the new cars as well as the steam coaches. ACF and Wason supplied the motor trucks. Motor car trailer trucks were remodeled from trucks previously used in steam operation. In addition, millwork for approximately 500 motorman's cabs for the conversion of steam trailers to motor cars was ordered from Wason. With a few exceptions these cars were all 6-4-6 cars. All

6000

work on the new and old cars was done at the Manhattan Railway Company's shops at Ninety-eighth Street and 129th Street, where 490 steam trailers were converted to multiple-unit (MU) motor cars, 378 steam trailers were wired for MU operation, and approximately 350 new cars were wired and fitted with controls and motors. The Ninth Avenue Line was the last to be electrified in April 1903.

The IRT was incorporated in 1902 as an operating company to run the new subway. Leasing the Manhattan Railway Company and ordering rolling stock for the new system were two priorities. Even though it had ordered two Composite samples and placed an order for 500 Composite cars, the IRT still pursued the idea of an all-steel car. Car 3342, the first steel car, finalized the decision to place an order for the Gibbs cars. The IRT ordered the Composites, Gibbs, Deck Roof cars, and Hedley cars as well as the 1915 series of Flivvers and high-voltage (HV) trailers under Contract Nos. 1 and 2. After the Composites had been rebuilt, they were classified as Elevated Extension cars and eventually assigned to the Manhattan Division. The 1915 Steinway cars as well as the 1916 and 1917 low-voltage (LV) cars were assigned to Contract No. 3. Starting with the 1922 trailers and continuing with the 1924 and 1925 motors, the cars were leased from the New York Trust Company. When the cars were paid off, they were assigned to Contract No. 3. The 1938 World's Fair cars, the last IRT cars, were financed by the Board of Transportation.

(*A Word about Dates:* IRT dates for the elevated cars were the years the cars arrived on the property and/or the order date. The subway car dates are the years the cars were ordered and/or delivered. All subway car dates shown were used by the IRT.)

---

### 1868 Manhattan Elevated Trailers

Manhattan Elevated Trailers (1–3)

In May 1868, car 1 was placed on the elevated structure of what would become part of the Ninth Avenue Elevated. The first three cable cars operated under the aegis of the West Side Patented Elevated Railway and later became the first cars of the New York Elevated Railroad. At first only two stations were constructed: Dey Street on the east side of Greenwich Street and Twenty-ninth Street on the west side of Ninth Avenue.

The elevated car was powered by a cable wound around a drum that was driven by a stationary steam engine, placed in a vault beneath the sidewalk under the elevated structure. The board of directors of the railroad were invited to a trial run on July 3, 1868. The car ran from Battery Place to Cortlandt Street.

Regular passenger service began on February 14, 1870, between Dey Street and Twenty-ninth Street. By November 15, 1870, the line had been closed because of the frequent breakdowns of the cable as well as financial difficulties. Under court proceedings, the sheriff sold the entire line to the bondholders at auction for $960. On April 20, 1871, operation was resumed with a steam dummy engine pulling these cars. A variation of this account states that three streetcars were used and a stereo viewing card from about 1872 shows a streetcar and one of the Shad Belly cars together on the Greenwich Street El structure.

### 1872, 1873, 1875 Shad Belly or Drop Center Cars

Manhattan Elevated Trailers (1–16)

This was the initial series of cars built for the New York Elevated Railroad. As soon as the first cars in this series arrived, the former cable cars were retired. The distinguishing feature of these cars was the depressed center, supposedly introduced to allay the fears of passengers about their height above the streets and also to maintain a low center of gravity. Cummings and Jackson & Sharp built a total of sixteen cars. In 1878, the floors of fifteen cars were raised to conform to new car deliveries. It is probable that they received a center door at this time. They also had their trucks rebuilt in conjunction with the track regauging to 4 feet 8.5 inches.

### 1875–1877 Side Door Cars

Manhattan Elevated Trailers (17–39)

Starting in 1875, a new car design arrived featuring a door on each side of the car body in addition to the usual end platforms. Gilbert & Bush and Jackson & Sharp were the builders. Aside from being slightly longer, not having the depressed floor, and having a different slope to the roofline at the end of the car, these cars were more or less comparable to the previous type. Several cars in this series were converted to work cars after their retirement from passenger service.

### 1877 Demonstrators

Manhattan Elevated Trailers (40, 41)

Jackson & Sharp designed and built these two cars. They were made without end platforms, so passengers would have to enter and exit through the side doors in the center of the car. These doors were much larger than those in the previous series of cars. The perforated veneer seats were arranged along the sides of the car. The plate glass windows were large, with a post between them. A novel feature with a beautiful effect was the use of tiles in the clerestory. The cars were painted a dark claret or maroon and overall were very attractive. They were placed in service early in 1878 with the intention of using them in extra-fare or first-class service.

## 1878–1879 Fourteen-Window Type

Manhattan Elevated Trailers (40–292)

There were three distinct styles of fourteen-window coaches. Gilbert & Bush and Wason built the first two. The basic difference in their appearance was the Gothic-style pointed windows of the Gilbert cars and the rounded-style windows of the Wason cars. Gilbert also built a short series of thirteen-window cars that had a shorter body length. By the 1890s, these cars had been assigned to the Second Avenue Elevated and lasted in this service until this line was electrified. As electrification proceeded, they were shifted to the Third, then Sixth, and finally the Ninth Avenue Elevated. They remained in service until the end of the steam operation. Several were converted to work cars (one in 1893 and others in 1904). The third series was a more basic design and was delivered beginning in 1879. They were built by Gilbert & Bush, electrified in 1907, and continued in service until at least 1940. All had been scrapped by the end of 1946.

M A N H A T T A N
152
152

284
284

THE NEW YORK

## 1878–1880 Metropolitan Elevated Railway

Manhattan Elevated Trailers (501–540, 541–580, 581–600, 601–699)

The initial deliveries of cars consisted of two types. Barney & Smith and Pullman built the first forty cars (twenty cars each), which had an eighteen-window arrangement on the car sides. These were referred to as the second-class cars. The next forty cars, built by Pullman, were the first-class or extra-fare cars. They had a combination of seven large and ten small windows in an alternating arrangement on the sides. Both types were noted for having light and airy interiors because of the large amount of glass used in the windows. The next

INTERBOROUGH
520

delivery of twenty cars reverted to the eighteen-window design and Pullman was once again the builder. The following Pullman order, starting in 1879, introduced the famous 6-4-6 window style, which continued as the regular design of the Metropolitan and later became the Standard of the Manhattan Railway Company (beginning with car 688). As an additional note, all Metropolitan cars and the first Manhattan cars had a deck roof that was later modified into a standard railroad clerestory roof.

## 1880–1882, 1885–1887, 1889–1891, 1893 Manhattan Railway

Manhattan Elevated Motors (370–500, 700–728, 791–1120)
Manhattan Elevated Trailers (1–39, 293–369, 729–790)

The Manhattan 6-4-6 design that evolved directly from the Metropolitan cars remained almost unchanged through the last delivery of cars in 1893. The only visible changes were the railroad clerestory-style roof and the simplified window frame and post design. The more ornate and simplified window frames and posts were used concurrently for a short time on the 6-4-6 cars starting in 1879. Both of these changes were in place prior to 1885. Bowers, Dure & Company, Gilbert & Bush, Pullman, and Wason were the car builders. Pullman and Gilbert & Bush built the Suburban Rapid Transit cars, which closely followed the Manhattan design. Many of these cars remained in service until subway unification in 1940. Circa 1940, however, they were mostly used in rush hour service on the Second, Third, and Ninth Avenue Elevated Lines. The downsizing of the elevated lines and other events such as World War II eventually affected the cars and the lines they traveled. Mayor Fiorello LaGuardia had a signifi-

MANHATTAN
703
703
703
703

INTERBOROUGH
415
415
Open Air Line

cant impact on them as well, as he disliked the appearance the elevated lines had on the cityscape.

The Sixth Avenue Elevated Line ceased operation on December 4, 1938. When the city took over the IRT on June 12, 1940, the Second Avenue Line north of Fifty-ninth Street was discontinued as was the entire Ninth Avenue Line except for the section north to The Bronx, which became the Polo Grounds Shuttle. The Manhattan Elevated cars were now surplus. Both motor and trailer cars were placed in storage. With the outbreak of World War II, the Office of Defense Transportation issued a directive that no railcar could be destroyed. In 1942, a number of these cars were taken over by the federal government and shipped around the country to various military facilities to be used as transport for war workers.

A large fleet of these cars went to Oakland, California, and were operated by the Key System (formerly the old East Bay Transport Company) and used by Shipyard Railway Company (a subsidiary of the Key System) throughout World War II. These cars were equipped with pantographs, and ran on their own power through overhead wires. Two still exist in a museum in Rio Vista, California—a long way from home. They were obviously built to last!

(*Note:* Some car builders used different names during the course of their existence. Gilbert & Bush used the names Gilbert Car Company, Gilbert, Bush & Company, and National Car Builders. Pullman used Pullman Palace Car Company, Pullman Company, and later Pullman Standard.)

---

### 1901–1904, 1907–1911 Manhattan Standard

Manhattan Elevated Motors (40–241, 1121–1218, 1255–1314, 1529–1672, 1753–1812)
Manhattan Elevated Trailers (11, 242, 1315–1528, 1673–1752)

These were the first cars built for electric service, and deliveries began in December 1901. The cars sported a new design, characterized by a six-pair or 4-4-4 window arrangement, and were referred to as

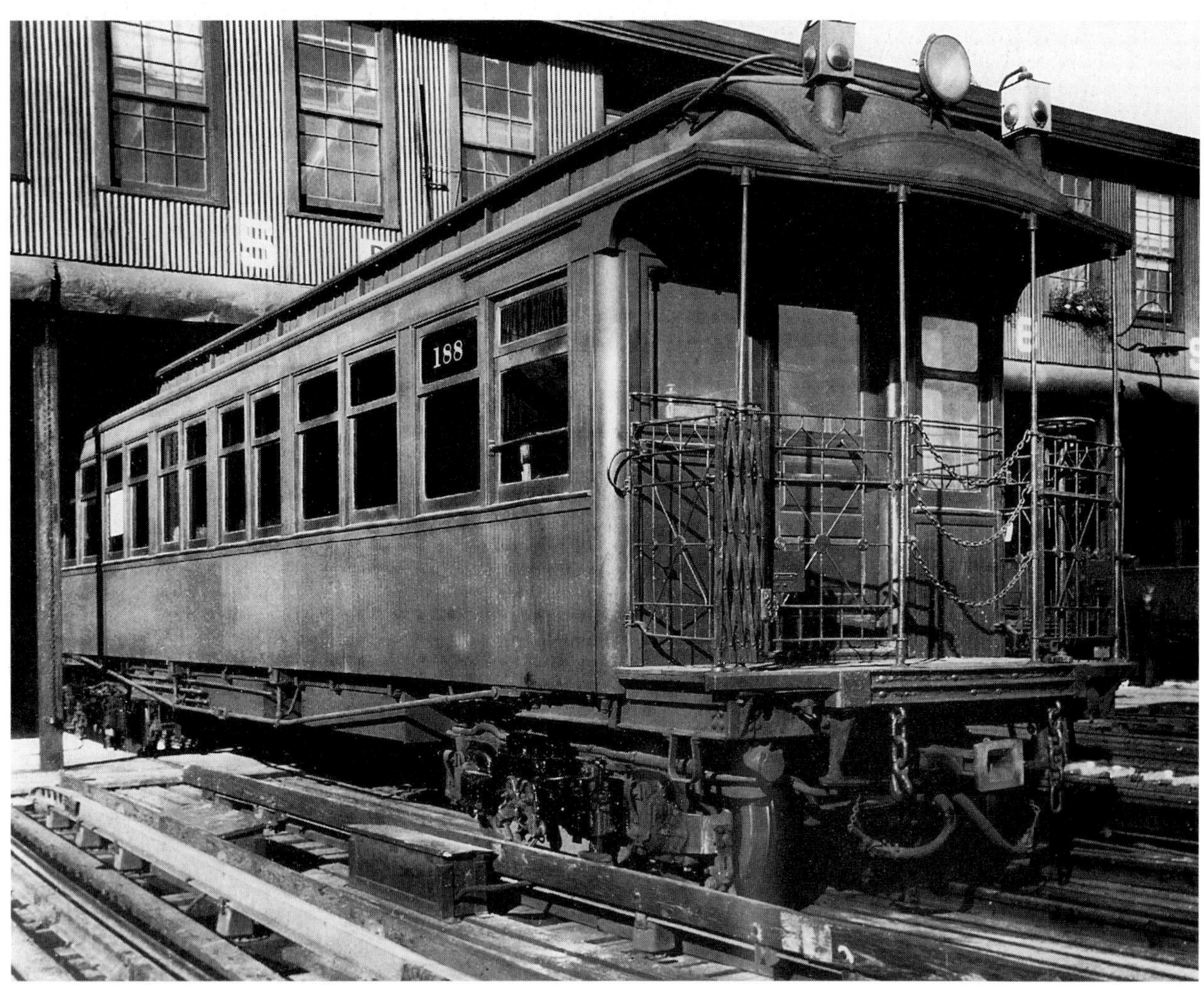

1447

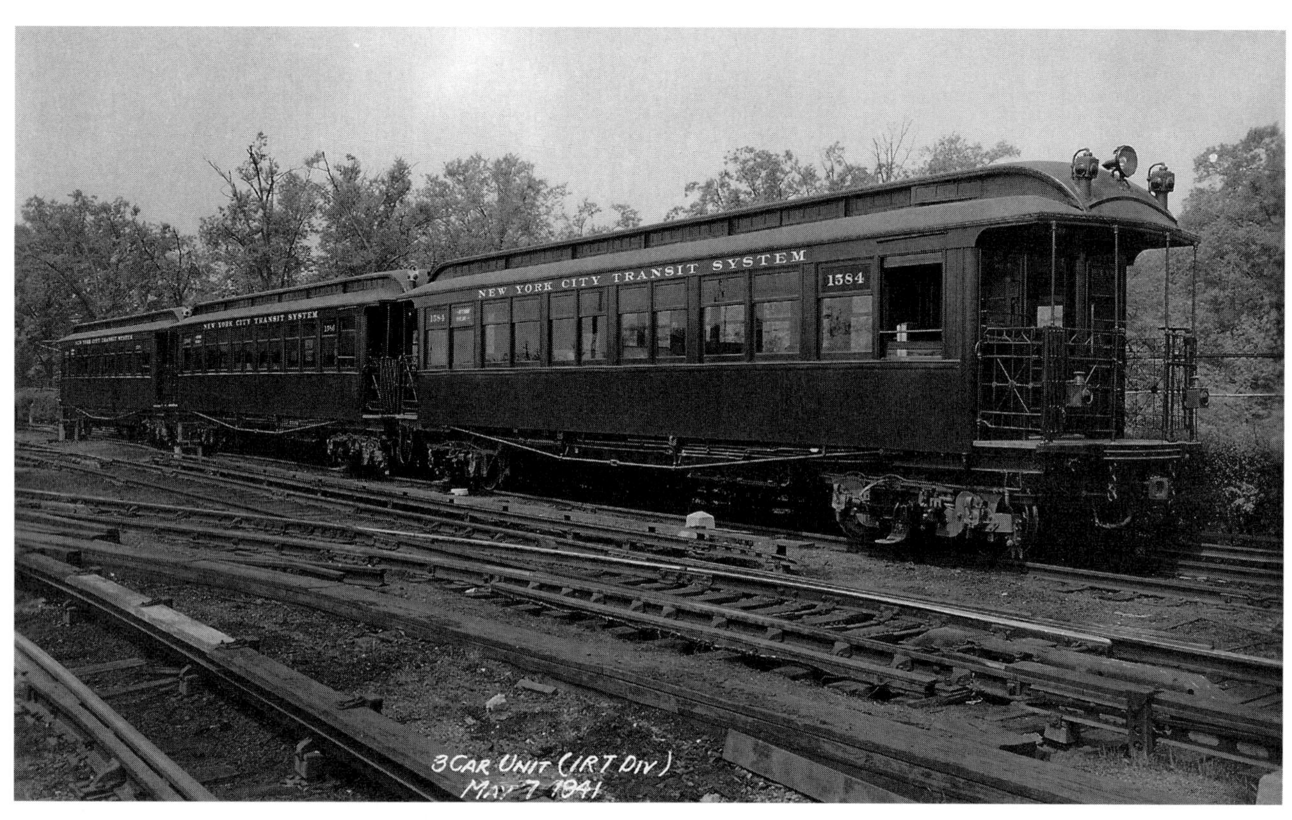

NEW YORK CITY TRANSIT SYSTEM
1584
3 CAR UNIT (IRT DIV)
MAY 7 1941

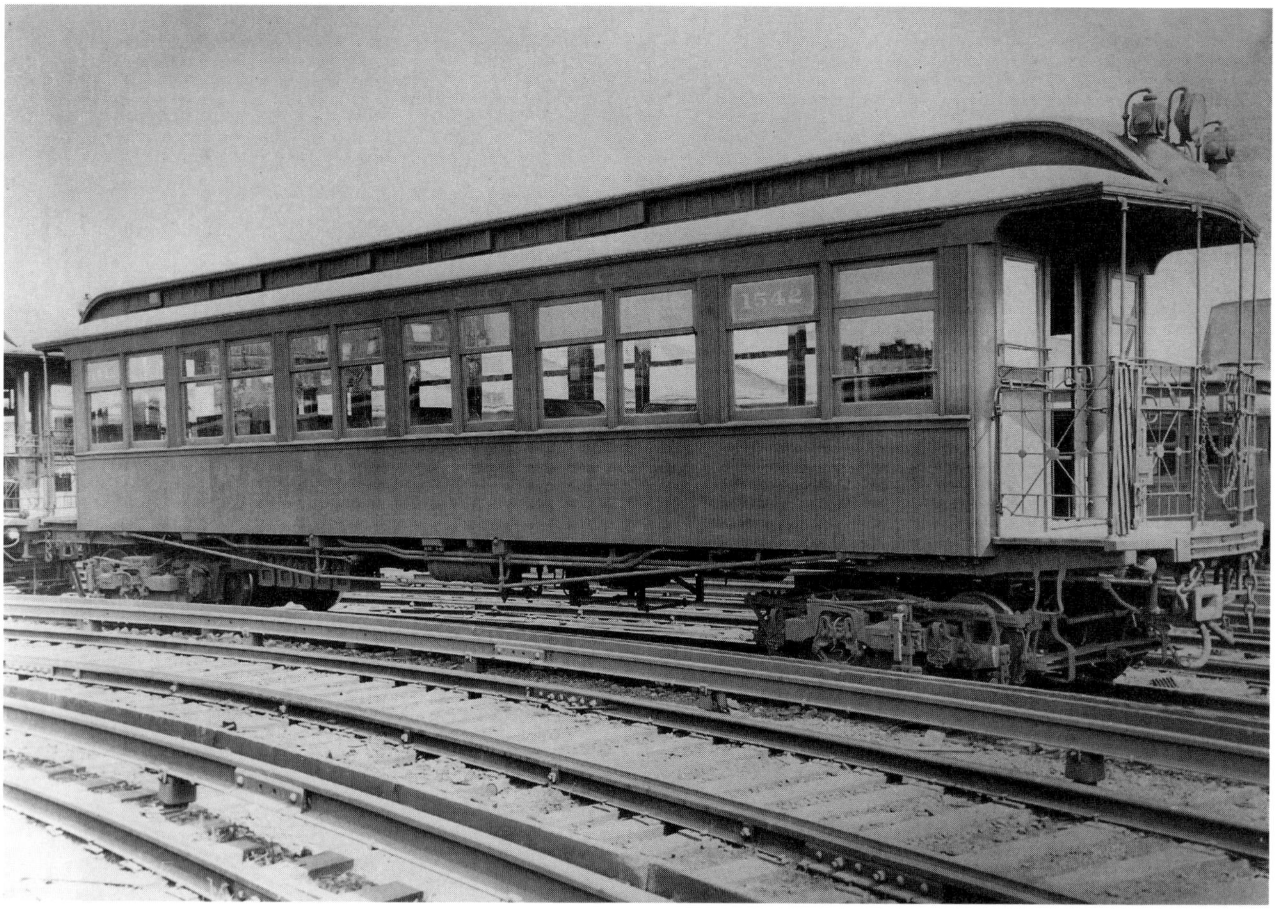
1542

Manhattan Standards. The cars had the usual Manhattan-style seating that accommodated forty-eight passengers. When the cab door was closed to protect the control apparatus, the motorman's seat was turned down, forming a seat for two passengers. The trailer cars had no cabs. Eventually the term *Manhattan Standard* was used to describe all gate cars (motors and trailers). They were slightly longer than the previous series of cars and easily distinguished by the larger windows and higher roof. Almost 900 cars were built following this design and the last 200 had steel window posts. The car builders were ACF, Barney & Smith, Cincinnati Car Company, Jewett Car Company, St. Louis Car Company, and Wason.

The initial deliveries of cars were lettered *Manhattan* on the letter board, but this was changed to *Interborough* within a short period of time. The old cars were mixed into train consists with the new ones, and trailers and motors were arranged in trains as follows:

| *Type* | *Motors* |
|---|---|
| six-car train | First, third, fourth, and sixth cars |
| four- or five-car train | First, third, and fourth or fifth cars |
| three-car train | First and third cars |

Eventually a seventh car was added by placing a motor car on the end of the train. Cars were selected from these groups to be converted from gate cars to MUDC in 1923–24.

## 1902 Manhattan Open Trailers

Manhattan Open Trailers (1219–1254)

Electrification of the Manhattan Elevated System began in 1901 and was completed by 1903. A series of open Manhattan trailers was built in 1902, but was short-lived. They were used exclusively on the Third Avenue Line, usually in local service, from City Hall to Bronx Park during the summer months. In the off-season the trailers were stored in sheds in the small yard directly under the Bronx Park spur. This yard could be reached via a small elevator from the El structure. Since these storage tracks were never electrified, a Forney steam engine was used to shunt the cars. These cars proved to be very popular. An additional order for sixty trailer cars was planned but was changed/cancelled at the last moment. The Manhattan open cars were last used in regular passenger service during the summer of 1917.

## 1902, 1903, 1907, 1908, 1909, 1910–1911, MUDC Cars: HV and LV

MUDC Cars: HV and LV Motor, HV and LV Trailer

Prior to 1923, cars with open platforms with gates for door operation were in use on the Manhattan El. Passengers boarded at the ends of cars through the gates, as they did on BRT's Brooklyn Union (BU) cars. In 1922 and 1923 all this began to change, as some of the 1910 Manhattan Elevated cars were reequipped for MUDC operation so that they could be operated by a single conductor. Unlike the BMT, which moved the motorman's cab to the end of the car, these cars were modified to have the ends enclosed, thus creating a vestibule. The motorman's cab remained in its original position and doors were built where gates had been. The motorman now looked through two sets of windows. With this construction it was possible, although not

NEW YORK CITY TRANSIT SYSTEM
227
227

permitted, for a passenger to walk in front of the motorman's view of the right of way.

The first car to be refitted was car 172, which presented a slightly different appearance from the others. Gate cars from several series and years were selected to be rebuilt to MUDC specifications. The Ninety-eighth Street, 129th Street, and 159th Street Shops of the Manhattan Division did the work. The MUDC cars (1910–11) were both motors and trailers. A small group of them were equipped for LV operation. These steel window post cars were never permitted to run on the Sixth Avenue Elevated or on the Ninth Avenue Elevated below Thirtieth Street. They were assigned as Second Avenue–Queens cars for years before eventually running on the Third Avenue Elevated.

However small in fleet composition, the LV group survived longer than any other cars—the LV Manhattan MUDC cars ran on the Third Avenue El in regular service until May 1955, when downtown service was discontinued. The cars had run from either South Ferry

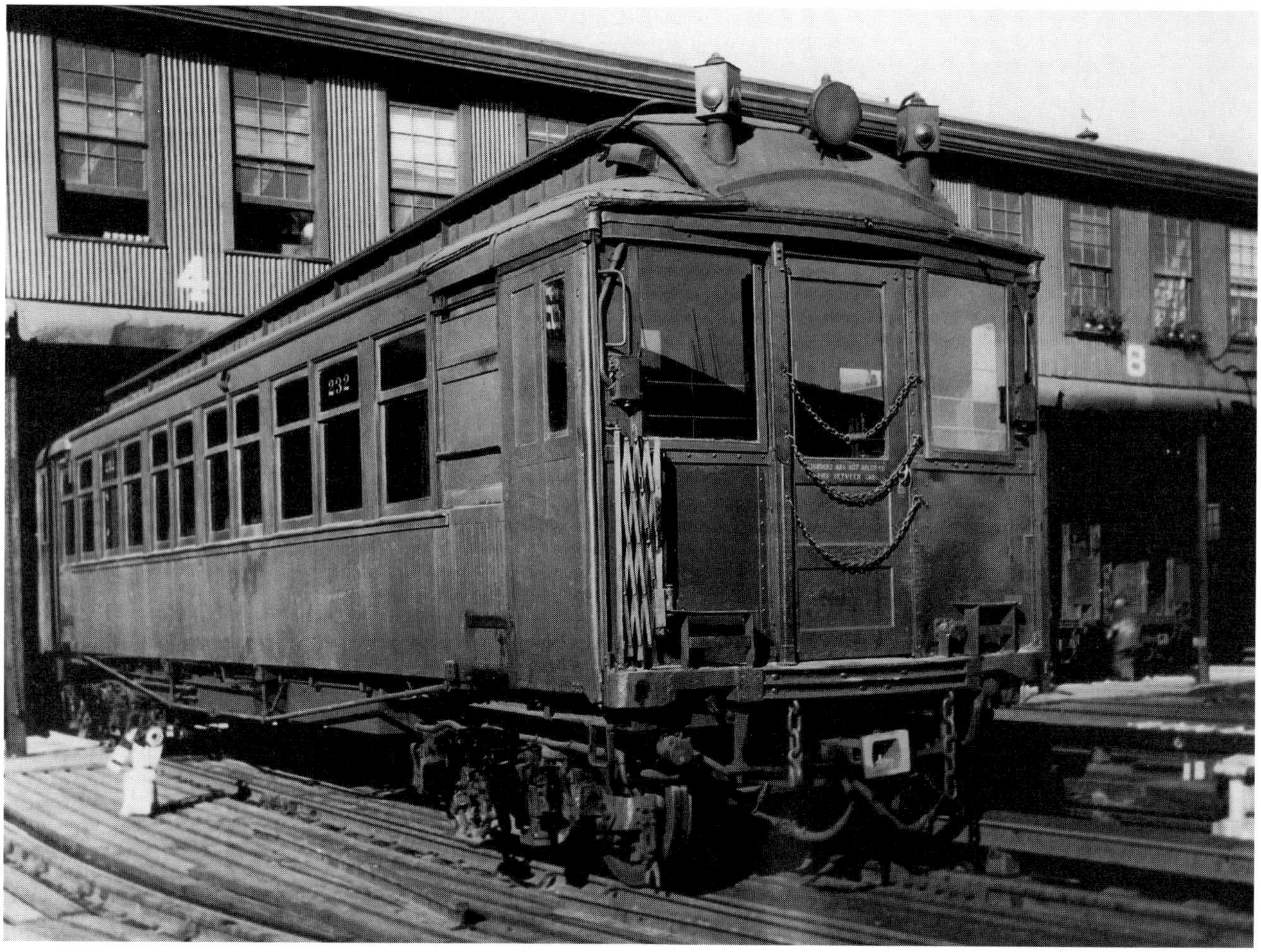

or City Hall in lower Manhattan on the Third Avenue El to one of several terminals in The Bronx: 241st Street–White Plains Road; the spur to Bronx Park (closed in 1951); or Freeman Street (until 1946). Service on the Third Avenue El's South Ferry spur ended on December 23, 1950, and was also cut back from 241st Street to Gun Hill Road. The City Hall to Chatham Square route was discontinued on December 31, 1953. On May 12, 1955, the El was cut all the way back to 149th Street. The LV Manhattan MUDC cars survived on the remaining portion of the Third Avenue El, running from 149th Street to Gun Hill Road, for approximately one more year, until December 15, 1956.

---

### 1904–1908 Q (Modified 1950)

Q Modified (1600–1629A, B, C)

Starting in 1949, several BMT Q types (three-car units) and one QX motor car (1634A) were sent to the IRT. The cars were replacements for the IRT gate cars and Composites used for the Third Avenue El Through Express service.

After tests using car 1634A, a few modifications were made. The marker lights were moved from outside the clerestory section of the roof to the inner part. This gave the cars a better appearance and provided sufficient clearance to the station canopies. Trucks from the Composites were placed under the A and C cars of the units to reduce weight.

All Q types were assigned to the Third Avenue Elevated and were placed in service between April 1 and September 15, 1950. The Qs were considered heavy and ran only in local express and through express service, just as the Composites had done. After the Manhattan section closed in 1955, they ran as locals in The Bronx until December 1956, when they were replaced by Steinway cars.

Car 1634A was used as a test car and later as a substitute revenue collection car when car G was not in service.

## 1902 Composite Samples

Composite Sample Motors (3340, 3341)

Wason built the August Belmont (Car 1) and the John B. McDonald (Car 2) according to the specifications of George Gibbs, consulting engineer of the IRT, and delivered them together in 1902. These cars served as test vehicles for the fleet of Composite cars to follow. Basic structural components in both cars were identical, but many details such as doors, windows, seats, lighting arrangements, and minor features differed materially. Several classes of furnishings were installed for purposes of comparison.

When delivered, the August Belmont was outfitted as an experiment for first-class service. More expensive lighting fixtures, Manhattan-style seating, and a half empire roof with a light-colored ceiling were only some of the appointments that distinguished this car from the John B. McDonald. The August Belmont had a white exterior while the John B. McDonald was painted chrome yellow. Car 1 had vestibule doors of the Gibbs type; Car 2 was fitted with the Gold,

Pitt, Duner, and Gibbs type. Car 1 had six clusters of five lights on the upper deck ceiling; Car 2 had a row of six single incandescent lamp fixtures down the center of the upper deck and five on each of the side deck ceilings. Headlinings in both cars were of triple veneer, the interiors were mahogany, and all the seats were finished in rattan. Both cars originally had headlights that were removed along with modification of the marker lights on both cars in 1903.

The August Belmont (Car 1) was renumbered 3340 approximately one year after delivery and became an instruction car based at the 147th Street Yard. The John B. McDonald (Car 2) was renumbered 3341 and was a pay car for many years (pay cars distributed the cash payroll to employees). Subsequently it was used as a time clock and canteen car. Later, in 1917, this car was refitted as a regular passenger car and became part of the rebuilt Composite fleet assigned to the Second and Third Avenue Elevated Lines (Manhattan Railway—Eastern Division).

## 1903–1904 Composite

The First IRT Subway Cars: Wood/Steel Composition Cars—
The Composites (2000–2159, 3000–3339)

Soon after construction of the subway began in 1900, IRT car engineers began to design suitable rolling stock capable of handling the heavy traffic they anticipated. The new car engineering work was finished in the spring of 1902. IRT management hoped to order an all-steel fleet because of the combustibility of wooden cars in the subway. The use of steel was a radical concept at the time; in fact, the only all-steel cars were experimental steam coaches built by the Pennsylvania Railroad in their own shops in Altoona.

When IRT started to place its order for cars, car manufacturers were busy and did not want to tie up their plants, so all refused to accept orders. Since time was running short (the line was scheduled to open in two and a half years), IRT management reconsidered their plan for an all-steel fleet. They went to work designing 500 wooden cars with enough metal and fireproofing to protect them from accidents and fire.

Two sample Composites were built in early 1902 and, after a series of successful tests, the final design was agreed upon. However, car manufacturers at the time were so busy that the IRT had difficulty obtaining contracts that offered delivery in time for the start of revenue

service in October 1904. The solution was to divide the order between four car manufacturers. In December 1902, contracts were awarded as follows: Jewett—100 cars, St. Louis Car Company—200 cars, Wason—100 cars, and John Stephenson Car Company—100 cars.

The IRT's car designers believed that they had ordered a fleet of cars that were stronger and safer than any other electric railway vehicle. This was a major concern since the IRT subway was the first heavy rail rapid transit underground operation. Until that time, all other subway operations were street railways that had been relocated underground. Speed and capacity of this first application of the heavy rail concept had to be matched with a vehicle that was faster, stronger, and safer than other electric railway vehicles built at the time.

The steel underframe of the car was reinforced with white oak timber and the side framing with steel. Car designers for the IRT wanted to make cars as fireproof as possible. The floor was double thickness, made of maple with asbestos fire-felt between the two lay-

ers. On the underside, it was completely covered with 0.25-inch asbestos transite board. Above the motor truck, the underside of the floor framing was protected by steel plates and 0.25-inch roll fire-felt. A layer of asbestos was placed on the outside of all wires, which were carried in ducts molded into asbestos. Switches, fuses, and high-voltage power wires were housed in an undercar steel compartment. The cars originally had copper sheathing on the sides, but it was removed from some cars during World War I.

When the orders were placed with the four car manufacturers, it was anticipated that two-thirds of the fleet would be motor cars and one-third trailers. Consequently, numbers 2000–2159 were assigned to the trailer cars and 3000–3339 to the motor cars. On September 14, 1903, a five-car demonstration train was run over the Second and Third Avenue Elevated Lines. Composite trains ran on both the Second and Third Avenue Lines carrying passengers for a period of time.

The cars were tested on the Second Avenue Elevated during the winter of 1903–4.

Car builders constructed only the car bodies and trucks. IRT car equipment personnel in IRT shops installed underfloor equipment. IRT engineers estimated that they required a 3:1 motor to trailer ratio. If the company had followed its original plan, it would have had an excess number of motor cars. Consequently, when the 3000 series cars arrived from the car builders, the IRT converted 78 of them to trailer cars. As more steel cars arrived, even more 3000s were converted: 73 in 1905, 10 in 1908, and 47 in 1910, for a total of 208 of the 340 cars.

The first all-steel cars were known as the Gibbs cars, named after George Gibbs, chief engineer of the Pennsylvania Railroad. These cars were put into revenue service toward the end of 1904 and in 1905, along with the subsequently ordered Deck Roof cars. IRT officials operated these cars in trains with Composite cars with great apprehension. They feared the consequences of operating mixed consists of wood/steel and all-steel in the same train. An accident would have been a disaster.

By July 1, 1916, the IRT had placed 490 new all-steel cars in service and as a result was able to transfer all the Composite cars from the subway to the elevated lines. When the Composites were new, one of the journals of the day had stated that "all the cars are adapted to run either on the elevated or subway lines, and are known as interchangeable equipment." However, the Composite cars were too heavy for the IRT's old elevated structures, which had been built in the nineteenth century. The IRT undertook extensive alteration work on these cars to reduce their weight at the New York Central's High Bridge Yard in The Bronx. The Composites originally had heavy GE-212 traction motors. In 1915–16, the original trucks were placed under newly arrived Flivver-type motor cars, and maximum traction trucks, which had larger-diameter wheels on the motor axle and smaller wheels on the trailer axle, were substituted. The motors used on these substitute trucks were of the lighter elevated type. All the Composite cars were motorized and the master controllers were changed from the HV type to the LV type. Cars equipped with HV controllers had 600 volts direct current (DC) power from the third rail fed directly into the master controller and then to relays that turned on the motors, accelerated them, and shut them off. When the cars were converted to use LV master controllers, battery power was fed into the master controller and then to relays that controlled the feed of 600 volts DC power to the traction motors.

When the work was completed, the 477 remaining Composite cars were assigned to the Second Avenue and Third Avenue Elevated Lines and to the East 180th Street–241st Street–White Plains Road Shuttle. Upon being transferred to the elevated lines, they became classified as "Elevated Extension cars." By 1916, twenty-three cars had been retired because of accidents and fires.

For the remainder of their lives, the Composite cars ran in service on those lines and some minor shuttle operations in the north Bronx. When service was reduced in the late 1920s and 1930s, some Composite cars were placed in storage. Sixty-nine of them were scrapped in 1938; by unification (the recapture of the transit properties of the IRT and BMT by the City of New York) on June 12, 1940, 401 of them had been sold to the City of New York. The fleet was further reduced with the closure of the Second Avenue Elevated Line, so that by September 1946, only 159 cars remained. The last Composite cars ran in Third Avenue Through Express service during the evening of April 28, 1950. The trucks were removed from sixty cars and placed under the Q type cars. The Composite car bodies were scrapped, and the Q cars provided Third Avenue Through Express service until closure of the Manhattan portion of the Third Avenue Line in 1955. The last two Composites were scrapped in 1953.

The IRT car designers did not realize that the traffic in the subway would become as heavy as it did. The Composite cars (and the original all-steel Gibbs cars) were constructed with doors at the ends of the cars and with eight transverse seats facing each other at the center of the car. There was no center door and the remainder of the seats were longitudinal. This arrangement slowed loading and did not provide enough room for standees. The next order of cars, the Deck Roof cars, were built with four transverse seats. While this design provided more room for standees, it still proved insufficient for handling the increased number of passengers. Since the Deck Roof cars were designed for center doors, they were among the first to have them installed. After the 1908 Gibbs modification was tested, a decision was made to remove all the transverse seats in all the subway cars and to install center doors in 1909 to speed loading. By the end of 1912, all the Composite cars had been converted.

(*Note:* IRT employees referred to the Composites as "copper sides." One of the curiosities relating to the Composites pertained to the removal of the copper sheathing on some of the cars. When the cars were being rebuilt for elevated service in 1915–16, the sheathing on some cars was damaged and was not reapplied.)

## 1903 First All-Steel Car

Steel Car Sample Motor (3342)

The design for a sample all-steel car was developed, and the car completed at Altoona by the Pennsylvania Railroad about fourteen months after the project had begun. It was designed by George Gibbs, primarily for use in the Pennsylvania–Manhattan Tunnel. Starting February 8, 1904, the car was subjected for several days to a thorough testing on the Second Avenue Elevated. When the car was in service, it was used as a trailer with no motors installed; it was equipped with motorman's cabs. The car proved satisfactory in every respect except weight, being about two tons heavier than the Composite subway cars. Deemed too heavy for passenger service, it was converted to a pay car at an early date; it was finally scrapped in 1956. This car paved the way for the first order of 200 steel cars from the ACF. Later, an add-on order upped this total to 300.

## 1904 *Mineola*

Director's Car—*Mineola* (3344)

Wason built the *Mineola,* August Belmont's private car, in 1904. This car was constructed of wood with elevated car dimensions except for the height, which was the same as that of the Composites, allowing it to run in the subway. The car had Van Dorn couplers, elevated-type trucks, and no MU connections: it also had a unique 14-point controller handle. The ends of the car were glass-enclosed, providing Mr. Belmont and his guests with an excellent view.

The car's exterior was finished in glossy maroon paint with golden leaf striping and lettering, while the interior trim was natural mahogany. There was a completely equipped steward's galley, as well as a lounge, lavatory, and office area with Belmont's rolltop desk. The floor had broadloom carpeting.

When the subway opened, the *Mineola* operated as a single car. As traffic increased, Belmont became concerned about the possibility of a collision between the *Mineola* and the steel subway cars. He eventually decided to couple a supply car at the rear of his car for protection. Either supply car 01 or 02 was used.

The *Mineola* was operated as a private car as late as 1919. Later, it was used for storage and construction. It was moved to the 147th Street Yard and stored in the shop. It was finally moved to a yard track and enclosed in a tarpaulin to protect it from the weather. After being sold for scrap in the late 1940s, the car was rescued several times and is now at the Shore Line (Branford) Trolley Museum in Connecticut. It is believed to be the only private subway car ever built.

### 1904–1905 Gibbs Cars

Gibbs Cars: HV Motor, Modified Gibbs (3350–3513, 3515, 3516), HV Motor, MUDC Gibbs (3514, 3517–3649)

Gibbs subway cars, beginning with car 3350, were the first steel subway cars built for the subway system. The order was originally for 200 cars, but was later upped to 300. The last 100 cars were slightly lighter and contained more aluminum. They were designed by George Gibbs, chief engineer of the Pennsylvania Railroad, and manufactured by ACF in 1904–5. They were built without center doors.

(*Note:* A group of cars with the same specifications was built for the Long Island Rail Road and ran over the Williamsburg Bridge to Chambers Street through a connection off Atlantic Avenue by Chestnut Street in Brooklyn. This connection was abandoned in 1917

and torn out during World War II, but one of the steel girders still stands on the property of the Columbia Wire and Cable Company.)

The first modification to these Gibbs cars was the addition of center doors by 1912. Some of these cars, termed *modified cars,* had hand-operated end doors and air-operated center doors. They also had a motorman's door indication light for other cars in the train. All HV cars with hand-operated doors were known as "pilot modified" (excluding the eight Gibbs work motors). The Gibbs cars were later used almost exclusively on the ends of the trains because they required a conductor to hand-operate the doors. The second group of cars was equipped for MUDC operation in 1936. New end and vestibule doors were installed. With air-operated doors in the vestibules and center, MUDC cars would be in the middle of the train and modified cars at the ends. At the same time, all Gibbs cars received the white line under the car numbers to indicate that they were to be used in local service. The Gibbs cars operated exclusively in local

Calvert
WRIGLEY'S

service, never on the express lines, probably because of their short platforms. They did not have large vestibules, like the Deck Roofs and later cars, or double-width doors on the ends (only at the center).

After World War II, five-car trains provided local service on the Broadway–Seventh Avenue, Seventh Avenue, and Lexington Avenue Lines. After World War II, platforms on the southbound lower Lexington Avenue Line were lengthened to accommodate ten-car trains, but the northbound local stations below Fourteenth Street were not lengthened until 1960–61.

Northbound platforms were lengthened to accommodate ten-car trains concurrent with the arrival of the R26/R28 cars. MUDC cars remained in local service, with modified Gibbs cars stationed at the ends of the trains, until just before they were retired (when they briefly appeared on the express service lines: the Broadway–Seventh Avenue Express, Lexington Avenue Express, and very rarely on a Seventh Avenue Express in the mid- to late 1950s). Modified cars were also assigned to the Polo Grounds Shuttle for a short time in the early

1950s. The modified cars were removed from service first; the last Gibbs MUDC cars ran in 1958 as the last of the R22s were being delivered.

IRT employees referred to the Gibbs cars as "Merry Widows" and "Battleship" cars. The term *Merry Widow* was derived from the parlor-type doors that closed off the vestibules. Later, the term *Battleship* was applied to the Deck Roof cars.

### 1908 Gibbs Modification

Modified Gibbs—Built in 1904 (3407, 3419, 3421, 3425, 3427, 3429, 3435, 3445)

Projections of passenger traffic on the subway were greatly underestimated. Thus, car design reflected IRT management's view of a moderately patronized line. Once the first subway opened, it became apparent that many more passengers had to be accommodated than had first been estimated. Most critical was the need to improve traffic flow to allow quicker boarding and exiting. The first attempt at

improved traffic flow was in the redesign of the door area in the second group of steel subway cars, the 1907 Deck Roof HV cars. The first steel cars, the Gibbs cars, and the original subway cars, the Composites, had 39-inch-wide doors located at the ends of the car. The Deck Roof cars were delivered with 50-inch-wide doors, but even that clearance proved insufficient.

In October 1907, a study was undertaken to address the problem. Bion J. Arnold, special consulting engineer, was engaged to examine the operation of the subway and make recommendations. His report to the Public Service Commission, dated February 18, 1908, proposed a modification to the cars already in service. This redesign included double doors near the car ends, with those nearest the end of the cars marked "Entrance" and those closer to the center marked "Exit." Eight cars were rebuilt and tested in passenger service in February 1909. Some difficulties were experienced with the performance of the additional doors.

A second experiment was conducted at almost the same time as that of Arnold's four-door-per-side modification. The new Deck Roof cars had a center door installed midway between the end doors, so that the space between the doors was reduced from approximately 40 feet to 18. More seating was thus retained. This experimental eight-car train outperformed the train modification of Arnold's proposal. As a result, all the existing subway fleet at the time was rebuilt with an added door on each side, and all future IRT cars were ordered with the same feature.

---

### 1907–1908 Deck Roof Cars

Modified Deck Roof HV Motor (3650–3699)

In 1907 and 1908, ACF manufactured fifty HV Deck Roof motor cars under the Gibbs design patents. Because of revised construction techniques, a weight reduction of 2,000 pounds per car was achieved. One reason advanced for the design of the roof was improved ventilation. The cars were built with the provision that center doors would be added, which was accomplished by early 1910. The cars were numbered 3650–3699; they were kept as modified cars, never having MUDC added. They were almost always used as end cars on a train, especially after 1936, when many Gibbs cars and the balance of the HV trailers received MUDC. The cars ran on the Broadway–Seventh

3666

Avenue Express along with other MUDC HVs and also on the Lexington Avenue Local between City Hall and Pelham Bay Park. At different times some cars were assigned to the Bowling Green and Times Square Shuttles as well as the Dyre Avenue Line.

## 1909 Hedley Motors

Modified HV (3700–3756, 3815, 3915)
MUDC HV (3757–3814, 3816–3914, 3916–4024)

In 1909, 325 additional HV motor cars were ordered (cars 3700–4024). Deliveries began in 1910 and were completed in 1911. These were the first standard-body IRT subway cars. ACF, Standard Steel Car Company, and Pressed Steel Car Company were the builders.

These cars were operated along with the Deck Roof, Gibbs, and Composite cars already in service and were intermixed in train consists. After the H System was inaugurated in 1918, the cars were eventually assigned to the Broadway–Seventh Avenue Express and Locals. The express ran to New Lots Avenue and the locals to South Ferry. Occasionally a few LVs were used in the express service as well. Cars

3700–3756, 3815, and 3915 had the hand or manual door controls. Cars 3757–4024 were equipped with MUDC, with the exception of cars 3815 and 3915. Several HV MUDC local cars were used in Times Square Shuttle service starting in the mid-1930s, replacing most of the Deck Roof cars used at the time. HV cars assigned to local service were identified by the white line under the exterior car numbers (cars 3951–4024 were HV MUDC local cars).

All HV subway cars had guide boards installed along the length of the side of the car body level with the door sills by 1916. All LV cars had them installed upon delivery. These helped to operate the gap fillers located at several of the IRT stations that were situated on sharp curves. The World's Fair Steinway cars were the only IRT cars without this feature. IRT employees referred to the 1909 series of HV motors as "Hedley Motors" (Frank Hedley was the General Manager and Superintendent of the IRT). They remained in service until 1958. Cars with the manual door controls were retired first, followed by the MUDC cars. The last of the HV cars were retired with the deliveries of the R22s.

## 1915 Steinway

The First Steinway Cars: "The Boilers" (4025–4036)

On April 3, 1913, the City of New York took ownership of the Steinway Tunnels and the IRT was selected as the operator of the line. The Steinway Tunnels were two single-track trolley tunnels built by the New York and Long Island Railroad Company between Lexington Avenue and Forty-second Street in Manhattan and Van Alst Avenue (near the present Hunters Point Avenue station) in Long Island City. After a brief consideration of the resumption of the trolley service (stopped on October 23, 1907, after legal complications), the IRT concluded it was best to convert the line to a heavy rapid transit style of operation. In order to have the tunnels in service as quickly as possible, only the reconstruction necessary for temporary rapid transit operation was carried out prior to the official opening. More extensive projects, such as platform lengthening, replacement of the old Van Alst Avenue trolley loop, and the extension to Queensboro Plaza, were completed later.

Standard IRT cars were impractical because of the steep grades in the tunnel. IRT car engineers redesigned the Standard car with a slightly lower weight and a gear ratio better suited for operating in this environment. These cars were also the first IRT LV cars. They were classified as Steinway-types and were not operated in the same consist with Standard IRT cars or trailers because of these differences.

Originally, the cars were referred to as Belmont Tunnel cars, but after a period of time they became known as the Boilers. Twelve cars were required for the initial subway service, which began on June 22, 1915, between Grand Central (formerly Lexington Avenue) and Jackson Avenue (now Vernon Boulevard–Jackson Avenue) stations. When the line opened, a single car was used except during rush hours, when a three-car train made the trip. These twelve cars, built in 1915 by Pressed Steel, were quite distinct from other Steinway-types and all Standard LV and HV cars that followed.

Six of the cars, 4026–4028 and 4031–4033, had GE PC-2 controllers and GE-240-C traction motors. Later PC-10-A controllers and GE-259-A traction motors were used. Cars 4025, 4029, 4030, 4034, and 4036 had Westinghouse Electric Company WH 214 controllers and car 4035 had a WH 214-B controller. These last six cars had WH 302-F2 traction motors. Two motors were located in the No. 2 truck of each car.

One of two features of the Steinway car may account for the name *Boiler:* the large air reservoir tank located in the motorman's cab or leaking air at the coupler head that made a hissing sound as the car operated.

---

## 1915 Flivver

The Flivver Class: An IRT Hybrid
Motors (4037–4160), Trailers (4161–4214)

From the start of service on the IRT subway until 1915, all IRT cars had HV electrical control equipment, which was state-of-the-art at the time. Frank Julian Sprague discovered that several motors could be turned on, accelerated, and turned off simultaneously from a single source. This was the basis for the first MU electric railcar, and most rapid transit cars followed this concept until 1915, when car builders found that battery power could be used to send electrical impulses between the master controller in the cab and the underfloor controller and that acceleration of the traction motors could be

achieved automatically. The IRT first used this concept on the twelve Steinway cars that were delivered early in 1915. The IRT would also apply it to the following group of subway cars, the Flivver class.

A large order for 478 car bodies was placed and these began arriving in the New York Central's High Bridge Yard in mid-1915. This was the location where the work would be done. Pullman Car Company was the builder. These cars were to replace the Composite cars in the subway by order of the Public Service Commission. Although billed as an exchange, the process was much more involved. When the new cars arrived, there were 124 Composite motors left in subway service. The trucks and many of the operating components were transferred to the steel bodies, resulting in 124 Flivver motors. Sixty-two Composite trailers underwent the same procedure and an identical number of Flivver trailers were created. At the same time, LV controls and equipment, previously ordered, were incorporated into the cars, making them the first IRT LV fleet (original Steinways excepted).

The Flivvers were true hybrid cars. They had all the pneumatic equipment that was used in the HV cars until that time, but in an LV configuration. The brake system was WH's Triple-R braking using its ME-21 brake valve. The master controller and control box were similar in outward appearance to those of the HVs, but operation was quite different. HV cars had manual acceleration; there were 10 points of power in the master controller, 5 in series and 5 in parallel. In addition, a heavy-duty spring prevented the motorman from accelerating too quickly. The Flivver's controller had only 3 points of power, as the automatic accelerator relay accelerated the cars to the proper switching, series, or parallel speeds. Because of these differences, Flivvers could not normally be operated in trains with other car classes. Pullman built all 186 cars and delivered them in 1915. In 1929, eight of the trailers (4215–4222) were converted to Steinway motor cars.

The Flivvers ran mostly on the Seventh Avenue Express (today's No. 2 Line) and, beginning in the 1950s, on the Lexington–Jerome Avenue and Lexington–White Plains Road Expresses (today's No. 4 and No. 5 Lines). Their last assignment was on the Lexington Avenue–White Plains Road Express, where they ran in ten-car trains (all motors, no trailers). They were removed from service in 1962. None survive today.

Three possible reasons have been given for their name: (1) the lighter weight of the cars; (2) the door control cranks at the car ends, suggestive of the crank on the automobiles of the time; (3) the behavior of the individual cars (cars with similar characteristics had to be assembled in the same train; if this was not done, the cars would buck and not behave uniformly in acceleration and braking, which gave a rough ride reminiscent of a Flivver automobile). The balance of 292 Composites in service gave up their trucks and other equipment to steel car bodies, which resulted in the creation of the same number of HV trailers.

### 1915, 1916 Steinway

Steinway Cars Converted from Other Car Classes
(4215–4222, 4555–4576)

The IRT was noted for its frugal operating practices, which were made necessary by ever-increasing costs and a refusal by the municipal government to permit fare increases. Traffic on IRT lines in The Bronx, Manhattan, and Brooklyn grew slowly because the areas that they served were already developed. However, traffic on the Flushing and Astoria Lines skyrocketed between 1917 and 1929 because of the construction of large apartment houses in Jackson Heights.

Faced with overcrowded trains on these lines and a lack of money to buy additional equipment, the IRT converted eight Flivver trailer cars (4215–4222) and twenty-two LV trailers (4555–4576) to Steinway motor cars in 1929. Pullman Company built the Flivver trailers in 1915 and the LV trailers in 1916. Both types of converted Steinway cars

4558

had GE PC-10-A controllers with GE 259 or WH 302-F motors, two per car in the No. 2 truck as usual. These Steinway-type motor cars had WH ME-30 brake valves.

When these and the other Steinway cars were transferred to the main line from Flushing and Astoria between 1948 and 1950, a red line was placed beneath the exterior car numbers to differentiate them from the LVs, Flivvers, and HVs. Cars 4215–4219 were the last group of cars assigned to the Polo Grounds Shuttle from mid-1956 until service on that line ended on August 31, 1958.

## 1915 HV Trailers

HV Trailers (4223–4514)

In conjunction with the 1915 Pullman order for the 478 car bodies, 292 HV trailers were created using the trucks and other equipment from the remaining Composite trailers. These HV trailers replaced the Composite trailers in subway service to keep pace with the expansion of the existing subway lines.

In the early 1920s, several groups in this series of cars received MUDC. In 1936 the balance were equipped. At the time a white line was placed under the car number on some of the cars to identify them as local service cars. Later, additional cars received this delineation.

In 1952 and again in 1955, several of these cars were converted to "blind" motors (motor cars without motorman's cabs). They were designated by a red *M* placed above the exterior car number. On the interiors, the red *M* was placed in front of the car numbers. The motors and other equipment were taken from Gibbs, Deck Roof, and Hedley HV cars with hand-operated doors that were being scrapped at the time.

## 1916 LV Motors and LV Trailers

LV Motors (4577–4699, 4719, 4772–4810)
LV Trailers (4515–4554, 4811–4825)

As the Composite cars were reequipped and transferred to the elevated lines, new cars were needed to fill the gap and provide equipment for planned extensions and new construction. In the interest of economy, some were trailer cars and some motor cars. This order of IRT cars would be known as the 311 type. The Pullman Company was the builder. The LV trailers (77 cars) were originally divided into 2 numerical groupings: 4515–4576 and 4811–4825. The LV motors (163 cars) were also divided into 2 groups: 4577–4699 and 4771–4810. The Steinways (71 cars) were originally numbered 4700–4770. The Steinways were delivered first, followed by the LV motors, and finally the LV trailers with some overlap between the last 2 groups. The LV trailers continued the numerical sequence from where the HV trailers

4531
4531
DANGER

ended. The first grouping in this series of trailers originally consisted of 62 cars. After the conversion of 22 cars to Steinway motors had taken place in 1929, the balance of 40 and the second group of 15 cars remained in main line operation until 1963–64.

Car 4719 was originally a Steinway and was converted to an LV; car 4771 was formerly an LV and converted to a Steinway. This was done because the Second Avenue Elevated was discontinued on June 13, 1942, and this had provided the track connection between the 147th Street Shop and the Flushing Line. Car 4719 was undergoing maintenance at the 147th Street Shop and was not going to be ready in time so the cars were switched instead.

Cars 4581 and 4583–4605 were transferred to the BMT Division and ran on the Culver and Franklin Shuttles in 1959 and 1960. Extension sills added to the sides of the cars compensated for the narrower width of the IRT cars.

---

## 1916 Steinway

Steinway Cars Built by Pullman (4700–4718, 4720–4771)

The Queensboro Subway, the first rapid transit line from Manhattan to Queens, opened on June 22, 1915. It provided service between Lexington Avenue (now Grand Central) and Jackson Avenue (now Vernon Boulevard–Jackson Avenue). After the opening, the IRT pushed to rebuild the Steinway trolley tunnels for rapid transit use and to extend the line to Queensboro Plaza, which occurred on November 5, 1916. Work was already in progress on further extensions: the Astoria Line to Ditmars Avenue (now Boulevard) and the Corona Line to 104th Street–Alburtis Avenue (now 103rd Street–Corona Plaza), which opened on February 1, 1917, and on April 21, 1917, respectively.

To provide equipment for service to these extensions, the IRT ordered seventy-one Steinway-type motor cars from the Pullman Company in 1916 in addition to other standard LV motor and trailer cars for the IRT main line extensions in The Bronx and Brooklyn. The seventy-one Steinway-type motor cars—4700–4770 (the original order), later 4700–4771 except 4719—featured the WH ABF Group model with model 214-B controllers, WH 302 traction motors, ME-23 brake valves, and CJ-131-8 master controllers. These seventy-one cars and the twenty-five additional Steinway cars in 1925 proved insufficient for the needs of the Astoria and Corona Lines, as traffic increased beyond the IRT's first expectations.

By 1923, ten-car trains were possible because of the elimination of the separation between IRT and BMT portions of stations on the Astoria and Corona extensions and the lengthening of subway station platforms in the Steinway Tunnel. To provide for ten-car train operation, twenty-two LV trailers, also built by Pullman in 1916, were converted to Steinway-type motor cars along with eight Flivver trailers.

## 1917 LV Motors and LV Trailers

LV Motor Cars (4966–5302), LV Trailer Cars (4826–4965)

These 1917 cars, known as the 477 type, consisted of 140 trailers numbered 4826–4965 and 337 motors numbered 4966–5302. The Pullman Company built all of them.

Several changes were incorporated into this group. All of the motor cars and twenty-eight of the trailers were equipped with brass window sashes, and all the cars had masonite ceilings. Some people thought these cars were cleaner in appearance since they did not

COUGH

have the customary row of lights down the center of the car body. The cars provided service on the Jerome Avenue, White Plains Road, and Pelham Bay Park extensions.

The IRT introduced a soundproof or noiseless train on December 4, 1933, and cars from this series were modified for this experiment. These same cars and a few others were part of another test in August 1941, when they were used in the air vent or blower train experiments.

Pullman converted the last car in the series, 5302, from a passenger car to a revenue collection car. It had wooden sashes similar to the 1916 LV cars and no center door. For security precautions, this car was designed to look as much as possible like a passenger car to detract attention. It was scrapped in 1979.

## 1922 LV Trailers

LV Trailers (5303–5402)

In 1922, it was decided to place the final order for LV trailers, a total of one hundred cars, with the Pullman Company. All were ordered, delivered, and placed in service in 1923. They were referred to in IRT records as 100 Lot Subway Trailers and 1922 Type Trailers. These cars had the traditional wooden sashes and the center row of lights down the middle of the car body. This was the first series of cars to have door guards on the exterior edges of the door pockets; all future IRT car orders would include them. Seventy-five of the cars had CP-28 compressors installed, the only trailer cars on the IRT system so equipped. They were also the only trailers to have third rail shoes on both trucks.

In 1924, after the IRT had decided that it would no longer run ten-car trains with four trailers, the compressors on the trailer cars became irrelevant. The IRT planned to remove the compressors and install them on the new motor cars arriving from ACF. This was not done and many trailers retained them until they were scrapped in the 1960s.

### 1924 LV Motors

ACF LV Cars (5403–5502)

The IRT placed the next to last order (the "100 lot order") of LV motor cars with the ACF in 1924. These cars were needed for added service on the last extensions under the IRT administration, the Nostrand and Livonia Avenue Lines. These cars had the traditional wooden sashes and other austerity features found on the 1922 Pullman trailers.

On the IRT lines, the trailers were incorporated into the trains of motor cars and generally operated as three trailers to a ten-car train. The trailer cars were placed as the second, ninth, and usually the fourth cars from the north end of the train. The reason for this was that if the train was cut at midnight, it would be divided into five-car trains. Alternate trains would be laid up (stored on tracks in specific locations when not in use [i.e., on weekends, after rush hour] on center tracks or yards). One train would have three motors and two trailer cars, and the other train would have four motors and one trailer car. After World War II, seven-car trains would often be run so that three cars—a motor car, a trailer car, and a motor car—could easily be dropped. The seven-car train was made up of five motors and two trailer cars. This was primarily done because a seven-car train required only one conductor. Under union agreement, a ten-car train

had to have two conductors: one conductor between the first and second cars and one between the ninth and tenth cars. This practice was maintained on IRT lines until the 1960s.

LV cars generally were used on the Seventh Avenue Express or the Lexington–White Plains and Lexington–Jerome Avenue Expresses. In fact, LVs rarely appeared in local service, except during the midnight hours and on weekends, when there were completely different routings/services. The LVs were considered safer as they did not have the 600 volt direct current (VDC) from the third rail in the master controller in the motorman's cab. The LV's master controller used power from a 32 volt (V) battery to operate the 600 VDC group switches under the car, reducing the risk of a crew member coming in contact with the 600 VDC supply.

### 1925 LV Motors

ACF LV Cars (5503–5627)

The IRT purchased 150 motor cars from the ACF of Berwick, Pennsylvania (the "150 lot order"), in 1925. Of these, twenty-five were of the Steinway-type configuration (see the following section) and 125 were of the LV design. The LVs were purchased to provide the additional trains needed for expansion of the subway system. Shuttle trains consisting of Composite cars were operated on lightly used portions of the outer ends of the system. The Livonia Avenue Line was nearing completion to New Lots Avenue, so the ACF LVs were used to provide a through service to all IRT terminals as well as the new service to New Lots Avenue.

Featured on these cars were GE PC-10-K-1 controllers with GE model 260-D traction motors, two per car in the No. 2 truck, as usual. These 125 LVs had WH ME-30 brake valves and WH D-3-F compressors—the first use of this type valve on the IRT along with the twenty-five Steinway-type cars that were purchased as part of this order of cars.

The 1925 LV cars had window sashes made of wood. The IRT had discovered it was more costly to maintain the brass sashes on older cars, so when the order for cars was placed with ACF, the IRT specified wooden window frames.

All standard LV cars operated interchangeably. Ten-car trains consisted of seven motor cars and three trailers. On weekends, when seven-car trains were operated, the trains consisted of five motor cars and two trailer cars. The cars saw service principally on the Lexington Avenue Express (today's No. 4 and No. 5 Lines). With the arrival of the R36 class cars, the 1925 LVs were retired by 1964 and were used in work service for many years.

Some of these cars have survived. Cars 5443 and 5483 along with two 1917 LVs, cars 5290 and 5292, are currently under restoration at Coney Island Yard. Car 5466 is preserved in the Shoreline Trolley Museum in Branford, Connecticut. Car 5600 is in Kingston with the Trolley Museum of New York, and Car 4902 is at the New York Transit Museum. Conceivably, if someone had the desire and the means, a seven-car weekend train from years gone by could be put together and run on the New York City transit system. World's Fair car 5655, which was converted to a standard LV when it was used as a work car, is also still in existence. So a standard eight-car Flushing Line train could be assembled and run today as well.

## 1925 Steinway

ACF (5628–5652)

These twenty-five Steinways, part of the 150-car lot purchased from the ACF, were acquired to provide the additional trains needed when the line was extended from the 104th Street–Alburtis Avenue (now 103rd Street–Corona Plaza) station to the current terminal at Main Street and westward under Forty-first Street to Times Square from Grand Central.

These cars featured PC-10-K-1 GE controllers with GE model 259-A traction motors, two per car in the No. 2 truck, as usual. These twenty-five cars had WH ME-30 brake valves; this was the first use of this type valve on IRT cars. Also, these cars were the last purchased by the IRT that were built to its Standard car body design.

In June 1942, just prior to the severing of the Flushing and Astoria Lines from the rest of the IRT system, thirty-five of the ACF LV motor cars (5593–5627) along with seven 1916 Pullman Company trailer cars were transferred from the IRT main line to the Corona Mainte-

nance Shop. These provided a sufficient number of spare cars for the additional time required to transfer cars from the Coney Island Overhaul Shops to the Corona Maintenance Shop. With the closing of the Second Avenue Elevated Line in 1942, the Queensboro Bridge connection to the IRT was lost. The Flushing and Astoria Lines were now isolated from the rest of the IRT system, including the 148th Street Overhaul Shops. These forty-two LV cars remained in the LV standard gear configuration; they were never converted to the Steinway-type gear ratio.

As the R12, R14, and R15 cars were placed in service between 1948 and 1950, Steinway cars were transferred to the IRT main lines. The above-mentioned ACF LV cars were the first cars to return and were quite distinguishable from other IRT equipment. They had not received the overhaul, rehabilitation, and modernization that all other IRT subway cars had received at the 148th Street Shops between 1941 and 1946. They still had the drop-seats near the center doors and an apple-green interior.

## 1938 World's Fair Steinway

World's Fair Steinway Cars: The Last IRT Company Cars (5653–5702)

In 1938, with the impending opening of the first World's Fair in Flushing Meadow Park and the proposed establishment of express service on the Flushing Line to handle increased traffic, the IRT purchased redesigned Steinway cars. Fifty cars were received from the St. Louis Car Company starting in September 1938, built to operate in trains of existing Steinway cars.

The IRT and the Board of Transportation designed these cars jointly, and as a result the World's Fair Steinways were a mix of characteristics of IRT and IND design practices of the time. They had a new door placement and an ogee roof. Each single-ended car was an independent unit, with the motorman's cab only at the No. 1 end and a conductor's position only at the No. 2 end. The cars were coupled in pairs at the No. 2 ends, facing each other. This arrangement allowed access to the conductor's door controls on both cars. With no cab at

the No. 2 end and a double passenger seat installed in the motorman's cab, which was accessible when the cab door was locked in the inside position, longitudinal seating capacity increased to forty-eight.

An unusual feature was a two-line, back-lighted route and destination front roll sign; these were the only cars so equipped in the IRT system. Introduced were the small-size sign plates and racks that were also retrofitted into the existing Steinways. This new arrangement facilitated quick sign changes at the terminals. The regular Steinway cars had the small signs and racks removed when they were transferred to the main line upon the arrival of the R12, R14, and R15 cars, but the World's Fair Steinways retained them until 1968. The cars were equipped with the WH ABF-UP-231-B control group, WH XM-129 master controller, AMUE brake equipment, with WABCO D-2-F compressors and ME-23 brake valves. Two WH 336-A1 motors were in the No. 2 truck only, standard practice on the IRT system. Another unusual feature was the 22 bulb, 30 V light circuit, arranged in two 600 V series circuits.

The original blueprints and the car mockup indicated that the cars would be numbered in the 6000 series. This was changed sometime

before delivery. When the cars initially ran on the Flushing Line, they were kept in separate trains and not mixed with other Steinways. Later they were occasionally mixed on the Flushing Line and on the Seventh Avenue Express. The cars ran in trains on the Flushing Line along with other Steinway cars until 1950, when they were replaced by the R15 class cars. All fifty World's Fair cars were reassigned to the Lexington Avenue Local (today's No. 6 Line), where they ran until the delivery of the R17 cars in 1955–56.

World's Fair cars were the only Steinway cars that never received the red line under the exterior car numbers. A test in the mid-1950s established that a six-car train of World's Fair cars would have door clearance problems on the Third Avenue Elevated in The Bronx. They were transferred to the Seventh Avenue Express (today's No. 2 and 3 Lines) and the Forty-second Street Shuttle (today's S Line). Further deliveries of new cars meant relegation to the Third Avenue Elevated by February 4, 1962, where they finished their passenger service on November 3, 1969. When running on the Third Avenue Elevated in The Bronx, they ran in five-car consists with an LV trailer at the center and two World's Fair cars at each end. Regular Steinways were substituted for them in these consists as necessary. All but one car was scrapped; car 5655 is under restoration at Coney Island Yard.

| Year Ordered and/or Built | Car Numbers | No. of Cars | Elevated Name/Designation | Description (page) | Technical Data (page) |
|---|---|---|---|---|---|
| 1884 | 1–51 | 51 | BU, Trailer | 110–11 | 394 |
| 1887, 1891, 1893 | 52–190 | 139 | BU, Trailer | 112 | 395 |
| 1888–89, 1893 | 191–271 | 81 | BU, Kings County Trailer | 113–14 | 396 |
| 1887 | 600, 601, 683 | 3 | BU, 600 series, Motor | 114–15 | 397 |
| 1891–93 | 602–619 | 18 | BU, 600 series, Motor | 116–17 | 398 |
| 1898 | 620–627 | 8 | BU, 600 series, Motor | 117–19 | 399 |
| 1901 | 633–682 | 50 | BU, 600 series, Motor | 119–20 | 400 |
| 1888 | 700–760 | 61 | BU, 700 series, Motor | 120–22 | 401 |
| 1884 | 800–832 | 33 | BU, 800 series, Motor | 122–23 | 402 |
| 1887 | 833–858 | 26 | BU, 800 series, Motor | 124–25 | 403 |
| 1898 | 900–936 | 37 | BU, 900 series, Motor | 126–27 | 404 |
| 1900 | 937–940 | 4 | BU, Brill 900 series, Motor | 128–29 | 405 |
| 1908 | 998 | 1 | BU, 900 series, Motor | 129–31 | 406 |
| 1903 | 1000–1119 | 120 | BU, Convertible 1000 series, Motor | 131–33 | 407 |
| 1904–5 | 1200–1299 | 100 | BU, 1200 series, Motor | 134–37 | 408 |
| 1908–14 | 1261, 1282, 1283, 1286, 1287 | 5 | BU, 1200 series, Motor | 134–37 | 409 |
| 1905–6 | 1300–1399 | 100 | BU, Convertible 1300 series, Motor | 137–41 | 410 |
| 1908 | 1400–1499 | 100 | BU, 1400 series, Motor | 141–43 | 411 |
| 1909 | 1448, 1482 | 2 | BU, 1400 series, Motor | 144–45 | 412 |
| 1893 & 1908 | 1500, 1501, A, B, C | 6 | C (Rebuilt 1923–25) | 146–49 | 413 |
| 1893 & 1908 | 1502–1526, A, B, C | 75 | C (Rebuilt 1925) | 146–49 | 414 |
| 1904–8 | 1600–1629 A, B, C | 90 | Q (Rebuilt 1938–41) | 149–53 | 415 |
| 1904–8 | 1600–1629 A, B, C | 72 | Q (Overhauled 1958 & 1962) | 149–53 | 416 |
| 1904–8 | 1630–1642 A, B | 26 | QX (Rebuilt 1939–40) | 149–53 | 417 |

| Year Ordered and/or Built | Car Numbers | No. of Cars | Subway Name/Designation | Description (page) | Technical Data (page) |
|---|---|---|---|---|---|
| 1914–17 | 2000–2399 | 400 | B, 2000 series, Motor | 154–57 | 418 |
| 1918 | 2400–2499 | 100 | B, BT, BX 2400 series, Motor | 154–57 | 419 |
| 1919 | 2500–2599 | 100 | B, 2500 series, Motor | 158–59 | 420 |
| 1920–22 | 2600–2899 | 300 | A, B, Motor | 159–61 | 421 |
| 1924 | 4000–4049 | 50 | AX, BX, Trailer | 161–62 | 422 |
| 1925, 1927–28 | 6000–6120, A, B, C | 121 | D, Triplex | 163–65 | 423 |
| 1934 | 7003 A, B, C, B1, A1 | 1 | MS, Multi-Section "Green Hornet" | 165–68 | 424 |
| 1936 | 7004–7013 A, B, C, B1, A1 | 10 | MS, Multi-Section | 168–71 | 425 |
| 1936 | 7014–7028 A, B, C, B1, A1 | 15 | MS, Multi-Section | 169–71 | 426 |
| 1934 | 7029 A, B, C, B1, A1 | 1 | MS, Multi-Section "Zephyr" | 166–68 | 427 |
| 1938 & 1940 | 8000–8005 A, B, A1 | 6 | Compartment "Bluebird" | 171–74 | 428 |
| 1925 | 2900–2924 | 25 | SIRT Motor | 174–77 | 429 |
| 1925, 1926 | 300–389, 500–509 | 100 | SIRT Motor & Trailer | 174–77 | 430 |

# VINTAGE BMT CARS

**The Brooklyn Rapid Transit Company and the Brooklyn–Manhattan Transit Corporation**

The Coney Island Elevated Railway, Brooklyn's first Elevated, was a mile-long structure completed in 1881. It ran from the Brighton Hotel to a terminal just west of the Culver depot in Coney Island. In 1886 the company was reorganized as the Sea View Railroad, and in 1897 BRT purchased and electrified it. During this period the New York and Brooklyn Bridge Railroad was formed. (The Brooklyn Bridge was originally known as the New York and Brooklyn Bridge, as Brooklyn was a separate city until 1898.) Its cars were larger than the standard elevated cars and were pulled by cables. The New York and Brooklyn Bridge Railroad began operation over the bridge on September 24, 1883. It was operated with cable although steam engines switched trains at the terminals and pulled passenger cars across if the cable was inoperative. The Brooklyn Bridge Railroad was electrified in November 1896, but cable remained the means of propulsion. The trackage was leased by the elevated companies in August 1897 and eventually through trains replaced the cable operation, which lasted until 1908.

In 1884 the Brooklyn Elevated Railroad Company was formed to operate an elevated railroad line from Fulton Ferry through various streets of downtown Brooklyn, to Lexington Avenue, and along Lexington Avenue to Broadway and Gates Avenue (all in Brooklyn). The lines were operated by steam locomotives pulling wooden coaches (or trailer cars).

NEW YORK
101
101

In 1887, the Union Elevated Railroad was formed to build and operate elevated railways along Myrtle Avenue, Broadway, and Fifth Avenue in Brooklyn. On March 28, 1890, a franchise was granted to the Seaside & Brooklyn Bridge Elevated Railroad Company to build a line connecting with the Union Elevated Railroad at Fifth Avenue and Thirty-seventh Street, continuing over private property, to Thirty-eighth Street and south along Third Avenue to the city line (Sixty-fifth Street).

The Brooklyn Elevated Railroad immediately acquired the controlling interest in the Union Elevated Railroad, the Seaside & Brooklyn Bridge, and additional cars were purchased. By 1899, the BRT, which was the holding company operating most of the street cars in Brooklyn, acquired the two elevated railroads and merged them into a single company known as the Brooklyn Union Elevated Railroad Company. These cars have since been known as BU cars, and the open platform models as gate cars.

The Kings County Elevated Railroad was also formed in 1887 to operate an elevated line along Fulton Street, Brooklyn. It placed an order with the Pullman Company in 1888 for sixty-one trailer cars. They were essentially similar to others in the Brooklyn fleet at the time. No doubt standardization of car design aided the ease of maintenance, something that is still in practice to this day.

Trains were initially hauled by "Puffing Billies," small locomotives built by H. K. Porter Company, Baldwin Locomotive Works, and others. Their main problems, as with their Manhattan counterparts, were smoke, cinders, and changing ends at the terminals. The high upkeep costs of boiler repair and coal-handling facilities also made them expensive to operate.

Electricity was still viewed as a marvel at the turn of the century, but nowhere was it more applicable than to mass transit. In a few years clean electric power virtually replaced the "Soot Spewers," as those who lived in proximity to El lines referred to these locomotives.

By the turn of the century, the BRT had acquired the Kings County Elevated Railroad Company and obtained rights to operate their trains over the Brooklyn Bridge. It immediately electrified the

track on the bridge and purchased the first electrically operated elevated cars in 1898. They were built with center doors to provide easy loading and unloading at Sands Street (Brooklyn) and Park Row (Manhattan). The original cable cars were too large for regular elevated operation and were retired as soon as new cars were acquired. In 1901, the BRT purchased fifty additional elevated electric cars.

Throughout the 1880s, both the Kings County and Brooklyn Union Companies purchased a number of cars to service the extensions to newly developing communities. During the late 1890s, the BRT acquired several steam-operated railroads to the seashore at Coney Island and Brighton Beach, and in 1906, the Canarsie and Rockaway Beach Railroad Company.

The Coney Island Lines included the Prospect Park and Coney Island Railroad Company (originally owned by Andrew Culver and known as the Culver Line), the Sea Beach Railroad Company, the

Brooklyn, Bath and West End Railroad Company, and the Brooklyn, Flatbush and Coney Island Railroad. All of these companies were incorporated into the BRT and turned over to its operating companies (as it was the controlling or holding company). The Culver Line was reorganized into the South Brooklyn Railroad Company shortly after the turn of the century and was assigned the freight operation of the BRT as well as the operation of the Culver Line.

696

BROOKLYN RAPID TRANSIT
697
697

Between 1900 and 1908, the BRT electrified these railroads, connected them with existing elevated lines, and purchased over 425 additional elevated cars to replace the old steam coaches. Some of the cars acquired from the railroad companies were fitted for electric operation and, renumbered in the surface series, were equipped with trolley poles and operated on the surface of the Sea Beach and West End Lines. Others were retained as work cars but their large size made them impractical for elevated line operation. They were all scrapped in 1905.

Many of the older steam coaches built for the original elevated lines were electrified. Because of the surface operations of the former railroads, as well as some of the extensions of the elevated lines into the suburban areas, virtually all of the electrified and newly acquired cars were equipped with a hinged platform that could be raised as well as steps under these platforms for surface loading. In addition, motor cars had trolley poles installed and brackets placed at the center of the roofs on each end of the cars for mounting carbon arc headlights. All cars had registers so conductors and trainmen could ring up fares collected on board the trains. The exception was the 900–935 group, which was never equipped for surface operation. All of this equipment was removed by 1923.

(*Note:* The steam coach and elevated car history for Brooklyn is very convoluted due to the mergers of various companies and the reconstruction of cars as the system moved from steam power to electric operation. Many records have been lost over the years, if indeed they even survived their own era. Rather than try to differentiate these cars into their "as delivered" categories as we have throughout most of this book, we organized the BU series by the cars' final numbering. Exceptions do exist, and some cars appear in more than one section because of the complex choreography of their history.)

## 1884 BU 1–51 Series

Brooklyn Union (1–51)

Built by the Pullman Company, these cars were virtually identical to those constructed for the Manhattan Railway Lines. They were originally included in the Brooklyn Elevated Railroad 100–189 group and were rebuilt for electric service between 1904 and 1906 at a cost of $3,000 per car. In 1923, shortly before its reorganization into the Brooklyn–Manhattan Transit Corporation, the BRT was granted rights to operate over the tracks of the IRT to Astoria and Flushing, and a number of these cars were transferred to that service. Most were retired in 1940, with a few surviving into World War II.

26

## 1887 BU 52–190 Series

Brooklyn Union (52–190)

These cars were built for the Brooklyn Elevated Railroad and were included in the 190–309 group. They were rebuilt for electric operation between 1904 and 1906. Cars from this group were selected to be included as the center cars of the reconstructed C units formed between 1923 and 1925. These rebuilt cars remained in service until 1956, while the others were retired during the 1940s.

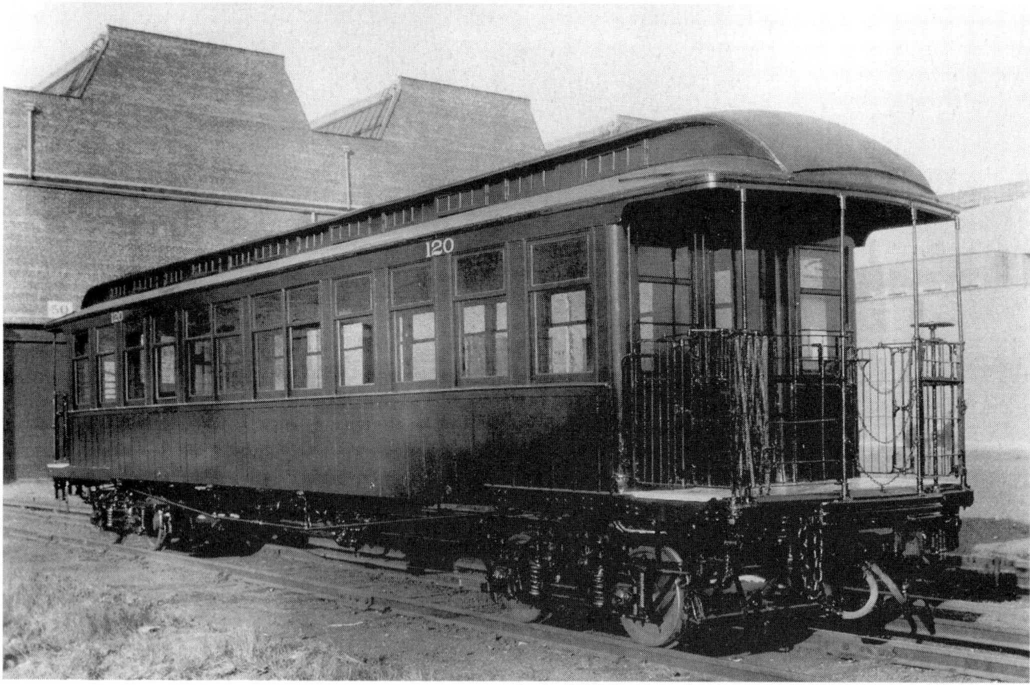

## 1887 and 1893 BU 191–271 Series

Kings County Elevated (191–271)

The Fulton Street Elevated Line was inaugurated on April 24, 1888, with cars being pulled by steam locomotives. The cars were later converted to electric operation between 1904 and early 1906. They generally remained in the Eastern Division of the BMT, although they were occasionally used in Southern Division service. The last of the cars were retired on June 26, 1952, when C units were assigned to replace the BU cars serving the West End Shuttle during rush hours. The Track Department at the Ocean Parkway station of the Brighton Line used car 218 as a storeroom until 1960. Car 197 was sold to the Branford Electric Railway Association at East Haven, Connecticut, in 1950.

### 1887 BU 600 Series

Brooklyn Union (600, 601, 683)

Originally these cars were Brooklyn Elevated Railroad cars 145, 190, and 188, respectively. All three cars had center doors added at an early date. Cars 600 and 601 were converted into motor cars with motorman's cabs constructed on the open platforms. In June 1906, they were reconstructed. The cabs were moved inside; the platforms, hoods, and seats were lengthened; and the cars were rewired and fireproofed, with wires running through conduits. The motors, controls, and air brakes were overhauled, and Gibbs suspension was provided on the motor trucks. At the same time, car 683 was reconstructed into a motor car and equipped with two WH 50-E motors, WH #131 control, and J. G. Brill Company motor and trailer trucks with air brakes. The wiring, trucks, and body received the same treatment as cars 600 and 601. Later, car 683 received Peckham-40 trucks. All three cars had been withdrawn from service by the time the city acquired the BMT on June 1, 1940, and remained in storage until they were scrapped in 1946.

600
600

## 1891–1893 BU 600 Series

Brooklyn Union (602–619)

Originally built as steam coaches for the Brooklyn Elevated Railroad and numbered in the 231–301 series, these cars were converted into motor cars with the electrification of the Els and equipped with outside motorman's cabs and Sprague-type controls. They were then renumbered 412, 413, and 415–431. Between February 21 and May 11, 1905, they were reconstructed at the cost of $5,000 per car. The platforms and hoods were lengthened, and new gates and inside motorman's cabs were installed. WH-type #131 controls were substituted for the Sprague-type, the trucks were overhauled and strengthened, new drawbars were installed, and the cars were rewired and fireproofed. Electric lamps and heaters were substituted for oil lamps and Baker heaters. The cars were renumbered 602–619. Cars 608, 616, and 618 were equipped with two WH 50-E motors each and

standard Christensen Engineering Company air brakes, thereby making them conform to the rest of the series. When the city acquired the BMT on June 1, 1940, the Fulton Street Elevated west of Rockaway Avenue was discontinued. All of the remaining cars in the series were placed in storage until they were scrapped in 1946. Fifth Avenue Elevated car 609 was assigned to the rubbish collection train, where it served until 1950.

## 1898 BU 600 Series

Brooklyn Union (620–627)

Cars 620–627 were the first electric cars on the Brooklyn Elevated system and were built with center doors for Brooklyn Bridge service. They were originally numbered 400, 404, and 406–411, respectively. In 1904 and 1905, their bodies were reconstructed and inside motorman's cabs were installed. The platforms and hoods were lengthened,

and new gates, new heaters, and drawbars were added. Oil lamps were replaced by electric lights and the Sprague control by a WH unit switch group type. All the rest of the equipment was removed, cleaned, repaired, and reinstalled and the cars were completely rewired and repainted. When the city acquired the BMT on June 1, 1940, the entire 600 series was placed in storage but these cars were later overhauled and returned to service by the spring of 1941. The cars were finally retired after the Lexington Avenue El was discontinued on October 13, 1950.

### 1901 BU 633–682 Series

Brooklyn Union (633–682)

Originally numbered 450–499, these cars were built as motor cars for the newly electrified elevated lines and included center doors for Brooklyn Bridge operation. The doors were permanently sealed and longitudinal seats placed across them during the 1930s. When the

city acquired the BMT in 1940, and discontinued the Fulton Street El west of Rockaway Avenue and the Fifth Avenue El, the cars were withdrawn from service with the intention of converting them to trailers as replacements for the older 100 and 200 series cars. However, when it was discovered that the wooden underframes of the 1000 series were beginning to deteriorate, the city decided to overhaul the 633–682 series as motor cars and they were returned to service beginning in May 1941. They continued to operate until they were retired in 1952.

### 1888 BU 700 Series

Kings County Elevated (700–758), Brooklyn Union (759–760)

The Kings County Elevated Railroad originally numbered these cars 100–159 when they were purchased for the new Fulton Street Elevated Line. Almost from the beginning, they were equipped with center doors for Brooklyn Bridge service. By the turn of the century, they were electrified. By 1905, they required extensive overhauling.

728

Two new WH 50-L motors were substituted for the four previously installed WH-8 motors. Trucks were equipped with Gibbs suspension and new driving axles, new drawbars were installed, and the brake apparatus was completely overhauled. The cars were rewired and fireproofed and the bodies were reconstructed and strengthened. New seats were installed across the side doors. After 1905, few changes were made to the cars other than the removal of trolley poles, steps, headlight brackets, and fare registers, which were required for surface operation.

Two manufacturers supplied the trucks: cars 701, 702, and 704 were equipped with Brill trucks; cars 700, 703, and 705–758 with Peckham-40 trucks. El-type third rail shoes and, later, tongue-shaped shoes were employed. Seating was longitudinal in style.

In 1930, cars 850 and 852 became 759 and 760. Many cars of the 700 series were converted to alcohol and work motors during the 1930s. What is most remarkable about the 700 series is its longevity. These cars had survived roughly seventy years on the system by the time they were scrapped in 1958. Superior maintenance played a major role in this feat.

---

### 1884 BU 800 Series

Brooklyn Union (800–832)

The Pullman Company built this fleet of cars in 1884, to be pulled by coal-burning steam locomotives. Kerosene lamps lit the interiors. Late in the nineteenth century, through the efforts of Frank J. Sprague, among others, MU motive power had become so effective that BRT management decided to implement the electrification of the system between 1898 and 1903. Locomotives were used for a short period because of fleet shortages. Center doors had been installed on these 800 series cars for Brooklyn Bridge service by the turn of the century.

As in other cars, seating was longitudinal. Cars weighed around 60,000 pounds so that lightweight El lines were not overstressed. Center side doors were sealed and seats were installed in the adjacent spaces. Platforms were also lengthened slightly.

The view from the cars was superior as there were more windows with slender columns in between. The arch-shaped side door had generous glass panes as well. Car body width was 8 feet 10.125 inches,

slightly narrower than the norm (8 feet 10.375 inches). Typically when lamping the interiors during electrification, high-glare nonfrosted bulbs were utilized. Later the lighting was improved by the use of frosted-type bulbs. Trucks were replaced with Peckham-40 trucks, retiring the trailer-type trucks. Destination signs were hung on the platform gates for passenger information.

In 1909 the Transit Development Company rebuilt car 817; its number was changed to 1282 in 1930. Its seating style was switched from longitudinal to cross with reversible seat backs. The year 1930 saw the conversion of cars 802 and 822 to work motors. They were renumbered 996 and 997. The rest of the fleet was scrapped just before the Great Depression, during 1924 through 1930.

## 1887 BU 800 Series

Brooklyn Union (833–858)

Gilbert constructed this fleet for the Union Elevated Railroad in 1887. Like most trailer coaches of this vintage, these cars were built to be hauled by small steam locomotives. The wooden bodies did not have side doors when originally built. Under steam operation only gates were needed for regular El service. The cars were not that distinguishable from those of other fleets operated by the BRT, having the usual ten windows per side with thick columns between each and clerestory vents. These cars, however, had four panes of glass on the vents rather than the usual three and seating was longitudinal with rattan cushioning.

The side doors with attractive arch glass were cut into the car body around the early 1900s. Steel and timber reinforcements were required to enhance the weakened area structurally. Side doors were needed for the Park Row terminal, which was located at the foot of

the Brooklyn Bridge on the Manhattan side. Park Row terminal was experiencing extreme crowding conditions and was expanded for longer trains. The new doors aided rush hour traffic immensely. Back then, BRT trains occupied the present left lane. Trolleys used the middle lane, and vehicles that were usually horse-drawn had the right lane. The Sands Street terminal on the Brooklyn side of the bridge accommodated rapid transit as well as trolley lines.

Steam power was reaching the end of its era, and this series of cars needed modification. During the 1898–1903 period, electrification took place and traction motors were added to make the cars self-propelled. Trucks were changed to Peckham-40 trucks on a 6-foot 4-inch wheelbase. Trailer axles were closer to the couplers.

In 1930, cars 850 and 852 were renumbered as 759 and 760. Car 843 became 995, a work motor. All cars were scrapped from 1924 through 1930.

## 1898 BU 900 Series

Brooklyn Union (900–936)

This handsome fleet of cars was designed for service mainly on the Els. At a relatively light weight of about 60,000 pounds, they were perfect for service on Els that had weight load restrictions, such as Fulton Street, Lexington Avenue, and portions of the Jamaica Line. Most of these lines had been constructed in the late 1880s through the early 1890s. Their latticework, longitudinal and cross girders, and support columns (at times mounted atop subterranean brick bases) could not support heavier rolling stock. Their construction, as well as other factors, led to their eventual demise.

Wason built all but one of the 900s (cars 900–935) in 1898. Pullman built car 936 in 1888. Baldwin supplied the trucks. The trailer had a wheelbase of 5 feet; the motor truck had a wheelbase of 5 feet 10 inches, and was made by Wason but was later replaced by the Peckham-40. Frames were prefabricated steel straps riveted into an archbar truss configuration.

The car body had clerestory vent flaps that contained four small window panes each. A manually operated side door was provided for ease of boarding. Top and bottom sills were reinforced for the opening; truss rods and turnbuckles provided additional strength. Metal route and destination signs were hung facing outward on the upper window pane opposite the motorman's cab. Of note, the Wason cars had a width of 8 feet 10.375 inches. The cars were extensively rebuilt in 1905, with improvements similar to other cars previously described.

In 1923, two cars, 913 and 919, were demolished in a crash on the El on Flatbush and Atlantic Avenues. A Pullman-built car constructed as a steam coach numbered 130 in 1888 was rebuilt as a motor car in 1905 and was assigned the number 936. This car was scrapped on November 7, 1929.

## 1900 BU 900 Series

Brooklyn Union (937–940)

In 1900, the Brooklyn Union Elevated Railroad Company purchased five semi-convertible electric motor cars and numbered them 436–440. They were exceptionally fine-looking cars with windows that dropped into the bodies, providing the effect of a half-open car. They were fitted with wood-slat, forward-facing seats that could be reversed.

In 1904 and 1905, the bodies were reconstructed and inside motorman's cabs were installed. The Sprague control was replaced by the WH unit switch group type. Oil lamps were replaced by electric ones; the platforms and hoods were extended; new gates were installed; and all the wiring, brake apparatus, and piping was overhauled. The cars were renumbered 628–632. During the 1920s, the

drop windows were replaced by standard double-hung sashes. On March 11, 1926, car 632 was scrapped, and in August and October 1930, the remaining cars were renumbered 937–940. During another overhaul in 1941, the wood-slat seats were covered with rattan. The cars remained in service through the early 1950s.

### 1908 BU 900 Series

Brooklyn Union (998)

Most people knowledgeable about rapid transit rolling stock never think of the gondola freight car as the design basis of subway transport. But that is what actually happened when the BRT management began thinking "subway." The wooden cars used and still ordered were fine for El service. Subways, however, would require the fireproof protection as well as the strength of steel cars. The IRT's success with the Gibbs car, named after the designer George Gibbs,

whose work in conjunction with Lewis B. Stillwell was a part of new technology in the making, did not go unnoticed by the BRT.

Car 998 was old technology in terms of being a wooden El car made of steel. As a developmental mule, it pointed in a direction the BRT management decided not to pursue for its subway car design. Later development decisions would dictate the innovative and well-received BMT 67-foot Standards. The BRT also took note of the Manhattan IRT MUDC cars, which lowered labor costs by using two-man crews to operate a consist.

Like a gondola car, construction of car 998 made use of side girders as well as a slender spine (center) girder. The advantages were significant. Electrical and brake equipment could be more safely hung beneath the car due to the floor's added strength. Framing from the lower window sills and up aided but was not necessary for car body strength. The seating was Manhattan-style. Known as an "easy access" car, it was normally teamed with gate cars from the 1200 and 1400 series. The car was also similar to those ordered for the Boston

Elevated Railway. El-type trolley poles and later tongue-shaped third rail shoes were used; steps were provided below the doors for low platform operation, although these were later removed along with the trolley poles.

Car 998 did not meet with great success, probably due to its narrow door openings. It ran in regular service only briefly. It performed shuttle service as a single car from the Broadway Ferry to Marcy Avenue on the Broadway–Brooklyn Line until 1916. This spur structure south of the Williamsburg Bridge was used for car storage between rush hours after 1916 and was finally demolished in 1941.

In 1924, car 998 was utilized as a mobile pay car for workers and for revenue collection service. Somewhere around this time, four small fish belly reinforcement plates were installed beneath the bottom sill below the second window columns on each side. In 1946 a center door was added for 998's new assignment as a newspaper bail (rubbish) car. It would survive at the Coney Island Overhaul Shop until its scrap date in 1960.

## 1903 BU Convertible 1000 Series

Brooklyn Union (1000–1119)

When one thinks of a convertible, it's usually an automobile, not a rapid transit car. But the BU convertible car was a hit with straphangers in the 1900s, who cooled off in the summer while riding the 1000 series. As rolling stock goes, this car was a maverick design, and not without problems. The BRT management thought that these cars, manufactured by J. Stephenson in 1903, would boost ridership, particularly during warmer weather on the routes to booming Coney Island and Canarsie Pier, as well as to racetracks in Sheepshead Bay and Gravesend on the Culver Line. Management was right on the money. Passengers usually flocked to these "open siders," as they were known.

The car bodies were constructed almost entirely of wood except for the hidden lower sill girder, which was truss rod and turnbuckle reinforced. This wood construction proved problematic. Due to the side profile of the wood fascia, water would collect and rot out hidden structural members. The 1000 series also was exceptionally vulnerable to electrical fires, caused by water-damaged components that shorted. These failings seemed to have been corrected on the later 1300 series convertibles, which were of a more stout design.

A-305

Unlike most other rolling stock, the 1000s sported arch, not clerestory, roofs on which fourteen ducts standing approximately 8 inches high were mounted. Inside bidirectional register flaps brought ventilation in as the train moved. It did not work as well as the proven clerestory setup. A version of these vents could be found on the BMT Standards, the Multi-Section cars, the 1938 World's Fair cars, and the R10, R12, and R14 cars.

The cars had removable side panels (thirteen per side) and two-window setups: nine had fixed window panes, upper and lower, and four had upper panels that could be lowered. With panels removed, wooden horizontal bars were installed for passenger safety since the window opening was huge. The two windows were removed when the wood sides were modified in the 1920s.

The platforms had split-style gates that folded in half during operation. The steps below the end platforms for low-level boarding were used when in surface service and removed as the Els appeared. The propulsion control cabinet enclosure was mounted behind the motorman's cab. Third rail shoes were U-shaped El-type and later tongue-shaped. Trolley poles were used on early ground-level operations and later removed. Route marker lights as well as headlights were employed. Headlight brackets and fare registers were removed during the 1920s. Seating was cross-style with reversible seat backs made of hardwood. Floors were poured Tucolith, a form of cement that acted as a seal for underfloor mechanisms in wet weather.

Trucks supplied by Peckham had one motored inner axle via spur gear units. These units served in single-car service on the Fifth Avenue El's Bay Ridge branch where the Brooklyn Queens Expressway (BQE) now exists. In 1910, car 1030 was rebuilt by the Transit Development Company, renumbered 1283 in 1930, and scrapped in 1955. Cars 1079 and 1081 were also rebuilt by the same firm in 1908; they were renumbered 1286 and 1287 in 1930. Car 1286 was scrapped by 1955 and car 1287 was scrapped on November 11, 1946. Cars 1093, 1096, 1097, and 1099 were converted to work motors and renumbered 988–991 in 1932 and 1933. By 1946–47, all other units had been scrapped. Today, none survive.

## 1904–1905, 1908–1914 BU 1200 Series

BRT (1200–1299) Series, Brooklyn Union (1261, 1282, 1283, 1286, 1287)

The 1200 series BU motor cars share many similarities with the later 1400 series in both design and components. Three car builders shared in the manufacture of this fleet: Osgood Bradley manufactured cars 1200–1234; Brill, cars 1235–1259; and Laconia Car Company, cars 1260–1299. All were built in 1903. That year was a busy one. The BRT had many low-level station platforms, as was customary with ground-level operations. The original configuration had platform steps built into the open ends. They were removed as Els were built. The cars were built as semi-convertibles with the windows dropping into the car bodies. Standard double-hung windows were installed by the 1920s. A reinforced bottom sill girder was utilized without the need for truss rods and turnbuckles. On the Brill cars, the truck suspension was unique, but was changed later. Bolster springs were metal leaves. The equalizer (primary) had coil springs, and there was an extra one right between the frame pedestals and journal boxes.

The ride quality must have been smooth albeit a bit bouncy over switches. American Locomotive motor trucks had a 6-foot 4-inch wheelbase and St. Louis Car Company trailer trucks had a 5-foot 6-inch wheelbase. Peckham-40 trucks replaced these trucks around 1921. Third rail shoes were U-shaped El-type and, later, modern tongue-shaped. All units utilized trolley poles. Seating was Manhattan-style with rattan cushioning. Floors were later replaced with poured cement, or Tucolith, rather than the more common wood strips originally used on these and other cars. Roofs had the usual marker lights with a provision for a carbon arc headlight that was installed only when cars were operated on surface lines at night.

Out of this fleet, twenty-seven remained as originally configured gate cars. The other seventy-three went on to become C type cars in 1925 and Q type cars in 1938–40. In 1950, car 1205 was converted into a trailer. In the early 1960s, the Branford Electric Railway of East Haven, Connecticut, acquired car 1227. There it has undergone some restoration. Q car 1622B which started life as car 1273, returned to its original gate stock appearance via overhaul at the Coney Island Shop in 1979 for the Diamond Jubilee celebration. It is now in the New York Transit Museum collection. Roof height reduction for tunnel clearances remained. It was also equipped with a modern-style coupler and third rail shoe beam assembly. All other cars were scrapped

DIETZ COAL
Chesterfield

in the 1950s. On an interesting note, while in Q type guise some of these cars were assigned as work equipment on the South Brooklyn Railway and received trolley poles again. Déjà vu!

The five other 1200 series BU motor cars were 1261, 1282, 1283, 1286, and 1287. These cars were rebuilt due to damage during their service on the BRT. Trailer car 82, built by Gilbert & Bush in 1887, damaged and rebuilt in motor car 684 in 1910, was renumbered 1261. Car 817, built by Pullman in 1884, was renumbered 1282. Cars 103, 1079, and 1081, built by J. Stephenson in 1903, were renumbered 1283, 1286, and 1287. These cars had been rebuilt after being damaged in accidents and retained their original numbers until 1930. At that time they were renumbered into the 1200 series to replace cars that had been converted into C units. The Transit Development Company rebuilt these cars sometime between 1908 and 1914. Seating was Manhattan-style on cars 1286 and 1287. Cars 1261, 1282, and 1283 retained the cross-style seating of noncushioned wood slatting with seat backs that were reversible. The upper sills of the car body had the Brooklyn Rapid Transit logo in bold serif lettering. Clerestory vents had two window panes to aid ventilation and illumination.

American Locomotive motor trucks had a wheelbase of 6 feet 4 inches. The trailer had a shorter wheelbase of 5 feet 6 inches and was made by St. Louis Car Company. The motor wheel diameter was 34 inches; the trailer wheel diameter was 31 inches. In 1930 trucks were replaced by Peckham-40 trucks. Since a good part of the BRT was at ground level, steps beneath open platforms were originally provided but were removed in the early 1920s.

Car 1287 was scrapped on November 11, 1946. The four remaining cars were scrapped in 1955 after many decades of service.

### 1905–1906 BU Convertible 1300 Series

BRT (1300–1399) Convertible Series (Brooklyn Union)

Prior to the early 1970s, most mass transit, not only in New York, but elsewhere, did not have air conditioning. Fans and vents circulated hot summer air. This is where the 1300 series really shone. Cincinnati built cars 1300–1349; Jewett, cars 1350–1374; and Laconia, cars 1375–1399. Construction occurred from 1905 through 1906.

Necessity is the mother of invention, as the saying goes. These convertibles embodied this motto. Ventilation was accomplished by

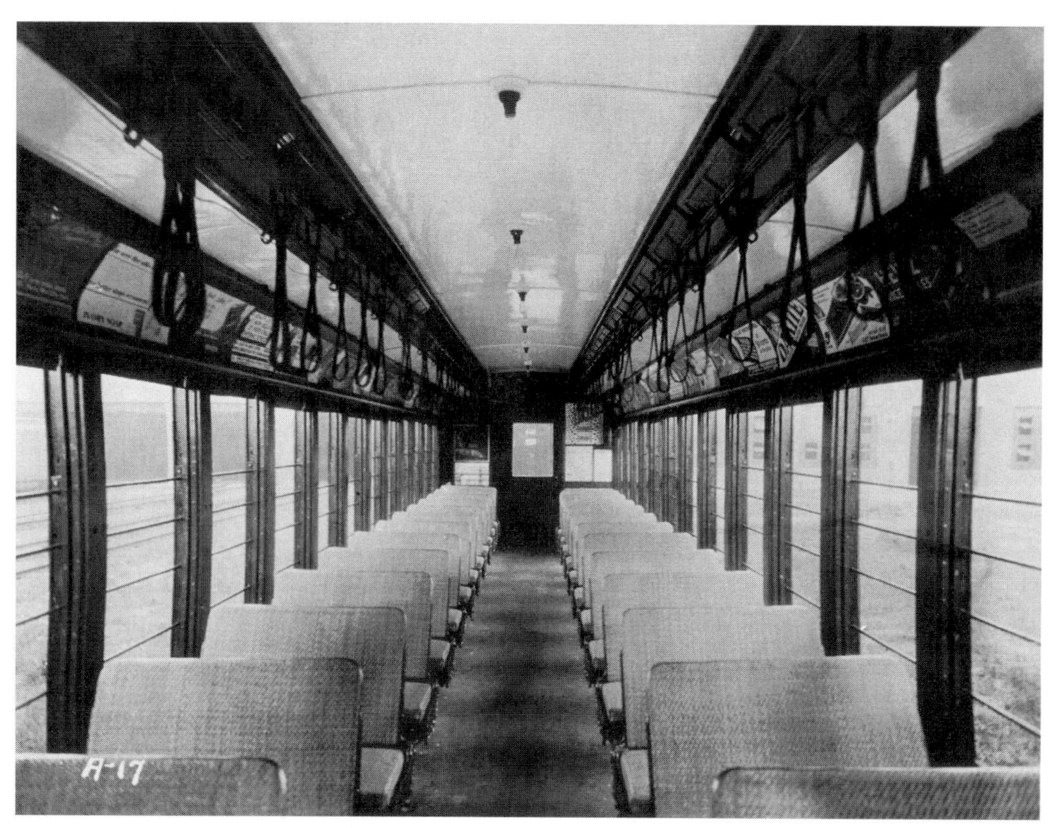
A-17

1312
1312

way of a "skeleton" car design. Removable panels that opened from beneath the roof sill to just below seat cushion level provided superior circulation of air. Twenty-eight single-hung window panels per car were the norm. Up to and throughout the summer of 1933, all panels were removed. From 1934 through 1946, eight panels on each side were taken out; they were reduced to six until 1950, when only four per side were removed. This practice continued until the cars were retired in 1958.

Aiding air movement, clerestory vents with glass panels lined the roof structure. The end doors remained open when the cars were out on various lines. Iron bars were installed horizontally between window columns for summer operation. As with any minimal load-bearing structure, reinforcement was required at times. On the convertibles, it came in the form of a huge horizontal riveted sill girder and platform decking with anticlimbers. Below the sill was a set of truss rods with turnbuckle tensioners.

The roof sported trolley poles for catenary power. It also had a provision for a carbon arc headlight, which was installed when the cars were assigned to lines operating on the surface during night hours. However, marker lights (four per car) as well as kerosene lamps were utilized. Fare registers were installed for use when the cars were on the surface portion of El lines. The trolley poles and fare registers were all removed in the early 1920s. U-shaped and tongue-shaped third rail shoes were also used. Cross-seating with rattan cushions and flip-over seat backs were employed.

Peckham supplied the trucks. They were prefabricated strap steel lengths riveted together in an arch bar (truss) configuration. WH traction motors ran through spur gears connected to the innermost axles. Axles near the couplers were unmotored. Peckham trucks were replaced with American Locomotive units and upgraded WH motors in 1950.

Convertibles were usually mixed with other BU equipment during normal runs over various BRT lines that included the Els on Fulton, Myrtle, Lexington, Fifth Avenue, Brooklyn Bridge, and Canarsie. Early Canarsie service was via the street and private right-of-way to the Canarsie amusement pier (where the Belt Parkway now runs at Rockaway Parkway).

Most railfans shunned the Q type cars as they brought the downfall of the 1300 series. Several cars survive: the Branford Electric Railway of East Haven, Connecticut, bought car 1349 in 1962 and car 1362 in 1964. Car 1365 went to the National Museum of Transport, St. Louis, Missouri, in 1964. All others were scrapped in 1962. They were the last of their type on the system and the last convertible and gate cars in North America.

### 1908 BU 1400 Series

BRT (1400–1499)

The BRT fleet of cars 1400–1499 had a somewhat normal beginning but varied endings. Jewett and Laconia manufactured the series from early 1907 to early 1908, in cooperation with the Transit Development Company. One differentiating factor of the fleet as far as the

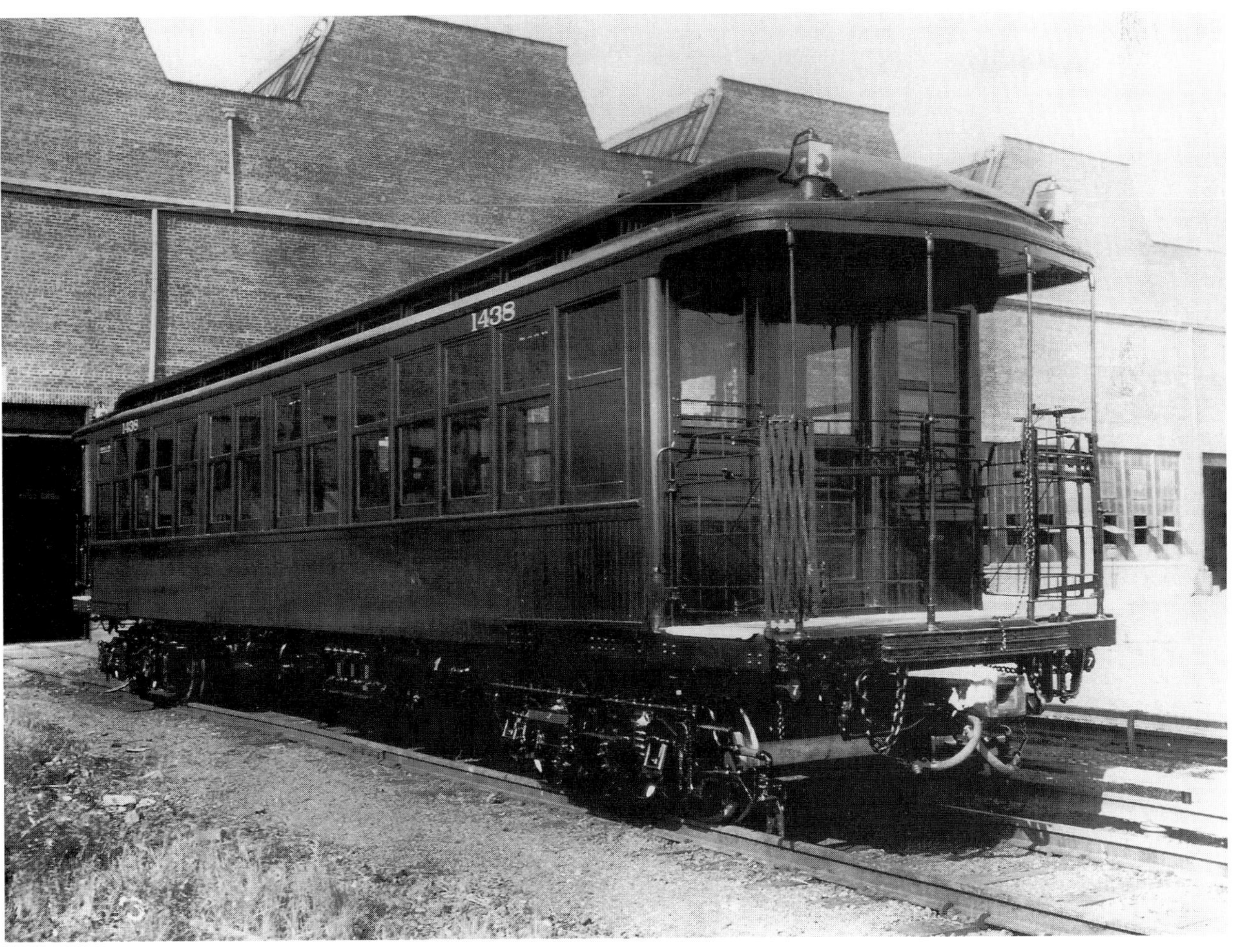

car body was concerned was thicker columns toward the center location. The vent flaps in the clerestory had twin glass panels to aid illumination while also making the interior more attractive. This had been an old practice on main line railroads and interurbans. Two roof-mounted marker lights were provided at each end of the cars as well as a bracket for a carbon arc headlight for night operation. The cars were almost identical to those of the 1200 series and were built to meet the service demands created by the trackage rights granted by the LIRR to the BRT for service to Rockaway Park.

Unlike cars in some other fleets, these were built with hidden sill girders, eliminating the need for a truss bar and turnbuckle setup. Platform ends were steel girder reinforced and were exterior extensions of the same hidden girder. Gates were manually operated. The width of the body varied slightly between cars assigned to Queens service and those in Brooklyn.

American Locomotive built trucks for the motored unit with a wheelbase of 6 feet 8 inches. Each axle had a gear unit (spur gears were used) and a traction motor. St. Louis Car Company constructed the trailer or unmotored truck with a shorter wheelbase of 5 feet 6 inches. Bolsters that supported the car body were mounted on leaf springs on the trucks. This was common practice at the time, and even the IND R1–R9 cars of the 1930s used it. El-type third rail shoes shaped like the letter U were used, and later changed to tongue-shaped modern-type shoes on those cars assigned to Fulton Street service and later to the Culver Line. Also utilized early in the careers of these cars were trolley poles installed on the roofs. When the BRT ordered these cars much of the system ran at ground level using catenary power. Also, railroad-style step plates were built in the open end platforms. Both trolley poles and step plates were removed as were the fare registers with which they were originally equipped when the El building boom commenced in the post–World War I period.

In the early 1920s a primitive form of MUDC was experimented with. Side girders at the lower sill were reinforced on three chosen cars. The gates had a "Dutch door" appearance, with wire mesh on the upper half and sheet steel on the lower half. The cars retained some of their open platform look. Though not successful during testing, MUDC was in future car plans. In 1923, the BMT gave it another go. The result was the C type cars discussed later in this book. For the C cars, MUDC was a success.

Of the BU 1400 series, cars 1448 and 1482 were rebuilt by the Transit Development Company in 1910 with cross-seating. (They are mentioned later in this book.) In 1917, car 1487 was destroyed in a derailment; the remains were cut up for scrap. Those cars not designated to become C cars wound up being rebuilt as Q/QX cars in late 1938 through mid-1940.

Two of the Q cars (1622A and 1622C, which began service as cars 1404 and 1407) were restored at the Coney Island Overhaul Shop to their original look. Aficionados will note that marker lights are in the relocated positions and the reduced high roof for tunnel clearance was retained. The couplers and shoe beam/third rail shoe assembly are of modern type. The end result is the beautifully restored consist on exhibit at the Transit Museum.

### 1909 BU 1400 Series

Brooklyn Union (1448, 1482) of the BRT and BMT

Two cars were singled out of the 1400 series car class for no apparent reason other than their existence on the system. Originally built for the Transit Development Company, the cars were typical of the rolling stock of the period. Their wooden car bodies had vented, clerestory roofs, and were built in 1907 as part of the regular 1400 series. They were damaged in an accident. In 1910, the Transit Development Company, a subsidiary of the BRT, rebuilt them to its own standards. Like other cars of the 1400 series, they were equipped with the usual trolley poles, headlights, and fare registers, all of which were removed by the early 1920s.

Steel was employed mainly as reinforcement material for truck mounting and open end platform locations. While not unique, the platform gating and supports were functionally handsome. Jewett manufactured car 1448 in 1907. Of note, this car had a seating arrangement atypical of others in the 1400 series. Seats had iron bases and were constructed of wood with no cushioning, just hardwood slatting. They were also reversible; at the end of the run the conductor would flip the seat backs to the new forward-facing position. Others in the BRT fleet had what was known as Manhattan-style seating: longitudinal seats toward the end of the car, with cross-seating at the middle.

Laconia built car 1482 in 1907 with a design similar to that of car 1448; they differed only in very minor details. In their original forms both cars employed truss rods with turnbuckles. The purpose was to slightly "spring" the bodies upward at the center when unloaded. As passenger loading occurred, the weight pushing downward normalized center body height.

Trucks had plain friction bearings, spur gear units, and one motored axle that faced the middle of the car. Wheels usually had holes cast into their fillets for weight saving and cooling. Both El-type and tongue-shaped third rail shoes were utilized. As with all wooden structures, timely and frequent paint cycles kept exteriors from deteriorating. By the time of the Great Depression, both cars had their original trucks, manufactured by American Locomotive Company and St. Louis Car Company, replaced with trucks made by Peckham.

Both cars remained in service until the mid-1950s, and were scrapped by 1955. They were fairly reliable workhorses during their lives on the system.

## 1893 and 1906, 1893 and 1908 C Type Cars

C (1500, 1501A, B, C), C (1502–1526A, B, C)

Unique is the best word to describe this fleet. Before significant changes in their appearance, they served mainly as El gate cars for the Brooklyn Union Elevated Railroad Company. The manufacturers were Bradley, Laconia, and Jewett. Construction spanned the years 1893 through 1908.

The C types represented a vigorous attempt on the part of the then new BMT to modernize and experiment with a portion of their revenue rolling stock. These cars were the mechanical guinea pigs that were a precursor of technology for future design. This early venture served as a developmental mule for later innovative subway car consists known as Triplexes, Multi-Sections, the Green Hornet, the Zephyr, and Bluebirds.

Cars were arranged in an A, B, C pattern, A and C being motored and B a trailer. They were not articulated as was originally proposed;

however, MUDC eliminated the need for a conductor between each car. Two-man crews could be utilized. The conductor's position was normally on step plates between cars. Doors were built by cutting into the side of the car body and adding sill reinforcing girders. Under the motored A and C cars, the girder was concealed. The B trailer girder could be seen and was known as a fish belly. Seating was longitudinal.

The first two C units had externally mounted, narrower doors similar to those used on the subway Standards with two window panes that slid into weatherproof pockets. The balance of the cars had larger doors with four window panes similar to those on the subway Triplex units. They were cut into the body away from the center, closer to the ends, and were actuated via an external crankarm—a problematic arrangement during inclement weather. The latter type of door arrangement was adopted as an economy measure because less reinforcement to the car bodies was required with the doors placed closer to the car ends.

The open ends of early Cs were enclosed with wood and steel and had external brake wheels. Later Cs had interior brake wheels and steel ends. Both types had railfan windows installed opposite the motorman's cab. The motorman's cab was moved forward just behind the anticlimber. Battery-operated running and taillights were mount-

Rinso
for Spaghetti-lovers
MUELLER'S SPAGHETTI
As a change from potatoes
Look!
FREE Today
RICE KRISPIES

311
C

ed below the conductor's step plate. The outer ones were white and the inner ones red. Marker lights retained their position. In between the cars, a steel portal replaced the original platforms, allowing passage in the three-car consist without having to go outside. The modern tongue-shaped third rail shoes were utilized.

Novel for the period, probably due to the success of broadcast radio, a PA system was tried but the technology of the period had not been developed to the point where understandable announcements could be made. Twin megaphones hung from center car body ceilings. This equipment was removed after only a few years.

The Fulton Street El was modified to handle cars 10 feet wide. All C types had their side sills rebuilt to accommodate the large platform gap. They remained in this service throughout their life. When originally built, however, they were operated on the Brighton–Franklin service for a short period. Several units were operated on West End Shuttle service between June 26, 1952, and December 10, 1953, as well as Culver El service between Kings Highway and Stillwell Avenue from June 25, 1953, until October 29, 1954. Unfortunately, no examples exist today of this daringly different car class. They were scrapped between 1955 and 1957.

## 1904–1908 Q and QX Type Cars

Q (1600–1629A, B, C), Q-Overhauled 1938–1941 (1600–1629A, B, C), Q-Overhauled 1958 and 1962 (1600–1629A, B, C), QX Overhauled 1938–1940 (1630–1642A, B)

These two car types were created by modifying an assortment of gate cars originally built from 1905 through 1908. The manufacturers were Jewett, Bradley, Brill, and Laconia. A Q car could be differentiated from a QX car by the A, B, C unit in the motor-trailer-motor configuration. The QX car was a two-car motor-control trailer setup. Other than that, the cars were essentially similar. The units started life as BU gate cars, numbered in the 1200 and 1400 series.

With the coming of the 1939–40 World's Fair, BMT management decided to modernize some gate cars to better serve passengers. The changes were many. Open-ended platforms were changed to enclosed steel vestibules and the motorman's cab moved forward to just behind the anticlimber. Two doors per side with high mounted double windows were installed on the car body. Exteriors were painted blue with bright orange trim, and the roofs were aluminum. The interior

layout remained the same, with longitudinal seats toward the end of the car and cross-seating in the center. Renumbering made these the 1600 series. MUDC was added, eliminating the need for a conductor between each car.

After World's Fair service, the Qs were assigned to the Astoria and Flushing Lines until 1949, when they were assigned to the Third Avenue El service (where they remained until December 1956 and were returned to the BMT). Steel door sill plates were installed to accommodate BMT platforms in 1958. In 1963, third rail shoes changed from El-type to tongue-shaped, as in use today. Canvas straps served as handholds. The modern term *straphanger,* denoting a standing passenger, was derived from the use of these straps.

Service on the Myrtle Avenue El came to an end on October 4, 1969, and most of the Qs were scrapped. This ended the age of the wood-body car. Of the Q cars saved, three were returned to their original gate car configuration (except for the high roof) for the Transit Museum. Car 1612C was restored as a Q car; car 1602A was sold to the Trolley Museum of New York at Kingston. Cars 1612B, 1630B, and 1636B were finally scrapped in February 2003 after being in storage for many years.

The original trucks were equipped with two motors on one truck, with the other serving as a trailer truck. They were replaced with trucks accommodating one motor, each salvaged from the IRT Com-

posite cars, which the Q cars had replaced. This was done for better weight distribution on the Third Avenue El. Upon their return to the BMT in 1957, the cars were repainted maroon and assigned to the Myrtle Avenue El. Because their height had been increased due to the Composite trucks and the DeKalb Avenue station had been rebuilt, it became clear that there would not be sufficient clearance to permit these cars to be transferred to and from Coney Island Shops. While the NYCTA considered reducing the height by placing these cars on the sturdier trucks under the 1300 series cars that were about to be scrapped, the necessary rewiring and relocation of other equipment under the cars that would be required made it more economical to lower the roofs by removing the clerestories with their ventilators. This solution was implemented beginning with car 1607. In order to provide air circulation, fans similar to those used on Multi-Section cars (on several units, including 1607, they were actually from Multi-Sections) were installed at the ends and in the center of the cars.

1640
A

## 1914–1918 AB Type Cars

B Motor (2000–2399), B, BT, BX Motor (2400–2499)

When the BRT decided to expand into subway operations, an extensive study was conducted beginning in 1910 to develop an entirely new type of car. The success of the IRT along with other considerations convinced the BRT to design a car that could accommodate more passengers through increased length and width. By 1913, a new car 67 feet long and 10 feet wide was proposed. A full-size wooden mockup was constructed and placed on display at the company's Thirty-ninth Street Shops. It received such rave reviews that the BRT placed an order for 500 cars of this design with the ACF. Delivery began in January 1915, and by 1918 all 500 cars were in service.

Originally equipped with a single row of fifteen lights down the center of the ceiling and no ceiling fans, the cars were found to be uncomfortably warm. In addition, ridership exceeded expectations, resulting in large numbers of standing passengers. The BRT placed an order for additional cars that would incorporate a number of changes based on experience and passenger complaints. Upon the arrival of the new cars in 1919, the original fleet was retrofitted to include these

2242

improvements. Lights were rearranged; fewer lights were placed down the center of the ceiling but others were installed over the seats, increasing the total number from fifteen to twenty. Four paddle fans were mounted to the ceiling, and handholds were provided over the longitudinal seats (many of these were later replaced with porcelain enameled railings).

Ultimately sporting nicknames such as "B types," "67-foot cars," "steels," and "(BMT) Standards," the design represented a significant advance in transit rail vehicles. From a passenger viewpoint, the car was almost all windows. The storm door pockets and the side door pockets were all double glass, although in later years they were either replaced with metal or painted over. The end (storm) doors could be operated pneumatically from the conductor's position, although like today's 75-foot cars, the end swings at curves and switches required them to be closed while in passenger service. Each car had seventy-eight seats in both cross and longitudinal positions. There were also seats that could be folded down across one door panel of each side door of the car during light riding periods. From an operating point of view, the conductor was inside the car (out of the weather). The running lights went from red to white as the reverser key was operated.

Originally, all motor cars came as single units—fully independent cars, with a cab at each end and conductor positions on the wide post between the center doors on each side of the car. The other side doors were separated by slender door posts. The conductor's controls were located on the wide post between the two center doors. Three opening buttons included two small buttons that controlled the storm doors to the front and rear of the positions, and a large center button that opened all the side doors. There were five closing buttons. The outer end buttons were used to close the storm doors, the button next to them was used to close the side doors on either side of the position, and the last button closed the center doors themselves. The conductor looked out from the center doors—he had no view of the train exterior after the doors were closed, although he did have an indication flag.

By 1921, the BRT decided that greater efficiency and cost savings would be realized if it coupled some of its cars into three-car B units. Cars 2000–2398 received this treatment. Car 2399 remained as a single, or A, unit and the 2400 series was joined to a fleet of trailers subsequently ordered. The motorman's controls in the four interior cabs were removed, the door of the cab was located in the forward position (against the windshield), and the seat was dropped. In addition, the conductor's positions in the end cars were deactivated, although the inert button board and the conductor emergency valve (CEV) remained. Conductors could now control the doors of all the cars on the train from one position.

Over the years, some cars were equipped with experimental components such as PA systems and special lighting. But by the late 1920s, all of this equipment had been removed, down to the unused window shades. In 1959 a modernization program was implemented that included the 2400 series cars. By virtue of renumbering, two cars from the earlier group were modernized as well: 2189, which became 2791, and 2351, which became 2576. Sealed beam headlights were added, unused motorman's cabs were removed, and some of the cross seats were replaced by longitudinal seats. Lighting was increased, new controllers were installed, and the cars were completely rewired. This program extended the life of the cars an additional ten years.

Today four of these cars survive: car 2204 is on display at the New York Transit Museum and cars 2390, 2391, and 2392 are stored at the Coney Island Shops.

## 1919 AB Type Cars

A Motor (2500), B Motor (2501–2599)

The one hundred cars of the 2500 series introduced modifications that would set the standard for the balance of the fleet. As previously described, lights over the seats as well as the center of the ceiling provided additional lighting, and ceiling fans were introduced. The 2500s were the last of the BMT Standards to be built by the ACF. They retained the empire roofs of their predecessors with larger ventilators installed in the raised sides of the roof. Larger, easier to read destination and line signs were relocated from the door pocket windows to the upper window sashes on each side of the center doors. The traditional window shades were not installed in these cars.

Car 2500 remained as a single, or A, car while the balance of the series—with the exception of car 2576, which had been switched with car 2351—was semi-permanently coupled into three-car B units. Car 2500 later replaced car 2006, which was damaged in an accident. It thus became part of the unit that included cars 2007 and 2008. It was scrapped in 1961. The balance of the series received extensive overhaul and rewiring during 1959 and 1960, and most of the cars survived until 1969.

## 1920–1922 AB Type Cars

A Motor (2600–2749), B Motor (2750–2899)

By 1919, the BRT realized that an additional 300 cars would be needed to meet service demands. Management had gained much insight from the operation of the cars already on the property. Therefore, more improvements were included in the car design with this contract. Brass window sashes were employed instead of wood; the traditional empire roof gave way to a more standard monitor or deck roof, which had a clerestory with adjustable ventilators along the sides. The order was placed with Pressed Steel, and new cars began to appear in 1920. A feature included as a part of all the Standards was milk-white shades that were fitted to all the light fixtures. They remained in place until about 1927.

The program of semi-permanently coupling the Standards into three-car units began in 1921. The BRT decided, though, that the first 150 of these new cars would remain as single A units. The BRT could then run trains of any length, ranging from one to eight cars. The BRT also realized that there would be occasions when one car of a three-car unit might require extensive repairs. Rather than lose all three cars of the unit while this work was being completed, a single A car could be substituted temporarily, thus reducing car shortages. Since the A units had motorman's controls on each end as well as conductor's controls in the center, these cars could be placed in any position of the unit.

2624
AA

A-44

Cars 2714 and 2741 were assigned to revenue collection service in the late 1950s. The remaining 2600 and 2700 series cars with a few exceptions were included in the overhaul program of 1959–60. Some of the 2600s were incorporated into B units to replace retired trailers. Car 2899 had been part of B unit 2899-2774-2775 and therefore was overhauled. The remaining cars of the 2800 series were not included in the program mainly because they used a different type of control switch unit. The 2800s were all out of service by the end of 1965, while the others remained in operation well into 1969. The Branford Electric Railway Association at East Haven, Connecticut, has preserved car 2775.

### 1924 AB Type Cars

BX Trailer (4000–4044), AX Trailer (4045–4049)

The final order of Standard cars was placed after the bankrupt BRT was reorganized into the BMT. In the interest of economy, the fifty cars would be trailers. By the time the contract was signed, the older cars were already being coupled as three-car B units.

Trailer cars 4000–4044 were originally coupled between a pair of 2400s, creating the desired three-car units. These sets were classified as BX units. The last five cars were used as single trailers, had motorman's cabs, and were designated AX units. The last ten 2400 series motor cars were made into two-car BT units. The AX units were added to the ends of trains only during rush hours. By 1933, this was no longer considered necessary; the cabs were removed and the cars placed between the 2400 series BT sets, thereby being reclassified as BX units.

The appearance of the cars was almost identical to that of the last motor cars built, although the line and destination signs had been moved over to the windows adjacent to the end side doors of the cars. One of the trailers, car 4036, was included in the first train to be overhauled in 1958. It was later decided not to use trailers, and 4000 series trailers were replaced by a 2600 series motor car. All of the trailers were retired by late 1960.

## 1925, 1927–1928 D Type Cars

D Triplex (6000–6120A, B, C)

The D types, also known as Triplexes, consisted of 121 articulated units, made up of three sections resting on four trucks. Each section had two large single-leaf doors per side. Conductor controls were on the ends of the car. Since there were two compressors, a single unit could operate by itself as was the case on the Culver Shuttle, which operated between Ninth Avenue and Ditmas Avenue.

The D types were probably the heaviest rapid transit cars ever built, weighing well over 100 tons. Despite their weight, the Triplexes did manage to exceed 50 mph when tested on the Sea Beach Line and were considerably faster than the AB types. Because of their weight, Triplexes were restricted to Southern Division use. The elevated structure along Fulton and Crescent Streets, built in 1893 and used by Jamaica Line trains, could not safely accommodate these heavy units.

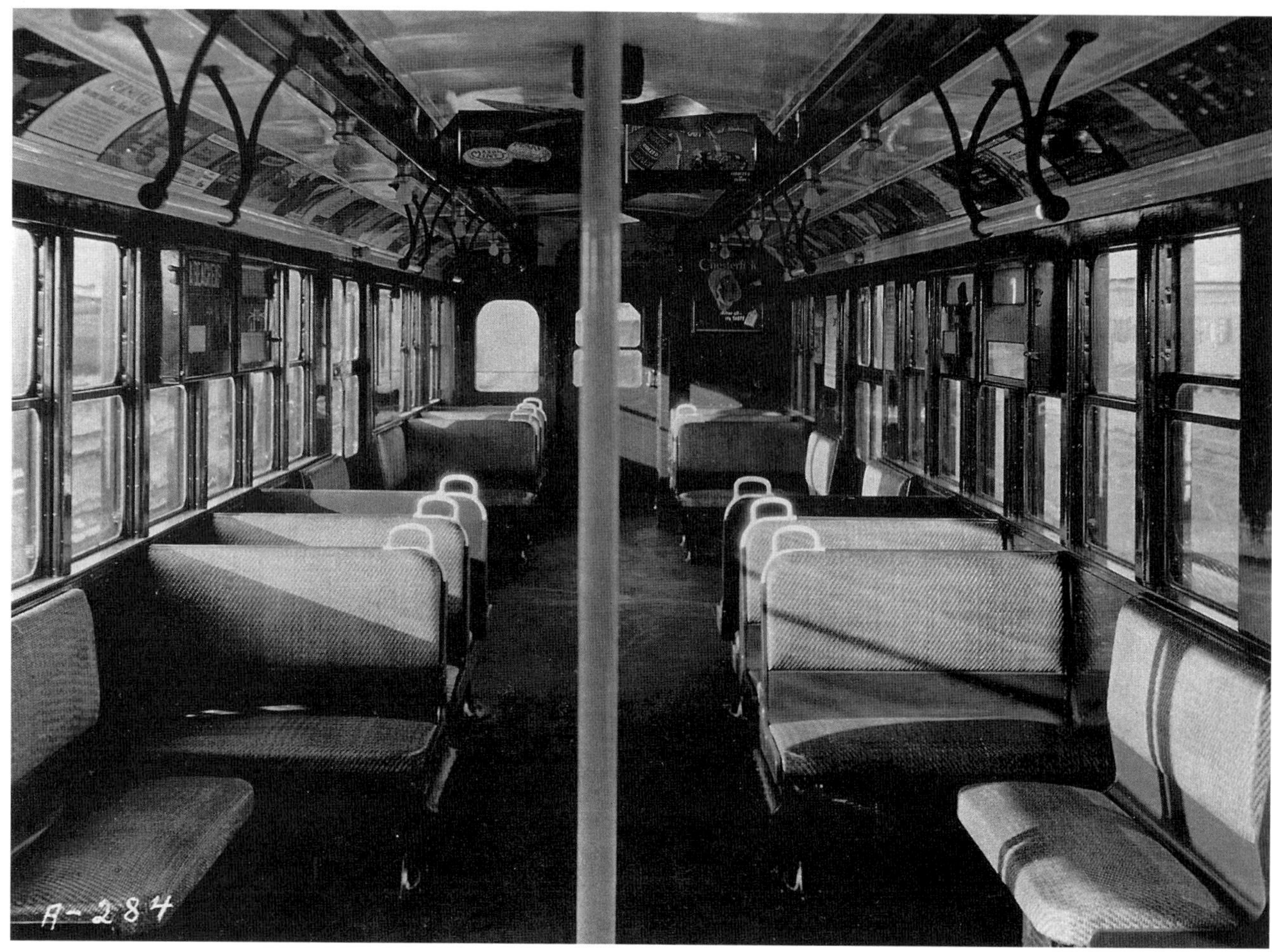


Because of their length (137 feet), only the two center trucks were equipped with contact shoes to avoid bridging third rail section breaks. The three sections were lettered A, B, and C.

In consists of two or more units, the conductor stood between the cars. MUDC was accomplished by using door control jumpers located at the stanchions of the end platforms. The cars had both interior and exterior conductor positions. They were the first New York City subway cars equipped with functioning dual controls. The interior control required a handle in a valve control fitting similar in appearance to a brake valve receptacle. The conductor would operate the car from either the trailing cab or the window opposite the cab. Because the large end platforms on the cars were an invitation for horseplay, the end doors were locked.

The D types were the first New York City subway cars to be equipped with lighted destination signs. The four-part side sign cluster had two destination rollers, one route roller, and "Bridge/Tunnel" glass that was color-coded indicating the route of the train. The ends of the cars carried a route number and a destination sign above the end door. Two sets of light bulbs were installed behind the sign, with one set placed behind a green lens. If the train was to be routed over the Manhattan Bridge, the light would shine through the green lens, giving the appearance of green letters on a black background. Montague Street Tunnel trains showed white letters on a black background. The side "Bridge" and "Tunnel" signs corresponded in color.

The four original units, built in 1925, were somewhat different in appearance from the others. This is because the balance of the Triplex production fleet was modified to include features missing in the first four cars; the original cars were later standardized to match their counterparts. The only exception to the retrofit was the Timken roller bearings that were introduced in the trucks of 6000–6003. They were retained but not applied to the rest of the fleet, except for car 6120, which was equipped with them to allow operation with the original four units.

The Triplexes were operated throughout their lives without any major modifications; they also required very little maintenance. The last of them ran in service as a West End Local on July 23, 1965. NYCT has retained three units, 6019, 6095, and 6112, for display in the New York Transit Museum and for use on the Nostalgia Special Train.

---

### 1934 Multi-Section Cars

Green Hornet (7003), Zephyr (7029)

The BMT was faced with a dilemma. After the notorious Malbone Street wreck on November 1, 1918, when 102 passengers were killed on what is now the Franklin Shuttle, wooden transit cars were phased out of passenger service in subway tunnels. Also, as the wooden car fleet aged, the BMT was faced with the need to have a car capable of running on elevated structures as well as entering the tunnel system. Compounding the problem was the fact that the older elevated lines (Fulton Street, Myrtle Avenue, etc.) could not handle the weight of the B and D types.

In 1934, the BMT took delivery of two prototype units: the Green Hornet (Car 7003) built by Pullman Standard, and the Zephyr (Car 7029) built by The Budd Company. The cars were five-section articulated units, measuring 179 feet, the longest articulated units ever operated on the New York City transit system. The five sections were

Westinghouse
New PEAKS

1
TIMES SQUARE
NYRT

lettered A, B, C, B1, and A1. Although 30 percent longer than a Triplex, the cars were light on their wheels, weighing under 90 tons. The Green Hornet was built of aluminum while the Zephyr was the first stainless steel car on the system. Both cars were designed for one-man operation. After evaluating the two special cars, the BMT ordered twenty-five more units: ten from the St. Louis Car Company and fifteen from Pullman Standard.

While the Green Hornet featured one double-leaf door per side per section, the Zephyr had two single-leaf doors per side per section (similar to the arrangement on the D types). The units were equipped with two air brake compressors; a single unit could operate as a train. Each Multi-Section had an interior sign to indicate the terminal. They were among the fastest accelerating cars, with a top speed in excess of 55 mph.

The Green Hornet (named after the radio character of the period) was also dubbed the "Blimp" by transit employees because of its curved sides. It boasted modern features, including chimes that would sound as the doors were being closed and light sensors that would automatically turn the lights on at dusk or when the train entered a tunnel. Traditional curtain signs including route number and

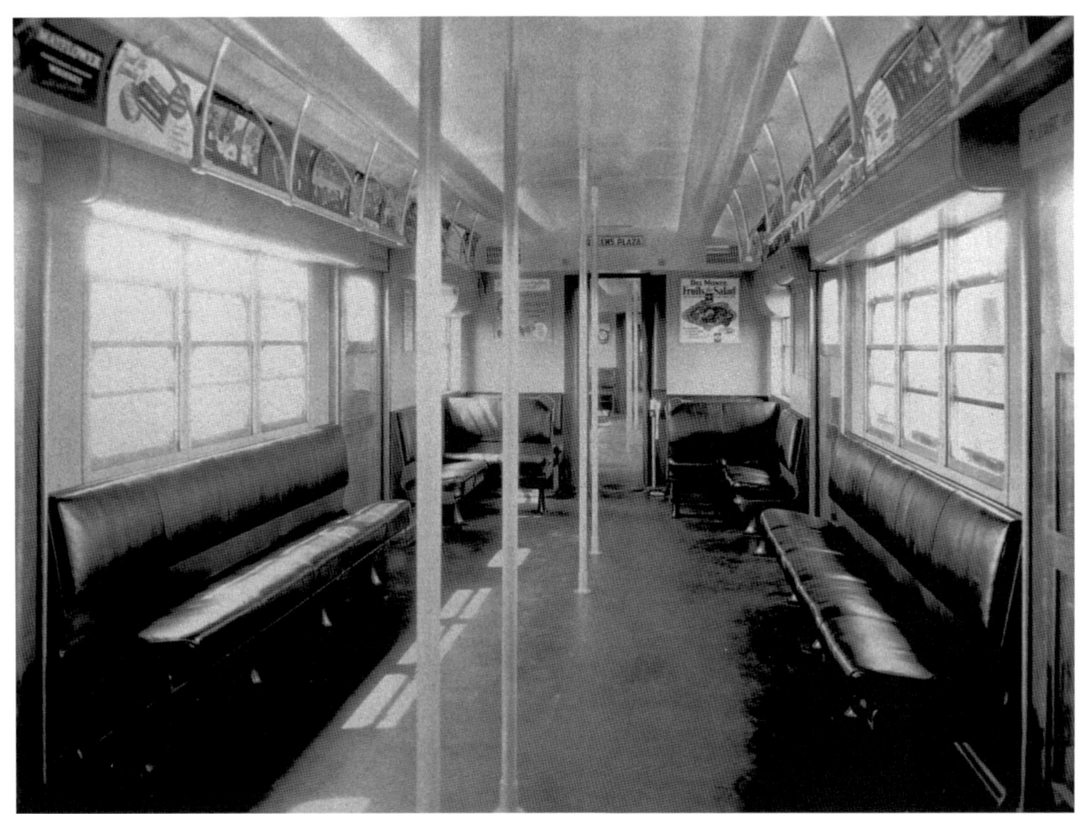

destination appeared on the ends of the cars, while new line and destination signs appeared on the sides and at the end interior bulkheads of the cars. These signs consisted of metal plates bearing the line name or destination affixed to a frame that was mounted on a cylinder. Each reading was numbered. The train crew would insert a key into a selection box and set all the signs simultaneously to the numbered reading desired.

The Green Hornet was plagued with many problems and saw little service, particularly after 1938. In 1942, the United States War Scrap Administration ordered the destruction of the car for its much needed aluminum. It was dismantled in early 1943.

The Zephyr was much more successful and remained in service on the Franklin Shuttle until shortly before it was retired on August 14, 1954. It was finally ordered scrapped in 1959.

IRT director's car *Mineola* (Car 3344) at the Shore Line Trolley Museum, East Haven, Connecticut (early 1990s).

BMT rebuilt BU cars 1404, 1273, and 1407 on an Electric Railroaders' Association (ERA) fan trip at Cortelyou Road on the Brighton Line (August 30, 1980).

IRT Deck Roof 3662 at the Shore Line Trolley Museum, East Haven, Connecticut (September 23, 1972).

IRT LV train on a farewell to the LV cars fan trip southbound on the Dyre Avenue Line (October 24, 1964).

BMT B Type 2390-2391-2392 on an ERA fan trip near 160th Street on the Jamaica Avenue Line (September 11, 1977).

BMT D Types 6019 and 6095 on an ERA fan trip at Prospect Park on the Brighton Line (September 20, 1970).

IND R6-2 1208 heading a train of R1/R9s passing Twentieth Avenue on the West End Line on a Nostalgia Special (May 7, 1977).

BMT Q Type 1622 at Central Avenue on the Myrtle Avenue Line (August 18, 1965).

BMT Bluebird and Zephyr at Fresh Pond Yard (April 22, 1956).

R11s at Coney Island Yard (October 26, 1963).

R15 6239 on a March of Dimes fund-raising trip passing Thirtieth Avenue (Grand Avenue) on the Astoria Line (August 23, 2003).

R17 6546 entering Queensboro Plaza on the Flushing Line (May 29, 1964).

R28 GOH Redbird 7902 (circa 1993).

R29s 8570-8571 brand-new at St. Louis Car Company (April 27, 1962).

R32s 3835-3834 at Coney Island Yard (September 4, 1965).

R36s 9628-9629 new at St. Louis Car Company (1963).

R46 Bicentennial cars 680 (1776) and 681 (1976) at Kings Highway on an ERA fan trip on the Brighton Line (February 26, 1977).

The Subway Series 2000 trains near Yankee Stadium in The Bronx on the Jerome Avenue Line (October 2000).

R142s on the Dyre Avenue Line south of Gun Hill Road (December 2002).

R143 8104 brand-new at Kawasaki, Kobe, Japan (2000).

BMT B Type 2390-2392-2392 and D Type 6112 Nostalgia Specials between Seaside and Rockaway Park on the Rockaway Line (October 23, 1977).

IRT Ticket Car G being delivered to the New York Transit Museum in company with R44s at 145th Street on the A-Washington Heights Line (November 1998).

IRT LVs and Redbirds at Westchester (Pelham) Yard on a fan trip (September 14, 1997).

R110A, the restored LVs, and R110B at Hammels Wye on the Rockaway Line (December 3, 1994).

## 1936 Multi-Section Cars

St. Louis Car Company (7004–7013),
Pullman Car Company (7014–7028)

Because of the popularity of the two experimental Multi-Section cars of 1934, the BMT decided to purchase twenty-five additional lightweight articulated units that could be used in both subway and elevated service. Orders were placed with St. Louis Car Company for ten units and with Pullman for fifteen units. At the time, St. Louis Car Company was building 101 Presidents' Conference Committee (PCC) trolley cars for the BMT (two of these cars were tag-on orders for Boston and Pittsburgh).

8th AV
7006

The new Multi-Section units were placed into service in 1936. They were quickly withdrawn when the BMT determined that modifications were needed on the trucks. The fleet was returned to service in 1937 and was assigned to the Eastern Division, specifically to the Fourteenth Street–Canarsie Line. The fact that their light weight permitted their use on elevated lines prompted the introduction of a new service, the Fourteenth Street–Fulton Line. The trains were operated during rush hours between Fourteenth Street and Eighth Avenue, Manhattan and Lefferts Avenue (now Boulevard) at Liberty Avenue, Queens, via the Fourteenth Street Subway and the Fulton Street Elevated, using the connection at Atlantic Avenue, East New York.

The cars retained many of the features of the two experimental units and were distinguished by their rapid acceleration and braking rate (4 mphps) as well as their high balancing speed (58 mph).

The cars remained in service until September 5, 1961. Their final assignment was to the Myrtle–Chambers Line.

## 1938 and 1940 Compartment Cars

Bluebirds (8000–8005)

The BMT did not let car design stand still. It established a reputation for innovative car design from the B types to the Multi-Sections. Its last design was another attempt to create lightweight cars for elevated/subway service—the "Compartment cars" or "Bluebirds."

Clark Equipment Company, a trolley car and motor truck manufacturer, built these three-section articulated cars. The Bluebirds were 80 feet long and weighed only 38 tons. Compare these statistics to a 67-foot-long B type, which was 12 feet shorter and weighed nearly 50 tons. The cars had one double-leaf door per side per section. Top rated speed was 39 mph, much slower than their companion cars (B, D, and MS), which could all reach 50 mph; with an acceleration of 4 mphps (the same as the MS), however, the Bluebirds could reach that top speed very quickly. A single-unit prototype was delivered in 1938. Five more were delivered in 1940. They were the last advanced design cars bought by the BMT.

Bluebirds were known as "Presidents' Conference Committee" (PCC) cars because of how they came to be developed. When bus lines started making serious inroads on trolley ridership, the heads of the street railways set up a design group to redesign the standard streetcar. The result was the PCC car design. (The bus manufacturers quickly copied the interior design themselves.)

The Bluebirds introduced new luxury that passengers in New York had never before experienced. The interiors were pale blue with mirrored panels at the ends of each section and featured mohair-covered seats. Bull's-eye lighting fixtures of the style adopted for PCC trolleys provided glare-free, soft, but pleasing lighting throughout the train. The cars rode on PCC Clark trucks identical to their trolley counterparts; the motorman's cab was equipped with a Westinghouse Air Brake Company (WABCO) Cineston controller, featuring one single handle for accelerating and braking the train.

The first unit was delivered in December 1938 and was tried out as an extra train on the Fulton Street El between Lefferts Avenue (Boulevard), Queens, and Park Row, Manhattan, via the Brooklyn Bridge, and also on the Fourth Avenue Subway between Ninety-fifth Street–Fort Hamilton, Brooklyn, and Queensboro Plaza, Queens. It also saw service on the Brighton–Franklin Line.

Having been built completely to PCC standards, virtually all the Bluebird's parts were interchangeable with those of the hundred PCC cars operated by the BMT and the 500 additional PCC cars that the company had planned to purchase. In view of this, the company placed an order with the Clark Equipment Company for fifty additional Bluebird cars. By the end of 1938, Mayor LaGuardia had informed the Federal Reconstruction Finance Corporation of his in-

tention to have the city purchase all the lines and assets of the BMT and advised them of his opposition to street cars. This resulted in the cancellation of the funding for the trolleys. The city acquired the BMT at midnight June 1, 1940, and immediately questioned the outstanding order for the fifty PCC subway units. Expressing his insistence on utilizing standard heavyweight subway cars, the engineer in charge of new car design for the Board of Transportation contacted Clark, instructing the company to complete the five units then under construction. The balance of the order was cancelled.

The five units were delivered on December 31, 1940, and were placed in service on the Fourteenth Street–Canarsie Line, where they remained until they were retired in 1956. They were occasionally operated over the Fourteenth Street–Fulton Line. Car 8000 was set aside awaiting new couplers for compatibility with the other units. The restrictions imposed by World War II prevented production of the couplers and the unit never operated again.

---

### 1925 Staten Island Cars

Motor (2900–2924), Motor and Trailer (300–389, 500–509)

What is now the Staten Island Railway, a wholly owned subsidiary of the Metropolitan Transportation Authority (MTA), New York City Transit, started its modern life as a subsidiary of the Baltimore and Ohio (B&O) Railroad. There were three main lines—the South Shore Line, the North Shore Line, and the South Beach Line. The lines operated both passenger and freight service, and included many level crossings with road traffic. Today, the South Shore Line (from the St. George Ferry to Tottenville) is the only operating segment. The North Shore Line exists but is currently dormant. It has been reduced to a single track and has been used only for freight for many years, but there are proposals to restore passenger service. The South Beach Line, which connected to the South Shore Line at Clifton, no longer exists. All that remains is some open cut areas and street bridges.

The B&O knew a good thing when they saw it. They copied the BMT's B type design for their electric cars when service shifted from steam to electric. The cars were basically the same as the B types with a slight change in appearance. While they retained the center two leaf doors for the conductor's position, the side doors near the ends of the cars were changed to LV type single-leaf doors at the end of the cars. Also, the cars were equipped with a large, roof-mounted headlight. The cars had to be designed and maintained according to

Interstate Commerce Commission/Federal Railroad Administration (ICC/FRA) Class I railroad standards.

The railway purchased ninety motor cars in 1925 and ten control trailers in 1926 to replace its steam coaches. Only two years later, in 1927, seven of the new cars were destroyed in a fire at Tottenville. On June 28, 1946, the St. George Ferry terminal burned to the ground, destroying eight more cars. On April 5, 1962, seven of the cars were lost in a fire at Clifton.

Because of the number of motor cars destroyed, the Staten Island Rapid Transit Railway Company (SIRT) converted five of the trailer cars to motors. While the motors had been numbered 300–389, the trailers were 500–509. The five converted trailers then became 390–394. With the elimination of passenger service on the South Beach and Arlington Lines in 1953, the railway found itself with a surplus of cars. The Board of Transportation had been making a number of improvements in service with the purchase of the R10 and R11 cars. In 1953, the Transit Authority was anticipating more improvements,

specifically the extension of the IND D Line to Coney Island via the Culver Line, the extension of the BMT Brighton Line to Forest Hills via the Sixtieth Street Tunnel connection, and the opening of the Rockaway Line. Even with the acquisition of the R16 cars, additional cars would be needed. The Authority purchased twenty-five motors and the five remaining trailers from SIRT. The trailers were intended for use as revenue collection cars, but were never converted. One was made into a Road Car Inspector's office at Fresh Pond Yard and renumbered 2925.

The motor cars were shopped at Coney Island. The headlights were removed; axiflow fans were mounted on the ceilings; BMT Standard door controls were installed between the center doors; BU type line and destination signs were placed at each end of the cars; the four end side windows were plated over for number plates; and the reversible seats were bolted into place, requiring half the passengers to ride backward. The interiors were painted gray and blue, and the exteriors a striking maroon and beige with black roofs and trim. The cars were numbered 2900–2924 and were usually assigned to Culver Nassau service between Ditmas Avenue and Chambers Street or on shuttle service nights and weekends. They also operated on the West End Nassau Local and on the Franklin Avenue Shuttle. By 1961, they were no longer needed and were finally scrapped in 1962.

The remaining forty-eight motor cars continued to serve Staten Island until 1972, when they were joined by five 1955 MP72 Long Island Rail Road cars. In 1973, the new MTA replaced all of these cars with new R44 SIR cars.

312
STATEN ISLAND
312

| *Year Built/ Re-Mfr.* | *Car Class/ Contract Number* | *Car Numbers* | *Description (page)* | *Technical Data (page)* |
|---|---|---|---|---|
| 1930–31 | R1 | 100–399 | 179–89 | 434 |
| 1932–33 | R4 | 400–899 | 179–89 | 435 |
| 1935 | R6–3 | 900–1149 | 179–89 | 436 |
| 1936 | R6–2 | 1150–1299 | 179–89 | 437 |
| 1936 | R6–1 | 1300–1399 | 179–89 | 438 |
| 1937 | R7 | 1400–1474 | 179–89 | 439 |
| 1937 | R7 | 1475–1549 | 179–89 | 440 |
| 1938 | R7A | 1550–1574, 1576–1599 | 179–89 | 441 |
| 1938/1947 | R7A | 1575 | 179–89 | 442 |
| 1938 | R7A | 1600–1649 | 179–89 | 443 |
| 1940 | R9 | 1650–1701 | 179–89 | 444 |
| 1940 | R9 | 1702–1802 | 179–89 | 445 |
| 1948–49 | R10 | 2950–3349 | 190–92 | 446 |
| 1949 | R11 | 8010–8019 | 193–94 | 447 |
| 1948 | R12 | 5703–5802 | 195–98 | 448 |
| 1949 | R14 | 5803–5952 | 195–98 | 449 |
| 1950 | R15 | 5953–5999, 6200–6252 | 198–200 | 450 |
| 1954–55 | R16 | 6300–6499 | 201–3 | 451 |
| 1955–56 | R17 | 6500–6899 | 203–10 | 452 |
| 1956–57 | R21 | 7050–7299 | 203–10 | 453 |
| 1957–58 | R22 | 7300–7749 | 203–10 | 454 |
| 1959–60 | R26 | 7750–7859 | 211–15 | 455 |
| 1985–87 | R26GOH | 7750–7859 | 211–15 | 456 |
| 1960–61 | R27 | 8020–8249 | 216–19 | 457 |
| 1960–61 | R28 | 7860–7959 | 211–15 | 458 |
| 1985–87 | R28GOH | 7860–7959 | 211–15 | 459 |
| 1962 | R29 | 8570–8685, 8688–8803 | 219–21 | 460 |
| 1962 | R29 | 8686–8687, 8804–8805 | 219–21 | 461 |
| 1985–87 | R29GOH | 8570–8805 | 219–21 | 462 |
| 1961–62 | R30 | 8250–8351, 8412–8569 | 216–19 | 463 |
| 1961 | R30A | 8352–8411 | 216–19 | 464 |
| 1985–86 | R30/30A GOH | 8250–8411 | 216–19 | 465 |
| 1964–65 | R32A | 3350–3649 | 222–27 | 466 |
| 1965 | R32 | 3650–3945 | 222–27 | 467 |
| 1965 | R32 | 3946–3949 | 222–27 | 468 |
| 1988 | R32GOH GE | 3594–3595, 3880–3881, 3892–3893, 3934–3937 | 222–27 | 469 |
| 1988–90 | R32GOH Phase I, Phase II excluding 10 GE cars | 3350–3949 | 222–27 | 470 |
| 1962–63 | R33 | 8806–9305 | 228–29 | 471 |
| 1986–91 | R33GOH | 8806–9305 | 228–29 | 472 |
| 1963 | R33S | 9306–9345 | 230–34 | 473 |
| 1985 | R33SGOH | 9307–9345 | 230–34 | 474 |
| 1964–65 | R34 | 8010–8019 | 235–36 | 475 |
| 1963–64 | R36 | 9346–9523, 9558–9769 | 230–34 | 476 |
| 1982–85 | R36GOH | 9346–9523, 9558–9769 | 230–34 | 477 |
| 1964 | R36 | 9524–9557 | 230–34 | 478 |
| 1982–85 | R36GOH | 9524–9557 | 230–34 | 479 |
| 1966–67 | R38 | 3950–4139 | 237–41 | 480 |
| 1967 | R38AC | 4140–4149 | 237–41 | 481 |
| 1987–88 | R38GOH | 3950–4149 | 237–41 | 482 |
| 1968–69 | R40 | 4150–4349 | 242–49 | 483 |
| 1987–89 | R40GOH | 4150–4349 | 242–49 | 484 |
| 1968–69 | R40 | 4350–4449 | 242–49 | 485 |
| 1988–89 | R40GOH | 4350–4449 | 242–49 | 486 |
| 1968–69 | R40 | 4450–4517 | 242–49 | 487 |
| 1968–69 | R40 | 4518–4549 | 242–49 | 488 |
| 1987–89 | R40GOH | 4450–4549 | 242–49 | 489 |
| 1969–70 | R42 | 4550–4949 | 250–53 | 490 |
| 1988–89 | R42GOH | 4550–4839 | 250–53 | 491 |
| 1988–89 | R42GOH | 4840–4949 | 250–53 | 492 |
| 1971–73 | R44 | 100–399 | 253–56 | 493 |
| 1991–92 | R44GOH | 5202–5479 | 253–56 | 494 |
| 1971–73 | R44 (SIR) | 400–435, 436–466 even only | 257–59 | 495 |
| 1988 | R44GOH (SIR) | 388–435, 436–466 even only | 257–59 | 496 |
| 1975–78 | R46 | 500–1278 | 260–63 | 497 |
| 1990–91 | R46GOH | 5482–6258 | 260–63 | 498 |
| 1983–85 | R62 | 1301–1625 | 264–66 | 499 |
| 1984–87 | R62A | 1651–2475 | 266–68 | 500 |
| 1986–88 | R68 | 2500–2924 | 268–70 | 501 |
| 1988–89 | R68A | 5001–5200 | 270–72 | 502 |
| 1992 | R110A | 8001–8010 | 273–76 | 503 |
| 1992 | R110B | 3001–3009 | 276–78 | 504 |
| 1999–2003 | R142 | 1101–1250, 6301–7180 | 278–81 | 505 |
| 1999–2005 | R142A | 7211–7810 | 278–81 | 506 |
| 2001–3 | R143 | 8101–8312 | 281–83 | 507 |
| 2005–8 | R160 | 8313–8972 | 281–83 | 508 |

# R TYPE NEW YORK CITY CARS

## R1/R4/R6-3/R6-2/R6-1/R7/R7A/R9 (IND)

Car Numbers: 100–399/400–899/900–1149/1150–1299/1300–1399/1400–1549/1550–1649/1650–1802

In 1925, construction was begun on a new municipal subway system that was later known as the IND. Compared to the other two divisions (BMT and IRT), the IND turned out to be a Spartan but very efficient engineering model. Stations were built on a larger scale with the entrances closer to both ends of a train, tiles were color-coded to help passengers identify their stops, and the spacing between stations was determined by actual engineering calculations.

Sharp curves force subway trains to operate at inefficiently slow speeds, so the new municipal lines were built with wide, high-speed curves. Costly "flying junctions," which increased the speed from 10 to 25 mph, were constructed so trains could proceed on diverging routes with minimum delay. In general, the physical plant of the new system was a glistening showpiece of a subway—a far cry from the older systems, which were starting to show not only their age but the limiting character of their construction.

Thanks to its well-built roadbed and efficient signal system, the IND subway featured a passenger capacity of 90,160 per hour per track. The BMT's was 73,680 and the IRT's 59,400.

Several other features of the IND are of interest. The older lines generally operated on conventional railroad-style roadbeds, with rails spiked to cross-ties sitting in stone ballast. The new system adopted a technique previously used only in stations. Ties were not embedded in ballast but in indentations in the concrete flooring, and half-ties instead of full-length ties were placed under each rail, with a drainage

ditch between them. The technical rationale for this design was that it resulted in a more secure roadbed without the distortions of ordinary ballast and ties.

The IND also began a new classification code for various car classes by ordering their first cars under contract R1. This code is still in use today and the latest procurement of new cars is under contract R160 (short for rolling stock contract R34160). This does not mean that 159 car classes of the R type have been running in New York City transit system—some R contracts were devoted to equipment and not necessarily the whole car. Starting in 1930, the Board of Transportation procured a total of 1,703 R1–R9 cars that were very similar to each other. These "city" cars, as they were called during their testing on the Sea Beach Line in 1931, started actual operation in 1932.

Before the initial Eighth Avenue subway lines were built, a Car Committee was established in the Board of Transportation, including representatives from the IRT, the BMT, and the Board Car Engineers. Input from car manufacturers and traction equipment suppliers was assimilated. A major decision regarding the length, width, and configuration of the new IND car had to be made.

### *Car Dimensions*

Since the City was constructing the new subways, and had previously built portions of the lines being privately operated in 1925–27, the committee's initial goal was to attempt to find a single private operator for all three systems (IRT, BMT, and IND). Accordingly, a serious and careful attempt was made to develop a composite car, capable of operating on all three divisions. It was thought that a 54-foot long car with 9-foot 6-inch floor width would do the job. Eleven-car trains were planned to fill the 600-foot IND station platforms during rush hour.

An IRT template car was modified and operated throughout the IRT lines. Tables of necessary clearances for a proposed steel composite car 54 feet 10 inches long over drawbar faces, 10 feet wide at the threshold plates, 9 feet 7 inches wide at the belt rails and IRT width at eaves, with 38-feet truck centers were tested. A table was developed showing the location of obstructions, type of obstructions, height of obstruction from top of rail, and additional clearances needed to accommodate this longer, wider car. Studies were also made for a car 61 feet 3 inches long, 10 feet wide at floors, 9 feet 9.5 inches wide at belt rail and eaves, with 47-foot truck centers. These tests appear to have been conducted prior to 1927.

Track centers were designed to provide maintenance space beside adjacent cars, though clearance at side wall safety niches could be as small as 2 feet 1 inch. Additional clearance was to be provided in the IRT Park Avenue Tunnel. The use of a 9-foot 6-inch wide "compromised" car would have required that the track centers be moved closer together. (They were then on 12-foot centers.) Other tight clearances on the IRT existed in the Harlem River Tunnel and the Steinway Tunnel, as well as at interlockings and special work (usually switches and crossovers).

After careful consideration, the Car Committee recommended that the composite car idea for all three divisions be abandoned. The structural costs involved, "shaving" of platform edges on the IRT, removing and realigning columns, problems in operation, the extent of structural work, and the resulting gaps that would exist if the narrower (8-foot 9-inch) standard IRT car were to be operated in the composite station platforms led the committee to this decision. They now concentrated their efforts on developing a car capable of operating only on the IND and BMT.

(*Note:* Before the three-division composite was rejected in favor of a car capable of operating only on the IND/BMT, configurations of cars 9 feet wide conforming to the IRT "envelope," 10 feet wide at the floor [threshold], but tapered in the upper portions to conform to the IRT roof and a 9-foot 6-inch floor width with appropriate taper were also considered.)

The possibilities for IND/BMT car configuration included the following 10-foot wide cars:

a. 54 feet long with three side doors
b. 54 feet 6 inches long with four side doors
c. 67 feet long (2000 series) BMT car
d. Three-car articulated D units, 137 feet long

The committee rejected the 67-foot BMT car due to "nosing" problems in the truck and the center-excess and end-excess clearance requirements. Though the Car Committee initially favored a 55-foot long, 10-foot wide car, and planned an eleven-car standard 600-foot train, a 60-foot 6-inch by 10-foot car eventually won out.

CONEY
ISLAND
999

D
1281

### *Seating*

To move rush hour crowds, the committee considered many permutations and combinations of seating and door arrangements. Ideally, rush hour capacity benefits most from fewer seats. During off-peak hours, ridership comfort benefits most from the maximum number of seats. For entering and leaving the train, longitudinal seating is preferable. A mix of cross and longitudinal seats minimally reduces standee capacity and slightly increases boarding and alighting time. The width of the door was increased from the then 32-inch standard to 48 inches. By using two independently operated leaves (halves) in each door opening, with no center post, and by using four doors per 60-foot car, studies showed that about 4 seconds per train per station loading/unloading operation would yield an additional train-per-hour operating capacity. This would permit 3,000 more passengers to be moved per hour than if there were only three doors per car. (Three doors per car became the standard on the IRT, superseding the former two doors per car.) The number of door openings approached, as close as practicable, a goal of having approximately 50 percent of the side of the car open to receive or detrain passengers. The double-leaf, center-part door with dependent halves operated twice as fast as single-leaf doors.

For seating arrangements, the goal was to speed boarding and exiting at the stations. The committee opted for a 17-inch wide seat (narrower than the IRT 18.5-inch wide seat) and a mix of cross and longitudinal seating.

The average length of a ride was 6 to 7 miles. In 1926 the IRT carried 5.82 passengers per each car-mile operated. The average car loading was 34 passengers per car, or 17 percent of the maximum crush load of 196 passengers per IRT 51-foot car.

### *Speed*

In 1927, 1.75 mphps was considered the maximum acceleration rate for passenger comfort. In subsequent car orders, this figure increased to 2.5 mphps.

The typical run for a train was considered to be 3,000 feet (average distance between stations was 2,640 feet). Energy consumption was planned with maximum passenger load; 72 watt-hours per ton-mile at 900 KWs per mile of single track was the standard. Station dwell times were 30 to 40 seconds in non-rush hours and 60 seconds during rush hours. Thirty trains ran per hour at an average speed of 18 mph.

In 1927, the rapid transit operation needed about 12 to 14 cars for every mile of revenue track. (Today's service density is not as great.)

*Passengers*

Two hundred passengers per IRT car (51 feet long and 9 inches wide) was considered the maximum carrying capacity at the time. Three hundred passengers could be crowded into the 67-foot BMT car. Each standee was allotted 1.4 square feet. Designs were based on an average load of 40 passengers traveling 6 to 7 miles per passenger-trip.

By October 11, 1927, the composite car idea, which would have permitted IRT/BMT/IND operation, was abandoned. The decision was made to go to a 60-foot car, with four doors per side, two cabs per single car.

*General*

Maximum efficiency of car utilization, designed to tailor service to actual ridership variations during the day, was predicated on using six-car articulated units (301 feet 3 inches long with three doors per car). Trains made up of an articulated unit could be extended with 60-foot single cars as required to best meet passenger loading needs.

Air-operated door engines were found to be superior to the electric door operators being used in Boston at the time. Outboard door controls offered the conductor more visibility but made the work more hazardous because he or she had to stand between cars. Traction power was provided by two 190 HP motors mounted on the No. 2 truck as on IRT cars. Controls were LV and much of the equipment was the same as that on other IRT/BMT cars. A significant safety deficiency was the lack of controller interlock to prevent the train from moving with side doors open. Later, in the 1980s the entire door safety issue including passengers being dragged became a major technical challenge to car equipment engineers.

Sixty thousand miles per car per year was planned. Draft gear, designed to absorb 200,000 pounds of buffing shocks, was provided. It was designed to move 1.75 inches on curves, and an allowance of 6 inches differential between car height ends was provided for vertical curves and bumps.

The Ohio Brass Company fought in the City Board of Aldermen for adoption of their shorter ball-joint coupler. The Board of Transportation rejected the shorter coupler for technical, operating cost, and safety reasons. With a short drawbar, the angle assumed by the draft gear was much more acute on tight curves than for a car with a long drawbar, which was the Board's standard. A short drawbar intensified the effects of buffing two cars together. Derailments would be more likely to occur at lower speeds than with a long drawbar.

During the development of the R6 cars in 1935, engineers considered eliminating one of the two cabs in each car. This was because both cabs were locked rather than being available to passengers when they were in the nonoperating position of the train. Four seats per car were thereby lost due to locked cabs. The initial design had been based on having two seats in each cab available for passenger use when the train operator's cab was not being utilized. Later, vandalism and unsanitary conditions made it necessary to lock unused cab doors. The elimination of a cab at one end of each car would have provided two more seats per car.

In conclusion, while the R1–R9 lacked some of the improvements incorporated into both the BMT Standard and the Triplex, they performed almost forty-five years of continuous, reliable service even without all those amenities that later became standard, including air conditioning, comfortable ride quality, and use of dynamic braking.

Several one-of-a-kind cars came about as retrofits. Car 1575 is the most famous of these. The R7A type car had the passenger area rebuilt by ACF as a prototype R10 (fluorescent lights, small window top roll signs, ogee roof, etc.). All car systems (except lighting) remained R9 type and the car always ran in train consists of R1–R9s (or R1/R9s).

## R10 (IND)

Car Numbers: 1803–1852*, 3000–3349 (*Renumbered 2950–2999)

In 1948 a new type of subway car appeared, the first of many similar car classes until the introduction of the R44s in 1972. The R10s offered little in the way of passenger comfort—they were noisy, dimly lit despite fluorescent lighting, and very hot during the summer. They are remembered as German "Panzer tanks" due to their durability.

The big changes in the R10s were related to the propulsion and braking systems. While the propulsion control system was similar to that which already existed, these cars were powered by a 100 HP traction motor on each of four axles, whereas most older cars had two large traction motors, either one per truck or both on one truck. The acceleration rate for the cars was 2.5 mphps as compared to only 1.75 mphps on the R1–R9 class.

The biggest change of all was in the braking system. These cars had both pneumatic and dynamic braking, and a more modern brake valve with a self-lapping feature. Whereas in earlier car classes the train operator would have to work the valve back and forth between service and lap or release and lap to gradually apply or release brakes, on a self-lapping brake valve a slight movement in either direction would control brakes as desired. This brake system was referred to as "SMEE" (short for straight air "ME" type valve self-lapping overlay) and was universally used up to the R44s, when the electronic P-wire was first introduced.

The inclusion of dynamic braking was the most significant improvement on the R10 and all rolling stock to follow. The amount of dynamic brake amperage is regulated by means of an actuator sensitive to increases or decreases in service braking via straight air pipe pressure. The traction motors, by having field polarities reversed, act as generators. The energy is dissipated as heat through electrical resistors, instead of heat generated by the friction of brake shoes on wheels. The beauty of this system is that brake shoe wear is reduced

tremendously, thereby decreasing maintenance costs. In the subway, it also results in much less brake shoe dust, which lowers the risk of signal failures, short circuits, and fires.

As a final note, 110 R10s went through a modified overhaul program in 1984 that was completely scoped, engineered, and performed by NYCTA personnel. The cars, painted silver (roof) and dark green (car body), provided several more years of solid, reliable service before finally being retired in 1989.

## R11 (BMT/IND)

Car Numbers: 8010–8019

In 1949, Budd manufactured a ten-car subway train for New York City. Budd was the industry pioneer in the production of shotwelded stainless steel, and the material offered a refreshing break from the grime-covered cars passengers had become so accustomed to. The trainset of R11s represented an NTTT that offered a sparkling new look, a host of fresh amenities, and the latest technical features best tried on a single train than across an entire fleet of cars. The train was a prototype for a planned 400-car fleet.

The stainless steel construction, window arrangement, and smooth roof design gave the train a futuristic appearance that was not to be duplicated in subway car design until 1963, when Budd would win an order for 600 R32s. Each R11 car cost $100,000, quickly giving the consist the nickname of "The Million Dollar Train." Aside from the new appearance, several passenger amenities were introduced. A PA system, germicidal lamps, forced air ventilation, and PCC-like crank-operated windows all made their initial appearance on the R11s.

Many new components were also introduced. Electric door motors replaced standard air engines, and disc brakes were installed instead of conventional wheel tread brakes. Unfortunately, like many other test cars, the results were marginal and few components were viewed as developed enough for inclusion in succeeding car orders. The 400-car fleet was never ordered. The R11s were eventually downgraded for use in either part-time service or included in consists of R16s. A midlife rebuild in 1964 allowed the cars (renamed R34) to soldier on until 1976. But once the engineers had squeezed every piece of knowledge they could out of them, the R11s (R34s) became unwanted orphans, requiring extra attention no one was willing to give them and special parts that were no longer available. The R11s, with the exception of car 8013, were scrapped in 1980.

## R12/R14 (IRT)

Car Numbers: 5703–5802/5803–5952

Except for the 1938 World's Fair cars, most of the IRT rolling stock looked pretty much the same. In the years immediately after the line's opening in 1904, the major car body design change was the addition of a center door to speed passenger flow. The first train of R12s placed into service in 1948 represented a revolutionary design change rather than an evolutionary one and was a real break with tradition.

Compared to then-current rolling stock, the modern-looking R12, manufactured by ACF, had six 12-inch bracket fans mounted on the stanchion poles, fluorescent lighting, quicker acceleration, a more efficient braking system, and front and side roll-sign windows for route identification. The car was light gray above the orange belt and dark gray below, and was 51 feet 4 inches long.

In 1948 the Board of Transportation took delivery of 100 single cars for the Flushing Line. The R12 was similar to the R10 in most of its equipment. Because it ran on the IRT system, it was shorter, narrower, and carried fewer passengers than the IND cars. All post-1946 cars were compatible as far as coupling, door controls, propulsion controls, and brake controls. The R12 was equipped with a solid maple or birch third rail contact shoe beam, whereas later cars had four-ply laminated maple or birch shoe beams.

The R12 cars were designed to operate in a maximum train unit of ten cars. Each car seated forty-four passengers. It had three double doors on each side of the car and an end door at each end. All side doors could be operated by the train crew from any assigned location between cars. No zone of control exceeded five cars. Steps were provided on the end sheets of the cars upon which the crew stood when operating the door control switches, which were located on the corner post. The R12 introduced route numbers to the IRT Division.

5
UTICA
AVE

With the exception of a simplified interior floor pattern with a color change, the R14 was an exact duplicate of the R12 and followed in production. It was the last New York City transit railcar to have exterior conductor's (door) controls.

---

## R15 (IRT)

Car Numbers: 5953–5999, 6200–6252

When the R15 cars were placed in service, passengers may have felt they were on an ocean liner with the innovative port-hole style windows on the side and end doors. The R11 BMT cars had the same window design. The R16 and R17 cars would keep the port-hole windows on the end doors, but the side door windows returned to the more conventional rectangular design.

The roof of the R15 was changed from ogee to turtleback construction, again concurrently with the R11 BMT cars. This type of roof construction remains to the present time. Axiflow fans were introduced for ventilation. The exterior color scheme was maroon

above and below a beige belt line. On the interior, the doors were dark blue, the walls gray, and the ceiling white.

The R15 cars had clasp brakes with automatic slack adjusters. This system remained on New York City transit cars until the 1970s, when package brakes were introduced on the R44 cars. The conductor's controls were moved inside the train operator's cab, eliminating the hazard of wet or icy steps.

The R15s entered service on March 6, 1950, inaugurating super express service between Times Square and Main Street on the Flushing Line. Car 6239 was used in the NYCTA's first attempt at air conditioning that began on the Flushing Line on September 8, 1955. This car has been preserved in the New York Transit Museum collection.

1
BROADWAY
SOUTH
FERRY
238TH STREET

## R16 (BMT/IND)

Car Numbers: 6300–6499

The R16 was no award winner in terms of beauty, but it was practical and functional and good enough to set the standard for 3,000-plus cars on the BMT/IND and the IRT. In fact, as recently as 2003, several hundred similarly designed cars were still operating on the IRT.

The R16s, manufactured before the NYCTA turned to stainless steel for car body construction, were real heavyweights. In fact, they outweighed every other R type car class that ever ran in the system except the 75-footers. As discussed earlier, even one extra pound carries with it a penalty in terms of car operation and maintenance. Extra weight translates into accelerated wear of the track and structure while requiring more energy for propulsion. It is estimated that each extra pound of weight costs the NYCTA as much as an extra dollar each year.

each pound must be paid for
each pound must be accelerated and braked
each pound must be maintained
each extra pound increases wheel and rail wear
each pound must be scrapped

The R16s were the first BMT/IND post–World War II cars to have the conductor's door controls inside the motorman's cab and a return to the large-size side signs that were on the R1/R9s. They also were the first cars to employ staggered side doors and to have tile floors. They began service on the Jamaica Line on January 11, 1955.

NYCTA mechanics will always fondly remember the R16s as the "money" cars. This nickname has to do with all the overtime spent trying to repair R16s equipped with the notorious GE propulsion equipment. The GE system consisted of flat bottom enclosed resistance grids cooled by forced air. This resulted in a whining noise during acceleration, a lot of dirt and debris accumulating and creating fires within the grids, and blasts of hot air at stations coming up be-

tween the car and the platform as the heat from the dynamic brake system was dissipated. This propulsion package was also used on R17/R21/R22 cars. It should be noted that later GE improved its performance record by providing very reliable propulsion equipment.

---

### R17/R21/R22 (IRT)

Car Numbers: 6500–6899/7050–7299/7300–7749

During the late 1940s and early 1950s, new IRT cars began to serve the Queens portion of the IRT, but it was not until the summer of 1955 that new rolling stock appeared to serve the main line, including both the west side and the east side. Three sister car classes (R17/R21/R22—a total of 1,100 cars), all built by St. Louis Car Company, started to replace the old HV equipment, including the 1904 Gibbs cars. The R21/R22 was a near perfect copy of the R17, including performance.

7
MAIN
STREET
6508

(*Note:* Even back then, there was no scientific car replacement program. Each car class has different engineering properties and, therefore, can perform at different standards for different periods of time. Likewise, the cost of maintaining and upgrading specific car classes varies according to the class design, equipment, and materials.)

Virtually the only differences between R17 and R21/R22 cars in terms of the body were a square, drop sash window in the storm door (the R17 had a round window) and reverse hinged cab doors (R21s). Additionally, the exterior color schemes differed. The R17s were maroon while the R21s and R22s were kale-green. The R22 fleet was the first to be delivered with sealed-beam headlights.

Ten R17 cars (6800–6809) were delivered with plastic seats and air conditioning. The air-conditioning experiments failed, however, and the cars were refitted with axiflow fans. R22 cars (7515–7524), dubbed R22 specials, featured plextone paint (green with black and white specks) over a rigidized aluminum interior, with pink molded fiberglass seats.

In a 1976 study conducted for the NYCTA by an engineering consultant, Wyer Dick and Company, the following point was made:

> It is generally assumed that effective life indicates the age when the car should be retired, reflecting the impact of various economic and intangible factors. However, none of these have been quantified in the arbitrary assignment of 35 years as the effective life. The precise origin of the 35 years life is lost in history. There is reason to believe that the 35 year life was originally related to the period of bonded indebtedness incurred when borrowing money for the purchase of new transit cars. It is fairly certain, however, that there was never any "engineered" life of precisely 35 years built into the vehicles.

Wyer Dick and Company studied other subway systems and identified a wide variance in the assumed useful life of rapid transit cars. The spectrum extended from twenty-five years (Chicago Transit Authority) to forty-five years (Stockholm, Sweden); from a Japanese practice of replacing standard railcars based on a twenty-year useful life to the Paris Metro historically using a sixty-year life. The only common denominator was that the "engineered" life of a railcar had definitely been extended with the introduction of stronger car bodies and that some transit properties prematurely retired cars only if they could achieve higher energy efficiency and/or passenger commodiousness levels.

The Wyer Dick report also observed that various series of New York City transit cars purchased between 1904 and 1940 were retired

at ages ranging from twenty-five years (the 1938 W-F series, which actually lasted thirty years in service) to fifty years (the 1904 HV series), with average retirement at about thirty-nine years for all pre–World War II series. Older cars were kept in service during World War II because of high transit ridership and because the war effort precluded the manufacture of domestic transportation vehicles. (*Note:* On December 23, 1946, an all-time record day was established when 8,872,244 passengers rode the subways during a twenty-four-hour period.) Cars might have been retired on an earlier schedule had the war not intervened.

(*Note:* In light of these facts, the objectives of a fleet maintenance program cannot be defined solely on the basis of age—either average age or maximum useful life. Rather, reliability, efficiency, and commodiousness of each car class enter into the decision as to when retirement occurs. Undoubtedly, a continuing, comprehensive overhaul program can extend the life of cars. Therefore, NYCT has recently extended the planning life of its subway fleet from thirty-five to forty years.)

R22 cars were the last single cars ordered for twenty-six years (except for the forty R33 single cars ordered to be the eleventh cars on Flushing Line trains). Three R22s had another unique role to play, however. About forty years ago, ATO was introduced on the Times Square and Grand Central Shuttle on Track 4. The automated shuttle was more evidence of the NYCTA's keen interest in modern transportation techniques. After all, since the beginning there had been a gradual automation of the trains. At one time brakes were set on each car by hand. The train operator then simply moved a handle to brake all the cars. Once, too, each door was swung shut with an arm, then all the doors were opened or closed by pushing a button.

In October 1960, the NYCTA demonstrated the feasibility of operating a fully automated, crewless, three-car (7513, 7509, 7516) subway train on a section of the Sea Beach Line, similar in profile and distance to the Grand Central–Times Square Shuttle track. After more than 1,500 test runs, this operation proved itself entirely safe

and practical and was demonstrated to the ICC and others. For the passengers, there was no appreciable difference between this and the ordinary shuttle train operation. Anyone boarding this automatic train rode just as safely, just as comfortably, and perhaps even a little faster.

A combination of electrical and electronic apparatus automatically provided all the commands, checks, safeguards, and controls required for the smooth and efficient movement of the train from one end of the shuttle line to the other and back again.

One of the key pieces of apparatus used in this operation was an electronic train dispatcher—a 35-millimeter, punched-film tape programmer that provided an automatic program for a twenty-four-hour period; the program could be varied to take care of irregular schedule days, such as Saturdays, Sundays, and holidays.

This automatic dispatcher read the train dispatching order punched on the tape and transmitted the order, via a wayside instrument case, to the running rails of the train in the form of electrical impulses. There was no physical connection between the automatic controls on the train and those on the wayside. Instead, receiver coils mounted on the truck of each lead car, just above the rails, picked up signal impulses and relayed them to amplifying equipment in the unoccupied cab. There the signal was fed to electrical/electronic circuits that interpreted the commands. The doors closed and the train started. The train picked up speed and accelerated smoothly to 30 mph.

As the train neared the end of its run, it passed a predetermined point, where the frequency of the impulses fed to the amplifier changed, causing the brakes to be gently applied. When train speed was reduced to approximately 5 mph, the train passed into a section of track fed with impulses of still another frequency. This commanded the electronic brain to slowly apply full brake pressure, bringing the train smoothly and accurately to a full stop. Regardless of load variations, platform stops were made within a tolerance of a few inches. The automatic equipment made a silent but certain check that all was safe and then opened doors, changed headlights to taillights and taillights to headlights, and reset the destination sign rolls to display the opposite terminal. After a desired period of time, preprogrammed on the tape of the automatic train dispatcher, the doors would close. A tape recording was used to alert the passengers that car doors were about to close, and a visual sign, with the same message, flashed on. Only then would the train begin its return trip—all automatically.

R-21

Each end car was equipped with automation control gear, only one set being in control of the train at a time, depending on direction of travel. An electronic axle-driven speed governor held train speed within prescribed limits. Information from the rail-carried electrical impulses, the speed-sensing governor, and brake-checking devices was delivered to the control equipment on the train. Here the information was interpreted and the car propulsion, braking, and door-opening circuits were ordered to respond when and as required. The electrical impulses fed into the running rails were automatically selected and controlled by apparatus in a steel cabinet on the platform at each station.

All of the equipment was of the fail-safe type. In case anything abnormal developed while the train was moving, such as excessive speed, a compensating device returned the speed to normal. If at any time the train failed to receive a continuous stream of electrical command impulses through the rails, the motors immediately shut off and the train automatically stopped.

A catch-up device maintained a rigid schedule. For example, if a passenger held a car door open beyond set time limits, the device correspondingly decreased the length of time the train remained at the opposite terminal in order to return it to schedule. Numerous electric scanning devices made a complete safety check before the train was allowed to start. The train was considered so safe, it came to stop within a few feet of the pedestrian bridge across Track 4 at Times Square.

Unfortunately, this major step toward automation came to an abrupt end on April 21, 1964, when a major fire destroyed the on-board automatic equipment and the car body of car 7740. The shuttle area at Grand Central also sustained heavy damage. Even though the cause of this fire was never determined, train operation went back to 100 percent manual—the way it still is today. But progress cannot be held back forever: NYCT has started a pilot program to demonstrate both communications-based train control (CBTC) and ATO.

## R26/R28 (IRT)

Car Numbers: 7750–7859/7860–7959

From 1959 through 1961, the ACF delivered 110 R26s and 100 R28s, which were the lightest cars ever built by ACF. They were also the last ACF cars for the NYCTA and were built at ACF's Berwick, Pennsylvania plant.

The most significant technical change for the IRT fleet was the fact that the R26 and R28 cars were built as married pairs—the even-numbered car had a motor generator and battery set to supply low voltage for control equipment of propulsion, doors, and the like, while the odd-numbered car had an air compressor to supply the air for the brake equipment. The NYCTA anticipated a significant reduction in maintenance costs with this arrangement. Although the cars had been designed and built to operate as married pairs, their marriage was achieved through a special version of the H2C coupler, so that they could easily be decoupled. Thus, they were actually semi-

permanent pairs. Later married pairs (R29/R32/R33/R36/R38/R40/R42) used a link bar and were considered permanent pairs due to the effort needed to split them. Rail buffs would refer to semi-permanent as Protestant marriages and permanent as Catholic marriages. Another characteristic of the married pair was the fact that the train operator's cab was only required at the No. 1 end of each car. The No. 2 end cars had only conductor's controls. Although referred to as "blind" ends, the No. 2 cab did have a vision glass so the conductor could observe the door zone light on the other half of the train. Also, after openable storm door windows on the R15/R16/R17/R21/R22, the R26/R28 returned to single, sealed, storm door windows.

The second important technical feature was the replacement of door control levers by single push-buttons and relays. Even today, decades later, the door system, which is the most critical in terms of safety and reliability, is controlled by relays, with microprocessors being the next logical sequence in the future.

Another interesting note about the R26s and R28s is the fact that they, along with the R29s, R33s, and R36s, are today referred to as the Redbirds due to the Fox-red color of the car body (similar to the Tuscan red used on the Pennsylvania Railroad electric locomotives). Throughout the years, the NYCTA had tried different color schemes, including 100 percent white, to combat graffiti but it was the Fox-red color that finally won the war on graffiti on May 12, 1989.

All the Redbirds, with the exception of the remaining 39 R33 single (R33S) cars, were retrofitted with air conditioning during the 1975–82 period, after more than two decades of engineering work to produce air-conditioning units small enough to fit IRT cars yet powerful enough to handle the summer cooling requirements of heavy passenger loading on rush hour trains.

The carbon steel structure of the Redbirds started to show fatigue as their planned life of thirty-five years was reached. A new generation of spot welded stainless steel cars with a specified life of forty years replaced these old but reliable warriors.

An interesting trend developed among the semi-permanent married pairs (R26/R28 and R27/R30/R30A) on the B Division. The A Division rarely split the cars, even if one was out of service for maintenance. The B Division, however, freely swapped mates, leading to the expression "Bigamy on the BMT" among rail buffs, who eagerly charted all the recouplings. This flexibility was eliminated at the time of the general overhaul (GOH) with the use of a link bar between the mates.

Florida
to face.
EASTERN

5
Dyre Avenue
7902
7824

## R27/R30/R30A (BMT/IND)

Car Numbers: 8020–8249/8250–8351, 8412–8569/8352–8411

These three car classes are grouped together because they are almost identical and St. Louis Car Company delivered all 550 cars during 1960–62. Their cosmetic features were similar to those of the IRT cars, with the obvious exception of BMT-IND cars, which were always longer and wider.

The cars mostly ran on BMT lines, especially the BMT Brighton Line, where their QT designation denoted service via tunnel to Manhattan–Queens (Astoria or Forest Hills) during weekdays while weekend and holiday service was designated as QB (via Manhattan Bridge). These letter designations replaced the previous numerical routes used on the BMT D types, Multi-Section, R11, and R16 fleets. After these cars, all the BMT lines were designated with letters from J through TT while IND routes used letters from A through HH. Today, IRT (A Division) lines still use numbers while BMT-IND (B Division) lines use various single letters after the operational standard of double letters to indicate local service was dropped.

When this fleet of 550 cars arrived, it allowed the NYCTA to retire all of the oddballs on the BMT and approximately 50 percent of the original BMT fleet. In 1991, a risk assessment analysis was performed to either scrap or mothball the fleet of 162 R30 GOH cars. Several factors contributed to the somewhat premature scrapping of the R30 fleet, notably their miserable mean distance between failures (MDBF), a steady decline in subway ridership, anticipated future demand, and budget considerations. Two more nails in the coffin were the facts that the cars did not have air conditioning and the determination that the B Division spare factor could be lowered to 12 percent with no adverse effect on service. Another consideration was that fact that this same R30 fleet had been temporarily removed from service in January 1990 due to a converter bracket failure and passenger service had continued on uninterrupted despite their absence. The last of the painted B Division subway cars would be retired after the R46 cars returned from their GOH program done in Hornell, New York, by Morrison Knudsen (MK).

S
FAR
8391

The decision to withdraw the R30s came in 1992, but some of their parts would live on in a sort of organ donor program. Some very valuable new components had been installed on the cars at the time of their overhaul, among them, expensive propulsion cam controllers. These nearly new parts were removed and installed on the R36 cars, which still had many older components. This project pumped some much needed new blood into the aging R36 fleet, helping to increase their MDBF and thus provide more reliable service to Flushing Line riders.

---

## R29 (IRT)

Car Numbers: 8570–8805

The R29s were basically a repeat of the R26/R28 car classes with a few differences. First, St. Louis Car Company, not the ACF, built 236 of these cars in 1962. As a matter of fact, St. Louis Car Company went on to construct another 2,636 subway cars for the NYCTA on its way to becoming the number one builder of transit passenger cars in the world. (Unfortunately, it was the last of the NYCTA contracts, the 352-car R44 order built during 1971–73, that contributed to the demise of St. Louis Car Company.)

The second significant deviation was that the R29s were built as permanent married pairs with link bars instead of the semi-permanent special couplers used for the R26/R27/R28/R30/R30A car fleets. The link bar/electrical connectors combination has since been modified to the standard method for marrying subway cars in pairs, triplets, and multicar combinations. The third and last new feature was related to the cosmetic appearance of the R29s: for the first time the exterior color scheme was an exciting bright tartar-red instead of the typical dull colors, and more pleasant colors were employed for the car interiors.

Four cars (8686–8687, 8804–8805) tested G70 inboard bearing trucks and were sometimes called the "roller skate" cars. The trucks were discarded in 1970.

## R29 PRE-GOH

## R29 GOH

## R32A/R32 (BMT/IND)

Car Numbers: 3350–3649/3650–3949

Back in the early 1960s, the New York City transit system lagged behind other railroads in the use of stainless steel passenger equipment. Although stainless steel was higher in initial purchase cost, the lighter cars built from it were equally as strong so power costs could be reduced; also the cars did not have to be painted. Therefore, the use of stainless steel construction would eventually recoup its cost and provide a savings besides. Budd had bid on stainless steel cars, and its bid was low enough to win without the aid of the price allowance for the 4,600 lb. weight saving of its design.

While the styling of these cars was quite similar to that of the R36 cars, their appearance was markedly different due to the use of fluted stainless steel in place of conventional steel. Body dimensions and basic outlines followed the established patterns set by the R27, and with a few insignificant refinements these cars were mechanically the same as their predecessors. A major difference was that the R32, and all married pair sets that followed, had solid drawbars between cars rather than couplers.

Collectively known as R32, there were two subcontracts to this fleet. The R32s were to be paid for by the NYCTA with the proceeds of a revenue bond. The R32As were to be paid for by the City of New York out of its 1963–64 capital budget. The R32s and R32As differed only in interior lighting. The R32As followed the established use of ribbed clear glass mounted beneath fluorescent lighting tubes, while the last 150 R32s had large plastic sheets, about a foot high, running the length of the lighting fixtures. Commercial advertisements for these cars were made of plastic rather than cardboard and were lighted.

When first delivered, the R32s had some growing pains. For one thing, the cars sat too high to clear subway tunnels safely because nobody thought of modifying the suspension to compensate for their lighter car body weight. Additionally, dynamic braking was exceptionally rough. These and other bugs were ironed out to some degree.

For years operating personnel thought of the R32s as the best-running cars. They jumped like jackrabbits and were fast; it stands to reason, as these cars had the same traction motors, controls, and braking equipment of their predecessors yet the R32 weighed about 70,000 pounds compared to the R27 at 80,000 pounds. Other rail

commuter equipment had the nickname "Silverliners"; the R32s got the moniker "Brightliners" when they were new because they were so much brighter than the rolling stock they were sharing trackage with.

The R32s were used to perform some important engineering tests. Four cars were equipped with Pioneer trucks using disc brakes. These test trucks were eventually replaced in 1976 and almost another twenty years passed before the NYCTA would again test disc brakes on the R110B. A Garrett flywheel energy storage unit was installed on two R32 cars in January 1974. After two years of extensive testing in Torrance, California, and Pueblo, Colorado, the two cars ran in New York City transit system passenger service for six months from February to August 1976. While the flywheel did show some potential energy savings, its very poor reliability and other negative fac-

tors convinced the NYCTA not to go ahead with a longer test sample of twenty cars in 1977.

This test of a different propulsion system is symbolic of the efforts of NYCTA engineers over the years to develop a more reliable and maintainable system than the standard cam controller: a WH chopper was tested on one R30 from 1966 to 1968, a State-of-the-Art-Car (SOAC) Garrett chopper was tested in October 1976, a GE chopper on two R46 cars was tested in 1979–80, alternating current (AC) propulsion by Strouberg/Garrett was tested on four R44 cars during 1986–88, a WH microprocessor controller was tested on an R36 in 1987–88, AC propulsion with AC traction motors AEG (now Bombardier) was tested on the R110A and GE on the R110B since 1992, and on an R38 test train (first with ALSTOM and later with MELCO propulsion). Furthermore, the new cars are equipped with AC propulsion, which also allows energy savings through a regenerative system active during dynamic brake applications.

Budd always built quality products and the R32s are excellent examples. Since they are lighter in weight, they have fast acceleration and good speed. They have seen use on virtually every B Division line and will easily surpass forty years in service. Only a few cars in this series have been retired.

3827

## R33 (IRT)

Car Numbers: 8806–9305

The R33 was virtually identical to the R29, except for the forty single cars ordered for the 1964–65 World's Fair (Corona) service. (See R36.) The R33 fleet was the first to go through the GOH program in-house. Periodic inspections and overhauls are the key foundations of a preventive maintenance program. After the NYCTA stopped its overhaul program at the Coney Island Overhaul Shop in 1974 and the following year at the 207th Street due to severe budget cuts during the city's financial crisis, the reliability of the cars started to deteriorate quickly. The key performance indicator, MDBF, fell to an all-time low of 6,700 miles (the twelve-month moving total for November 2003 was 140,479 miles) and car availability dropped to less than 70 percent. The worst single day in New York City transit system history was January 13, 1981, when 2,117 cars, or one out of three, sat idle, waiting to be repaired.

In order to prevent the collapse of the subway system, a budget close to $2 billion was allocated to perform a major GOH of 4,177 cars during the first two five-year capital programs (1982–86 and

1987–91). While most of the overhauls were performed by outside vendors, with MK taking the lion's share, all 494 R33 cars were overhauled by NYCTA labor during 1986–91. The significance of this successful program relied on the fact that for the first time in many years an NYCTA team made up of managers, analysts, engineers, and maintainers was able to turn out a competitive product maybe even of a higher quality than that of outside vendors.

The tremendous success of the R33 GOH can be easily measured in terms of economic and car performance results, and it especially stands out when compared with the GOH of 110 R42 cars at Coney Island: that program had cost overruns of about 50 percent; it probably would have been cheaper for the NYCTA to scrap the twenty-year-old cars and buy brand-new ones. The reason for the R42 fiasco was weak project management by a rookie team recently hired from bankrupt Pan Am and the fact that the R42 GOH was being performed while the Coney Island facility was in the midst of heavy reconstruction.

As a final note on the in-house GOH programs, the R44 fleet was later divided between MK and Coney Island for overhaul work during 1991 and 1992, and this time the Coney Island team proved that it could perform the same quality of workmanship as either a vendor or the 207th Street Shop.

## R33S/R36WF/R36ML (IRT)

S-Single, WF-World's Fair, ML-R36 Main Line
Car Numbers: 9306–9345/9346–9523, 9558–9769/9524–9557

The No. 7 Flushing Line is the only subway line where an eleven-car train is operated. The train consist was made up of ten R36 cars (five married pairs) and one single car, the R33S. The forty R33Ss were the only cars that were not air-conditioned due to the lack of physical space and did not run during the summer months in their final years. Historically, this fleet, which was maintained by the Corona Shop, was one of the most reliable. The cars were delivered during 1963–64 to provide service for the 1964–65 New York City World's Fair event, including super express service between Times Square and the Fair. Therefore, the cars received special cosmetic attention in terms of radical paint schemes and picture windows, as well as extra work in terms of quality control to minimize equipment failures. (*Note:* Thirty-four cars [9524–9557] had original drop sash windows and were assigned to the main line to "make up" for forty R33S cars.) With the delivery of these cars, every pre–World War II IRT car was retired except for the fleet of LVs and Steinways used on the Third Avenue El remnant in The Bronx.

Later, in the 1970s, the cars started a steady decline until they were selected as the first cars in 1982 to be overhauled during the GOH program. While the GOH provided a boost in their performance, the cars did not receive as complete a workscope as the other car classes that were later enrolled in the GOH. In fact, the GOH program, which covered 4,000 cars at a total cost of $1.7 billion, eventually became a comprehensive remanufacturing program.

When the cars approached their thirtieth birthday, their performance again started dipping and this seriously impacted the on-time performance of this critical IRT line, which connects the boroughs of Queens and Manhattan and serves 250,000 passengers daily as it runs from Flushing to Times Square. A corrective action program was implemented that included the upgrade of key components and involved a partial replacement of propulsion equipment. The upgrade and the ongoing scheduled maintenance system (SMS) program have helped to breathe new life into these old cars, which would be retired after the R142/R142A cars arrived. The last Redbird train in scheduled service ran as a No. 7 Flushing Local from Times Square to Willet's Point Boulevard on the morning of November 3, 2003. A brief ceremony was held to commemorate the event with speeches from the MTA New York City Subways President Lawrence Reuter and Chairman Peter S. Kalikow.

One of the historical curiosities about the R33S/R36 fleet is the original nickname "Silver Foxes," which they were given following the GOH program, during which they were painted Fox-red with a silver roof. Fox-red is a practical color because it hides stains caused by metal dust; it became the standard color for all non-stainless steel cars. A Silver Fox Flushing train and a new Kawasaki R62 (stainless steel car shells) No. 4 Line train were the first soldiers in a long but victorious war against graffiti.

In the 1960s, New York's colorful subway cars had become a symbol of the thwarted artistic ambitions of inner-city kids, who covered the entire exterior and interior car surface with spray-painted tags and cartoon characters. Over the next twenty years, graffiti got uglier and started blanketing even the ceilings and floors. The indecipherable black scrawls made the passengers feel vulnerable to crime because they conveyed the message that "nobody is in charge here."

Starting in May 1984, the new Kiley-Gunn administration declared war on graffiti. A policy was established that no car in the "clean car program" could remain in service with graffiti on it for more than twenty-four hours. The program would be implemented on a train-by-train, line-by-line basis. The goals were to restore public confidence, attract new passengers, and establish an acceptable work environment for transit employees.

The program was a success from the beginning because speedy erasure of graffiti discouraged vandals, who could no longer see their tags traveling throughout the city. Nevertheless, it required a constant effort and a sustained top management focus and deployment of interdepartmental resources. The war finally ended on May 12, 1989, when the last graffiti-covered train was removed from passenger service. This war had cost plenty of valuable resources, including those of the transit police. According to some critics, it was money ill spent since "passengers would stand up to their knees in garbage if they can ride safely" and many had that attitude of "get me there on time and I don't care what hell the train looks like."

It is ironic that today some of these same critics are crediting the 1984–89 NYCTA graffiti war as the catalyst for lowering crime and improving the quality of life not only in New York's subways but throughout the entire city.

## R34 (BMT/IND)

Car Numbers: 8010–8019

In 1964, when the R11s were only fifteen years old, the NYCTA rebuilt all ten cars under contract R34 at a cost of $50,000 per car—half the price of a brand-new car. The R34 represents the only car class that changed its original designation after the cars were rebuilt. Fuses were replaced by circuit breakers and disc brakes by tread brakes, new fans were installed, the original velon seats were replaced with vandal-resistant fiberglass material, and the middle stanchions were eliminated. Unfortunately, even after this rebuilding program, the R34s, like many other orphan test cars, ended their short life as three-car trains running on the Franklin Avenue Shuttle.

One of the items replaced during their rebuilding, the stanchions, deserves some discussion. Many New Yorkers prefer these upright metal posts, usually located right in the middle of a car and anchored to the ceiling and to the floor, because they provide support for standing passengers regardless of height or size. However, stanchions restrict passenger flow inside a car and definitely impede the maneuverability of a wheelchair. For these reasons, stanchions were re-

moved on the R34s, nearly removed on the R62s, minimized on the R110As, and never installed on the R110Bs. What is the future of the vertical stanchion? While the Swedish are exploring a new twist, literally a spiral pole, for their cars in order to provide more gripping spaces and to avoid passengers' fingers making contact with each other, New York City Transit designs are concentrating on stanchions attached to the passenger seats, as they are on buses, instead of in the middle of the car.

Car 8013 is preserved in the New York Transit Museum.

## R38 (BMT/IND)

Car Numbers: 3950–4149

The R38s looked much like the R32s. One telltale feature that distinguishes this car class from the R32s is the car body siding, which is fluted only on the lower half instead of being fluted all the way up to the roof. The R38s were the first stainless steel cars built by St. Louis Car Company, and they were delivered at the rate of five cars per week instead of the traditional ten due to a strike at the plant that slowed down the production process.

As far as special equipment, the last ten cars (4140–4149) were delivered with air conditioning instead of fans. This experiment was very successful and later air conditioning became standard equipment on all New York City transit cars. Today, the entire fleet is air-conditioned.

It is interesting to note that four (4140, 4141, 4144, 4145) of the original ten cars were retrofitted with a Toshiba roof-mounted unitized heating, ventilating, and air-conditioning (HVAC) system in 1985. The main advantage of the Toshiba unit was that it came all in one package like a window room air conditioner instead of the traditional split evaporator/blower assembly up in the car roof and the compressor/condenser assembly under the car. A split system is definitely more prone to leaks and the undercar environment is hotter and dirtier than it is close to the roof. While the Toshiba test had mixed results, the roof-mounted unit is alive and well today on new car procurements.

## R38 AC

4144

## R38 GOH

## R40 (BMT/IND)

Car Numbers: 4150–4549

The R40s will always be remembered for two main features: their lack of standardization and their crashworthiness. The first factor relates to the three different types of the R40 fleet as delivered: 200 slant-end without air conditioning, 100 slant-end with air conditioning, and 100 straight-end with air conditioning. The R40 straight-end cars are nearly identical to the R42s; however, an astute observer can quickly tell them apart by the notch at the belt line and the narrow windows on the No. 2 end storm doors.

In the slant-end cars, the slant consisted of a slope at the No. 1 end of each married pair. This slope, which was later abandoned, gave the R40s a futuristic and aerodynamic space age look. Most railroaders want to give credit for this futurist look to former New York Mayor John V. Lindsay, who had always enjoyed a reputation for the fine arts and wanted a more stylish body than the old-fashioned "boxy" look. This design had no significant impact on their operation and maintenance with the single exception that the sloped ends created a huge

gap for passengers walking between cars. Due to the safety issue, hardware was added to the slanted ends, completely ruining the aesthetics of these cars. The last 100 cars of the R40 fleet were built with the conventional straight end and later, after the GOH program, were renamed "modified."

To the delight of the mystically inclined, the area within a ten-mile radius of the East New York Shop, where some of the R40s were assigned, was once labeled a "Bermuda Triangle" by a newspaper due to a spate of accidents and mishaps. Which other transit system in the world can make the same claim?

Some of the most important technical innovations in the R40s were air conditioning and semiautomatic door controls. In fact, both the R40 and R42 car classes did not have individual door zone switches, and the trainline door unlocking and the door opening wires were zoned through push button relays. By upgrading door controls from manual to semiautomatic, this feature allowed a conductor to set up door operation in the entire train consist in a timely fashion.

The R40s also introduced for the first time, on only thirty-two cars, package brakes with composition brake shoes. This type of friction brake equipment later became standard on both new and existing cars. The advantages of composition over cast iron brake shoes were many: less wheel thermal checks and cracks, fewer flat wheels, lighter weight, better emergency stops at higher speeds, and reduced iron dust in the tunnels (excess dust contributes to signal failures, short circuits, and track fires).

The R40s introduced color-coded signage on both the end and side signs and marker lights were eliminated starting with this order of cars. Finally, special side signs were used. The large, one-piece sign curtain displayed the route and north/south destinations, and on the inside, a strip route map showed all stops and transfers. The curtains were so bulky, however, that the entire B Division could not fit on one curtain, so cars were restricted to certain lines and lacked the flexibility of the three-sign design used previously. Although popular with passengers, constant route changes became very expensive, and the design (also used in R42/R44/R46 cars) reverted. R40s and R42s got three-piece signs, and R44/R46 cars were equipped with electronic LCD designs.

The R40 car numbers were switched around shortly after delivery, puzzling many who look at early pictures of them.

| *Body Style* | *Original #* | *New #* | *Air Comfort* | *Propulsion* |
|---|---|---|---|---|
| Slant | 4150–4249 | 4150–4249 | Fan | WH |
| Straight | 4250–4349 | 4450–4549 | A/C | WH |
| Slant | 4350–4449 | 4250–4349 | Fan | GE |
| Slant | 4450–4549 | 4350–4449 | A/C | GE |

F
4253

## R40 PRE-GOH

## R40 "MODIFIED"

## R42 (BMT/IND)

Car Numbers: 4550–4949

When the R42s were first introduced in 1969, many NYCTA engineers were reluctant to decide which term would best apply to them: "electronic lemons" or "electronic wonders." After all, the R42s were first to introduce new features both obvious and subtle to the riding public: air conditioning, a solid-state converter instead of a motor generator to provide LV power supply, and a solid-state inverter to provide interior lighting. The R42s were also the first car class in many years to be powered by propulsion equipment provided by one vendor (WH) instead of being split fifty-fifty between WH and GE, a significant fact since the propulsion equipment represents the single most expensive system on a subway car.

In 1977, an R42 test train was the first to be modified (no field shunting of the traction motors) in order to reduce energy consumption during the days of power shortages and long lines at gas stations. Today, no field shunting allows the entire fleet to operate with a larger safety margin for the signal system following the 1995 Williamsburg Bridge accident.

With the R42 class came a change in the existing policy of assigning a new car fleet to a single maintenance shop. The assignment of a car fleet to one shop meant that all the documentation, drawings, special tools and gages, and spare parts were in the same place; training and maintenance work could be performed in one location by the same personnel. Further, while most car classes can be mixed and can operate together, it is a proven fact that they perform better in a train consist when the cars are all of the same class. Nevertheless, the R42 was the first fleet to be split and assigned to various shops in the B (BMT/IND) Division of the New York City transit system. Why?

The answer was very simple. It was politically important to introduce air-conditioned cars to millions of voting passengers in all boroughs of the City of New York.

The mid-1980s were focused on door safety programs to minimize incidents of doors flying open and passengers being dragged ("Years of the Door"); the early 1990s homed in on the passenger environment and customer-friendly features. Ten R42s were also retrofitted in 1995 to test a customer strip alarm system similar to the one installed on the R110A/B cars. While the alarm system was eventually removed on the ten cars due to the high number of false activations, all new cars have an intercom system to allow passengers to communicate with the train crew in case of an emergency.

A final curiosity about the R42 fleet was the fact that two cars were shipped to and extensively tested at the U.S. Department of Transportation test facility in Pueblo, Colorado. The benefits of this test were mostly for the U.S. government and not for NYCTA engineers.

### R44 (BMT/IND)

Car Numbers: 100–399

Sleek and silvery with a polished stainless steel exterior, dark blue belly band, and pastel-colored interior, the R44s were the first subway cars to sport the Metropolitan Transportation Authority's "corporate" look.

High-tech, high-speed, and designed and built with a far higher level of passenger comfort than any subway car before it, the R44 was the harbinger of a new age of subway cars. The handsome air-conditioned subway cars were delivered by St. Louis Car Company in 1971–73 after a three-year development period. (*Note:* The R44s were the last subway cars built by the company, which afterward went out of business.) The fleet, powered by a WH propulsion system, consisted of 300 cars divided into four-car units. The unit mixed cab cars with non-cab cars, and trains were designed to run in an A-B-B-A configuration. Relying heavily on electronic control and the first New York City subway car to be equipped with ATO hardware, the R44 was a high-tech design that proved to be too much too soon. As delivered, each car had couplers on both ends, but hostling individual cars (bringing them to the shop and delivering them to the road crew) was very difficult and the requisite safety measures could bring a yard to a halt.

With a length of 75 feet, the R44s were a full 15 feet longer than car series R1 through R42 but 10 feet shy of the 85-foot length of the Penn Central and Long Island commuter railcars they so closely resembled. The decision to go to a longer car was made largely as a potential cost-savings measure. While an eight-car train of R44s is the same length as a ten-car train of R42s—600 feet—fewer cars trans-

lated into fewer components. Fewer cars, however, also translated into fewer doors—sixty-four, compared with eighty door panels on each side of a ten-car train of 60-foot cars.

The R44 has the distinction of holding the speed record for subway trains, an astonishing 87.75 mph. A second run was made with the motors cut out on two cars to simulate a fully loaded train. This time, the R44s managed a still respectable 77 mph. The high-speed tests were carried out on a 5.9-mile stretch of Long Island Rail Road track between Jamaica and Woodside on January 31, 1972.

One of the many technical curiosities associated with the R44s was the fact that for the first time on New York City subways, eight cars came equipped with a hydraulic (made by Abex) instead of air system to provide a quicker, more reliable brake. After many fluid leaks, the test was terminated and the eight cars were converted to the standard WABCO air system.

The side roll signs, similar to the R40, were motorized and the crew could resign the entire train to another service at the touch of a button. (These signs were replaced with LCD displays during GOH.)

Except for the last twelve cars, the fleet had WH propulsion. After GOH, the twelve GE cars (388–399) were reassigned to Staten Island, which also was GE equipped. Cars 100–387 were assembled into four-car units (A-B-B-A) with link bars and renumbered 5202–5479 (not in numerical order). Ten R44s removed from passenger service prior to the GOH program were not included in the renumbering.

A

## R44 SIR

Car Numbers: 400–435 Consecutive, 436–466 Even Only
Differing Characteristics between R44 Main Line and R44 SIR

When the R44 class cars were ordered, 300 cars (288 WH, 12 GE, evenly divided A and B units) were for the main line and 52 cars (all GE, 34 A, 18 B) were for Staten Island. Staten Island at the time was running three-car trains (A-B-A or A-A-A) and was under ICC/FRA rules, since it was considered to be a railroad. Major differences at delivery from the main line cars included:

- FRA grab irons at all doorways (passenger and end doors)
- No trip cocks
- Headlight dimmers (for use when entering stations/passing oncoming trains)

After GOH, the twelve main line GE cars were assigned to Staten Island because powerhouse upgrades made it possible to run four-car trains (A-B-B-A or A-A-B-A or A-A-A-A). Differences now include:

- Staten Island Railway (SIR) cars all GE, main line all WH
- Side designation signs removed
- FRA grab irons removed from passenger doors, added to cab windows
- All cars remain single cars (couplers at each end) [B cars cannot move by themselves]
- 12 KW converters (from 6 KW converters)

The installation of threshold plate heaters and door pocket heaters to prevent ice and snow buildup on the plates in winter has been completed in approximately 40 percent of the cars (2003).

## R46 (BMT/IND)

Car Numbers: 500–1227 Consecutive, 1228–1278 Even Only

The R46 was the sequential brother of the R44. Just like second-year new model automobiles, the R46s had many improvements based on the operational experience and test results of the R44s. Pullman Standard did a much better job in the actual manufacturing and systems integration than St. Louis Car Company which had, by this time, gone out of business. (*Note:* The R46s were the last subway cars built by Pullman Standard, which afterward built only freight equipment.)

The propulsion, the brakes (still controlled by the electrical analog P-wire), and the train control proved to be more reliable and maintainable than the equipment on the R44s. Still, many transit officials rate the R46s as the worst lemons ever in the history of the New York City transit system's subway car fleet. Why? Simply because everybody remembers the R46 Rockwell truck.

The Rockwell articulated truck, an unproven design on NYCTA property, was selected by top management in spite of many technical reservations. On March 27, 1977, the first cracked Rockwell truck was found. This prompted a series of events, including routine inspections, a limited mileage program, and eventually after many cracks and several attempted fixes, the replacement by 1983 of all R46 trucks with a total of 1,548 new truck castings of conventional design.

As a matter of information, the NYCTA did win a financial litigation against Rockwell. While the NYCTA and the riding public suffered through many years of hardship due to this problem, it did teach NYCTA management an important lesson: never accept cars in large numbers unless the design has been proven in the tough New York City transit environment—regardless of any successful testing anywhere else in the world. (The Rockwell truck design is still used today by other transit systems.)

562

E

During the 1976 U.S. Bicentennial celebration, cars 680 and 681 were renumbered 1776 and 1976 and had starred bands on the car body. Another historical curiosity about the R46 was its limited use on train consists dedicated to "The Train to the Plane"—one of many unsuccessful attempts to provide subway service to Kennedy Airport. The main reason why this experiment did not work was because it did not provide a direct link to the airport. A bus and a train were required to make the trip between Manhattan and Kennedy. Three cars assigned to this service had a special "Train to the Plane" logo on their sides.

After GOH, the cars were assembled into four-car A-B-B-A units with link bars and renumbered as 5482–6258. The extra A units are paired and currently make up six-car trains on the G Line.

G
5542

## R62 (IRT)

Car Numbers: 1301–1625

In 1980, the New York City transit fleet reliability reached one of the lowest levels ever (MDBF of less than 10,000 miles). Only two out of three cars were available for service due to the old age of the vehicles and poor maintenance. A top management level Bond Car Steering Committee was established to direct the Bond Car Projects which, at the time, consisted of new R62 IRT cars and the R67 fabricated trucks. In view of the deteriorated fleet, shops, and operational factors, the committee specified single cars instead of the long-standing policy of purchasing married pairs. The single-car design was based on a recommendation from a management consultant (The Emerson Consultants) in 1978 for single, double-ended (a cab at each end) cars. The technological level of the R62s and later cars went back twenty years to the reliable but obsolete design of the R36s based on the poor performance of the R44/R46 classes.

The first R62s arrived from Kobe, Japan, in 1983 and from the beginning proved to be reliable performers. The only negative comment was the one from the public and the news media about the narrow width of the seats, which was wrongly associated with the smaller anatomical features of the Japanese people. In reality, the narrow seats were the result of a high-level directive to specify a minimum number of seats, which simply could not be installed within the space constraints of the car. Despite the seats, the R62 soon became the flagship of the fleet; it also received special attention during the SMS program.

In 1989, Car Equipment developed the SMS workscope for the R62s at a time when the fleet was rapidly approaching a state of good repair and precise and standardized maintenance procedures were being followed in the shops. The justification for the need of a single car was no longer valid, yet its existence required the maintenance and overhaul of many redundant components. By marrying ten single cars into two five-car units, only those components still functional after the marriage went through SMS, while superfluous and trouble-prone components were removed to save energy and weight. (The five-car unit is created by converting the two end cabs into cross-cabs [this feature was part of the design of the car] and activating the off side conductor's controls. The cars are connected with link bars, and can be run as a single unit or two units.) Not only did this provide substantial economic savings but it also resulted in an MDBF of almost 150,000 miles, the highest of all car classes in the fleet.

The R62 and R62A cars have consistently shown themselves to be solid performers right down to the present day. The R62s have been assigned to the Lexington–Jerome Avenue Express (the No. 4 Line) for almost twenty years and are about at the midpoint of their lives. They are still regarded as excellent cars and have the service records to prove it.

---

## R62A (IRT)

Car Numbers: 1651–2475

The development of stainless steel railcars resulted from the efforts of American companies like Boeing and Budd, which initially developed stainless steel technology for the aircraft industry. Budd built a stainless steel aircraft after World War II and went on to utilize this technology in the design and manufacture of stainless steel railcars. Other car manufacturers from that time forward bought the license and the technology from Budd, Boeing, or Pullman. The process of purchasing the license and technology to produce stainless steel cars has continued throughout the history of the industry in the United States.

Historically, all contractors who have won contracts in the United States to supply stainless steel coaches initially subcontracted the construction of the car shells to specialized manufacturers of stainless steel car shells. Contractors such as Kawasaki Heavy Industries (KHI), Bombardier, Sumitomo, and a few others initially purchased the license for a design, and through a program of technology transfer, began manufacturing their own car shells under supervision. In addition to Mafersa of Brazil and Comenge from Australia, the small, specialized group of designers and builders of stainless steel car shells includes Sorofame of Portugal and Sofretu of France. Japanese manufacturers have also built stainless steel, bi-level, high-speed coaches for the Japanese Railway.

1984
3
7814
NEW YORK CITY
M
TRANSIT

KHI entered the stainless steel railcar business by purchasing the license and technology from Budd. KHI had successfully manufactured 325 R62 cars; however, it did not pursue the following procurement for 825 R62A cars due to a combination of factors, including production concerns and limiting its exposure to risk in the North American market. This opened the door for Canadian Bombardier, which subsequently entered the stainless steel car business in 1982 on the R62A project by purchasing the design and manufacturing technology from KHI. For the R62A project, Bombardier assigned several people to KHI's factory in Kobe to learn the stainless steel manufacturing process firsthand. Afterward, KHI assigned several people to Bombardier's manufacturing facility to assist in the manufacture of the R62A cars. Bombardier also purchased ten car shells from KHI, which were sent to Canada to serve as production samples. The R62A project was successfully completed on time. The main difference between the R62 and R62A, in addition to different system suppliers, is that today all the R62s are unitized, while slightly over 250 R62As are still single cars. Today, R62As provide all service on the No. 1 Broadway Line and No. 7 Flushing Line (2004).

---

### R68 (BMT/IND)

Car Numbers: 2500–2924

The R68 is one of the most handsome cars in the fleet, with its gleaming stainless steel exterior, powerful air conditioning, comfortable seating, and bright interiors. While it is beautiful to look at, it has a number of deficiencies. In a way, this was quite a surprise. After all, the R68s were built in France and French railroads have always had the reputation of being the best in the world—at times even better than German and Japanese rail systems. The French Train à Grande Vitesse (TGV), the Spanish Alta Velocidad Española (AVE), and the upcoming American Flyer are prime examples of the French genius in the railroad industry. So why is the R68 such a marginal performer and one of the heaviest and most expensive cars?

A combination of factors explains the R68's lackluster performance. The technical specifications had contradictory requirements with regard to car body weight and maximum wheel loads; manufacturing car shell methods had not been well defined; configuration control was not enforced; little emphasis was placed on doing first article inspection (FAI) of equipment; quality assurance/quality control (QA/QC) activities should have been proactive rather than reactive; not enough attention was given to preliminary and final design reviews; and a consortium (WH-Amrail) instead of a single manufacturer had the contract to build these 425 cars. Both the design and production suffered from the constant lack of coordination among the consortium members. A typical delay was caused by the fact that the steel to be used for the end underframe assembly was chosen by the firm responsible for the car shell and not by the people who actually manufactured the assembly.

The bottom line is that for many years after their delivery, the NYCT engineering team has been going back to the drawing board,

trying to make these cars as smart as their Canadian and Japanese counterparts. In recent years, the R68 and R68A have improved their performance considerably and now have the best MDBF of any of the B Division cars. This has been the result of several modifications in the propulsion system as well as the linking of the cars into four-car units.

---

### R68A (BMT/IND)

Car Numbers: 5001–5200

The R68A was the last car class of the "new car" fleet that consisted of 1,775 R62/R62A/R68/R68A cars at a capital expense of about $2 billion. While the 200 KHI R68As were quite an improvement over their predecessors, the French R68s, because of the program management and selection of systems' suppliers, still suffer from their heavy weight and all other constraints associated with the operation of a 75-foot-long car on a transit system designed and built for shorter cars. In a way, the problems with the 75-foot cars are similar to the ones a driver would encounter maneuvering a full-size automobile up and down a narrow driveway and into a small garage built fifty years ago for a different generation of people and automobiles.

The cost-savings factors that decided the length of the R44s are presently being challenged, and it could very well be that the R68A will be the last car of this particular length. Today, the design is gradually shifting back to 60-foot-long cars like the R30s, R32s, R38s, R40s, and R42s. Why? In the first place, the 60-footer is the largest car that can physically fit on all BMT and IND routes, while there are restrictions for both 67- and 75-foot cars. Because there are more doors in a 60-foot train consist, there is a better passenger flow. The cars are lighter, have a lower center of gravity, are faster and stop better, offer better visibility due to more glass, and have none of the

other problems associated with 75-foot cars, such as stress on body structure and trucks and higher wheel loadings. Passengers also dislike the inability to walk between cars. The end doors must be kept locked for safety because the end doorways are not aligned when the train is negotiating curves or switches.

On December 2, 1989, sixteen R68As were involved in a flood similar to the previous one in a Harlem station where 18 million gallons of water had flowed into the tunnels. In both events, the only way to reach the roof of the trains was to swim straight down; therefore, some people started calling the cars the silver submarines. As in the case of other car classes, the R68A survived this flood relatively unscathed; some damaged floor panels eventually had to be replaced.

## R110A (IRT)

Car Numbers: 8001–8010

The ten-car R110A is one of the two NTTTs delivered in October 1992 representing the Tomorrow Trains, a glimpse at twenty-first-century New York City subway travel. Each train also represents a $40 million investment ($20 million from NYCTA, $20 million from various suppliers) to introduce and qualify new technology, including AC propulsion with regeneration, microprocessor-controlled doors and brakes, roof-mounted hermetic air-conditioning units, and fabricated trucks with air bags suspension. A lot of design effort was expended to make passengers feel safer and more secure while riding these trains. The door edges act like elevator doors and reopen when something gets stuck between them. There are passenger emergency intercoms for contacting train crews, passenger alarm strips to press in case of an emergency, improved lighting, lots of glass all around to see into other cars and the platform, and computerized announcements for improved audio clarity.

The KHI R110A train received special attention with regard to interior design. Famous Italian designer Massimo Vignelli was hired to

design the car interior under the direction of the Metropolitan Transportation Authority Arts for Transit program. The R110A has very bright colors with speckled black floors and speckled gray walls. The seats are also in different bright colors. Unnecessary edges were removed from stanchions, poles, and bars to create a smoother, cleaner appearance. All these efforts were rewarded with the 1995 national award for transportation design by the U.S. Department of Transportation National Endowment for the Arts.

In order to improve passenger traffic flow in and out of the cars, the door openings were increased from 50 to 64 inches and a space niche was added next to each door to prevent passengers from leaning on door panels. Also, cross-sectional seats were introduced for the first time in many years on these IRT cars. One of the unfortunate results was a significant reduction in seats, from a total of 440 for a previous IRT R62A train down to 264 for the R110A (on the other hand, number of standees went up, from 1,332 to 1,684). This became the number one complaint from the public and transit communities. In view of this criticism and further studies, the NYCTA restored most of the seats on the IRT R142/R142A procurement.

In the future, the R110A and R110B test trains will be used to qualify new vendors and new products.

(*Note:* Although commonly called the "R110A" fleet, the cars were actually purchased with money from contract R130.)

## R110A New Technology Test Train Cab (KHI)

The cab console includes an automatic signage and announcement programming keypad (left side), a built-in radio (dedicated speaker and controls located on left side), a microprocessor screen with selection buttons on each side, and a single-handle controller and communication handset (left of console). Below the console is an access cover for windshield washer fluid (left side) and windshield wiper control (right side). In the center of the picture is an adjustable, ergonomic cushioned seat, and to the right are a cup holder and a small bag rack. The left side of the picture shows a sun visor, a manual door lock (one of two, lower lock not shown), and a schedule holder.

## R110B (BMT/IND)

Car Numbers: 3001–3009

The nine-car R110B consists of three triplets, each made up of three cars 67 feet long, a concept going all the way back to the BMT Standards of the 1920s. The R110B has all the innovative communications features of the R110A, plus electronic strip maps displaying each station on the A Line where the train operated. Unfortunately, the R110B has the same deficiencies as the R110A: passengers criticize the lack of seats, handholds, and poles, which makes it difficult to ride inside these cars during rush hour. Interestingly, this arrangement was the result of passenger focus groups consulted during the train's design phase. One of the top priorities they identified was the need to enter and exit trains quickly during rush hours. The eventual design of the R110Bs resulted in fewer seats. Designers would be going back to the drawing board when it was time to design the new R142/R142A cars.

The R110B performance has also been slightly better than that of the R110A, mostly due to the following factors: (1) the extensive use

207TH
SHOP
A

of a high-speed test track in Canada, which allowed the engineers to do a lot of debugging before shipping the cars to New York City; and (2) the management decision to design and build some of the most critical car systems internally instead of relying on outside suppliers. The Canadians were able to perform a better systems integration at the beginning, which gave them a head start in this friendly competition even though the Japanese, relying on their own resiliency, eventually did catch up with their competitors. Meanwhile, NYCTA engineers have drawn some very important conclusions from these tests, which will decide the design of twenty-first-century subway cars.

(*Note:* Although commonly called the "R110B" fleet, the cars were actually purchased with money for contract R131.)

---

## R142/R142A (IRT)

With the exception of nineteen New Technology Test cars delivered in October 1992, it had been more than ten years since the last fleet of 200 R68A cars arrived in New York from Japan. Meanwhile, New York City Transit experienced a constantly rising ridership due to the introduction of the MetroCard, which allowed free transfers between bus lines and the subway, increased service reliability, and the fact that IRT Redbirds' fleet of roughly 1,400 cars was reaching the end of its thirty-five-year useful life. This prompted the procurement of a new fleet of high-technology cars that soon were dubbed the "New Millennium Trains."

With the assistance of LTK & Associates Engineering Consultants, the R142 technical specifications outlined a rigorous process of design reviews and testing to assure new standards in safety, reliability, and maintainability. At the same time, active participation of a myriad of external groups (customers, peers, industry, news media, transportation committees, professional societies, associations, etc.) during many focus group presentations and open houses, and the use of actual size mock-ups assured that, like never before, every single detail related to customer environment and the ergonomic design of the crew cabs would be addressed.

The original plan was to purchase only 740 cars, but fierce competition and effective value engineering during the request for proposal process drove prices down to such an affordable level that an additional 340 cars could be funded within the original budget. Bombardier had the lion's share of a split award (R142) and was designated as

the lead designer, while KHI had the balance of the contract (R142A) with the requirement that both car classes would look the same and be operationally compatible. Subsequently, Bombardier was awarded the original 680 cars in 1997, with a later award of 200 option cars plus an additional 150 in 2001, for a total of 1,030 cars. KHI was awarded 400 base cars in 1997 and 120 option cars in 2001, and won a competitive bid for eighty (R34142S) cars in 2003 for a total of 600 cars. This fleet of 1,630 cars has allowed NYC Transit to retire all of the Redbirds from passenger service and satisfy additional ridership demand.

Shortly after the start of the delivery of the New Millennium Trains in December 1999, the cars experienced many technical and "cultural" difficulties due to a combination of revolutionary technology (for NYCT), infant mortality, and hundreds of design modifications. Nevertheless, the seasoned project management under the leadership of President Lawrence Reuter faced and methodically resolved these human and technical challenges. Today, the entire program is being recognized as a great accomplishment, with New York State resources responsible for most of the project. MTA Chairman

E. Virgil Conway has called it "a victory for the workers at the Kawasaki plant in Yonkers and Bombardier's facility in Plattsburgh," referring to the two assembly plants.

The R142As are now breaking new MDBF records. They have become the number one choice for train crews and subway riders because all of the amenities successfully tested on the R110A/R110B cars were further refined and implemented, especially the addition of seats and hand poles. In addition to a smoother and quieter ride, passengers now hear automated public information announcements from familiar voices donated by Bloomberg Radio and Television. Under the direction of Metropolitan Transportation Authority Arts for Transit, interior design consultant Antenna Design New York created a rider-friendly, bright, and at the same time soft atmosphere that includes a nostalgic nod to the departing Redbirds with some red speckles sprinkled in the colors of the cars. Both the R142 and R143 received very favorable reviews by the *International Design Magazine,* with the R142 as a "Concept Design Distinction" winner of the 45th Annual Design Review.

New York State Governor Pataki best summed it up: "These trains will set the tone for a new century where we will have a transit system more rider-friendly than ever before."

## R143/R160 (BMT/IND)

With all the fanfare surrounding the introduction of the New Millennium Trains R142/R142A for the IRT (A) Division, the BMT/IND (B) Division was practically overlooked. Yet not only was it preparing for its own share of New Millennium Cars, but it was advancing the technology level to an even higher bar. In December 1998, KHI was awarded the R143 contract for 100 cars, followed two months later

by an option for 112 cars. In addition to the well-known physical differences dictated by clearance requirements (IRT versus BMT/IND), the R143s differed from the R142/R142A in one major way: these are the first cars designed and equipped to operate with CBTC, originally known as moving block signal. The key to this new technology is the use of microprocessors on the trains and in wire loops on the wayside for high-speed, two-way communications. When NYCT replaces its century-old fixed block signal system, it will take CBTC to a new level, pioneering the use of radio communication that eliminates wires on the track. With CBTC, the signal block moves along with the train, allowing more trains to be packed closer together, thus reducing overcrowding. Naturally, as with any other new technology, this process will not happen overnight but through a long and often painful evolution of hardware, software, and "humanware." CBTC will debut on the L (Canarsie) Line, which today exclusively uses the R143 fleet.

Another significant development related to the R143s is the use of four trains completely dedicated to qualifying new major suppliers for the R160 procurement. In addition to the original CBTC test train, Fuji Doors, Panasonic Communications, and Siemens Propulsion are in the final stages of qualification tests. This will improve the limited pool of suppliers for new car orders not only for the New York City transit system but for the entire transit world. Being able to demonstrate a successful trial in the toughest transit environment, New York, is a coveted benchmark.

In December 2000, NYCT announced a request for proposal for the procurement of as many as 1,700 BMT/IND cars. This is most probably the largest order of new subway cars in the entire world. Proposals were received from Bombardier (R142 car builder), KHI (R142A and R143 car builder), and ALSTOM Transportation Inc. (supplier of the R142 propulsion equipment). While KHI received the highest technical ranking, ALSTOM submitted the lowest bid. Since the R143 had to look the same and be compatible with the R160 cars, it was well recognized that any car builder other than KHI would face a greater risk in attaining R143 operational compatibility. In order to mitigate this risk, ALSTOM was requested to revise its proposal. Subsequently, ALSTOM enhanced its proposal by committing to form a contractual relationship with KHI to produce a single design by an integrated team. This was achieved under a technology transfer by KHI that will maximize the use of common suppliers and ensure compatibility at the best price.

In October 2002, ALSTOM was awarded the federally funded base contract for 660 new cars, with KHI taking approximately 40 percent of the contract value. After the award, ALSTOM entered into a limited liability company with KHI to form ALSKAW LLC. Today, engineering and manufacturing efforts are under way with a target delivery of July 2005 for the pilot cars, followed one year later by the production cars.

At a public board meeting, MTA Chairman Peter Kalikow proudly stated, "I think today we have done something that twenty-five to thirty years from now, people in New York will be grateful for and thank us for." And with this statement, we have finally arrived at the terminal of the nostalgic trip that we undertook with so many friends riding every single type of rolling stock that so elegantly and efficiently glided over the century-old New York City subway rails.

# Appendix A

*Subway Car Fundamentals*

## Power System Overview

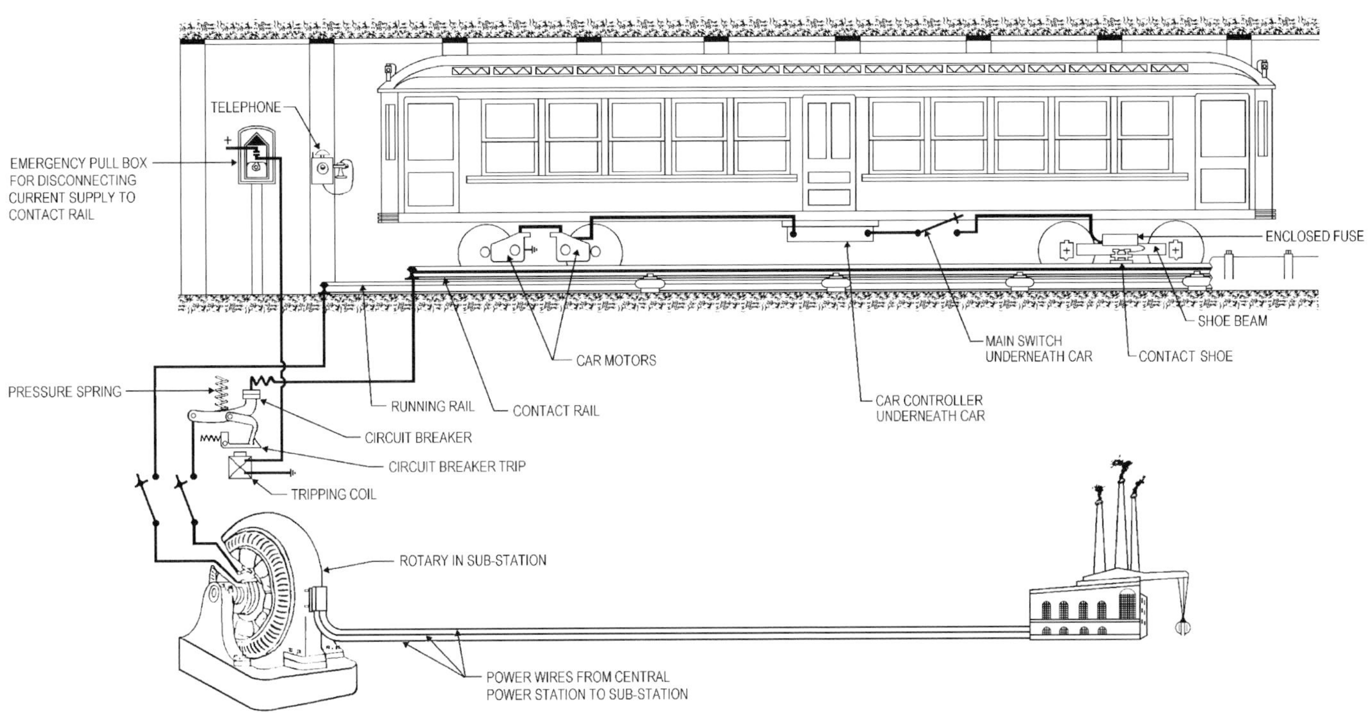

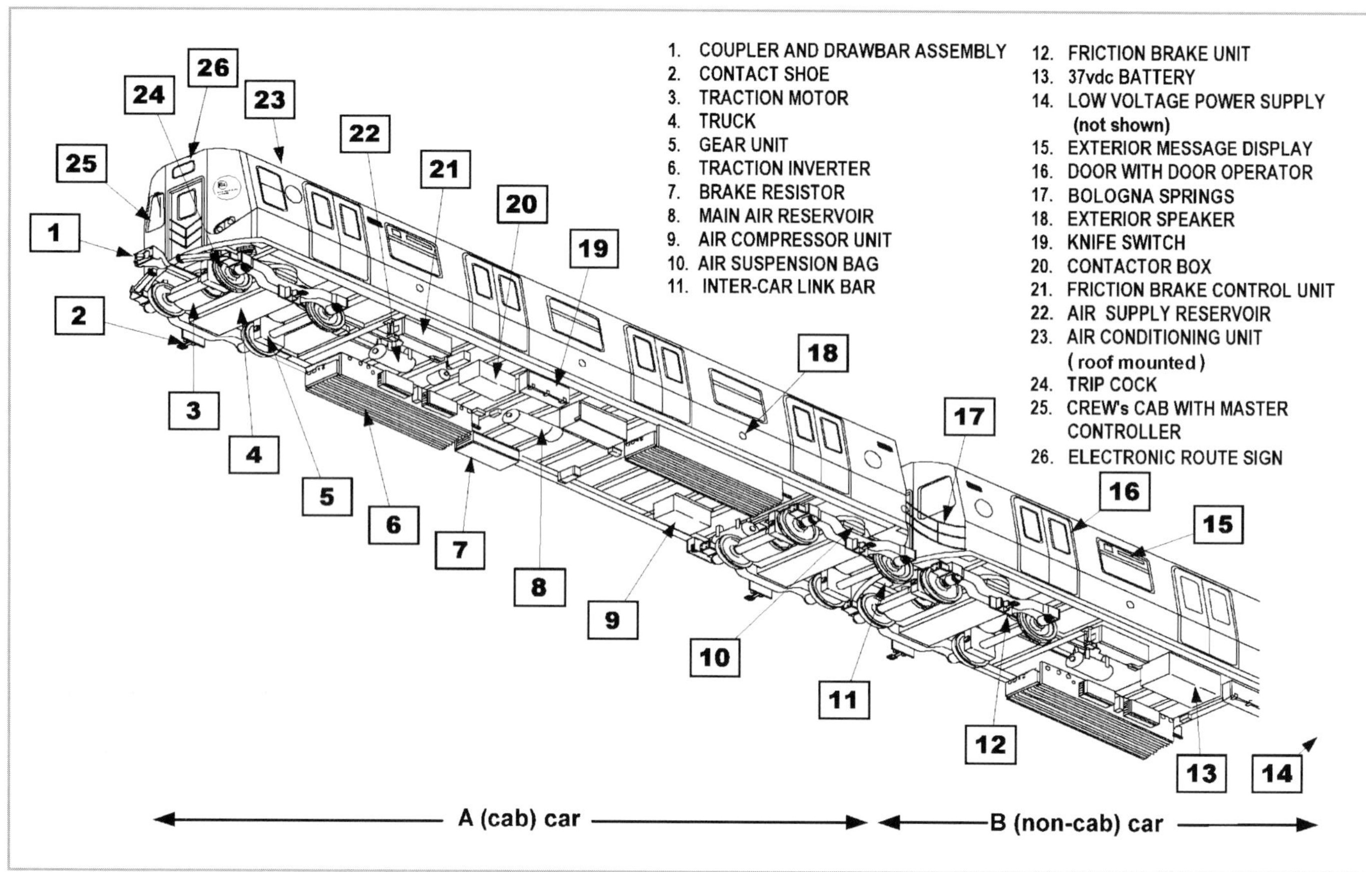


### What Makes a Subway Car Run?

If a rapid transit system wishes to attract and hold riders, it must provide service that is both swift and reliable. The following generic description outlines what makes a subway car run. Please refer to the corresponding numbers listed in the drawing above.

A third rail supplies 600 VDC electric power to operate and provide heat, air conditioning, and light for the car. This 600 VDC electric power is picked up from the third rail via the contact shoe (2) to the knife switch (19). From the knife switch, 600 V power is distributed to traction inverters (6) and, through the contactor box (20), to other high-powered systems, such as the two roof-mounted air conditioners (23) and the air compressor unit (9).

Cars are assembled into multicar units with intercar link bars (11) that carry the air and electric connections. B (non-cab) cars are sandwiched between A (cab) cars. Then, the units are coupled together to form the train with automatic couplers (1).

Electric power is used to operate the air compressor unit (9). The air compressor unit provides pressurized air to fill the main air reservoir (8) and the air supply reservoir (22) for operation of the friction braking system and car body air suspension system.

To provide speed and reliability, the latest generation of New York City transit cars uses four three-phase asynchronous traction motors (3), each connected to the axle through a gear unit (5). The train operator controls the speed using the master controller handle located in the cab (25). Powering and braking commands are sent to each traction inverter (6) and friction brake control unit (12) of every car in the train over a digital computer network. In powering, the traction inverter converts 600 VDC current into three-phase AC current using Insulated Gate Bi-Polar Transistor (IGBT) semiconductor switches. The control computer assures that right amount of three-phase current is delivered to the traction motors to provide smooth and rapid acceleration.

In braking, the traction motors act as alternators, converting the car's kinetic energy into three-phase AC current electric energy. The traction inverter converts this energy back into 600 VDC current that is "regenerated" back into the third rail. Any excess energy that cannot be consumed by the third rail is diverted into the brake resistor (7). The air-operated friction brakes are automatically activated for a final stop to hold the car at a standstill. They also serve as a backup system in case of electric brake failure. In addition, the trip cock (24) is activated when the train hits debris on the roadbed or a stop arm (red signal), causing the train to come to a full stop.

The doors (16) are opened and closed electronically utilizing a door operator. A door obstruction sensing system prevents the train from moving while an object is stuck in the doors. Upon detection of obstruction, the doors will automatically open and then reclose, allowing the passenger to retrieve the stuck object.

Two heating and air-conditioning units (23) on each car keep the passengers cool in the summer and warm in the winter.

The air suspension system (10) provides a smooth ride whether the car is nearly empty or fully loaded with passengers.

To help passengers find their way in the vast New York City subway, an automatic system announces the next station stop, while the colorful electronic passenger information signs (15, 26) and computerized strip maps indicate train route, transfer information, current time, and even advertisements.

The LV power supply (14) converts 600 VDC to 37.5 VDC to charge the battery (13) and to power the lights, door operators, train control computers and command networks, PA system, and other functions. Batteries provide standby control power when 600 VDC is not available.

And last, but certainly not the least, are the bologna springs (17) between the cars. They provide an additional safeguard to prevent passengers from trying to board the subway car from station platforms and falling to the roadbed. In addition, these barriers help to guide people with visual impairments away from the area and toward doors of the car.

# Appendix B

*Historic Subway Cars of the New York Transit Museum*

| Car Number | Year | Builder | Type | Class |
|---|---|---|---|---|
| 100 | 1930–31 | American Car & Foundry | IND | R-1 |
| 484 | 1932–33 | American Car & Foundry | IND | R-4 |
| 1273 | 1904–5 | Laconia Car | BMT | BU (1200 Series Trailer) |
| 1404 | 1908 | Jewett | BMT | BU (1400 Series Motor) |
| 1407 | 1908 | Jewett | BMT | BU (1400 Series Motor) |
| 1575 | 1938/1947 | Pullman Standard/ American Car & Foundry | IND | R-7A / R10 Prototype |
| 1612C | 1908 | Jewett | BMT | Q |
| 2204 | 1916 | American Car & Foundry | BMT | B BMT Standard |
| 4902 | 1917 | Pullman | IRT | 1917 LV Trailer |
| 5655 | 1938 | St. Louis Car | IRT | 1938 World's Fair, Steinway |
| 5760 | 1948 | American Car & Foundry | IRT | R-12 |
| 5782 | 1948 | American Car & Foundry | IRT | R-12 |
| 5871 | 1949 | American Car & Foundry | IRT | R-14 |
| 6019 | 1927 | Pressed Steel Car | BMT | D Triplex |
| 6095 | 1928 | Pressed Steel Car | BMT | D Triplex |
| 6112 | 1928 | Pressed Steel Car | BMT | D Triplex |
| 6239 | 1950 | American Car & Foundry | IRT | R-15 |
| 6387 | 1954–55 | American Car & Foundry | BMT/IND | R-16 |
| 6609 | 1955–56 | St. Louis Car | IRT | R-17 |
| 6895 | 1955–56 | St. Louis Car | IRT | R-17 |
| 7267 | 1956–57 | St. Louis Car | IRT | R-21 |
| 7486 | 1957–58 | St. Louis Car | IRT | R-22 |
| 8013 | 1949 | Budd | BMT/IND | R-11 / R-34 |
| 8506 | 1961–62 | St. Louis Car | BMT/IND | R-30 |
| 9306 | 1963 | St. Louis Car | IRT | R-33S |
| 9542–9543 | 1964 | St. Louis Car | IRT | R-36 |
| 51050 | 1957–58 | St. Louis Car | Money Train | R-22 |

# Appendix C

## *Specials*

Special trains have been running on New York's transit lines since the first elevated lines began providing service. As with any transit system, specials are used for anything from commemorating the opening of a new line or the closing of one, to inaugurating a new service, to showing off a new type of railcar, to just running a unique service. There have been shoppers specials, theater specials, and business specials as well as limited express services, summer specials, racetrack specials, baseball specials, railfan specials, and a legion of others. When there are temporary service changes, special trains can be called upon to bridge gaps and unify operations. The special provides a flexible way to fill a rail system's "special" needs.

On December 1, 1955, Miss Subways, at the front door of a train of B Types, cuts the ribbon to inaugurate service between the IND's Queens Plaza Station and the BMT's Lexington Avenue station via the Eleventh Street connection in Long Island City. Brighton locals, operating weekdays between Coney Island and Forest Hills via the Montague Street tunnel, provided this service.

R16s on the inaugural trip over the new Rockaway Line on June 28, 1956.

A summertime round-trip excursion service that ran daily between the lower level of Forty-second Street–Eighth Avenue station and Playland in Rockaway with an intermediate stop at Hoyt–Schermerhorn Street in Brooklyn. R16s were usually used in this service, but this 1958 photo shows R10s.

In 1960 and 1961, New York Titans Football Specials were run to bring football fans from 169th Street or 179th Street Jamaica to the Polo Grounds. These R1/R9 trains ran nonstop to the lower level of Forty-second Street–Eighth Avenue. After this one stop, the motorman reversed ends south of the station and ran nonstop to 155th Street on the Concourse Line.

This eight-car train of R32As (equipped with modified third rail shoes) operated under its own power from the Mott Haven Yard of the New York Central Railroad to Grand Central terminal, where the train was displayed for several hours on September 9, 1964. The run demonstrated that rapid transit equipment could run on electrified commuter railroads.

## Subway Series 2000

In the fall of 2000, the New York Yankees and the New York Mets met in the World Series, making a "Subway Series" event for the first time since 1956, when the Yankees had played the Brooklyn Dodgers. To celebrate the occasion, two special trains were decked out with the logos of the New York Yankees and the New York Mets as well as banners proclaiming "Subway Series" affixed to the car sides.

On Saturday evening, October 21, 2000, a train consisting of ten new R142 cars (ends covered in Yankee pinstripes) ran on the No. 4 Line from Grand Central to 161st Street. The governor, the mayor, and other dignitaries were present as well as their families. A train of R33S/R36 cars painted in New York Mets colors operated on the No. 7 Line to Shea Stadium for the third game of the World Series, the first game at the Mets' home stadium on Tuesday, October 24, 2000. Both trains ran for a considerable time afterward in regular service on their respective lines before the color schemes were removed. (See color gallery.)

To celebrate the New York Yankees' victory over the New York Mets in the 2000 World Series, the City of New York sponsored a parade on Monday, October 30. Several players rode aboard R62 car 1415, which was decorated with Yankee logos. Placed on a flatbed truck, it traveled the entire parade route.

# Appendix D

## *The State-of-the-Art Car of 1974*

During the 1970s the federal Urban Mass Transportation Administration sponsored the design of a new generation of urban mass transit cars known as the State-of-the-Art Car (SOAC). The operation of the SOAC cars on existing rapid transit systems allowed firsthand evaluation of the best in current technology by transit and public officials, the community, and the riding public. The two SOAC cars operated in New York City, Boston, Philadelphia, Cleveland, and Chicago. The SOAC cars made a demonstration run for the press and invited guests on May 13, 1974. They were placed in passenger service on May 17, 1974, in New York City, where they ran through July 19, 1974, on the A, D, E, and N Lines.

The features of the SOAC cars included lightweight air suspension trucks with rubber chevron primary springs, solid-state DC chopper control, AC auxiliaries, stainless steel body construction, reduced noise level, separately excited field DC motors, styled exterior, human-engineered passenger-oriented interiors, and a high flow capacity AC air-conditioning system. The SOAC cars were capable of reaching 80 mph in less than 60 seconds. Braking from 80 mph could be achieved with any of the three brake systems (blended dynamic/air, dynamic only, and air only) in less than 2,000 feet. Yet, the acceleration/deceleration rates were well within the limits of passenger comfort, contributing to the excellent ride quality of the cars. The improved truck and suspension system design incorporating air bags achieved a ride quality as good or better than that of any other rail transit car.

HIGH DENSITY CAR

Special emphasis was placed on the control of interior and wayside noise. The interior SOAC noise levels were considerably less than that experienced on current cars. The SOAC HVAC system was designed for all-season passenger comfort. Each car had two independent 8 ton air-conditioning systems, each separately controlled by its own temperature control panel and thermostats. Smaller size AC motors resulted in lower weight, improved reliability, and reduced maintenance.

Many of the features of the SOAC cars were later incorporated in cars purchased by the NYCTA.

# Appendix E

## *Redbird Reefings*

Because of the increased costs associated with the scrapping of obsolete subway cars, the Redbirds would eventually become artificial reef material on the ocean floor. The operation began in August 2001 with the first shipment of cars to Delaware. Additional shipments of virtually all of the R26, R28, R29, R33, and R36 cars were made to Delaware, South Carolina, Georgia, Virginia, and New Jersey. Reefings were finally completed by 2004.

The trucks, doors, windows, and other parts were removed and the car bodies cleaned of grease, oil, and other contaminants. They were then loaded on barges and delivered to the locations where the reefs were to be created and sunk in approximately 80 to 90 feet of water. The artificial reefs will provide a protective habitat for fish and other marine animals.

WEEKS 246
7850

7851
2 0 2

# Glossary

*Note:* Some definitions are standard, while others refer only to New York City Transit railcars.

**A Car**—Usually denotes the end car in a multiple-car coupled unit; typically has a train operator's cab.

**A Division**—Previously known as the Interborough Rapid Transit (IRT); now refers to the part of the system used by the numbered lines 1–9 and Forty-second Street Shuttle Line.

**AC Propulsion**—Propulsion system using alternating current (AC) traction motors.

**Acceleration**—Rate of change of velocity with respect to time, usually expressed in mphps.

**Actuator**—Portion of a device that converts an input into mechanical motion.

**Air Bag Suspension**—Inflatable resilient bag in the suspension system on a truck to improve ride quality.

**Air Compressor**—A device used to compress air for the operation of air brakes and other air-actuated equipment; also called an air supply unit.

**Air Condenser**—A device used for removing moisture from compressed air by cooling the air.

**Air Conditioning**—Any system used to modify air qualities, but usually refers to a system to cool and dehumidify air.

**Anticlimber**—End-of-car underframe having two or more parallel horizontal plates protruding from the end sill to prevent one car from overriding the next car in the event of a collision.

**Articulated Car**—A unit where the ends of two car sections share a common truck.

**Automatic Train Control (ATC)**—The system for automatically controlling train movement, enforcing train safety, and directing train operations; must include automatic train protection (ATP) and may include automatic train operation (ATO) and/or automatic train supervision (ATS).

**Automatic Train Operation (ATO)**—That subsystem within the automatic train control system that performs any or all of the functions of speed regulation, programmed stopping, door control, performance level regulation, or other functions otherwise assigned to the train operator.

**Automatic Trip Arm**—Automatic stop that provides a positive means of train control by enforcing obedience to a stop signal.

**AW0 Load (Empty)**—Empty car weight with neither crew nor passengers aboard.

**AW1 Load (Normal)**—Car weight with crew and all passenger seats occupied.

**AW2 Load (Full)**—Car weight with crew, all passenger seats occupied, and an equal number of passengers standing.

**AW3 Load (Crush)**—Car weight with crew, all passenger seats occupied, and the largest number of passengers that can occupy the car (crush load).

**Axle**—The shaft on which the car wheels of the truck are mounted; it holds the wheels to gage and transmits the load from the journal bearings to the wheels.

**B Car**—Usually denotes a non-cab car located between A cars.

**B Division**—Previously known as the Brooklyn–Manhattan Transit (BMT) and Independent System (IND); now refers to the part of the system served by the lettered lines.

**Bad Order**—A car that needs mechanical attention or repairs.

**Ballast [Electrical]**—Electrical device to obtain electrical conditions necessary to start and maintain illumination in fluorescent lights.

**Ballast [Stone]**—Cut stone applied to rails and ties on the roadbed to prevent movement.

**Barn**—Facility used for inspection, service, and running repair; today called a maintenance shop.

**Bologna Springs**—Large-diameter springs connected to ends of cars to prevent passengers from attempting to board trains from station platforms and falling to the roadbed.

**Brake Pipe**—Trainlined air pipe used by the brake system.

**Brake Rigging**—A system of levers and associated hardware for transmitting forces developed from the brake cylinders to the brake shoes on all wheels.

**Brake Shoe**—A device that is pushed against a wheel to slow or stop the train.

**Brake Valve**—A device used by the train operator to control the train's brakes.

**Buff Load**—Horizontal compressive load applied at the coupler transmitted to the car body; the opposite of draft.

**C Car**—Usually denotes a non-cab car located between A cars; similar to a B car. *See* B Car.

**Cab**—An enclosed area at an interior end of a subway car where equipment to operate the train or doors is located.

**Cam Controller**—A device to regulate traction motor current by varying motor circuit resistance.

**Cam Switch**—High-voltage (600 VDC) power switches used on main power circuit controllers for motoring and braking.

**Camber**—A slight deviation from a horizontal or vertical straight line in an arc between two points on the line. Cars have a slightly positive camber built in to allow for the deflection of the car under load.

**Car**—Individual vehicle in a train consist; also called a railcar. (A car built to carry customers is called a passenger car, otherwise it is called a work car.)

**Car Body**—The car exclusive of the trucks and exterior mounted components.

**Car Body Bolster**—A transverse member on the underframe over the truck that transmits the load to the truck bolster center plate.

**Carline**—Longitudinal roof framing members from end to end of a car.

**Center Pin**—Pin designed to align the car body to the truck while allowing rotation.

**Center Plate Casting**—A pair of plates that support the car body on the truck with a center pin.

**Center Sill**—Central longitudinal member of the underframe of a car.

**Chevron Spring**—Sandwiched rubber and steel plate fabricated into a V form used in the primary suspension on a truck.

**Chime**—A two-note sound that indicates that the doors are closing.

**Chopper**—Electrical device providing variable direct current voltage from a fixed supply.

**Circuit Breaker**—A protective device that automatically interrupts the circuit when excessive current is detected.

**Clasp Brake**—Brake arrangement with two brake shoes on a wheel.

**Communications-Based Train Control (CBTC)**—A continuous automatic train control system utilizing high-resolution train location determination, independent of track circuits; continuous, high-capacity, bidirectional train-to-wayside radio frequency (RF) data communications; and trainborne and wayside vital processors capable of implementing vital functions.

**Composite Car**—Car constructed of wood and steel.

**Consist (Car Consist)**—The makeup or composition of individual cars in a train. *See* Unit (Operating).

**Console**—Control panel containing major controls for the train operator to operate the car/train.

**Contact Rail**—Rail adjacent to running rails to supply electric power to car (also called the third rail).

**Controller**—*See* Group Box.

**Converter**—Solid-state device that converts 600 VDC to 37.5 VDC for control voltage.

**Coupler**—Device located at a car end to provide a mechanical connection to another car.

**Coupler Centering Device**—An arrangement for maintaining the coupler at the centerline of the car.

**Coupler Head**—The portion of the coupler that houses the locking mechanism.

**Cross Bearer**—Transverse member of the underframe.

**Deadman Control**—An alertness device that activates an emergency brake and releases the master controller handle if the train operator becomes incapacitated.

**Deck Roof**—Structure above the roof arranged in the form of a deck to provide ventilation from the sides.

**Derailment**—Any time the wheel(s) of a car is/are either on top of the head of a rail/switch point or suspended above a rail.

**Destination Sign**—A sign for passenger information showing the train's destination.

**Disc Brake**—Discs mounted on axle with pads arranged to squeeze the disc for braking (as opposed to tread brakes).

**Door Hanger**—Device to suspend the sliding door panel from the top.

**Door Operator**—An electric motor or pneumatic engine used to open and close the door panel(s).

**Draft Gear**—An arrangement of rubber and metal working together during compression and tension to form a shock-absorbing action.

**Draft Load**—Horizontal tensile load applied at the coupler transmitted to the car body.

**Drawbar**—Bar located on the car body undercarriage: pivots laterally from side to side, allowing one car to follow another around curved tracks.

**Dwell Time**—Total time from the instant that a train stops in a station until the instant it resumes moving.

**Dynamic Brake**—A system to brake a train by using the traction motors as generators and dissipating the energy electrically through resistors.

**Dynamic Outline**—The greatest dimensional cross-section of a moving train accounting for all possible car body positions due to rocking and the like.

**Electric Portion**—Attachment to the mechanical coupler through which trainline electrical circuits pass to the other coupled cars.

**Emergency Brake**—Fail-safe, open-loop braking to a complete stop with an assured maximum stopping distance considering all relevant factors; once the brake application is initiated, it is irretrievable, that is, it cannot be released until the train has stopped or a predetermined amount of time has passed.

**End Destination Sign**—Signs on both ends of the train showing the train's final destination.

**End Door**—Sliding or hinged door at ends of a car (also called storm door).

**End Frame**—Car frame that forms the end of the car.

**End Sill**—Transverse member of the underframe extending across the ends of all the longitudinal members (sills).

**Fabricated Truck**—A truck with main components that are welded rather than cast.

**Fish-Belly Side**—A car side that has an outward curvature at the level of the passenger seats in order to allow more standing room.

**Floor Height**—Distance from top of the running rail to top of car floor.

**Flying Junction**—(Track) switch arrangement that avoids grade crossing of opposing moves (for maximum flexibility).

**Flywheel Energy Storage**—Method of storing potential energy from braking effort into electrical power for later use.

**Gear Ratio**—Ratio of rotations of motor shaft to wheel axle.

**Gear Unit (Gear Case)**—Housing that encloses the gear and pinion installed on the truck to reduce the high rotational speed of the traction motor to the low rotational speed of the truck axle.

**Germicidal Lamp**—Lamp installed in the ventilation system of a car to provide germ-free air (used on R11 prior to overhaul).

**Grids**—Resistances introduced in electric circuits converting electrical energy into heat, thereby controlling the current in the circuit.

**Group Box**—A large fabricated enclosure box containing a number of cam-operated power switches and interlock switches for traction motor control.

**Guard Light Circuit**—In-car circuit to indicate doors are open or unlocked.

**Guard Rail**—A rail or other structure laid parallel to and inside the running rails to reduce the risk of derailment when negotiating a curve or switch by restraining the wheel flange.

**Hand Bar**—Horizontal bar above the edge of seats to provide support for standing passengers.

**Hand Brake**—Wheel or handle that is attached by linkage to the brake system to apply a force to the brake shoe manually regardless of the air brake system's state.

**Headway**—Time interval between two successive trains moving in the same direction on the same route.

**Heater**—Electrical elements to heat the air space within the car body.

**High Voltage Car (HV)**—Cars where the master controller operated directly at 600 V (*see also* Low Voltage Car [LV]).

**HVAC**—Heating, ventilating, and air-conditioning system to provide a comfortable air environment for the passengers and train crew.

**IGBT**—Insulated Gate Bi-Polar Transistor inverter controlled (AC propulsion) system.

**Interlock**—A device activated by the operation of some other device to make or break an electrical circuit.

**Inverter**—Solid-state device to convert direct current (DC) to alternating current (AC).

**Inverter Ballast**—A combination device that inverts a DC supply voltage to AC voltage and then performs the job of a fluorescent light ballast.

**Journal Bearing**—Bearing that transmits the load between the axle and the frame of a truck.

**Leaf**—A door panel.

**Left (B) Side of Car**—The side of the car to the left when facing the No. 1 end of the car.

**Line Breaker**—An electrically or pneumatically operated device used to open and close power coming into a main circuit.

**Link Bar**—A permanent connection between cars in lieu of a coupler.

**Load Weigh**—A weight-sensing system to control the tractive and braking effort for a constant effort-to-weight ratio despite changing passenger loads.

**Low Voltage Car (LV)**—Cars with low-voltage master controllers (*see also* High Voltage Car [HV]).

**Manhattan Seating**—*See* Manhattan Style.

**Manhattan Style**—An arrangement of longitudinal seats except for four cross-seats on each side of the aisle of which two seats face each other and are located at the center of the car length.

**Marker Lights**—A pair of colored changeable lamps mounted on the roof of the car facing forward to indicate the train's route to towers and platform conductors.

**Married Pairs**—Cars that can only be operated as a linked unit of two cars (shared equipment).

**Master Controller**—A device used by a train operator to control speed and direction of the train; some models also control the brake.

**Motor-Generator (MG Set)**—Mechanically paired motor and generator to change voltage (superseded by converter).

**Motor Truck**—A powered truck with traction motors under a railcar.

**Motorman's Indication**—Trainline circuit that indicates to the motorman (train operator) whether or not all the doors in the train are closed and locked.

**Multi-Section Car**—Five-section lightweight articulated car for use on BMT elevated lines.

**Multiple Unit Door Control (MUDC)**—System allowing operation of doors in a train from one control position.

**No. 1 End**—End of car body where the hand brake is located.

**No. 2 End**—End of car body opposite to No. 1 end.

**Offside**—The side of the train not facing the platform.

**Overhaul Shop**—Facility used for heavy maintenance (not within the capacity of a maintenance shop).

**P-Wire**—A trainline wire that carries an electrical signal indicating power or brake demand (only brake for certain New York City transit cars).

**Package Brake**—*See* Tread Brake.

**Pantograph Gates**—Spring-loaded gates used between cars to prevent passengers from attempting to board trains from station platforms and falling to the roadbed.

**Parking Brake**—A means that supplies static braking forces to maintain a vehicle or train in a no motion state.

**Platform Side**—The side of the train facing the platform.

**Pneumatic**—Operated by air pressure.

**Propulsion Control**—System used to control the acceleration and maintenance of train speed.

**Public Address (PA)**—System used to make audible announcements throughout the train and to station platforms via external speakers.

**Purlin**—Longitudinal roof frame member extending over the carlines from one end to the opposite end.

**Regenerative Braking**—A form of dynamic brake in which the electrical energy generated by braking is returned to the power supply line, provided to on-board loads, or a combination thereof during the braking cycle instead of being dissipated in resistors.

**Reverser**—Electromechanical device used by a train operator to select the direction of train movement.

**Right (A) Side of the Car**—The side of the car to the right when facing the No.1 end of car.

**Rolling Stock**—Generic term for rail vehicles.

**Rotary Converter**—*See* Motor-Generator (MG Set).

**Route Sign**—Sign indicating route of a train.

**Running Lights**—Illumination source located near platform height in front of the car for visibility (prior to sealed beam lamp).

**Safety Chains**—Devices attached between cars adjacent to the end doors to permit passenger movement between cars.

**Sash**—Vertical or horizontal sliding window frame.

**Service Brake**—A nonemergency brake application.

**Shoe Beam**—A wooden beam used for electrical insulation.

**Side Bearing**—A local bearing component, located on either the truck or body bolster, to absorb vertical loads arising from the rocking motion of the car.

**Side Destination Sign**—A sign on the side of the car telling the train's final destination.

**Side Door**—Door installed on the side of the car.

**Sill**—The general term to describe main structural member of a car underframe.

**Single Car**—(1) a car with couplers at both ends or (2) a car that is capable of independent movement (e.g., not part of a married pair).

**SMEE**—Straight-air ME type brake valve self-lapping electric overlay.

**Stanchion**—A vertical tube or pipe fastened between the floor and ceiling for standing passengers to hold.

**Standard Gage**—The standard distance between rails of most railways, being 4 feet 8.5 inches measured between the inside faces of the rail heads.

**Storm Door**—*See* End Door.

**Subway**—An urban rapid transit service using below-ground right-of-way; a transportation service constructed beneath the ground.

**Switch**—A device for controlling the direction of a train, either diverting it to another track or allowing it to continue on the track it is occupying.

**Template Car**—Dynamic outline of the vertical profile of the car mounted to an existing car to check clearance. *See* Dynamic Outline.

**Threshold Plate**—Floor plate at the bottom of a doorway.

**Track**—Two parallel lines of rails spaced nominally 4 feet 8.5 inches apart measured from the inside surfaces of the two running rails.

**Traction Motor**—Electric motor mounted on a truck for propulsion.

**Tractive Effort**—The useable force exerted by the wheels of a locomotive at the rails for pulling a train.

**Trailer Truck**—An unpowered truck (no traction motors) under a railcar.

**Train**—A consist of one or more operating units.

**Trainline**—A continuous circuit (wire or bus) that transmits signals through each car of the train.

**Tread Brake**—(1) a system using brake shoes that press on wheel treads (as opposed to disc brakes) or (2) a single acting brake shoe on a wheel.

**Trip Cock**—Mechanical device located on a truck which, when struck by a trip arm adjacent to rail, causes an emergency brake application (signal system enforcement).

**Trolley Pole**—Moveable, spring-loaded pole installed on roof of car that collects power from an overhead wire.

**Truck**—A rail vehicle component that consists of a frame, normally two axles, brakes, suspension, and other parts, which supports the vehicle body and can swivel under it on curves; if powered, it may also contain traction motors and associated drive mechanisms. Called a bogie in European railways.

**Truck Bolster**—Component part of truck distributing one half of the weight of the car body.

**Truck Centers**—The distance between the centers of the two trucks of a car.

**Truck Frame**—The major structural component of the truck.

**Underframe**—The bottom structure of the car body.

**Unit (Operating)**—A group of cars, either within a train, or a complete "locked-in" train consist.

**Vertical Curve**—Measure of the transition from level track to a grade.

**Vestibule**—The area at ends of cars delineated by a partition between the door openings and the seats.

**Wheel Load**—Distribution of car weight as transmitted through the wheels at a point of contact on the rail in relation to a measured reference point.

**Windscreen**—Screen located near the side door opening as a weather shield for passengers.

**Yoke**—A metal housing that surrounds the draft gear.

**Zone**—A section of the train to the conductor's left or right side as he or she looks out the side window from his or her operating position.

# Index to Cars

## IRT CARS

| Elevated Name/Designation | Year Ordered and/or Built | Car Numbers | No. of Cars | Description (page) | Technical Data (page) |
|---|---|---|---|---|---|
| Manhattan Elevated, Trailer | 1868 | 1–3 | 3 | 30–31 | 332 |
| Manhattan Elevated, Trailer | 1872, 1873, 1875 | 1–16 | 16 | 32 | 333 |
| Manhattan Elevated, Trailer | 1875–77 | 17–39 | 23 | 33 | 334 |
| Manhattan Elevated, Trailer | 1877 | 40–41 | 2 | 34 | 335 |
| Manhattan Elevated, Trailer | 1878, 1879 | 40–119, 150–205 | 136 | 35–37 | 336 |
| Manhattan Elevated, Trailer | 1878–79 | 120–149, 206–242 | 67 | 35–37 | 337 |
| Manhattan Elevated, Trailer | 1885 | 1–39 | 39 | 41–44 | 338 |
| Manhattan Elevated, Trailer | 1907 | 11 | 1 | 44–47 | 339 |
| Manhattan Elevated, Motor | 1901–2 | 40–241 | 202 | 44–47 | 340 |
| Manhattan Elevated, Trailer | 1907 | 242 | 1 | 44–47 | 341 |
| Manhattan Elevated, Trailer | 1879 | 243–292 | 50 | 35–37 | 342 |
| Manhattan Elevated, Trailer | 1882, 1885 | 293–369 | 77 | 41–44 | 343 |
| Manhattan Elevated, Motor | 1886–87 | 370–500 | 131 | 41–44 | 344 |
| Manhattan Elevated, Trailer | 1878 | 501–540 | 40 | 38–41 | 345 |
| Manhattan Elevated, Trailer | 1878 | 541–580 | 40 | 38–41 | 346 |
| Manhattan Elevated, Trailer | 1879 | 581–600 | 20 | 38–41 | 347 |
| Manhattan Elevated, Trailer | 1879–80 | 601–699 | 99 | 41–44 | 348 |
| Manhattan Elevated, Motor | 1880 | 700–728 | 29 | 41–44 | 349 |
| Manhattan Elevated, Trailer | 1880–81 | 729–790 | 62 | 41–44 | 350 |
| Manhattan Elevated, Motor | 1881, 1885–87, 1889–91, 1893 | 791–1120 | 330 | 41–44 | 351 |
| Manhattan Elevated, Motor | 1902–3 | 1121–1218 | 98 | 44–47 | 352 |
| Manhattan Elevated, Trailer | 1902 | 1219–1254 | 36 | 48 | 353 |
| Manhattan Elevated, Motor | 1903 | 1255–1314 | 60 | 44–47 | 354 |
| Manhattan Elevated, Trailer | 1904 | 1315–1414 | 100 | 44–47 | 355 |
| Manhattan Elevated, Trailer | 1907–8 | 1415–1528 | 114 | 44–47 | 356 |
| Manhattan Elevated, Motor | 1907 | 1529–1612 | 84 | 44–47 | 357 |

| Elevated/Subway Name/Designation | Year Ordered and/or Built | Car Numbers | No. of Cars | Description (page) | Technical Data (page) |
|---|---|---|---|---|---|
| Manhattan Elevated, Motor | 1910 | 1613–1672 | 60 | 44–47 | 358 |
| Manhattan Elevated, Trailer | 1909, 1911 | 1673–1752 | 80 | 44–47 | 359 |
| Manhattan Elevated, Motor | 1911 | 1753–1812 | 60 | 44–47 | 360 |
| Q (Modified 1950) | 1904–8 | 1600–1629 A, B, C | 90 | 53–54 | 361 |
| Composite, LV Motor | 1903–4 | 2000–2159, 3000–3339 | 500 | 56–60 | 362 |
| Composite samples, Motor | 1902 | 3340, 3341 | 2 | 55–56 | 363 |
| Steel car sample, Motor | 1903 | 3342 | 1 | 61 | 364 |
| Mineola (Director's Car), Motor | 1904 | 3344 | 1 | 62–63 | 365 |
| Modified Gibbs, HV Motor | 1904 | 3350–3513, 3515, 3516 | 166 | 63–67 | 366 |
| Modified Gibbs, HV Motor | 1904 | 3407, 3419, 3421, 3425, 3427, 3429 3435, 3445 | 8 | 67–68 | 367 |
| MUDC Gibbs, HV Motor | 1904–5 | 3514, 3517–3649 | 134 | 63–67 | 368 |
| Modified Deck Roof, HV Motor | 1907–8 | 3650–3699 | 50 | 68–70 | 369 |
| Modified Hedley, HV Motor | 1909 | 3700–3756, 3815, 3915 | 59 | 71–72 | 370 |
| MUDC Hedley, HV Motor | 1909 | 3757–3814, 3816–3914, 3916–4024 | 266 | 71–72 | 371 |
| Steinway, LV Motor | 1915 | 4025–4036 | 12 | 73–74 | 372 |
| Flivver, LV Motor | 1915 | 4037–4160 | 124 | 74–77 | 373 |
| Flivver, LV Trailer | 1915 | 4161–4214 | 54 | 74–77 | 374 |
| Steinway, LV Motor | 1915 | 4215–4222 | 8 | 78–80 | 375 |
| HV Trailer | 1915 | 4223–4514 | 292 | 80–81 | 376 |
| LV Trailer | 1916 | 4515–4554 | 40 | 82–84 | 377 |
| Steinway, LV Motor | 1916 | 4555–4576 | 22 | 78–80 | 378 |
| LV Motor | 1916 | 4577–4699, 4719 | 124 | 82–84 | 379 |
| Steinway, LV Motor | 1916 | 4700–4718, 4720–4771 | 71 | 84–85 | 380 |
| LV Motor | 1916 | 4772–4810 | 39 | 82–84 | 381 |
| LV Trailer | 1916 | 4811–4825 | 15 | 82–84 | 382 |
| LV Trailer | 1917 | 4826–4965 | 140 | 86–88 | 383 |
| LV Motor | 1917 | 4966–5302 | 337 | 86–88 | 384 |
| LV Trailer | 1922 | 5303–5402 | 100 | 89–90 | 385 |
| LV Motor | 1924 | 5403–5502 | 100 | 90–92 | 386 |
| LV Motor | 1925 | 5503–5627 | 125 | 92–94 | 387 |
| Steinway, LV Motor | 1925 | 5628–5652 | 25 | 95–96 | 388 |
| World's Fair Steinway, LV Motor | 1938 | 5653–5702 | 50 | 97–99 | 389 |

## BMT CARS

| Elevated Name/Designation | Year Ordered and/or Built | Car Numbers | No. of Cars | Description (page) | Technical Data (page) |
|---|---|---|---|---|---|
| BU, Trailer | 1884 | 1–51 | 51 | 110–11 | 394 |
| BU, Trailer | 1887 1891, 1893 | 52–190 | 139 | 112 | 395 |
| BU, Kings County Trailer | 1888–89, 1893 | 191–271 | 81 | 113–14 | 396 |
| BU, 600 series, Motor | 1887 | 600, 601, 683 | 3 | 114–15 | 397 |
| BU, 600 series, Motor | 1891–93 | 602–619 | 18 | 116–17 | 398 |
| BU, 600 series, Motor | 1898 | 620–627 | 8 | 117–19 | 399 |
| BU, 600 series, Motor | 1901 | 633–682 | 50 | 119–20 | 400 |
| BU, 700 series, Motor | 1888 | 700–760 | 61 | 120–22 | 401 |
| BU, 800 series, Motor | 1884 | 800–832 | 33 | 122–23 | 402 |
| BU, 800 series, Motor | 1887 | 833–858 | 26 | 124–25 | 403 |
| BU, 900 series, Motor | 1898 | 900–936 | 37 | 126–27 | 404 |
| BU, Brill 900 series, Motor | 1900 | 937–940 | 4 | 128–29 | 405 |
| BU, 900 series, Motor | 1908 | 998 | 1 | 129–31 | 406 |
| BU, Convertible 1000 series, Motor | 1903 | 1000–1119 | 120 | 131–33 | 407 |
| BU, 1200 series, Motor | 1904–5 | 1200–1299 | 100 | 134–37 | 408 |
| BU, 1200 series, Motor | 1908–14 | 1261, 1282, 1283, 1286, 1287 | 5 | 134–37 | 409 |
| BU, Convertible 1300 series, Motor | 1905–6 | 1300–1399 | 1001 | 137–41 | 410 |
| BU, 1400 series, Motor | 1908 | 1400–1499 | 100 | 141–43 | 411 |
| BU, 1400 series, Motor | 1909 | 1448, 1482 | 2 | 144–45 | 412 |
| C (Rebuilt 1923–25) | 1893 & 1908 | 1500, 1501 A, B, C | 6 | 146–49 | 413 |
| C (Rebuilt 1925) | 1893 & 1908 | 1502–1526 A, B, C | 75 | 146–49 | 414 |
| Q (Rebuilt 1938–41) | 1904–8 | 1600–1629 A, B, C | 90 | 149–53 | 415 |
| Q (Overhauled 1958 & 1962) | 1904–8 | 1600–1629 A, B, C | 72 | 149–53 | 416 |
| QX (Rebuilt 1939–40) | 1904–8 | 1630–1642 A, B | 26 | 149–53 | 417 |

| Subway Name/Designation | Year Ordered and/or Built | Car Numbers | No. of Cars | Description (page) | Technical Data (page) |
|---|---|---|---|---|---|
| B, 2000 series, Motor | 1914–17 | 2000–2399 | 400 | 154–57 | 418 |
| B, BT, BX 2400 series, Motor | 1918 | 2400–2499 | 100 | 154–57 | 419 |
| B, 2500 series, Motor | 1919 | 2500–2599 | 100 | 158–59 | 420 |
| A, B, Motor | 1920–22 | 2600–2899 | 300 | 159–61 | 421 |
| AX, BX, Trailer | 1924 | 4000–4049 | 50 | 161–62 | 422 |
| D, Triplex | 1925, 1927–28 | 6000–6120 A, B, C | 121 | 163–65 | 423 |
| MS, Multi-Section "Green Hornet" | 1934 | 7003 A, B, C, B1, A1 | 1 | 165–68 | 424 |
| MS, Multi-Section | 1936 | 7004–7013 A, B, C, B1, A1 | 10 | 169–71 | 425 |
| MS, Multi-Section | 1936 | 7014–7028 A, B, C, B1, A1 | 15 | 169–71 | 426 |
| MS, Multi-Section "Zephyr" | 1934 | 7029 A, B, C, B1, A1 | 1 | 165–68 | 427 |
| Compartment "Bluebird" | 1938 & 1940 | 8000–8005 A, B, A1 | 6 | 171–74 | 428 |
| SIRT Motor | 1925 | 2900–2924 | 25 | 174–77 | 429 |
| SIRT Motor & Trailer | 1925, 1926 | 300–389, 500–509 | 100 | 174–77 | 430 |

## R TYPE CARS

| *Car Class/ Contract No.* | *Year Built/ Re-Mfr.* | *Car Numbers* | *Description (page)* | *Technical Data (page)* |
|---|---|---|---|---|
| R1 | 1930–31 | 100–399 | 179–89 | 434 |
| R4 | 1932–33 | 400–899 | 179–89 | 435 |
| R6–3 | 1935 | 900–1149 | 179–89 | 436 |
| R6–2 | 1936 | 1150–1299 | 179–89 | 437 |
| R6–1 | 1936 | 1300–1399 | 179–89 | 438 |
| R7 | 1937 | 1400–1474 | 179–89 | 439 |
| R7 | 1937 | 1475–1549 | 179–89 | 440 |
| R7A | 1938 | 1550–1574, 1576–1599 | 179–89 | 441 |
| R7A | 1938 | 1575 | 179–89 | 442 |
| R7A | 1938 | 1600–1649 | 179–89 | 443 |
| R9 | 1940 | 1650–1701 | 179–89 | 444 |
| R9 | 1940 | 1702–1802 | 179–89 | 445 |
| R10 | 1948–49 | 2950–3349 | 190–92 | 446 |
| R11 | 1949 | 8010–8019 | 193–94 | 447 |
| R12 | 1948 | 5703–5802 | 195–98 | 448 |
| R14 | 1949 | 5803–5952 | 195–98 | 449 |
| R15 | 1950 | 5953–5999, 6200–6252 | 198–200 | 450 |
| R16 | 1954–55 | 6300–6499 | 201–3 | 451 |
| R17 | 1955–56 | 6500–6899 | 203–10 | 452 |
| R21 | 1956–57 | 7050–7299 | 203–10 | 453 |
| R22 | 1957–58 | 7300–7749 | 203–10 | 454 |
| R26 | 1959–60 | 7750–7859 | 211–15 | 455 |
| R26GOH | 1985–87 | 7750–7859 | 211–15 | 456 |
| R27 | 1960–61 | 8020–8249 | 216–19 | 457 |
| R28 | 1960–61 | 7860–7959 | 211–15 | 458 |
| R28GOH | 1985–87 | 7860–7959 | 211–15 | 459 |
| R29 | 1962 | 8570–8685, 8688–8803 | 219–21 | 460 |
| R29 | 1962 | 8686–8687, 8804–8805 | 219–21 | 461 |
| R29GOH | 1985–87 | 8570–8805 | 219–21 | 462 |
| R30 | 1961–62 | 8250–8351, 8412–8569 | 216–19 | 463 |
| R30A | 1961 | 8352–8411 | 216–19 | 464 |
| R30/30A GOH | 1985–86 | 8250–8411 | 216–19 | 465 |
| R32A | 1964–65 | 3350–3649 | 222–27 | 466 |
| R32 | 1965 | 3650–3945 | 222–27 | 467 |
| R32 | 1965 | 3946–3949 | 222–27 | 468 |
| R32GOH GE | 1988 | 3594–3595, 3880–3881, 3892–3893, 3934–3937 | 222–27 | 469 |
| R32GOH Phase I, Phase II excluding 10GE cars | 1988–90 | 3350–3949 | 222–27 | 470 |
| R33 | 1962–63 | 8806–9305 | 228–29 | 471 |
| R33GOH | 1986–91 | 8806–9305 | 228–29 | 472 |
| R33S | 1963 | 9306–9345 | 230–34 | 473 |
| R33SGOH | 1985 | 9307–9345 | 230–34 | 474 |
| R34 | 1964 | 8010–8019 | 235–36 | 475 |
| R36 | 1936–64 | 9346–9523, 9558–9769 | 230–34 | 476 |
| R36GOH | 1982–85 | 9346–9523, 9558–9769 | 230–34 | 477 |
| R36 | 1964 | 9524–9557 | 230–34 | 478 |
| R36GOH | 1982–85 | 9524–9557 | 230–34 | 479 |
| R38 | 1966–67 | 3950–4139 | 237–41 | 480 |
| R38AC | 1967 | 4140–4149 | 237–41 | 481 |
| R38GOH | 1987–88 | 3950–4149 | 237–41 | 482 |
| R40 | 1968–69 | 4150–4349 | 242–49 | 483 |
| R40GOH | 1987–89 | 4150–4349 | 242–49 | 484 |
| R40 | 1968–69 | 4350–4449 | 242–49 | 485 |
| R40GOH | 1988–89 | 4350–4449 | 242–49 | 486 |
| R40 | 1968–69 | 4450–4517 | 242–49 | 487 |
| R40 | 1968–69 | 4518–4549 | 242–49 | 488 |
| R40GOH | 1987–89 | 4450–4549 | 242–49 | 489 |
| R42 | 1969–70 | 4550–4949 | 250–53 | 490 |
| R42GOH | 1988–89 | 4550–4839 | 250–53 | 491 |
| R42GOH | 1988–89 | 4840–4949 | 250–53 | 492 |
| R44 | 1971–73 | 100–399 | 253–56 | 493 |
| R44GOH | 1991–92 | 5202–5479 | 253–56 | 494 |
| R44 (SIR) | 1971–73 | 400–435, 436–466 even only | 257–59 | 495 |
| R44GOH (SIR) | 1988 | 388–435, 436–466 even only | 257–59 | 496 |
| R46 | 1975–78 | 500–1278 | 260–63 | 497 |
| R46GOH | 1990–91 | 5482–6258 | 260–63 | 498 |
| R62 | 1983–85 | 1301–1625 | 264–66 | 499 |
| R62A | 1984–87 | 1651–2475 | 266–68 | 500 |
| R68 | 1986–88 | 2500–2924 | 268–70 | 501 |
| R68A | 1988–89 | 5001–5200 | 270–72 | 502 |
| R110A | 1992 | 8001–8010 | 273–76 | 503 |
| R110B | 1992 | 3001–3009 | 276–78 | 504 |
| R142 | 1999–2003 | 1101–1250, 6301–7180 | 278–81 | 505 |
| R142A | 1999–2005 | 7211–7810 | 278–81 | 506 |
| R143 | 2001–3 | 8101–8312 | 281–83 | 507 |
| R160 | 2005–8 | 8313–8972 | 281–83 | 508 |

# A Few Words about This Edition

The concept of this book was born about seventeen years ago, when it became clear that no one had yet put together a comprehensive listing and description of all New York City Transit passenger cars. This updated volume continues the most ambitious attempt yet to present a continuous history of all New York City heavy rail transit vehicles from the very beginning up to the present day. While every effort has been made to ensure accuracy, verifying even recent data can present problems, not to mention validating information that exceeds one hundred years of age. Many sources of information have been accessed and the result is this new centennial edition.

As with any other publication, this book would not have been possible without the efforts of many people who, often on their own time and at their own expense, made a tremendous contribution in researching and locating data thought to be lost. Among the many people, the following eight made the most significant contributions to the original edition: Donald Harold, founder of the Transit Museum, Michael Hanna, "jack of all trains," and Arthur "Interborough" Murphy, who were responsible for the accuracy of the information on the older cars; Superintendent William Wall, who found and corrected many of the errors describing the old equipment; the late Frank Scimone, the project engineer, who focused on the engineering data associated with all the cars; Charles Seaton, the in-house editor, who reviewed and corrected all of the writing. Two individuals who made this publication happen are Vincent Lee, the project manager, who provided most of the staff resources, and Dr. Kathleen Collins, former New York City Transit's archivist, who arranged for the actual printing and publication of the first edition. Finally, special thanks to Debby Lee and Kenneth Teu who converted 137 years of documents into a readable format.

The following personnel are among those who have contributed information, photographs, and their time to producing this book: Murad Alikhan, Eric Barthell, Sam Basilious, Ray Berger, Penny Brackett, Lorraine Brillante, Rudolph Brooks, Joe Burgin, Jerome Capoccia, David Chan, Brian Cohen, Kathleen Collins, Christopher Creed, Edward R. Crew, Joseph Cunningham, Wayne Galante, Nate Gerstein, Jerrold Gross, Douglas Grotjahn, George Hanna, Michael Hanna, Donald Harold, Elijah Hilbert, Stanley Kwa, Debby Lee, Vincent Lee, Osborne Maitland, Omar A. F. Messado, Jason Moscowitz, Peter Muller, Arthur Murphy, Steve Neufeld, Barbara Orlando, Paul Prasek, Rob Rones, Victor Rucklin, Bronique Sanders, Erwin Schaefer, Frank Scimone, Charles Seaton, Richard Stewart, Don Teichman, Kenneth Teu, Masamichi Udagawa, William Wall, George Watson, Shirley Wiggins, Mark Wolodarsky, Nick Yong, Bill Zucker, and many more too numerous to mention.

Key contributors to the centennial edition include Joseph Cunningham, Jerrold L. Gross, Douglas Grotjahn, Mike Hanna, Stanley Lui, Raymond Mercado, Bill Zucker, and especially Don Harold. Brian Cohen, our digital image editor, restored all the photographs to their original state and made modifications to the line drawings. Paul Scarpone also contributed an extraordinary effort as the assistant editor of the technical specifications. Additional thanks are also due to Amy Hausmann of the New York Transit Museum and Robert J. Brugger of the Johns Hopkins University Press.

GENE SANSONE

VINCENT LEE

ARTHUR MURPHY

# Metropolitan Transportation Authority

### Board of Directors

Peter S. Kalikow, Chairman
David S. Mack, Vice Chairman
Edward R. Dunn, Vice Chairman

Andrew Albert*
John H. Banks III
James F. Blair*
Nancy Shevell Blakeman
Anthony J. Bottalico*
Michael J. Canino*
Barry Feinstein
Lawrence W. Gamache***
James H. Harding Jr.
Susan L. Kupferman
Mark D. Lebow
James L. McGovern**
Mark Page
Ernest J. Salerno***
Andrew M. Saul
James L. Sedore Jr.***
James S. Simpson
Edward A. Vrooman***
Ed Watt*
Alfred E. Werner

* Nonvoting Member
** Alternative Nonvoting Member
*** Casts One Collective Vote

### MTA Management Team

Peter S. Kalikow, Chairman
Katherine N. Lapp, Executive Director
Thomas J. Savage, Chief Operating Officer
Stephen L. Kessler, Chief Financial Officer

Maureen Boll
Linda G. Kleinbaum
Catherine A. Rinaldi
Christopher P. Boylan
William A. Morange
Paul Spinelli

**MTA New York City Transit/Subways/Car Equipment**

Lawrence G. Reuter, President

Michael Lombardi, Senior Vice President, Subways

Carlo Perciballi, Chief of Operations, Rolling Stock and Maintenance of Way

Hank Insinna, Chief Mechanical Officer

Anthony Gagliardi, Assistant Chief Mechanical Officer, A-Division Maintenance Shops

Gene Sansone, Assistant Chief Mechanical Officer, Car Equipment Engineering and Technical Support

Robert Smith, Assistant Chief Mechanical Officer, B-Division Maintenance Shops

Richard Sowa, Assistant Chief Mechanical Officer, Overhaul Shops

# About the Author

Gene Sansone began his career with MTA New York City Transit in 1973 as an analyst with the Executive Office of Operations and Maintenance. Subsequently, he was appointed to various management positions within the Division of Car Equipment. His current title is Assistant Chief Mechanical Officer of Car Equipment Engineering and Technical Support, the subdivision in which all engineering and maintenance support issues concerning the 6,400 passenger and 450 work cars are addressed.

Concurrent with his employment with MTA New York City Transit, Gene is an adjunct professor at Polytechnic University of New York, teaching a graduate course titled "Management of Transit Maintenance and Operations." He is also a NYC College of Technology advisory commission member.

Prior to joining MTA New York City Transit, Gene was a field service engineer with Westinghouse Electric Corporation, where he provided engineering, maintenance, and installation support of various industrial equipment, primarily associated with rail transportation.

Gene, born in Naples, Italy, received his bachelor of science in electrical engineering from Kansas State University and his MBA from Iona College. He has published and presented technical papers at conferences hosted by the Federal Transit Administration, the American Public Transit Association, the Transportation Research Board, and the UITP International Metropolitan Railways Committee. In 2003 Gene received the Engineer of the Year award by the New York chapter of the Institute of Electrical and Electronics Engineers (IEEE). He is a resident of Queens, New York, where he lives with his wife and two grown children.

# TECHNICAL DATA

1584

SOUTH FERRY
9TH AVE
INTERBOROUGH

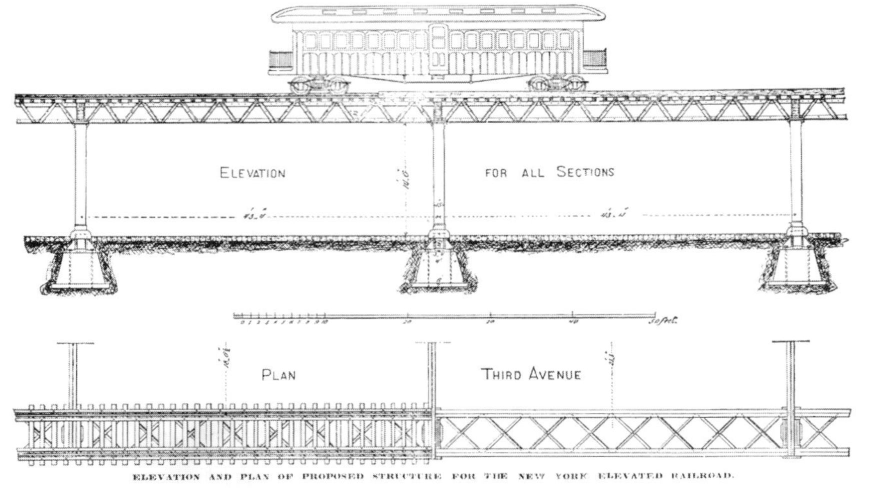


ELEVATION AND PLAN OF PROPOSED STRUCTURE FOR THE NEW YORK ELEVATED RAILROAD.

SEPT. 28, 1877.

(191) CON 3-0-5 DELANCY + CHRISIE
7.12.1907

CAR NUMBERS: 1 - 3
TOTAL: 3 CARS
BUILT BY: UNKNOWN

DATE: 1868
AVERAGE COST PER CAR: N/A

# 1868 (IRT)

MANHATTAN ELEVATED, TRAILER

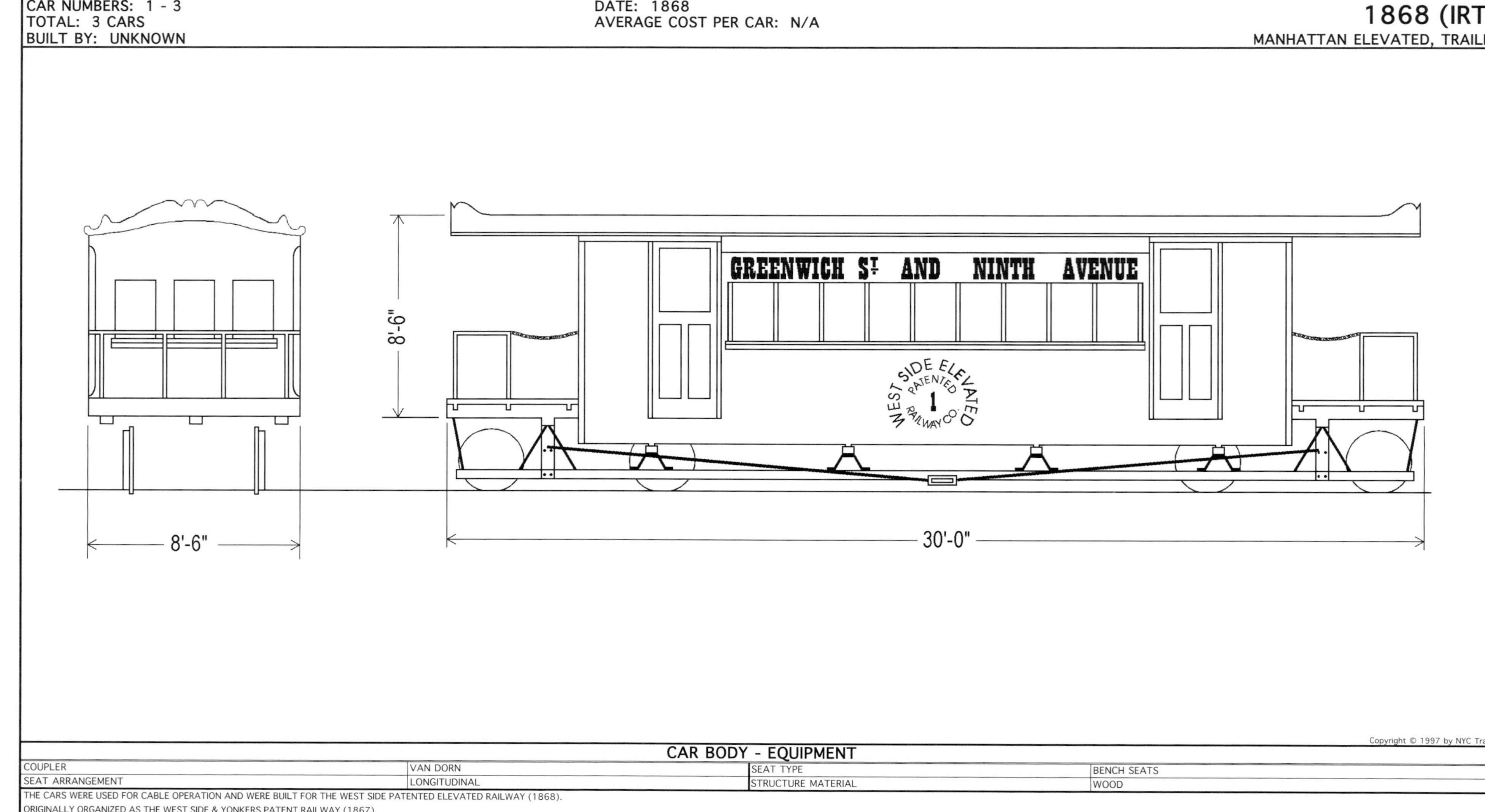


Copyright © 1997 by NYC Transit

## CAR BODY - EQUIPMENT

| | | | |
|---|---|---|---|
| COUPLER | VAN DORN | SEAT TYPE | BENCH SEATS |
| SEAT ARRANGEMENT | LONGITUDINAL | STRUCTURE MATERIAL | WOOD |

THE CARS WERE USED FOR CABLE OPERATION AND WERE BUILT FOR THE WEST SIDE PATENTED ELEVATED RAILWAY (1868).
ORIGINALLY ORGANIZED AS THE WEST SIDE & YONKERS PATENT RAILWAY (1867).
THE CARS OPERATED ON THE GREENWICH STREET AND NINTH AVENUE LINES AND LATER THEY WERE DRAWN BY THE FIRST PURCHASED STEAM DUMMY (EARLY 1871).
ORIGINALLY, CARS HAD NO COUPLERS.
THE CARS WERE REMOVED FROM SERVICE WHEN THE NEW YORK ELEVATED CARS WERE DELIVERED (1872).
ALL CARS WERE SOLD OR SCRAPPED.

CAR NUMBERS: 1 - 16
TOTAL: 16 CARS
BUILT BY: CUMMINGS, JACKSON & SHARP (SEE NOTE 1)

DATE: 1872, 1873, 1875 (SEE NOTE 1)
AVERAGE COST PER CAR: SEE NOTE 1

# 1872, 1873, 1875 (IRT)

MANHATTAN ELEVATED, TRAILER

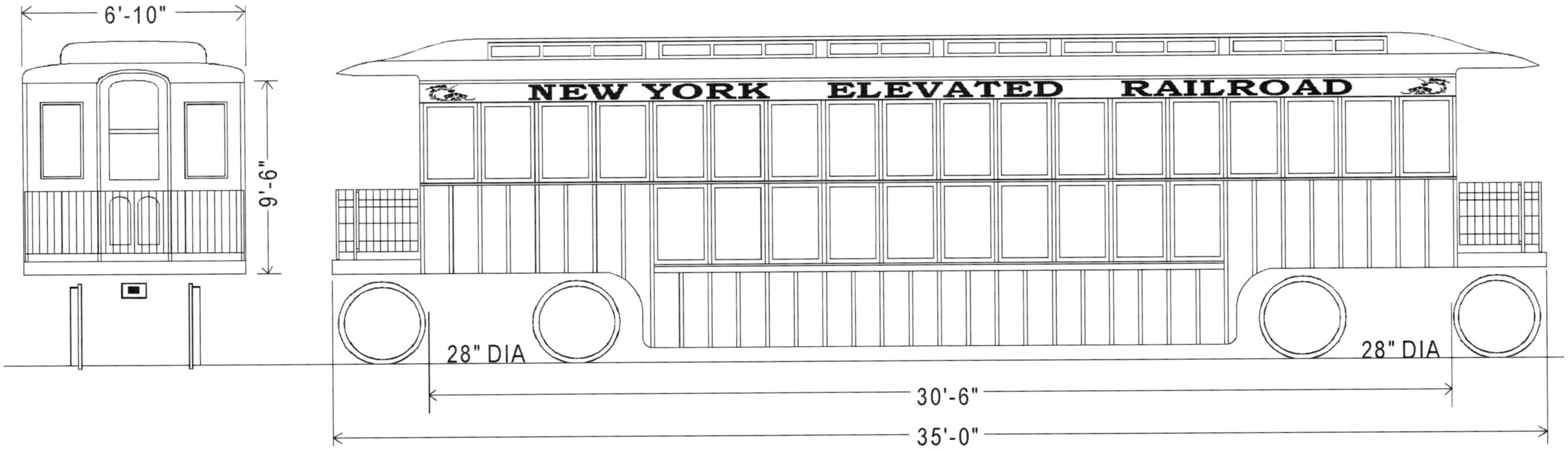

6'-10"
9'-6"
NEW YORK ELEVATED RAILROAD
28" DIA
28" DIA
30'-6"
35'-0"

Copyright © 1997 by NYC Transit

| CAR BODY - EQUIPMENT | | | | |
|---|---|---|---|---|
| COUPLER | | VAN DORN | SEATING CAPACITY | CARS 1-4 [40]; CARS 5 - 16 [44] |
| END DOOR | | SWING | | |
| **TRUCK** | | | | |
| JOURNAL BEARING | | FRICTION | | |
| **GENERAL** | | | | |
| CAR LIGHT WEIGHT | CUMMINGS | 14,000 LBS | | |
| | JACKSON & SHARP | 12,600 LBS (AVG.) | | |

NOTE 1:

| CARS | CAR BUILDER | YEAR BUILT | AVERAGE COST PER CAR |
|---|---|---|---|
| 1 - 4 | JACKSON & SHARP | 1872 | $3,518 |
| 5 - 8 | | 1872, 1873 | $3,279 |
| 9 - 10 | CUMMINGS | 1873 | $3,768 |
| 11 - 16 | | 1875 | $3,237 |

ORIGINALLY OWNED BY NEW YORK ELEVATED RAILROAD.
FLOOR RAISED ON 15 CARS (1878).
TRACK GAUGE OF 4' 10 1/2" WAS CHANGED TO 4' 8 1/2" (1875).
ALL CARS WERE SOLD OR SCRAPPED BY 1885.

CAR NUMBERS: 17 - 39
TOTAL: 23 CARS
BUILT BY: GILBERT & BUSH, JACKSON & SHARP (SEE NOTE 1)

DATE: 1875, 1876, 1877 (SEE NOTE 1)
AVERAGE COST PER CAR: SEE NOTE 1

# 1875, 1876, 1877 (IRT)

MANHATTAN ELEVATED, TRAILER

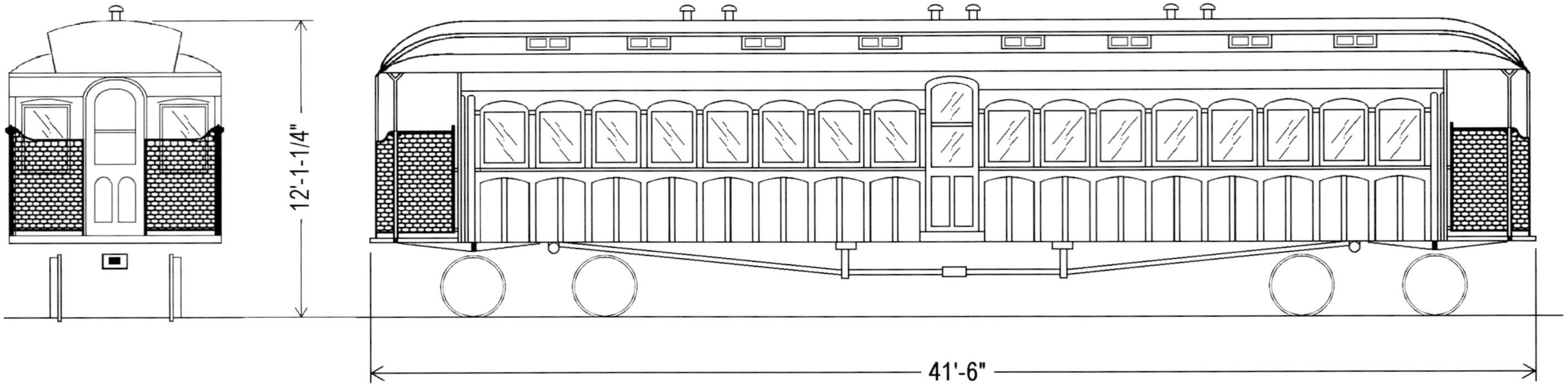

12'-1-1/4"

41'-6"

Copyright © 1997 by NYC Transit

## CAR BODY - EQUIPMENT

| | | | |
|---|---|---|---|
| COUPLER | VAN DORN | STRUCTURE MATERIAL | WOOD |
| END DOOR | SWING | SEATING CAPACITY | CARS 17 -21 [44]; CARS 22 - 39 [46] |

## TRUCK

| | | | |
|---|---|---|---|
| JOURNAL BEARING | FRICTION | | |

## GENERAL

| | | | | |
|---|---|---|---|---|
| CAR LIGHT WEIGHT | JACKSON & SHARP | 14,330 LBS (AVG.) | | |
| | GILBERT & BUSH | 16,000 LBS (AVG.) | | |

NOTE 1:

| CARS | CAR BUILDER | YEAR BUILT | AVERAGE COST PER CAR |
|---|---|---|---|
| 17 - 18 | JACKSON & SHARP | 1875 | $3,294 |
| 19 - 21 | | 1876 | $3,103 |
| 22 - 39 | GILBERT & BUSH | 1876 - 1877 | $2,346 |

ORIGINALLY OWNED BY NEW YORK ELEVATED RAILROAD.
CARS 28 AND 39 CONVERTED TO SUPPLY CAR "E" AND TICKET CAR "F" RESPECTIVELY (1887).
CAR 21 CONVERTED TO SUPPLY CAR "B" (4/1880).
ONE CAR CONVERTED TO SUPPLY CAR "A" (CA. 1880).
ALL CARS WERE SOLD OR SCRAPPED.

CAR NUMBERS: 40, 41
TOTAL: 2 CARS
BUILT BY: JACKSON & SHARP

DATE: 1877
AVERAGE COST PER CAR: N/A

## 1877 (IRT)

MANHATTAN ELEVATED, TRAILER

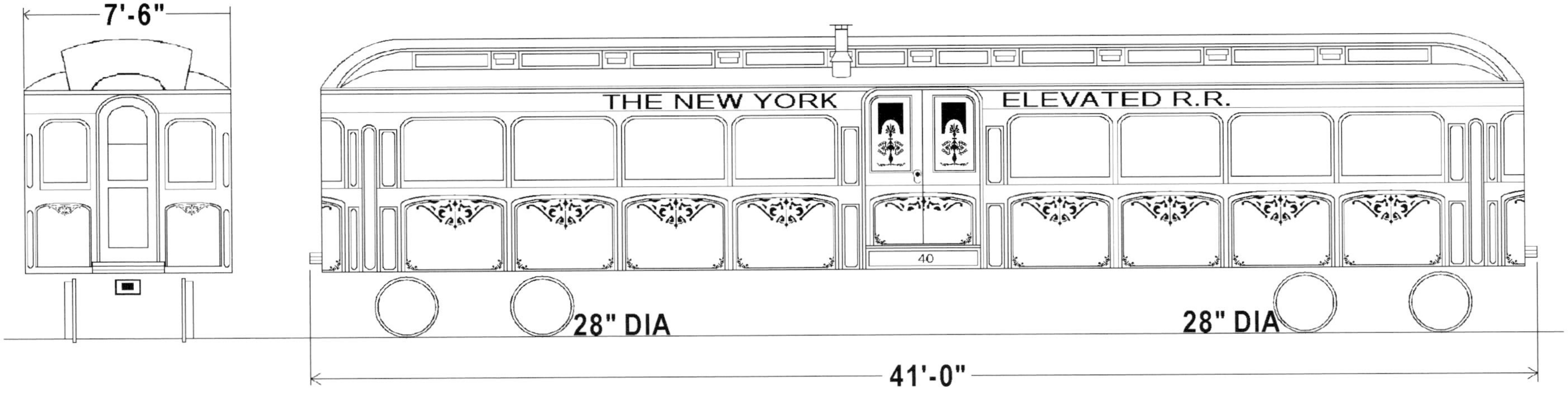

7'-6"

THE NEW YORK ELEVATED R.R.

40

28" DIA

28" DIA

41'-0"

Copyright © 1997 by NYC Transit

| CAR BODY - EQUIPMENT | | | |
|---|---|---|---|
| COUPLER | VAN DORN | SEAT ARRANGEMENT & CAPACITY | [48], SEE NOTE 1 |
| END DOOR | SWING | STRUCTURE MATERIAL | WOOD |
| TRUCK | | | |
| JOURNAL BEARING | FRICTION | | |
| GENERAL | | | |
| CAR LIGHT WEIGHT | 16,000 LBS (APPROX.) | | |

NOTE 1: LONGITUDINAL SEATING EXCEPT THAT SEATS WERE CURVED NEAR CAR ENDS.

ORIGINALLY OWNED BY NEW YORK ELEVATED RAILROAD AND WERE DEMONSTRATORS.
REMOVED FROM SERVICE AND BELIEVED TO BE RETURNED TO BUILDER. CAR NUMBERS WERE APPLIED TO NEXT ORDER OF CAR PURCHASES (SEE PAGE 336).
ALL CARS WERE SCRAPPED.

CAR NUMBERS: 40 - 119, 150 - 205 (SEE NOTE 1)
TOTAL: 136 CARS
BUILT BY: GILBERT & BUSH

DATE: 1878, 1879
AVERAGE COST PER CAR: (SEE NOTE 2)

# 1878, 1879 (IRT)

MANHATTAN ELEVATED, TRAILER

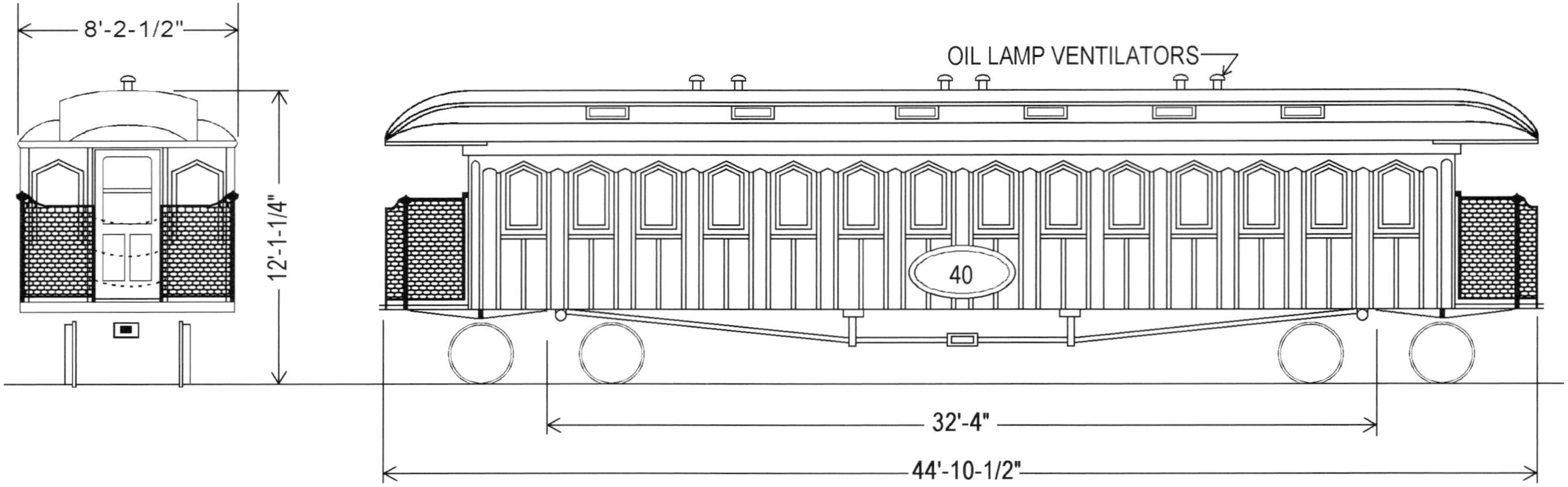

8'-2-1/2"

12'-1-1/4"

OIL LAMP VENTILATORS

40

32'-4"

44'-10-1/2"

Copyright © 1997 by NYC Transit

## CAR BODY - EQUIPMENT

| | | | |
|---|---|---|---|
| COUPLER | VAN DORN | SEATING CAPACITY | [48] |
| END DOOR | SWING | STRUCTURE MATERIAL | WOOD |

## TRUCK

| | | | |
|---|---|---|---|
| JOURNAL BEARING | FRICTION | | |

## GENERAL

| | | | | |
|---|---|---|---|---|
| CAR LIGHT WEIGHT | CARS 40-119, 168-205: | 22,620 LBS | | |
| | CARS 150-167: | 21,080 LBS | | |

NOTE 1: CARS 150-167 WERE 42'-3-5/8" OVER PLATFORMS, SEATING CAPACITY 44, AND HAD 13 SIDE WINDOWS.

NOTE 2:

| CARS | YEAR BUILT | AVERAGE COST PER CAR |
|---|---|---|
| 40 - 59 | 1878 | $2,216 |
| 60 - 94 | | $2,238 |
| 95 - 119 | 1878 - 1879 | $2,811 |
| 150 - 155 | 1878 | $2,100 |
| 156 - 167 | 1879 | $2,312 |
| 168 - 205 | | $1,798 |

ORIGINALLY OWNED BY NEW YORK ELEVATED RAILROAD.
CAR 41 CONVERTED TO TICKET CAR "G" (7/9/1893) AND IS CURRENTLY AT BRANFORD MUSEUM, CT.
CAR 43 CONVERTED TO SUPPLY CAR "H" (1904). CAR 52 CONVERTED TO SUPPLY CAR "O" (1904).
CAR 53 CONVERTED TO BAGGAGE CAR "Q" (1904) AND LATER CONVERTED TO SUPPLY CAR "01".
CARS 57, 62, 68, 70, 71 CONVERTED TO BAGGAGE/EXPRESS CARS "K", "P", "L", "M", "N" RESPECTIVELY (1904).
CARS 54 AND 55 CONVERTED TO SUPPLY CARS "I" AND "J".
CAR 72 CONVERTED TO BAGGAGE CAR "R" (1904) AND LATER CONVERTED TO SUPPLY CAR "02".
CARS 40 AND 41 WERE REPLACEMENTS FOR CARS OF BUILDER JACKSON & SHARP (SEE PAGE 335).
ALL OTHER CARS WERE SOLD OR SCRAPPED (BEFORE 1928).

CAR NUMBERS: 120 - 149, 206 - 242
TOTAL: 67 CARS
BUILT BY: WASON (SEE NOTE 1)

DATE: 1878, 1879
AVERAGE COST PER CAR: (SEE NOTE 1)

# 1878, 1879 (IRT)

MANHATTAN ELEVATED, TRAILER

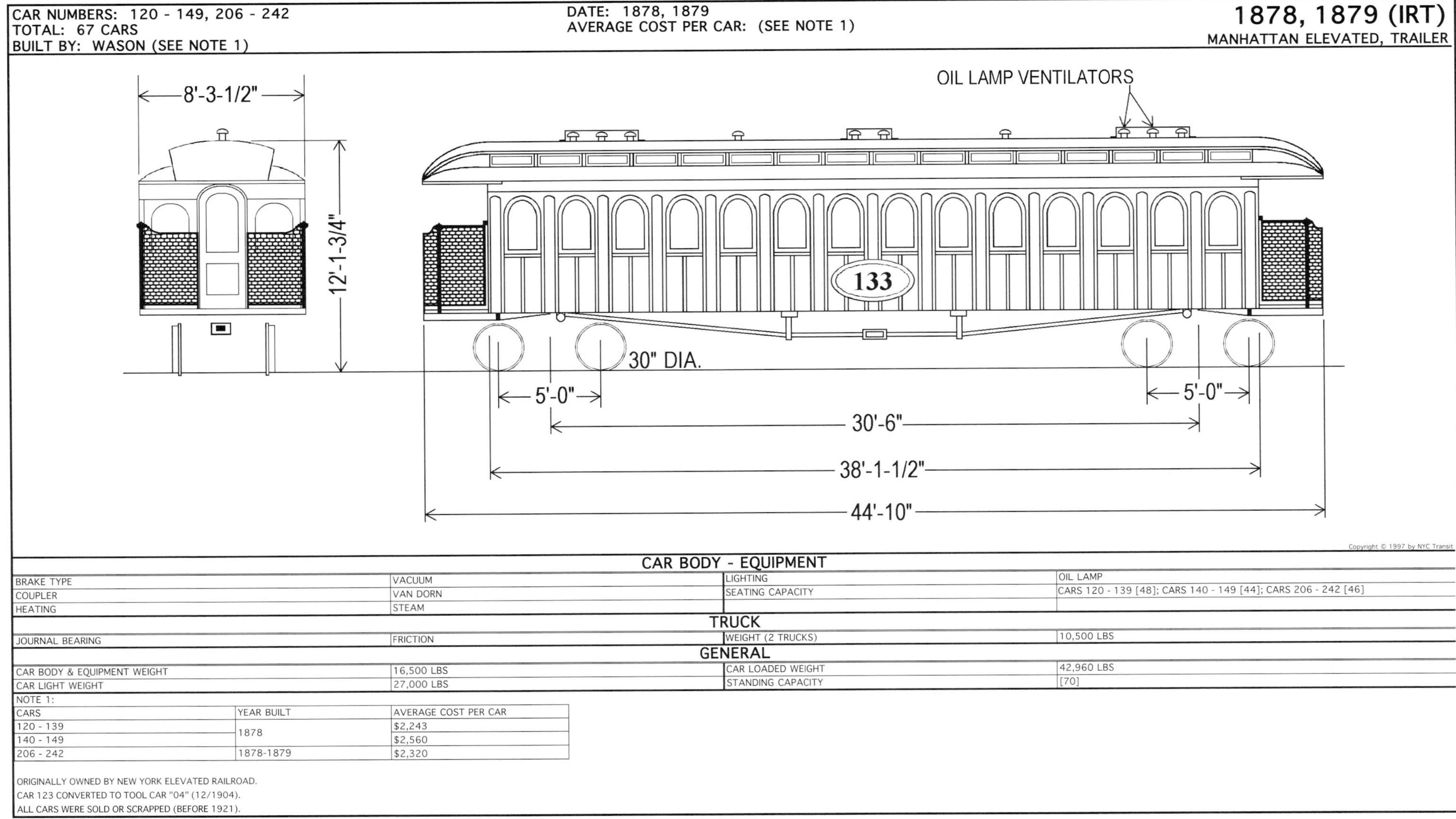


Copyright © 1997 by NYC Transit

## CAR BODY - EQUIPMENT

| | | | |
|---|---|---|---|
| BRAKE TYPE | VACUUM | LIGHTING | OIL LAMP |
| COUPLER | VAN DORN | SEATING CAPACITY | CARS 120 - 139 [48]; CARS 140 - 149 [44]; CARS 206 - 242 [46] |
| HEATING | STEAM | | |

## TRUCK

| | | | |
|---|---|---|---|
| JOURNAL BEARING | FRICTION | WEIGHT (2 TRUCKS) | 10,500 LBS |

## GENERAL

| | | | |
|---|---|---|---|
| CAR BODY & EQUIPMENT WEIGHT | 16,500 LBS | CAR LOADED WEIGHT | 42,960 LBS |
| CAR LIGHT WEIGHT | 27,000 LBS | STANDING CAPACITY | [70] |

NOTE 1:

| CARS | YEAR BUILT | AVERAGE COST PER CAR |
|---|---|---|
| 120 - 139 | 1878 | $2,243 |
| 140 - 149 | | $2,560 |
| 206 - 242 | 1878-1879 | $2,320 |

ORIGINALLY OWNED BY NEW YORK ELEVATED RAILROAD.
CAR 123 CONVERTED TO TOOL CAR "04" (12/1904).
ALL CARS WERE SOLD OR SCRAPPED (BEFORE 1921).

CAR NUMBERS: 1 - 39
TOTAL: 39 CARS
BUILT BY: PULLMAN

DATE: 1885
AVERAGE COST PER CAR: $2,838

# 1885 (IRT)

MANHATTAN ELEVATED, TRAILER

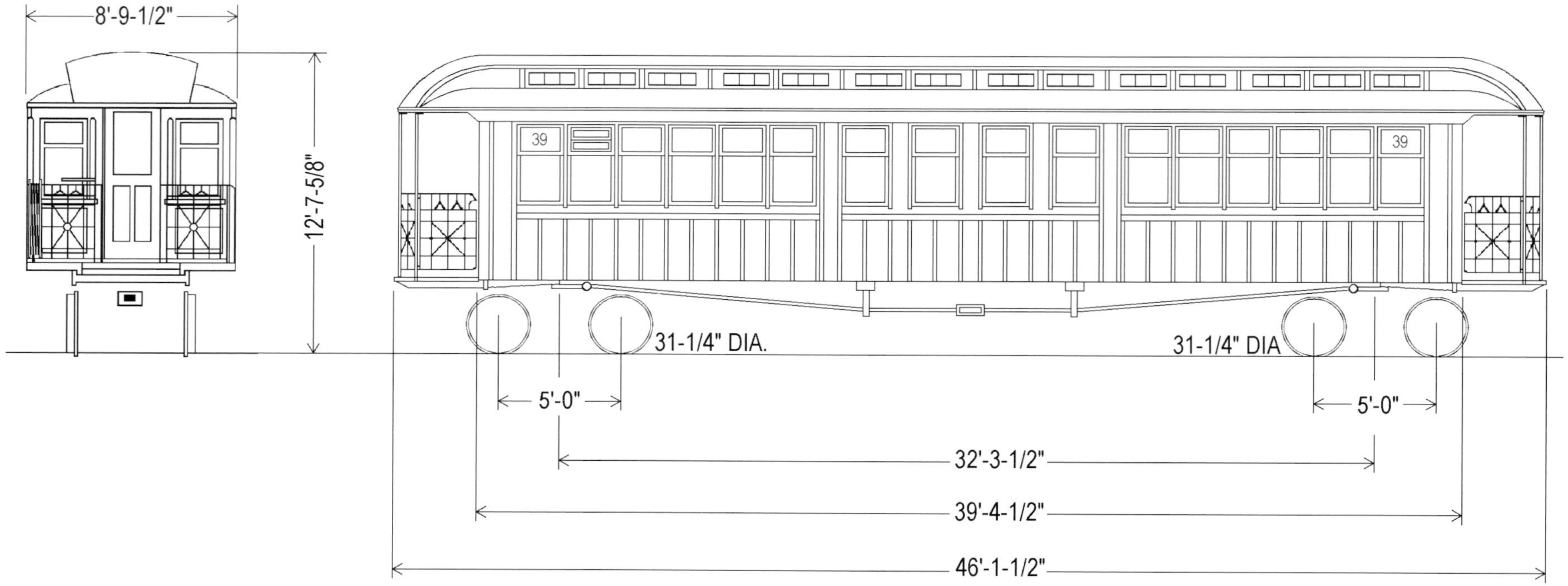

8'-9-1/2"

12'-7-5/8"

31-1/4" DIA.

31-1/4" DIA

5'-0"

5'-0"

32'-3-1/2"

39'-4-1/2"

46'-1-1/2"

Copyright © 1997 by NYC Transit

| CAR BODY - EQUIPMENT | | | |
|---|---|---|---|
| COUPLER | VAN DORN | SEAT MATERIAL | RATTAN |
| END DOOR | SWING | SEAT TYPE | INDIVIDUAL CUSHION |
| SEAT ARRANGEMENT & CAPACITY | LONGITUDINAL [48], SEE NOTE 1 | STRUCTURE MATERIAL | WOOD |
| **TRUCK** | | | |
| JOURNAL BEARING | FRICTION | | |
| **GENERAL** | | | |
| CAR LIGHT WEIGHT | 34,280 LBS | STANDING CAPACITY | [77] |

NOTE 1: LONGITUDINAL SEATING EXCEPT FOUR CROSS SEATS ON EACH SIDE OF AISLE TWO OF WHICH FACED EACH OTHER AND WERE LOCATED AT CENTER OF CAR LENGTH.

REPLACEMENT FOR THE OLDEST NEW YORK ELEVATED RAILROAD CARS IN NUMBERING SEQUENCE (SEE PAGES 333 & 334).
CARS 7 AND 10 HAD CONTROLS AT ONE END OF CAR AND WERE USED ON THE 42ND STREET ELEVATED SHUTTLE FROM THIRD AVENUE TO GRAND CENTRAL (UNTIL DECEMBER 1923).
CAR 11 DESTROYED IN COLLISION (4/1905).
CARS ELECTRIFIED (1902, 1903).
ALL CARS WERE SOLD OR SCRAPPED (1930's - 1940's).

| CAR NUMBER: 11<br>TOTAL: 1 CAR<br>BUILT BY: ST. LOUIS | DATE: 1907<br>AVERAGE COST PER CAR: N/A | **1907 (IRT)**<br>MANHATTAN ELEVATED, TRAILER |
|---|---|---|

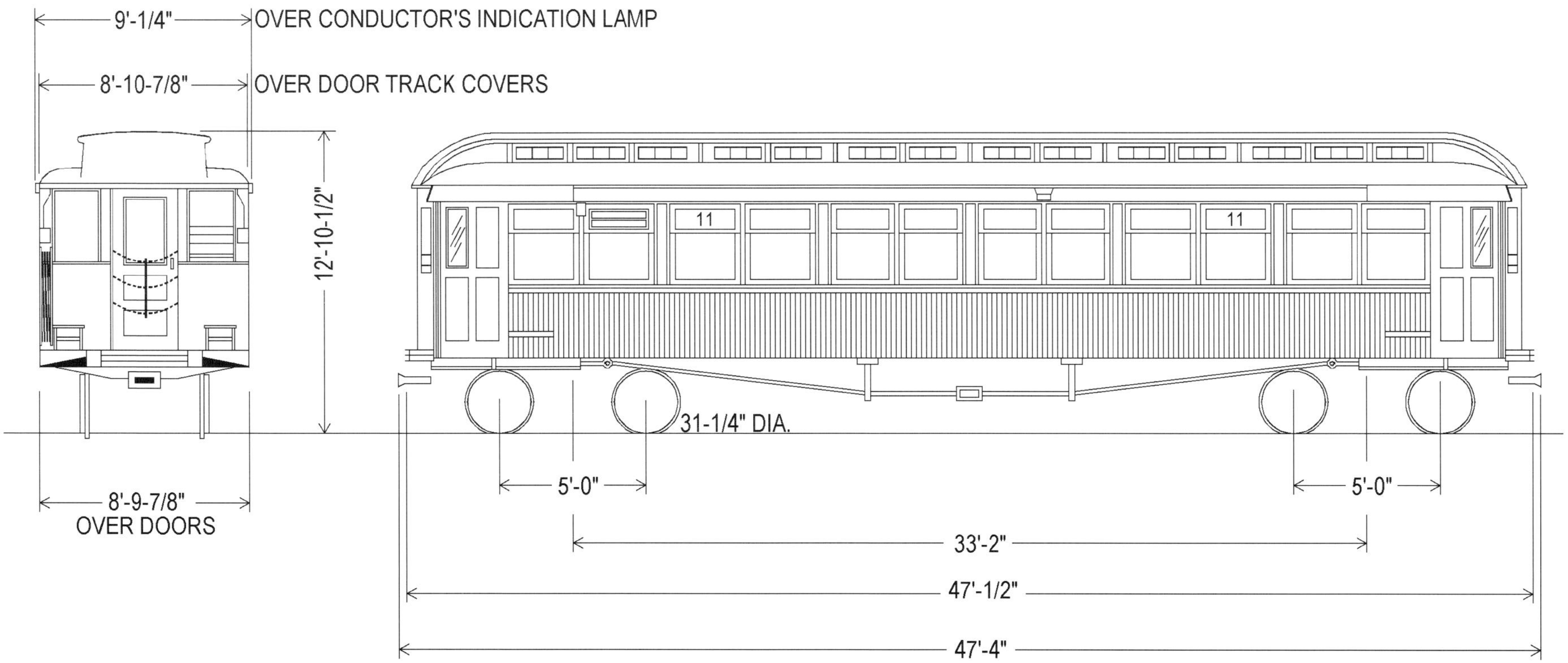

9'-1/4" OVER CONDUCTOR'S INDICATION LAMP

8'-10-7/8" OVER DOOR TRACK COVERS

12'-10-1/2"

8'-9-7/8" OVER DOORS

11

11

31-1/4" DIA.

5'-0"

5'-0"

33'-2"

47'-1/2"

47'-4"

Copyright © 1997 by NYC Transit

| CAR BODY - EQUIPMENT | | | |
|---|---|---|---|
| COUPLER | VAN DORN | SEAT MATERIAL | RATTAN |
| END DOOR | SLIDE | SEAT TYPE | INDIVIDUAL CUSHION |
| SEAT ARRANGEMENT & CAPACITY | LONGITUDINAL [48], SEE NOTE 1 | STRUCTURE MATERIAL | WOOD |
| **TRUCK** | | | |
| JOURNAL BEARING | FRICTION | WEIGHT (2 TRUCKS) | 16,480 LBS |
| **GENERAL** | | | |
| CAR BODY & EQUIPMENT WEIGHT | 26,600 LBS | CAR LOADED WEIGHT | 60,580 LBS |
| CAR LIGHT WEIGHT | 43,080 LBS | STANDING CAPACITY | [77] |

NOTE 1: LONGITUDINAL SEATING EXCEPT FOUR CROSS SEATS ON EACH SIDE OF AISLE TWO OF WHICH FACED EACH OTHER AND WERE LOCATED AT CENTER OF CAR LENGTH.

CAR IS REPLACEMENT FOR CAR 11 BY BUILDER PULLMAN (SEE PAGE 338).
CONVERTED TO MUDC OPERATION (1924).
CAR WAS SCRAPPED (4/16/1951).

CAR NUMBERS: 40 - 241
TOTAL: 202 CARS
BUILT BY: AMERICAN CAR FOUNDRY, WASON (SEE NOTE 2)

DATE: 1901, 1902 (SEE NOTE 2)
AVERAGE COST PER CAR: $9,303

# 1901, 1902 (IRT)

MANHATTAN ELEVATED, MOTOR

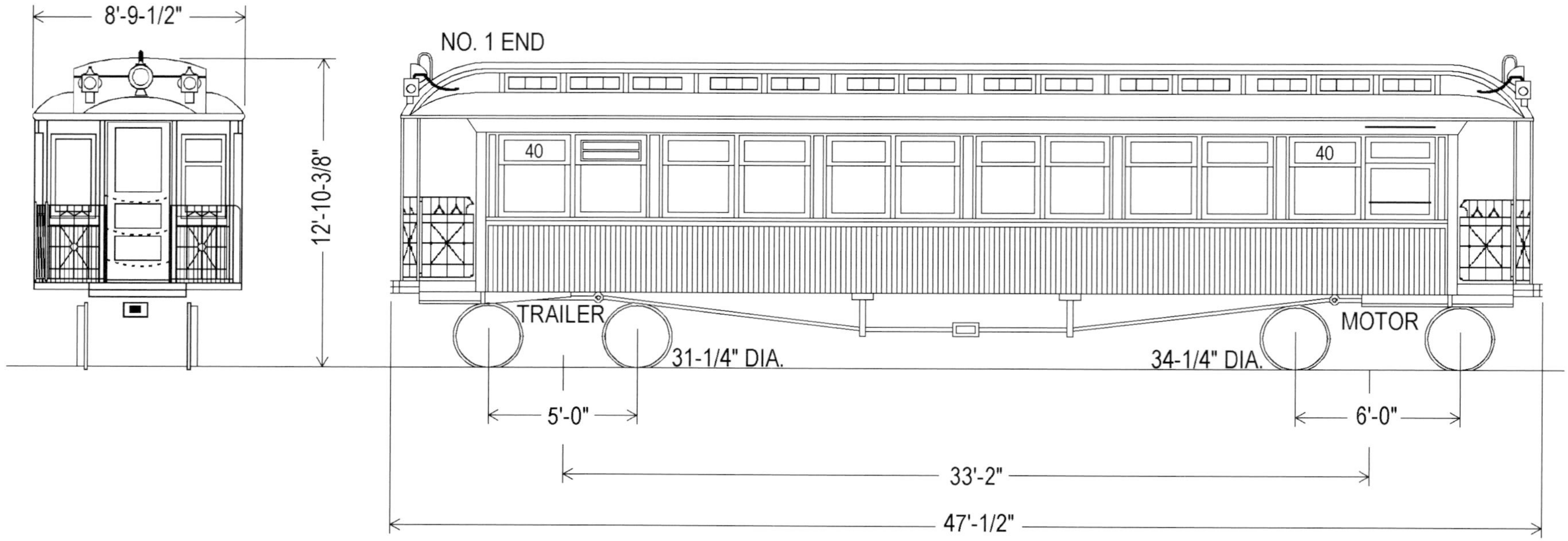

8'-9-1/2"
12'-10-3/8"
NO. 1 END
40
40
TRAILER
MOTOR
31-1/4" DIA.
34-1/4" DIA.
5'-0"
6'-0"
33'-2"
47'-1/2"

Copyright © 1997 by NYC Transit

## CAR BODY - EQUIPMENT

| | | | |
|---|---|---|---|
| COUPLER | VAN DORN | SEAT MATERIAL | RATTAN |
| END DOOR | SLIDE | SEAT TYPE | INDIVIDUAL CUSHION |
| HEAD LIGHT | ELECTRIC LAMP | STRUCTURE MATERIAL | WOOD |
| MARKER LIGHT | ELECTRIC LAMP | TAIL LIGHT | KEROSENE SINGLE LENS RED LAMP |
| SEAT ARRANGEMENT & CAPACITY | MANHATTAN [44], SEE NOTE 1 | | |

## TRUCK

| | | | | |
|---|---|---|---|---|
| AXLE LOAD | MOTOR | 17,520 LBS | JOURNAL BEARING | FRICTION |
| | TRAILER | 11,730 LBS | WEIGHT (2 TRUCKS) | 28,740 LBS |

## GENERAL

| | | | |
|---|---|---|---|
| CAR BODY & EQUIPMENT WEIGHT | 29,760 LBS | CAR LOADED WEIGHT | 72,620 LBS |
| CAR LIGHT WEIGHT | 55,120 LBS | STANDING CAPACITY | [77] |

NOTE 1: LONGITUDINAL SEATING EXCEPT FOUR CROSS SEATS ON EACH SIDE OF AISLE TWO OF WHICH FACED EACH OTHER AND WERE LOCATED AT CENTER OF CAR LENGTH.

NOTE 2:

| CARS | CAR BUILDERS | YEAR BUILT |
|---|---|---|
| 42 - 91 | AMERICAN CAR FOUNDRY | 1901 - 1902 |
| 40, 41, 92 - 241 | WASON | 1902 |

CARS 40 - 241 ARE REPLACEMENTS FOR THE OLDER NEW YORK ELEVATED RAILROAD CARS (SEE PAGES 336 & 337).
CAR 107 USED AS DYRE AVE. SNOWPLOW.
CARS 161, 162, 172, 189, 191, 194, 211, 212, 216, 218, 220, 222-225, 227, 229-241 CONVERTED TO MUDC OPERATION (1924) (SEE PAGE 352 & PAGES 49 - 53).

CAR NUMBER: 242
TOTAL: 1 CAR
BUILT BY: ST. LOUIS

DATE: 1907
AVERAGE COST PER CAR: N/A

# 1907 (IRT)

MANHATTAN ELEVATED, TRAILER

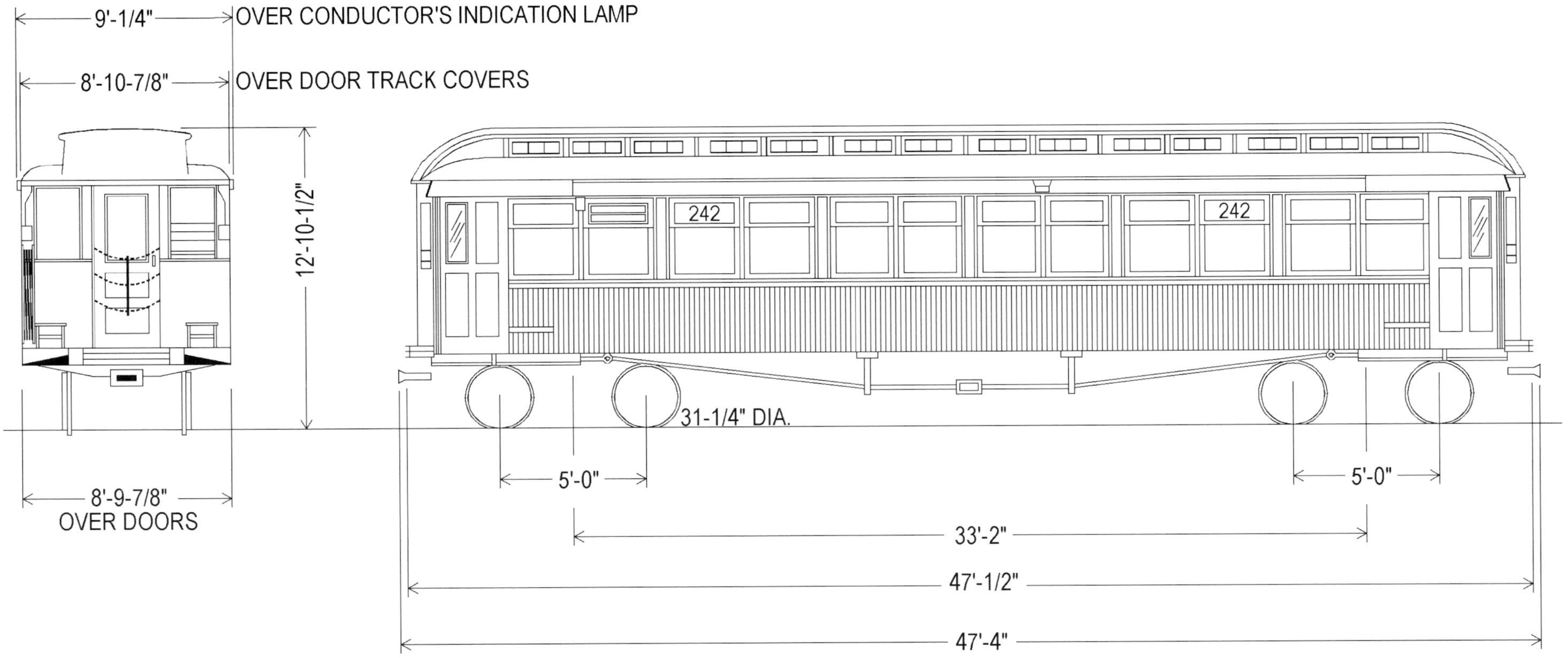

Copyright © 1997 by NYC Transit

| CAR BODY - EQUIPMENT | | | |
|---|---|---|---|
| COUPLER | VAN DORN | SEAT ARRANGEMENT & CAPACITY | MANHATTAN [48], SEE NOTE 1 |
| END DOOR | SLIDE | | |
| **TRUCK** | | | |
| JOURNAL BEARING | FRICTION | WEIGHT (2 TRUCKS) | 16,480 LBS |
| **GENERAL** | | | |
| CAR BODY & EQUIPMENT WEIGHT | 26,600 LBS | CAR LOADED WEIGHT | 60,580 LBS |
| CAR LIGHT WEIGHT | 43,080 LBS | | |

NOTE 1: LONGITUDINAL SEATING EXCEPT FOUR CROSS SEATS ON EACH SIDE OF AISLE TWO OF WHICH FACED EACH OTHER AND WERE LOCATED AT CENTER OF CAR LENGTH.

CAR CONVERTED TO MUDC OPERATION (1924).
CAR WAS SCRAPPED (12/29/1952).

CAR NUMBERS: 243 - 292
TOTAL: 50 CARS
BUILT BY: GILBERT & BUSH

DATE: 1879
AVERAGE COST PER CAR: $2,233

# 1879 (IRT)

MANHATTAN ELEVATED, TRAILER

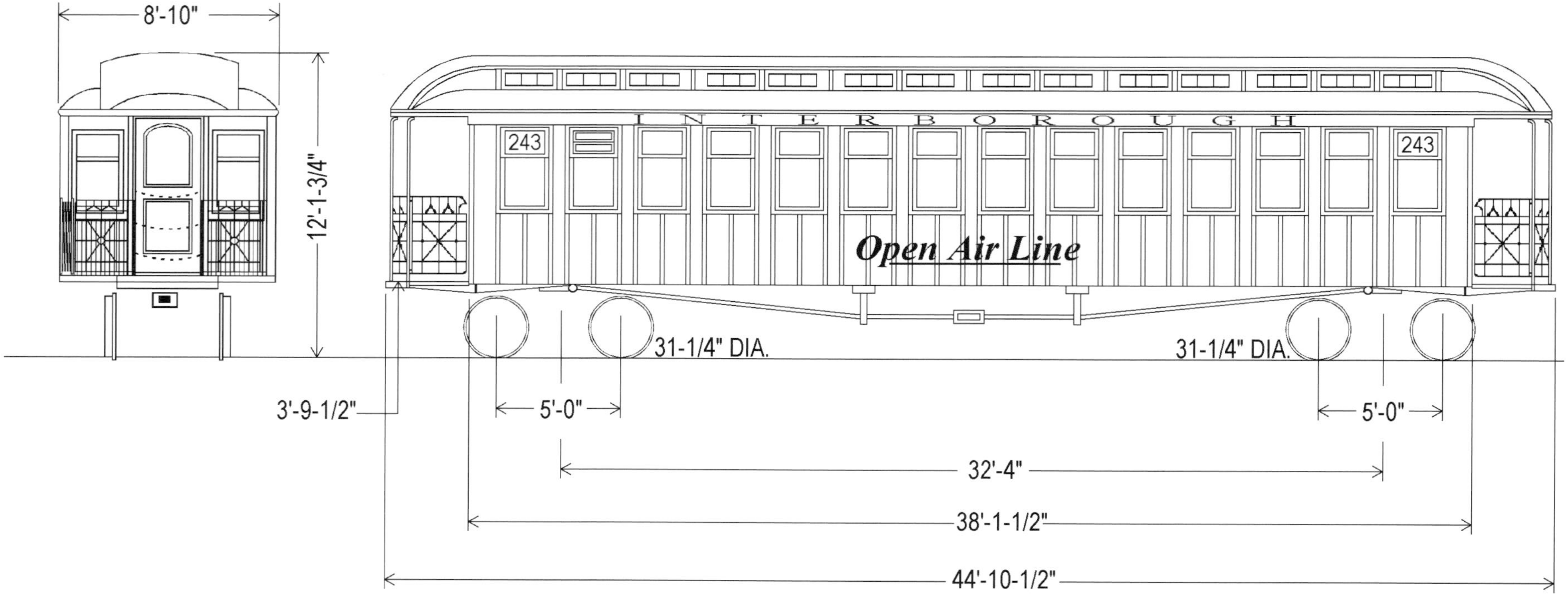

8'-10"
12'-1-3/4"
3'-9-1/2"
243
INTERBOROUGH
Open Air Line
31-1/4" DIA.
31-1/4" DIA.
5'-0"
5'-0"
32'-4"
38'-1-1/2"
44'-10-1/2"

Copyright © 1997 by NYC Transit

## CAR BODY - EQUIPMENT

| | | | |
|---|---|---|---|
| COUPLER | VAN DORN | SEAT MATERIAL | RATTAN |
| END DOOR | SWING | SEAT TYPE | INDIVIDUAL CUSHION |
| SEAT ARRANGEMENT & CAPACITY | LONGITUDINAL [44] | STRUCTURE MATERIAL | WOOD |

## TRUCK

| | | | |
|---|---|---|---|
| AXLE LOAD | 7,260 LBS | WEIGHT (2 TRUCKS) | 11,400 LBS |
| JOURNAL BEARING | FRICTION | | |

## GENERAL

| | | | |
|---|---|---|---|
| CAR BODY & EQUIPMENT WEIGHT | 17,640 LBS | CAR LOADED WEIGHT | 46,260 LBS |
| CAR LIGHT WEIGHT | 29,040 LBS | STANDING CAPACITY | [79] |

ORIGINALLY OWNED BY NEW YORK ELEVATED RAILROAD.
CARS ELECTRIFIED (1907).
ALL CARS WERE SOLD OR SCRAPPED (1930's - 1940's).

CAR NUMBERS: 293 - 369
TOTAL: 77 CARS
BUILT BY: BOWERS, DURE, PULLMAN (SEE NOTE 3)

DATE: 1882, 1885 (SEE NOTE 3)
AVERAGE COST PER CAR: SEE NOTE 3

# 1882, 1885 (IRT)

MANHATTAN ELEVATED, TRAILER

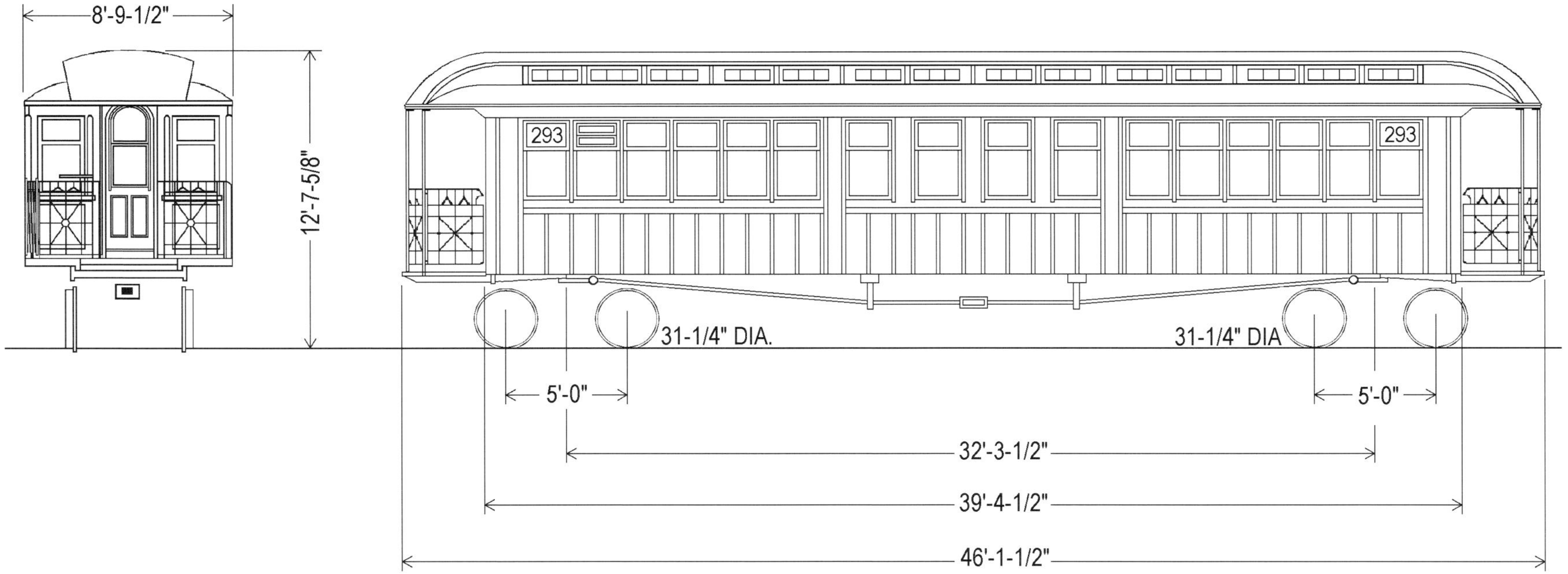

8'-9-1/2"
12'-7-5/8"
293
293
31-1/4" DIA.
31-1/4" DIA
5'-0"
5'-0"
32'-3-1/2"
39'-4-1/2"
46'-1-1/2"

Copyright © 1997 by NYC Transit

## CAR BODY - EQUIPMENT

| | | | |
|---|---|---|---|
| COUPLER | VAN DORN | SEAT MATERIAL | RATTAN |
| END DOOR | SWING | SEAT TYPE | INDIVIDUAL CUSHION |
| SEAT ARRANGEMENT & CAPACITY | MANHATTAN [48], SEE NOTE 1 | STRUCTURE MATERIAL | WOOD |

## TRUCK

| | | | |
|---|---|---|---|
| JOURNAL BEARING | FRICTION | WEIGHT (2 TRUCKS) | 13,700 LBS |

## GENERAL

| | | | |
|---|---|---|---|
| CAR BODY & EQUIPMENT WEIGHT | 20,580 LBS | STANDING CAPACITY | [77] |
| CAR LIGHT WEIGHT | 34,280 LBS | | |

NOTE 1: LONGITUDINAL SEATING EXCEPT FOUR CROSS SEATS ON EACH SIDE OF AISLE TWO OF WHICH FACED EACH OTHER AND WERE LOCATED AT CENTER OF CAR LENGTH.
NOTE 2: CARS HAD TWO TYPES OF WINDOW POSTS AND FRAMES.
NOTE 3:

| CARS | CAR BUILDERS | YEAR BUILT | AVERAGE COST PER CAR |
|---|---|---|---|
| 293 - 364 (SEE NOTE 2) | BOWERS, DURE | 1882 | $3,465 |
| 365 - 369 | PULLMAN | 1885 | $2,838 |

CARS 293 - 364 ORIGINALLY HAD DECK ROOF MODIFIED TO EXISTING ROOF (EARLY 1880'S).
CAR 293 MOTORIZED BY SPRAGUE FOR ELECTRICAL TESTS IN 1886 ONLY.
CARS ELECTRIFIED (1902-1903).
ALL CARS WERE SOLD OR SCRAPPED (1930's - 1940's).

CAR NUMBERS: 370 - 500
TOTAL: 131 CARS
BUILT BY: GILBERT & BUSH

DATE: 1886, 1887 (SEE NOTE 2)
AVERAGE COST PER CAR: $2,720

# 1886, 1887 (IRT)

MANHATTAN ELEVATED, MOTOR

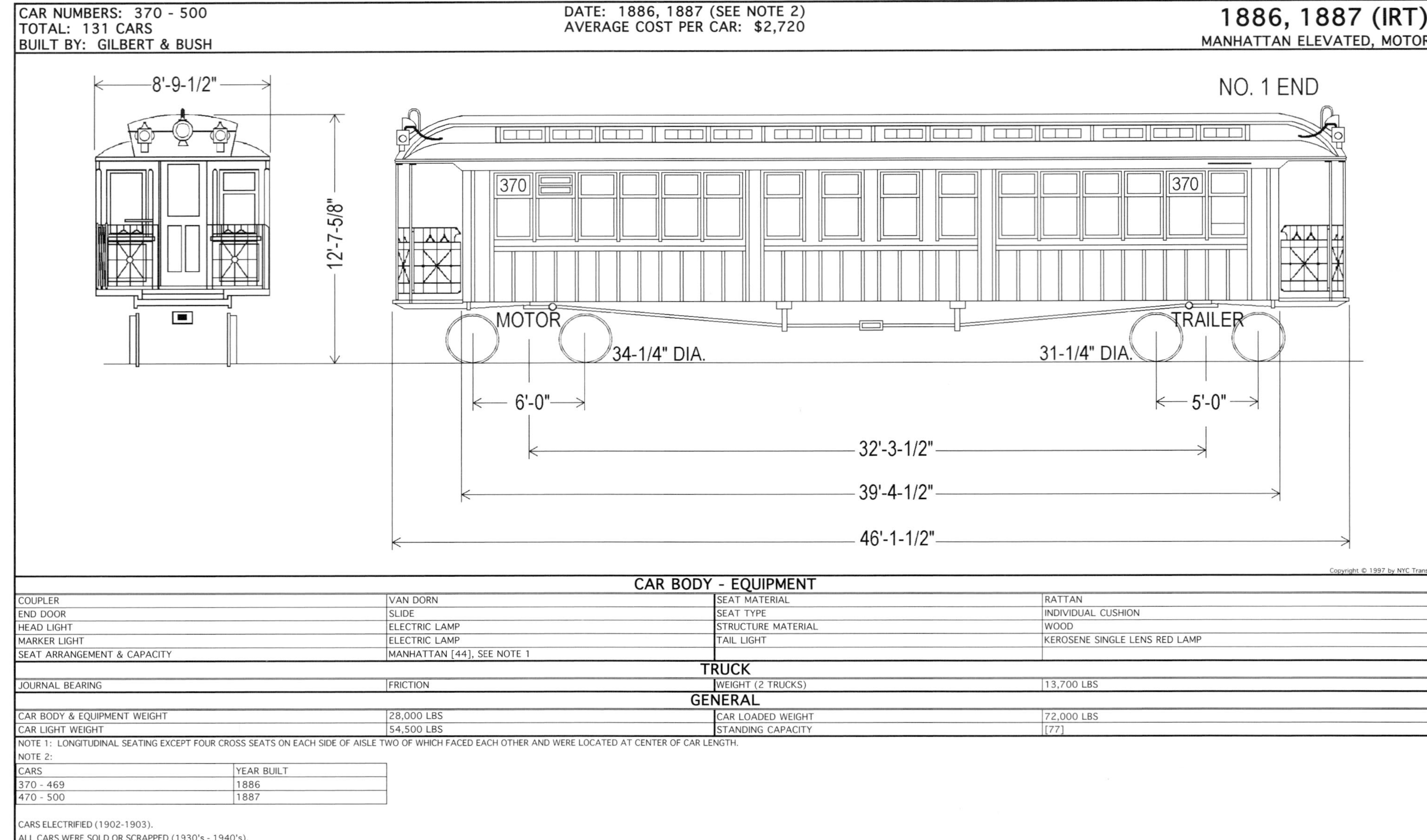


Copyright © 1997 by NYC Transit

## CAR BODY - EQUIPMENT

| | | | |
|---|---|---|---|
| COUPLER | VAN DORN | SEAT MATERIAL | RATTAN |
| END DOOR | SLIDE | SEAT TYPE | INDIVIDUAL CUSHION |
| HEAD LIGHT | ELECTRIC LAMP | STRUCTURE MATERIAL | WOOD |
| MARKER LIGHT | ELECTRIC LAMP | TAIL LIGHT | KEROSENE SINGLE LENS RED LAMP |
| SEAT ARRANGEMENT & CAPACITY | MANHATTAN [44], SEE NOTE 1 | | |

## TRUCK

| | | | |
|---|---|---|---|
| JOURNAL BEARING | FRICTION | WEIGHT (2 TRUCKS) | 13,700 LBS |

## GENERAL

| | | | |
|---|---|---|---|
| CAR BODY & EQUIPMENT WEIGHT | 28,000 LBS | CAR LOADED WEIGHT | 72,000 LBS |
| CAR LIGHT WEIGHT | 54,500 LBS | STANDING CAPACITY | [77] |

NOTE 1: LONGITUDINAL SEATING EXCEPT FOUR CROSS SEATS ON EACH SIDE OF AISLE TWO OF WHICH FACED EACH OTHER AND WERE LOCATED AT CENTER OF CAR LENGTH.

NOTE 2:

| CARS | YEAR BUILT |
|---|---|
| 370 - 469 | 1886 |
| 470 - 500 | 1887 |

CARS ELECTRIFIED (1902-1903).
ALL CARS WERE SOLD OR SCRAPPED (1930's - 1940's).

CAR NUMBERS: 501 - 540
TOTAL: 40 CARS
BUILT BY: BARNEY & SMITH, PULLMAN (SEE NOTE 2)

DATE: 1878
AVERAGE COST PER CAR: $3,500

# 1878 (IRT)

MANHATTAN ELEVATED, TRAILER

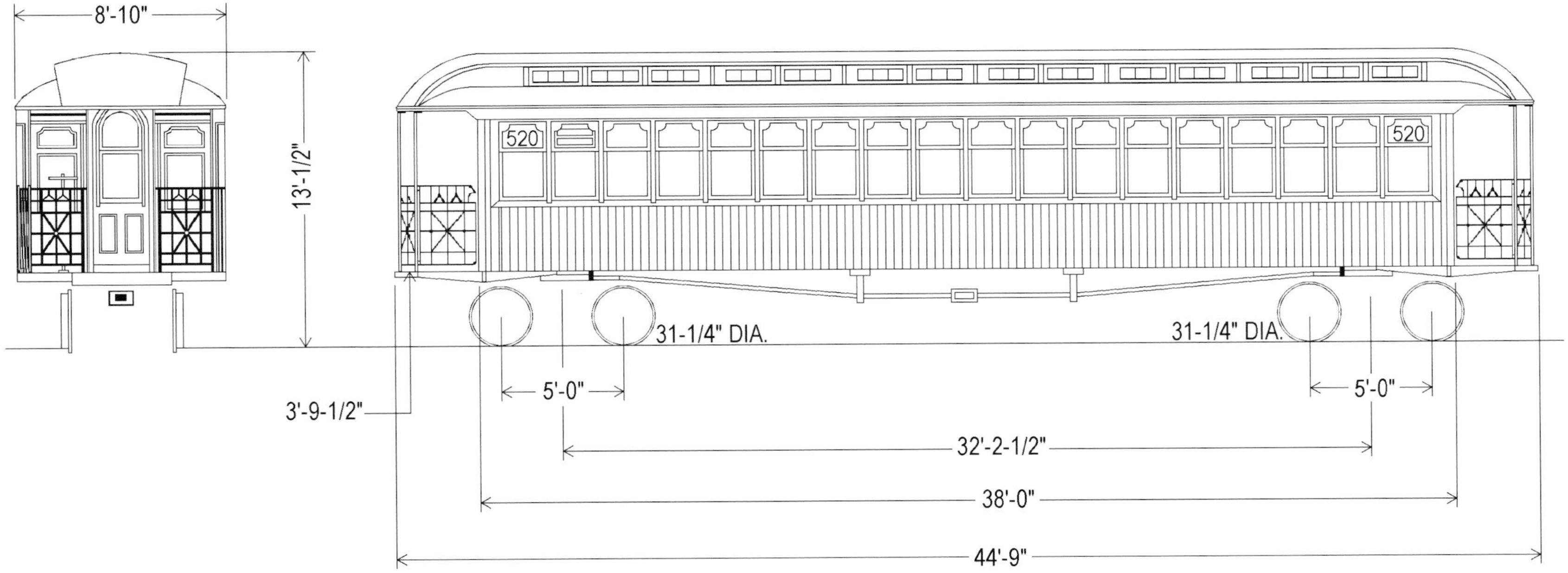

8'-10"
13'-1/2"
3'-9-1/2"
520
31-1/4" DIA.
31-1/4" DIA.
5'-0"
5'-0"
32'-2-1/2"
38'-0"
44'-9"

Copyright © 1997 by NYC Transit

## CAR BODY - EQUIPMENT

| | | | |
|---|---|---|---|
| COUPLER | VAN DORN | SEAT MATERIAL | RATTAN |
| END DOOR | SWING | SEAT TYPE | INDIVIDUAL CUSHION |
| SEAT ARRANGEMENT & CAPACITY | MANHATTAN [48], SEE NOTE 1 | STRUCTURE MATERIAL | WOOD |

## TRUCK

| | | | |
|---|---|---|---|
| AXLE LOAD | 8,570 LBS | WEIGHT (2 TRUCKS) | 15,000 LBS |
| JOURNAL BEARING | FRICTION | | |

## GENERAL

| | | | |
|---|---|---|---|
| CAR BODY & EQUIPMENT WEIGHT | 19,280 LBS | CAR LOADED WEIGHT | 51,780 LBS |
| CAR LIGHT WEIGHT | 34,280 LBS | STANDING CAPACITY | [77] |

NOTE 1: LONGITUDINAL SEATING EXCEPT FOUR CROSS SEATS ON EACH SIDE OF AISLE TWO OF WHICH FACED EACH OTHER AND WERE LOCATED AT CENTER OF CAR LENGTH.

NOTE 2:

| CARS | CAR BUILDERS |
|---|---|
| 501 - 520 | BARNEY & SMITH |
| 521 - 540 | PULLMAN |

ORIGINALLY OPERATED ON METROPOLITAN ELEVATED RAILWAY AS CARS 1-40.
ORIGINALLY HAD DECK ROOF MODIFIED TO EXISTING ROOF (EARLY 1880'S).
CARS 501 & 502 CONVERTED TO PAY CARS (1887 AND 1892 RESPECTIVELY). CAR 501 MOTORIZED (EARLY 1900's).
CARS HAD TWO TYPES OF WINDOW POSTS AND FRAMES.
ALL CARS WERE SOLD OR SCRAPPED (1930's - 1940's).

CAR NUMBERS: 541 - 580
TOTAL: 40 CARS
BUILT BY: PULLMAN

DATE: 1878
AVERAGE COST PER CAR: $4,300

# 1878 (IRT)

MANHATTAN ELEVATED, TRAILER

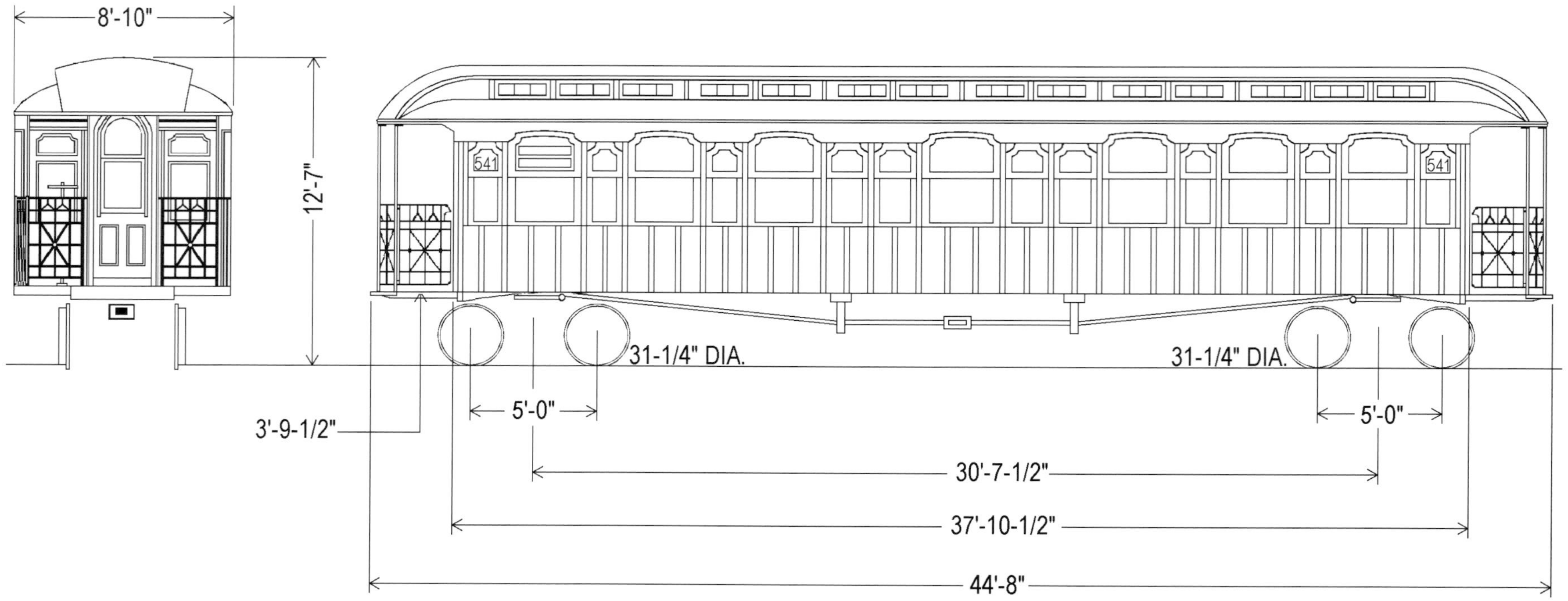

8'-10"
12'-7"
541
3'-9-1/2"
31-1/4" DIA.
31-1/4" DIA.
5'-0"
5'-0"
30'-7-1/2"
37'-10-1/2"
44'-8"

Copyright © 1997 by NYC Transit

| CAR BODY - EQUIPMENT | | | |
|---|---|---|---|
| COUPLER | VAN DORN | SEAT MATERIAL | RATTAN |
| END DOOR | SWING | SEAT TYPE | INDIVIDUAL CUSHION |
| SEAT ARRANGEMENT & CAPACITY | MANHATTAN [44], SEE NOTE 1 | STRUCTURE MATERIAL | WOOD |
| **TRUCK** | | | |
| AXLE LOAD | 8,575 LBS | WEIGHT (2 TRUCKS) | 15,000 LBS |
| JOURNAL BEARING | FRICTION | | |
| **GENERAL** | | | |
| CAR BODY & EQUIPMENT WEIGHT | 19,300 LBS | CAR LOADED WEIGHT | 49,925 LBS |
| CAR LIGHT WEIGHT | 34,300 LBS | STANDING CAPACITY | [81] |

NOTE 1: LONGITUDINAL SEATING EXCEPT FOUR CROSS SEATS ON EACH SIDE OF AISLE TWO OF WHICH FACED EACH OTHER AND WERE LOCATED AT CENTER OF CAR LENGTH.

ORIGINALLY OWNED BY METROPOLITAN ELEVATED RAILWAY AS CARS 41-80.
ORIGINALLY HAD DECK ROOF MODIFIED TO EXISTING ROOF (EARLY 1880'S).
CARS ELECTRIFIED (1902-1903).
ALL CARS WERE SOLD OR SCRAPPED (1930's - 1940's).

CAR NUMBERS: 581 - 600
TOTAL: 20 CARS
BUILT BY: PULLMAN

DATE: 1879
AVERAGE COST PER CAR: $3,500

# 1879 (IRT)

MANHATTAN ELEVATED, TRAILER

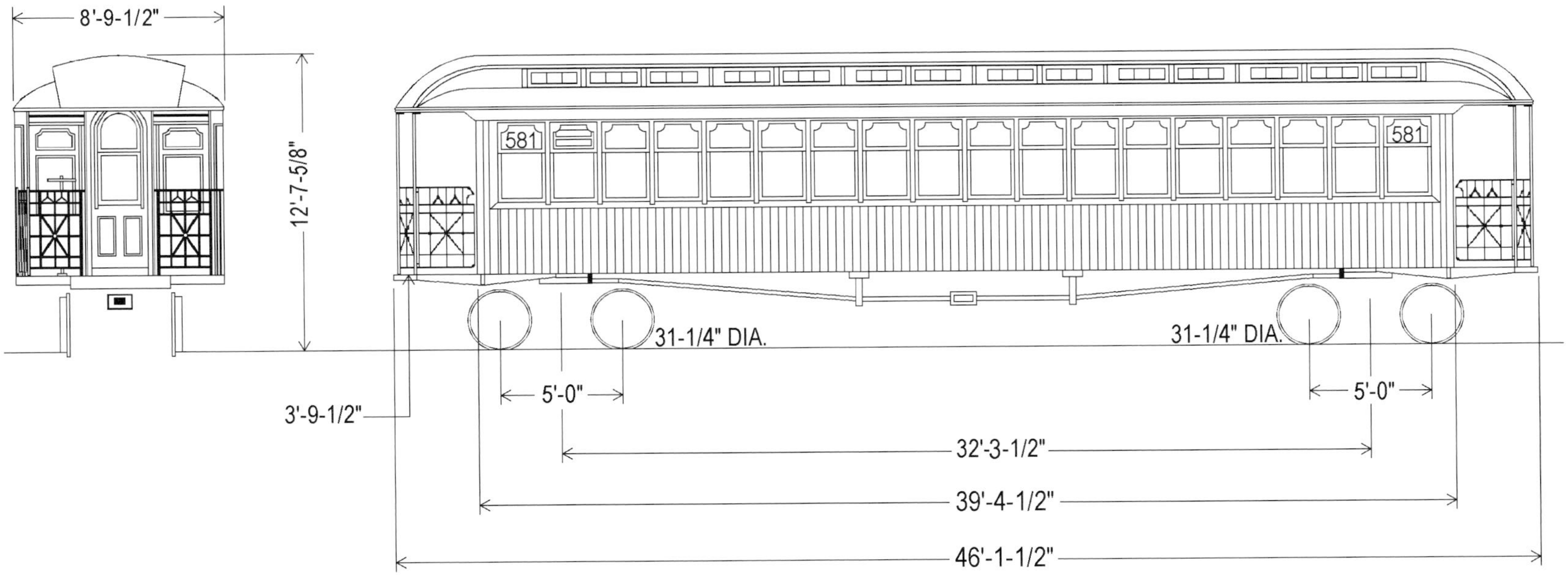

8'-9-1/2"
12'-7-5/8"
3'-9-1/2"
581
31-1/4" DIA.
31-1/4" DIA.
5'-0"
5'-0"
32'-3-1/2"
39'-4-1/2"
46'-1-1/2"

Copyright © 1997 by NYC Transit

| CAR BODY - EQUIPMENT | | | |
|---|---|---|---|
| COUPLER | VAN DORN | SEAT MATERIAL | RATTAN |
| END DOOR | SWING | SEAT TYPE | INDIVIDUAL CUSHION |
| SEAT ARRANGEMENT & CAPACITY | MANHATTAN [48], SEE NOTE 1 | STRUCTURE MATERIAL | WOOD |
| **TRUCK** | | | |
| AXLE LOAD | 8,570 LBS | WEIGHT (2 TRUCKS) | 15,000 LBS |
| JOURNAL BEARING | FRICTION | | |
| **GENERAL** | | | |
| CAR BODY & EQUIPMENT WEIGHT | 19,280 LBS | CAR LOADED WEIGHT | 51,780 LBS |
| CAR LIGHT WEIGHT | 34,280 LBS | STANDING CAPACITY | [77] |

NOTE 1: LONGITUDINAL SEATING EXCEPT FOUR CROSS SEATS ON EACH SIDE OF AISLE TWO OF WHICH FACED EACH OTHER AND WERE LOCATED AT CENTER OF CAR LENGTH.

NOTE 2: ALTHOUGH DIMENSIONS SHOWN ON THIS PAGE ARE FROM MANHATTAN RAILWAY COMPANY RECORDS, THERE IS THE POSSIBILITY THAT THESE CARS HAVE THE SPECIFICATIONS OF THE 1878 TYPE (SEE PAGE 345).

ORIGINALLY OPERATED ON METROPOLITAN ELEVATED RAILWAY AS CARS 81-100.
ORIGINALLY HAD DECK ROOF MODIFIED TO EXISTING ROOF (EARLY 1880'S).
CARS HAD TWO TYPES OF WINDOW POSTS AND FRAMES.
ALL CARS WERE SOLD OR SCRAPPED (1930's - 1940's).

CAR NUMBERS: 601 - 699
TOTAL: 99 CARS
BUILT BY: PULLMAN

DATE: 1879, 1880 (SEE NOTE 2)
AVERAGE COST PER CAR: $3,500

# 1879, 1880 (IRT)

MANHATTAN ELEVATED, TRAILER

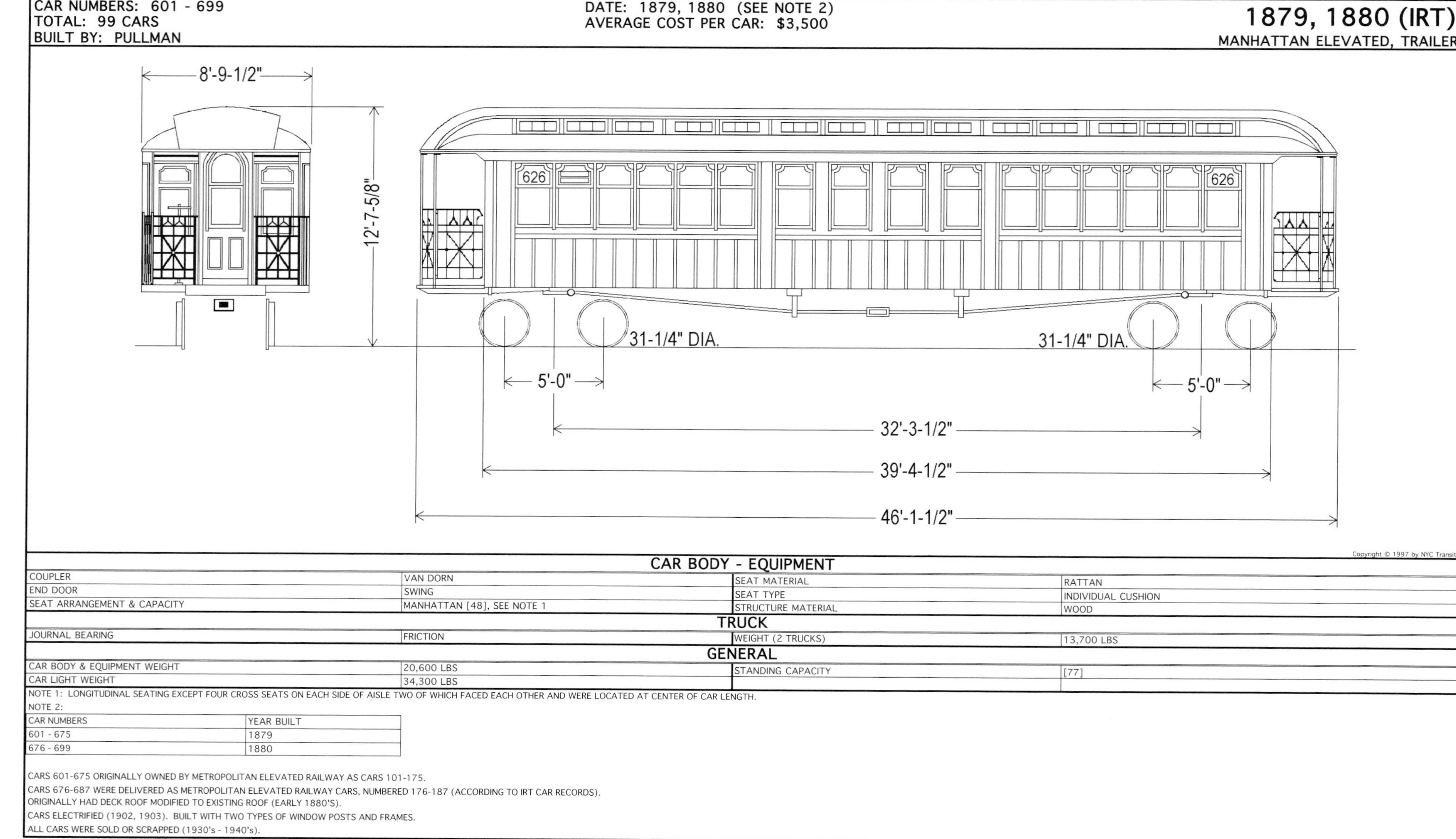


Copyright © 1997 by NYC Transit

## CAR BODY - EQUIPMENT

| | | | |
|---|---|---|---|
| COUPLER | VAN DORN | SEAT MATERIAL | RATTAN |
| END DOOR | SWING | SEAT TYPE | INDIVIDUAL CUSHION |
| SEAT ARRANGEMENT & CAPACITY | MANHATTAN [48], SEE NOTE 1 | STRUCTURE MATERIAL | WOOD |

## TRUCK

| | | | |
|---|---|---|---|
| JOURNAL BEARING | FRICTION | WEIGHT (2 TRUCKS) | 13,700 LBS |

## GENERAL

| | | | |
|---|---|---|---|
| CAR BODY & EQUIPMENT WEIGHT | 20,600 LBS | STANDING CAPACITY | [77] |
| CAR LIGHT WEIGHT | 34,300 LBS | | |

NOTE 1: LONGITUDINAL SEATING EXCEPT FOUR CROSS SEATS ON EACH SIDE OF AISLE TWO OF WHICH FACED EACH OTHER AND WERE LOCATED AT CENTER OF CAR LENGTH.

NOTE 2:

| CAR NUMBERS | YEAR BUILT |
|---|---|
| 601 - 675 | 1879 |
| 676 - 699 | 1880 |

CARS 601-675 ORIGINALLY OWNED BY METROPOLITAN ELEVATED RAILWAY AS CARS 101-175.
CARS 676-687 WERE DELIVERED AS METROPOLITAN ELEVATED RAILWAY CARS, NUMBERED 176-187 (ACCORDING TO IRT CAR RECORDS).
ORIGINALLY HAD DECK ROOF MODIFIED TO EXISTING ROOF (EARLY 1880'S).
CARS ELECTRIFIED (1902, 1903). BUILT WITH TWO TYPES OF WINDOW POSTS AND FRAMES.
ALL CARS WERE SOLD OR SCRAPPED (1930's - 1940's).

CAR NUMBERS: 700 - 728
TOTAL: 29 CARS
BUILT BY: PULLMAN

DATE: 1880
AVERAGE COST PER CAR: $3,120

# 1880 (IRT)

MANHATTAN ELEVATED, MOTOR

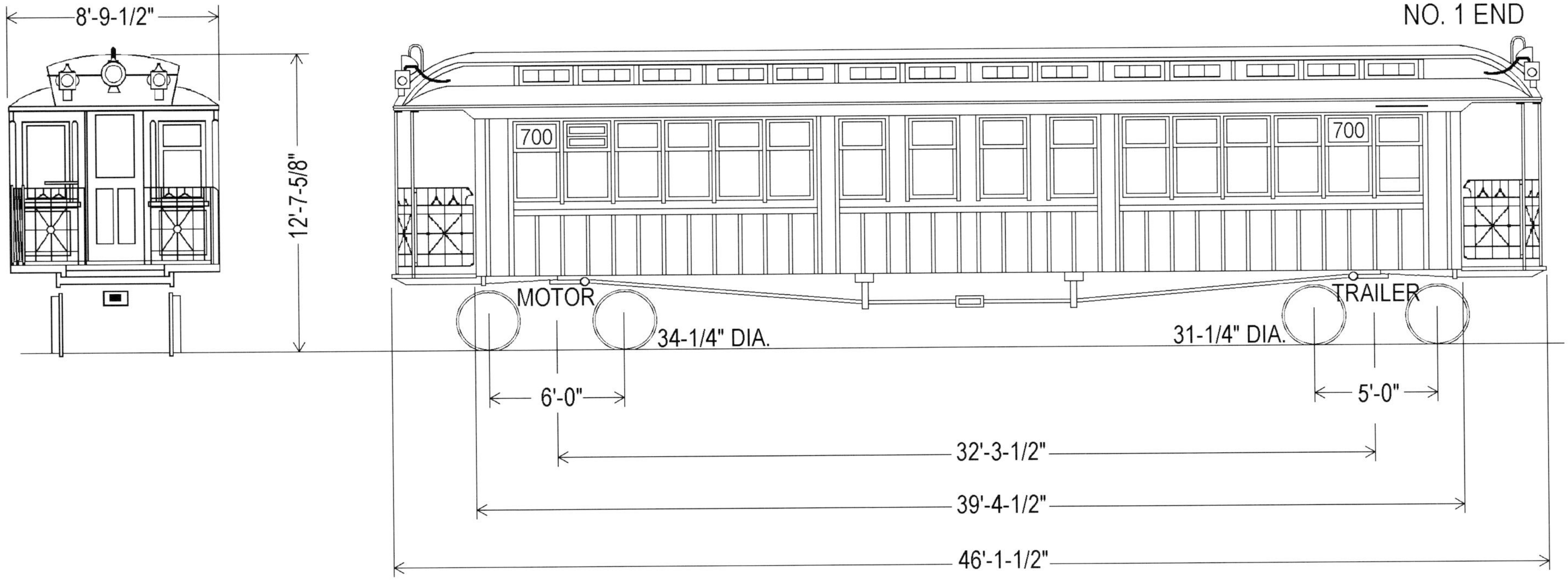

NO. 1 END

8'-9-1/2"
12'-7-5/8"
700
700
MOTOR
TRAILER
34-1/4" DIA.
31-1/4" DIA.
6'-0"
5'-0"
32'-3-1/2"
39'-4-1/2"
46'-1-1/2"

Copyright © 1997 by NYC Transit

| CAR BODY - EQUIPMENT | | | |
|---|---|---|---|
| COUPLER | VAN DORN | SEAT ARRANGEMENT & CAPACITY | MANHATTAN [44], SEE NOTE 1 |
| END DOOR | SLIDE | SEAT MATERIAL | RATTAN |
| HEAD LIGHT | ELECTRIC LAMP | SEAT TYPE | INDIVIDUAL CUSHION |
| MARKER LIGHT | ELECTRIC LAMP | TAIL LIGHT | KEROSENE SINGLE LENS RED LAMP |
| **TRUCK** | | | |
| WEIGHT (2 TRUCKS) | 13,700 LBS | | |
| **GENERAL** | | | |
| CAR BODY & EQUIPMENT WEIGHT | 33,740 LBS | STANDING CAPACITY | [77] |
| CAR LIGHT WEIGHT | 52,940 LBS | | |

NOTE 1: LONGITUDINAL SEATING EXCEPT FOUR CROSS SEATS ON EACH SIDE OF AISLE TWO OF WHICH FACED EACH OTHER AND WERE LOCATED AT CENTER OF CAR LENGTH.

CARS 703 & 704 MODIFIED WITH SPRAGUE MU ELECTRIFICATION EQUIPMENT FOR EXPERIMENTAL TESTS (11/21/1900).
CARS 703 & 704 WERE MOTORS, CARS 705, 706, 707 & 708 WERE TRAILERS.
ORIGINALLY HAD DECK ROOF MODIFIED TO EXISTING ROOF (EARLY 1880's).
BUILT WITH TWO TYPES OF WINDOW POSTS AND FRAMES.
CARS ELECTRIFIED (1902-1903).
ALL CARS WERE SCRAPPED (1930's - 1940's).

CAR NUMBERS: 729 - 790
TOTAL: 62 CARS
BUILT BY: PULLMAN

DATE: 1880, 1881 (SEE NOTE 2)
AVERAGE COST PER CAR: $2,065

# 1880, 1881 (IRT)

MANHATTAN ELEVATED, TRAILER

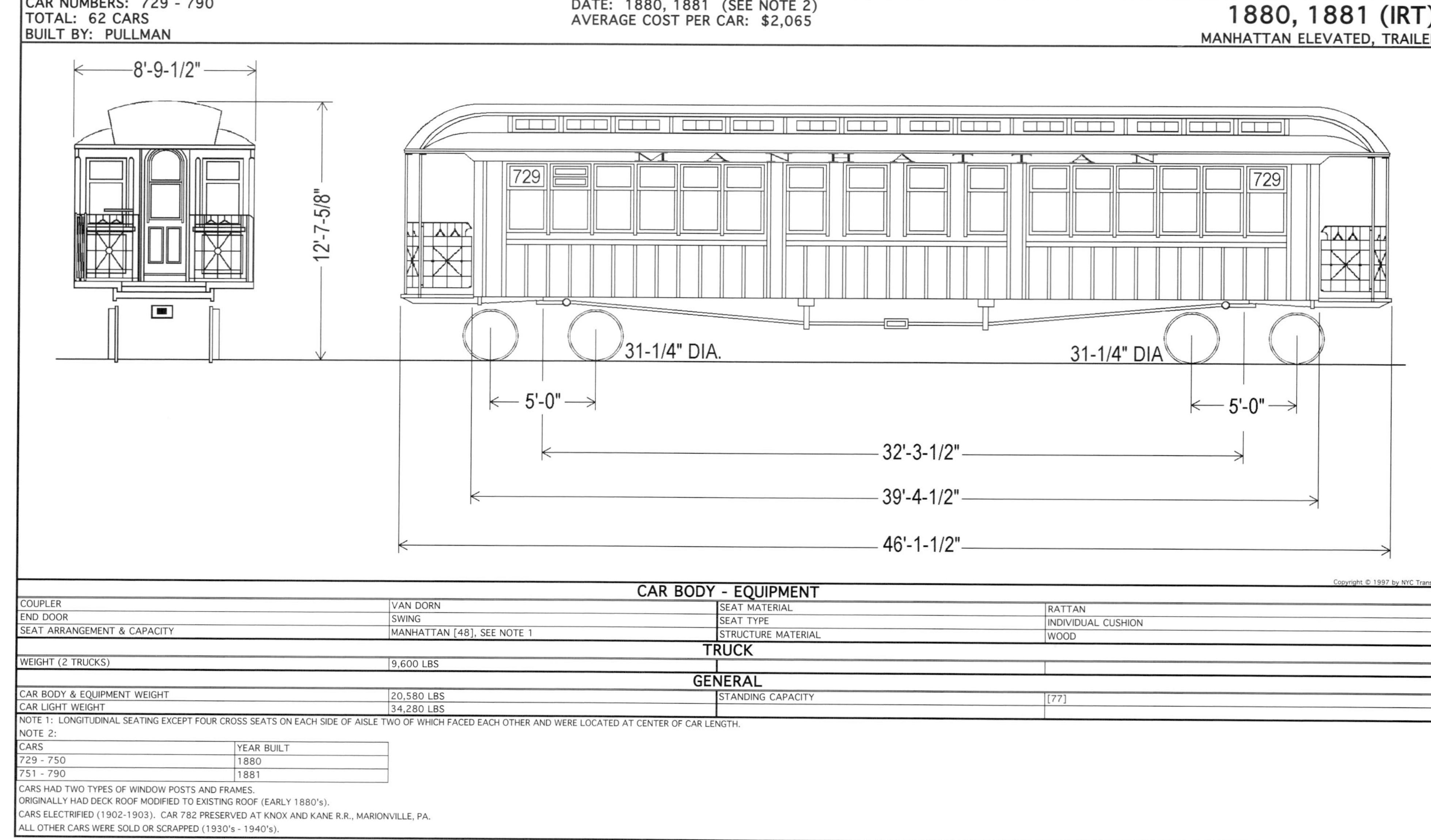


Copyright © 1997 by NYC Transit

## CAR BODY - EQUIPMENT

| | | | |
|---|---|---|---|
| COUPLER | VAN DORN | SEAT MATERIAL | RATTAN |
| END DOOR | SWING | SEAT TYPE | INDIVIDUAL CUSHION |
| SEAT ARRANGEMENT & CAPACITY | MANHATTAN [48], SEE NOTE 1 | STRUCTURE MATERIAL | WOOD |

## TRUCK

| | | | |
|---|---|---|---|
| WEIGHT (2 TRUCKS) | 9,600 LBS | | |

## GENERAL

| | | | |
|---|---|---|---|
| CAR BODY & EQUIPMENT WEIGHT | 20,580 LBS | STANDING CAPACITY | [77] |
| CAR LIGHT WEIGHT | 34,280 LBS | | |

NOTE 1: LONGITUDINAL SEATING EXCEPT FOUR CROSS SEATS ON EACH SIDE OF AISLE TWO OF WHICH FACED EACH OTHER AND WERE LOCATED AT CENTER OF CAR LENGTH.

NOTE 2:

| CARS | YEAR BUILT |
|---|---|
| 729 - 750 | 1880 |
| 751 - 790 | 1881 |

CARS HAD TWO TYPES OF WINDOW POSTS AND FRAMES.
ORIGINALLY HAD DECK ROOF MODIFIED TO EXISTING ROOF (EARLY 1880's).
CARS ELECTRIFIED (1902-1903). CAR 782 PRESERVED AT KNOX AND KANE R.R., MARIONVILLE, PA.
ALL OTHER CARS WERE SOLD OR SCRAPPED (1930's - 1940's).

CAR NUMBERS: 791 - 1120
TOTAL: 330 CARS
BUILT BY: GILBERT & BUSH, PULLMAN, WASON (SEE NOTE 2)

DATE: 1881, 1885 - 1887, 1889 - 1891, 1893 (SEE NOTE 2)
AVERAGE COST PER CAR: SEE NOTE 2

# 1881, 1885 - 1887, 1889 - 1891, 1893 (IRT)

MANHATTAN ELEVATED, MOTOR

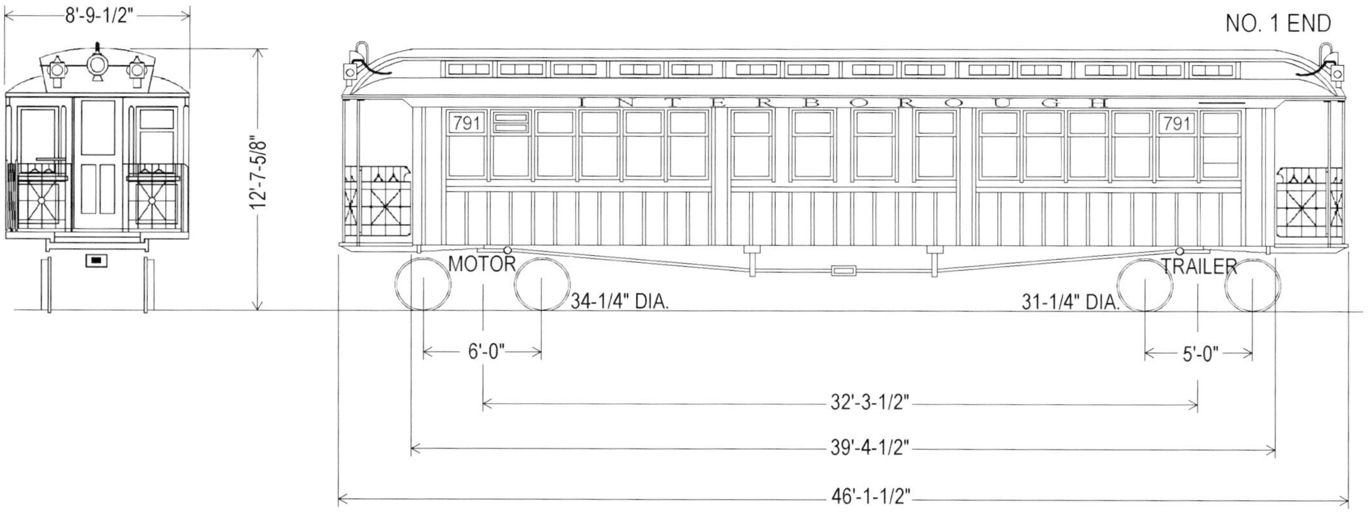

8'-9-1/2"
12'-7-5/8"
NO. 1 END
INTERBOROUGH
791
MOTOR
TRAILER
34-1/4" DIA.
31-1/4" DIA.
6'-0"
5'-0"
32'-3-1/2"
39'-4-1/2"
46'-1-1/2"

Copyright © 1997 by NYC Transit

## CAR BODY - EQUIPMENT

| | | | |
|---|---|---|---|
| COUPLER | VAN DORN | SEAT MATERIAL | RATTAN |
| HEAD LIGHT | ELECTRIC LAMP | SEAT TYPE | INDIVIDUAL CUSHION |
| MARKER LIGHT | ELECTRIC LAMP | STRUCTURE MATERIAL | WOOD |
| SEAT ARRANGEMENT & CAPACITY | MANHATTAN [44], SEE NOTE 1 | TAIL LIGHT | KEROSENE SINGLE RED LENS LAMP |

## TRUCK

| | | | |
|---|---|---|---|
| JOURNAL BEARING | FRICTION | WEIGHT (2 TRUCKS) | 13,700 LBS |

## GENERAL

| | | | |
|---|---|---|---|
| CAR LIGHT WEIGHT | CARS 791 - 825: 52,940 LBS; CARS 826 - 919: 54,540 LBS; CARS 920 - 1094: 55,100 LBS; CARS 1095 - 1120: 54,500 LBS | STANDING CAPACITY | [77] |

NOTE 1: LONGITUDINAL SEATING EXCEPT FOUR CROSS SEATS ON EACH SIDE OF AISLE TWO OF WHICH FACED EACH OTHER AND WERE LOCATED AT CENTER OF CAR LENGTH.

NOTE 2:

| CARS | CAR BUILDERS | YEAR BUILT | AVERAGE COST PER CAR |
|---|---|---|---|
| 791 - 825 | PULLMAN | 1881 | $3,200 |
| 826 - 919 | GILBERT & BUSH | 1887 | $2,830 |
| 920 - 969 | WASON | 1890 | $2,995 |
| 970 - 1019 | | 1891 | $2,995 |
| 1020 - 1094 | | 1893 | $2,898 |
| 1095 - 1098 | PULLMAN | 1885 | $3,016 (APPROX.) |
| 1099 - 1114 | GILBERT & BUSH | 1886 | $2,807 (APPROX.) |
| 1115 - 1120 | | 1889 | $2,855 |

CARS 791-825 HAD TWO TYPES OF WINDOW POSTS AND FRAMES. CARS ELECTRIFIED (1902-1903).
CARS 791 - 825 ORIGINALLY HAD DECK ROOF MODIFIED TO EXISTING ROOF (EARLY 1880'S).
CARS 1095-1120 ORIGINALLY OPERATED ON SUBURBAN RAPID TRANSIT AS CARS 1-26.
CAR 805 CONVERTED TO DRILL FLAT CAR (1905). CAR 824 CONVERTED TO INSTRUCTION CAR (1902) AND IS CURRENTLY AT THE BRANFORD ELECTRIC RAILWAY MUSEUM, CT. CAR 873 CONVERTED TO COLLECTION CAR (12/17/1942).
CARS 844 & 889 AT WESTERN RAILWAY MUSEUM , RIO VISTA, CA. OF THE BAY AREA ELECTRIC R.R. ASSOCIATION (LATE 1940's).
ALL OTHER CARS WERE SOLD OR SCRAPPED (1930's - 1940's).

CAR NUMBERS: 1121 - 1218
TOTAL: 98 CARS
BUILT BY: WASON

DATE: 1902, 1903
AVERAGE COST PER CAR: N/A

# 1902, 1903 (IRT)

MANHATTAN ELEVATED, MOTOR

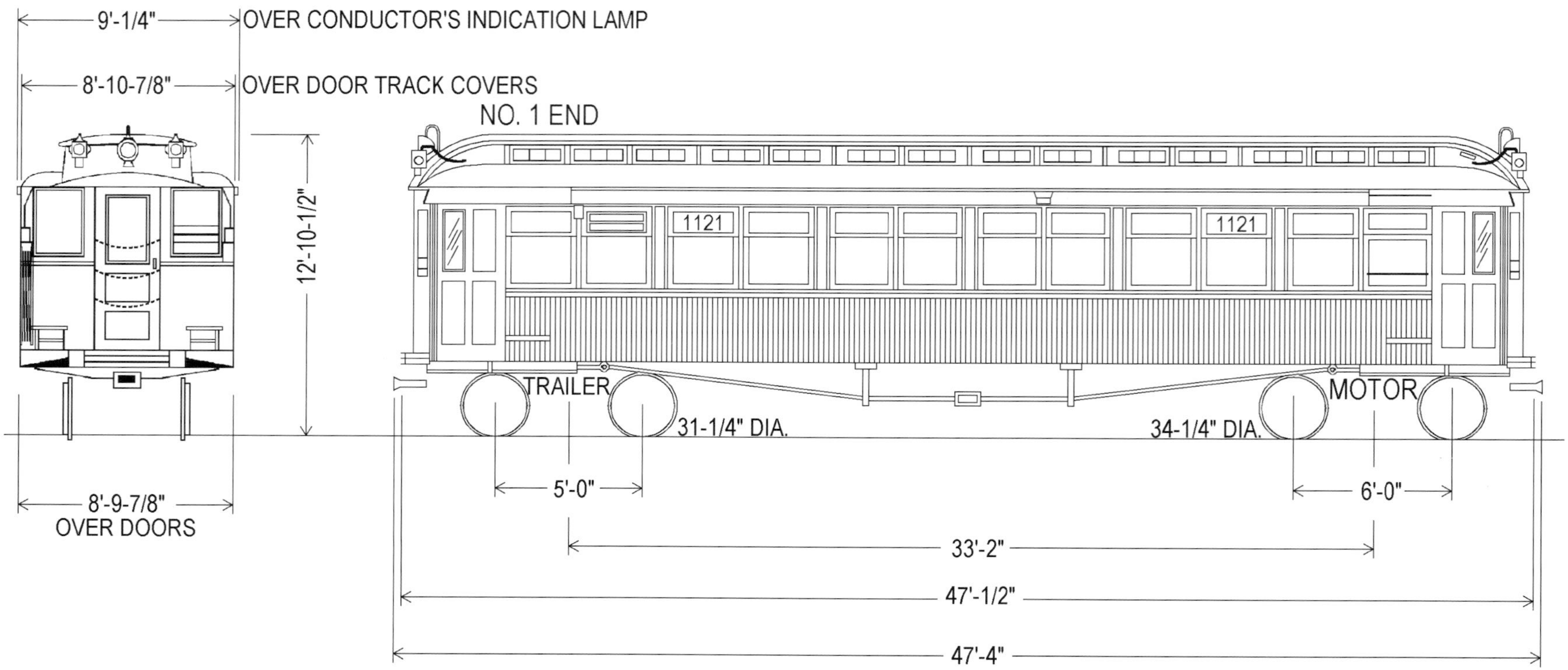

9'-1/4" OVER CONDUCTOR'S INDICATION LAMP
8'-10-7/8" OVER DOOR TRACK COVERS
NO. 1 END
12'-10-1/2"
8'-9-7/8" OVER DOORS
1121
1121
TRAILER
MOTOR
31-1/4" DIA.
34-1/4" DIA.
5'-0"
6'-0"
33'-2"
47'-1/2"
47'-4"

Copyright © 1997 by NYC Transit

| CAR BODY - EQUIPMENT | | | |
|---|---|---|---|
| COUPLER | VAN DORN | SEAT MATERIAL | RATTAN |
| END DOOR | SLIDE | SEAT TYPE | INDIVIDUAL CUSHION |
| HEAD LIGHT | ELECTRIC LAMP | STRUCTURE MATERIAL | WOOD |
| MARKER LIGHT | ELECTRIC LAMP | TAIL LIGHT | KEROSENE SINGLE LENS RED LAMP |
| SEAT ARRANGEMENT & CAPACITY | MANHATTAN [44], SEE NOTE 1 | | |
| **TRUCK** | | | |
| JOURNAL BEARING | FRICTION | WEIGHT (2 TRUCKS) | 27,180 LBS |
| **GENERAL** | | | |
| CAR BODY & EQUIPMENT WEIGHT | 31,940 LBS | CAR LOADED WEIGHT | 76,620 LBS |
| CAR LIGHT WEIGHT | 59,120 LBS | STANDING CAPACITY | [77] |

NOTE 1: LONGITUDINAL SEATING EXCEPT FOUR CROSS SEATS ON EACH SIDE OF AISLE TWO OF WHICH FACED EACH OTHER AND WERE LOCATED AT CENTER OF CAR LENGTH.

ALL CARS EXCEPT 1130, 1144, 1185, 1191, 1193, 1201 WERE CONVERTED TO MUDC OPERATION (1924).
ALL CARS WERE SOLD OR SCRAPPED (LATE 1940's - 1950's).

CAR NUMBERS: 1219 - 1254
TOTAL: 36 CARS
BUILT BY: AMERICAN CAR FOUNDRY, JEWETT (SEE NOTE 2)

DATE: 1902
AVERAGE COST PER CAR: N/A

# 1902 (IRT)

MANHATTAN OPEN, TRAILER

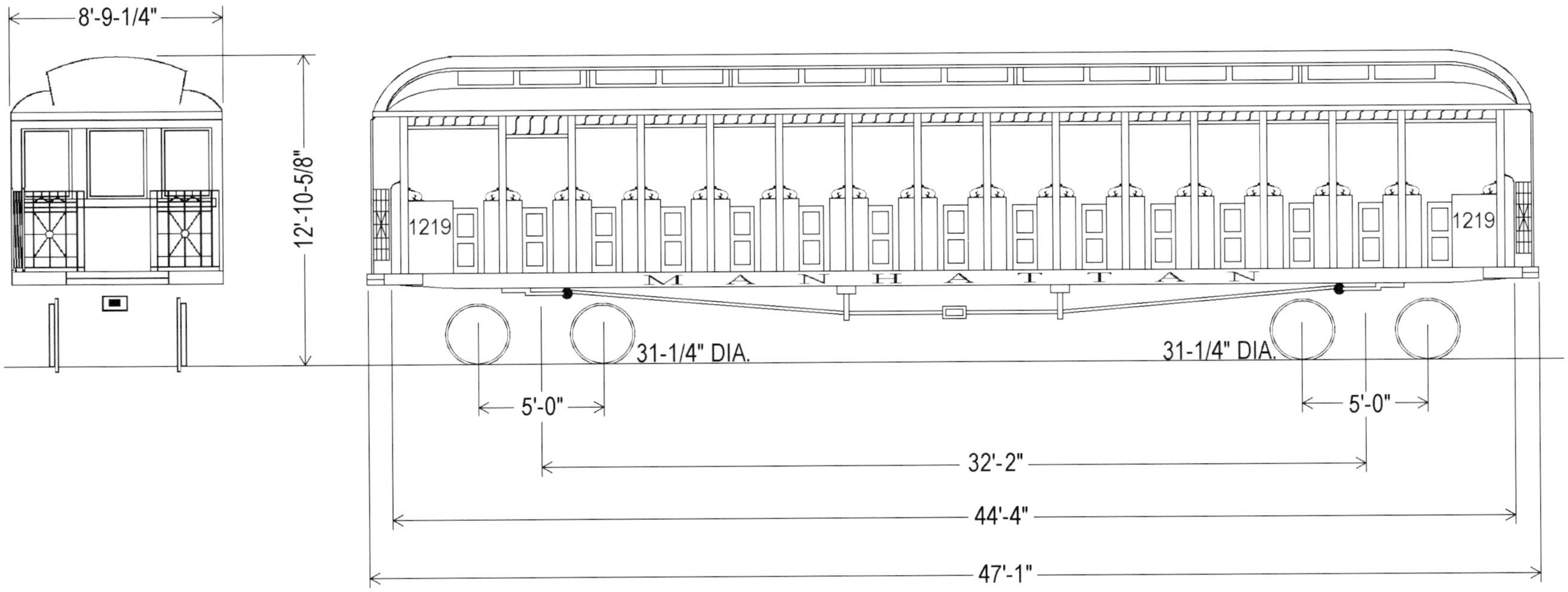

8'-9-1/4"
12'-10-5/8"
1219
MANHATTAN
31-1/4" DIA.
31-1/4" DIA.
5'-0"
5'-0"
32'-2"
44'-4"
47'-1"

Copyright © 1997 by NYC Transit

## CAR BODY - EQUIPMENT

| | | | |
|---|---|---|---|
| COUPLER | VAN DORN | SEAT MATERIAL | WOOD |
| SEAT ARRANGEMENT | SEE NOTE 1 | SEAT TYPE | FIVE PASSENGER |
| SEATING CAPACITY | [80] | STRUCTURE MATERIAL | WOOD |

## TRUCK

| | | | |
|---|---|---|---|
| AXLE LOAD | 8,280 LBS | WEIGHT (2 TRUCKS) | 20,800 LBS |
| JOURNAL BEARING | FRICTION | | |

## GENERAL

| | | | |
|---|---|---|---|
| CAR BODY & EQUIPMENT WEIGHT | 21,720 LBS | CAR LOADED WEIGHT | 44,120 LBS |
| CAR LIGHT WEIGHT | 33,120 LBS | STANDING CAPACITY | NONE |

NOTE 1: SEATS ARRANGED LATERALLY (CROSS SEAT) WITH REVERSIBLE BACK.

NOTE 2:

| CARS | CAR BUILDERS |
|---|---|
| 1219, 1245 - 1254 | AMERICAN CAR FOUNDRY |
| 1220 - 1244 | JEWETT |

CARS OPERATED ONLY ON 3RD AVENUE LINE WITH A MAXIMUM OF TWO (USUALLY ONE) PER TRAIN CONSIST (FROM 1902 - 1917).
CARS ORIGINALLY HAD NO ANTICLIMBERS.
CAR 1219 WAS DELIVERED AS CAR 142 (RENUMBERED SHORTLY AFTER DELIVERY).
CARS 1221, 1228, 1240, 1245, 1247 WERE CONVERTED TO FLAT CARS (1925).
THREE CARS WERE SCRAPPED (10/1938). BALANCE OF CARS WERE SOLD (1918).

CAR NUMBERS: 1255 - 1314
TOTAL: 60 CARS
BUILT BY: WASON

DATE: 1903
AVERAGE COST PER CAR: N/A

# 1903 (IRT)

MANHATTAN ELEVATED, MOTOR

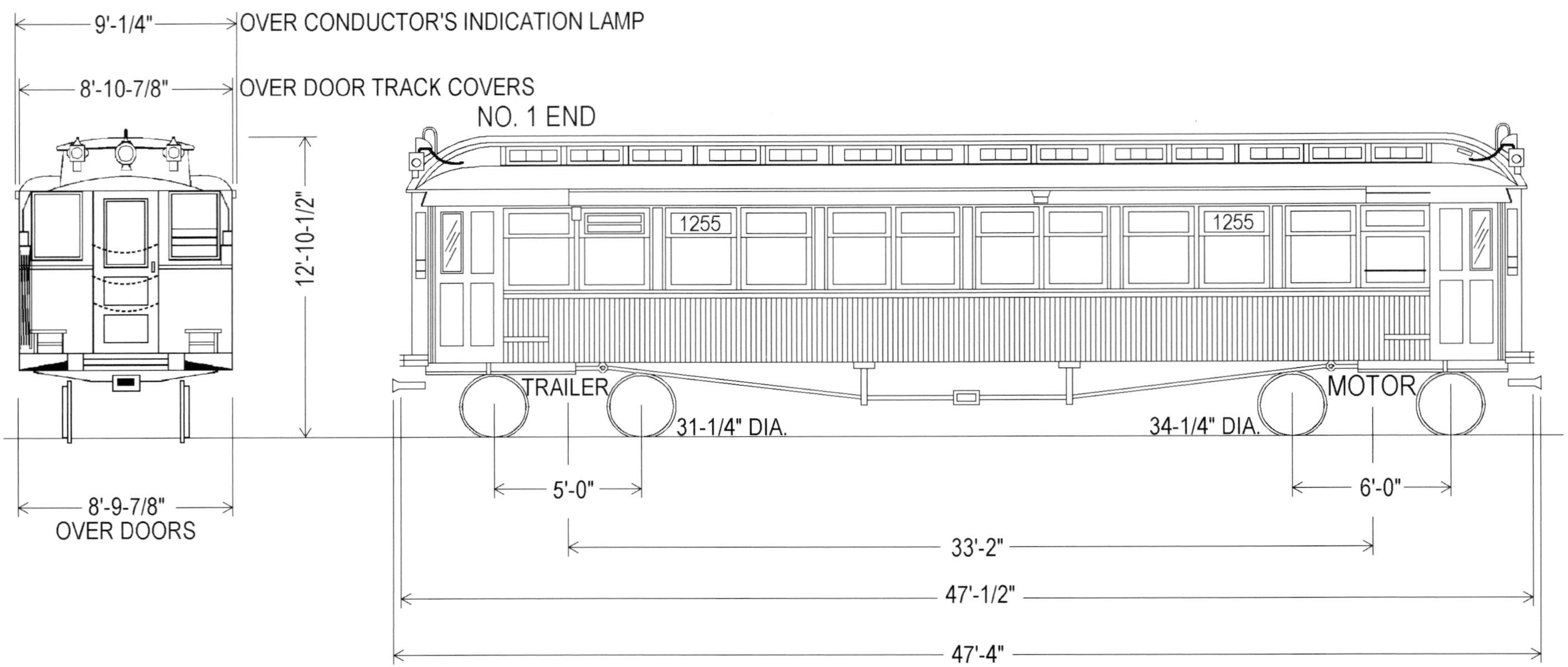

9'-1/4" OVER CONDUCTOR'S INDICATION LAMP
8'-10-7/8" OVER DOOR TRACK COVERS
NO. 1 END
12'-10-1/2"
1255
1255
TRAILER
31-1/4" DIA.
MOTOR
34-1/4" DIA.
8'-9-7/8"
OVER DOORS
5'-0"
6'-0"
33'-2"
47'-1/2"
47'-4"

Copyright © 1997 by NYC Transit

| CAR BODY - EQUIPMENT | | | |
|---|---|---|---|
| COUPLER | VAN DORN | SEAT MATERIAL | RATTAN |
| END DOOR | SLIDING | SEAT TYPE | INDIVIDUAL CUSHION |
| HEAD & MARKER LIGHTS | ELECTRIC LAMP | STRUCTURE MATERIAL | WOOD |
| SEAT ARRANGEMENT & CAPACITY | MANHATTAN [44], SEE NOTE 1 | TAIL LIGHT | KEROSENE SINGLE LENS RED LAMP |
| TRUCK | | | |
| JOURNAL BEARING | FRICTION | WEIGHT (2 TRUCKS) | 27,180 LBS |
| GENERAL | | | |
| CAR BODY & EQUIPMENT WEIGHT | 31,940 LBS | CAR LOADED WEIGHT | 76,620 LBS |
| CAR LIGHT WEIGHT | 59,120 LBS | STANDING CAPACITY | [77] |

NOTE 1: LONGITUDINAL SEATING EXCEPT FOUR CROSS SEATS ON EACH SIDE OF AISLE TWO OF WHICH FACED EACH OTHER AND WERE LOCATED AT CENTER OF CAR LENGTH.

ALL CARS EXCEPT 1258, 1292, 1307, 1314 WERE CONVERTED TO MUDC OPERATION (1924).
ALL CARS WERE SCRAPPED (1940's - 1950's).

CAR NUMBERS: 1315 - 1414
TOTAL: 100 CARS
BUILT BY: ST. LOUIS, WASON (SEE NOTE 2)

DATE: 1904
AVERAGE COST PER CAR: N/A

# 1904 (IRT)

MANHATTAN ELEVATED, TRAILER

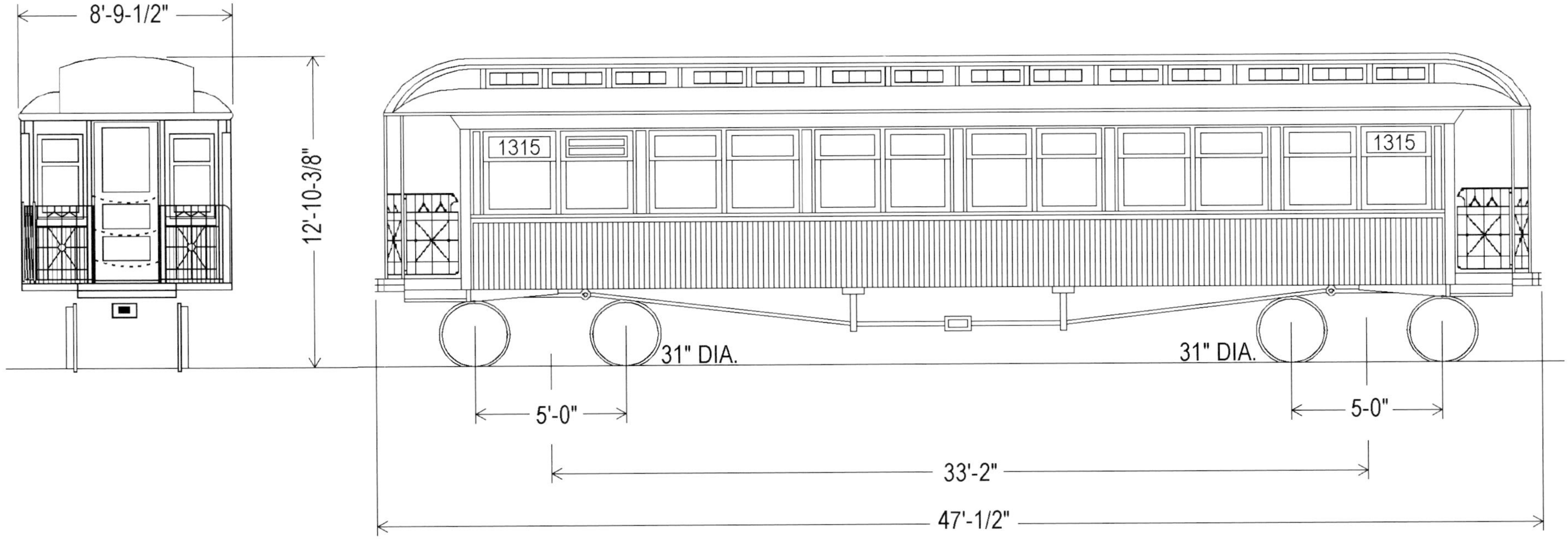

8'-9-1/2"

12'-10-3/8"

1315

31" DIA.

5'-0"

5-0"

33'-2"

47'-1/2"

Copyright © 1997 by NYC Transit

## CAR BODY - EQUIPMENT

| | | | |
|---|---|---|---|
| COUPLER | VAN DORN | SEAT MATERIAL | RATTAN |
| END DOOR | SLIDING | SEAT TYPE | INDIVIDUAL CUSHION |
| SEAT ARRANGEMENT & CAPACITY | MANHATTAN [48], SEE NOTE 1 | STRUCTURE MATERIAL | WOOD |

## TRUCK

| | | | |
|---|---|---|---|
| JOURNAL BEARING | FRICTION | WEIGHT (2 TRUCKS) | APPOX. 16,000 LBS |

## GENERAL

| | | | |
|---|---|---|---|
| CAR BODY & EQUIPMENT WEIGHT | 24,080 LBS | STANDING CAPACITY | [77] |
| CAR LIGHT WEIGHT | 40,080 LBS | | |

NOTE 1: LONGITUDINAL SEATING EXCEPT FOUR CROSS SEATS ON EACH SIDE OF AISLE TWO OF WHICH FACED EACH OTHER AND WERE LOCATED AT CENTER OF CAR LENGTH.

NOTE 2:

| CARS | CAR BUILDERS |
|---|---|
| 1315 - 1364 | WASON |
| 1365 - 1414 | ST. LOUIS |

ALL CARS WERE SOLD OR SCRAPPED (1940's - 1950's).

CAR NUMBERS: 1415 - 1528
TOTAL: 114 CARS
BUILT BY: ST. LOUIS

DATE: 1907, 1908
AVERAGE COST PER CAR: N/A

# 1907, 1908 (IRT)

MANHATTAN ELEVATED, TRAILER

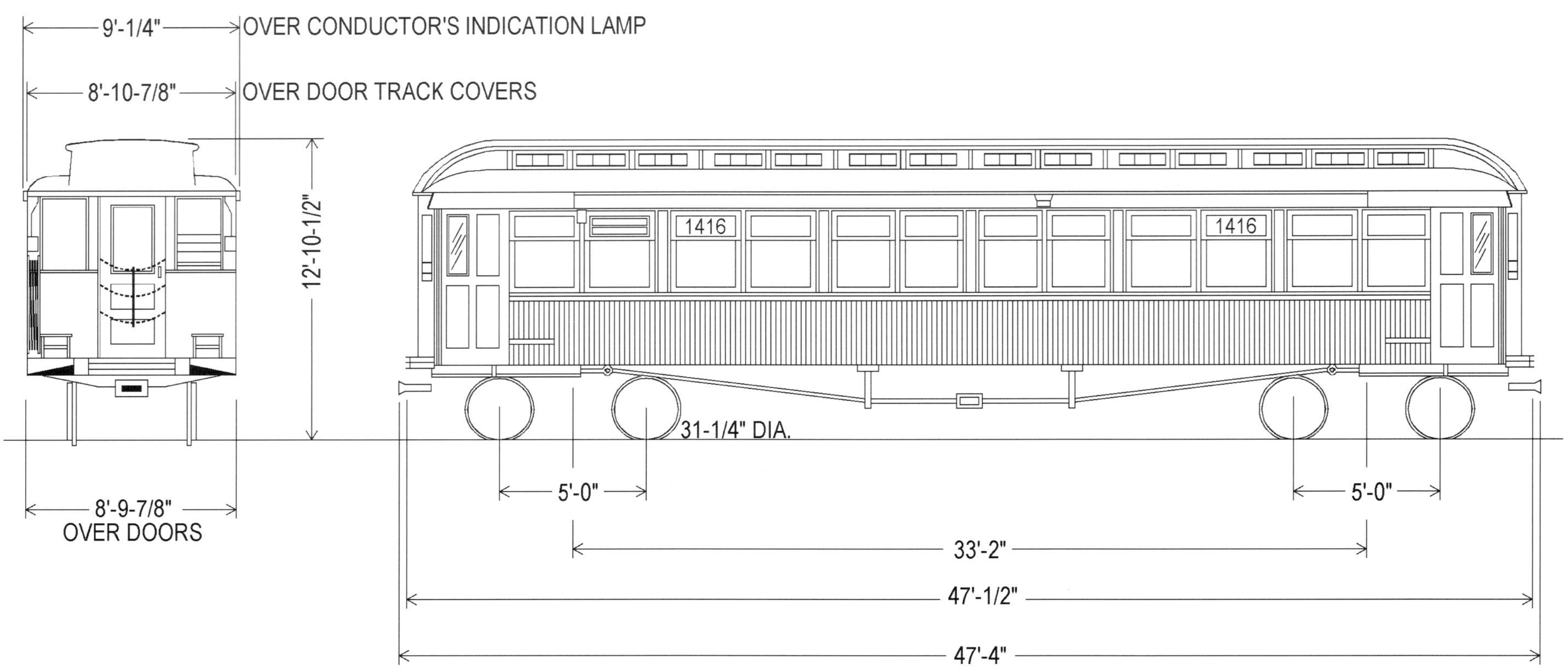

9'-1/4" OVER CONDUCTOR'S INDICATION LAMP

8'-10-7/8" OVER DOOR TRACK COVERS

12'-10-1/2"

8'-9-7/8" OVER DOORS

1416

31-1/4" DIA.

5'-0"

5'-0"

33'-2"

47'-1/2"

47'-4"

Copyright © 1997 by NYC Transit

| CAR BODY - EQUIPMENT | | | |
|---|---|---|---|
| COUPLER | VAN DORN | SEAT MATERIAL | RATTAN |
| END DOOR | SLIDING | SEAT TYPE | INDIVIDUAL CUSHION |
| SEAT ARRANGEMENT & CAPACITY | MANHATTAN [44], SEE NOTE 1 | STRUCTURE MATERIAL | WOOD |
| **TRUCK** | | | |
| JOURNAL BEARING | FRICTION | WEIGHT (2 TRUCKS) | 16,480 LBS |
| **GENERAL** | | | |
| CAR BODY & EQUIPMENT WEIGHT | 26,600 LBS | CAR LOADED WEIGHT | 60,580 LBS |
| CAR LIGHT WEIGHT | 43,080 LBS | STANDING CAPACITY | [77] |

NOTE 1: LONGITUDINAL SEATING EXCEPT FOUR CROSS SEATS ON EACH SIDE OF AISLE TWO OF WHICH FACED EACH OTHER AND WERE LOCATED AT CENTER OF CAR LENGTH.

ALL CARS EXCEPT 1415, 1447, 1459, 1515 WERE CONVERTED TO MUDC OPERATION (1924).
ALL CARS WERE SOLD OR SCRAPPED (1940's - 1950's).

CAR NUMBERS: 1529 - 1612
TOTAL: 84 CARS
BUILT BY: WASON

DATE: 1907
AVERAGE COST PER CAR: N/A

# 1907 (IRT)

MANHATTAN ELEVATED, MOTOR

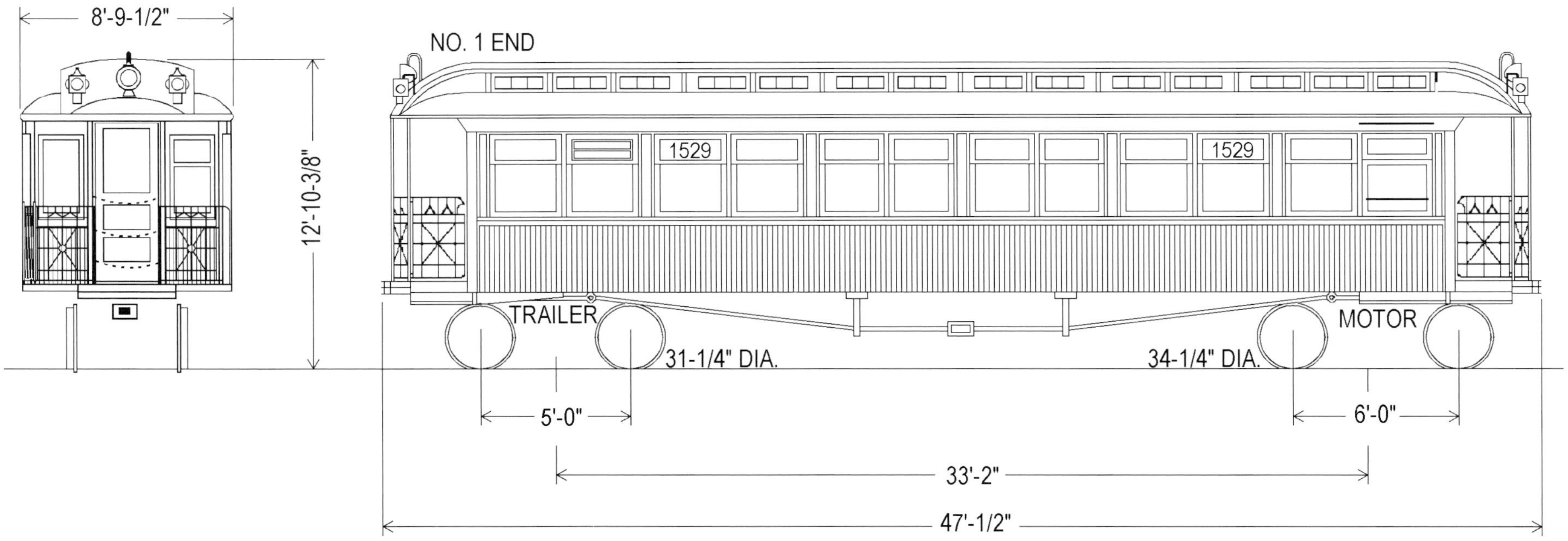

Copyright © 1997 by NYC Transit

| CAR BODY - EQUIPMENT | | | |
|---|---|---|---|
| COUPLER | VAN DORN | SEAT MATERIAL | RATTAN |
| END DOOR | SLIDING | SEAT TYPE | INDIVIDUAL CUSHION |
| HEAD LIGHT | ELECTRIC LAMP | STRUCTURE MATERIAL | WOOD |
| MARKER LIGHT | ELECTRIC LAMP | TAIL LIGHT | KEROSENE SINGLE LENS RED LAMP |
| SEAT ARRANGEMENT & CAPACITY | MANHATTAN [44], SEE NOTE 1 | | |
| **TRUCK** | | | |
| JOURNAL BEARING | FRICTION | WEIGHT (2 TRUCKS) | 20,800 LBS |
| **GENERAL** | | | |
| CAR BODY & EQUIPMENT WEIGHT | 37,700 LBS | STANDING CAPACITY | [81] |
| CAR LIGHT WEIGHT | 58,500 LBS | | |

NOTE 1: LONGITUDINAL SEATING EXCEPT FOUR CROSS SEATS ON EACH SIDE OF AISLE TWO OF WHICH FACED EACH OTHER AND WERE LOCATED AT CENTER OF CAR LENGTH.

CAR 1580 CONVERTED TO SNOW PLOW (1940's).
CAR 1588 TO NYC DEPARTMENT OF SANITATION (1942).
CARS 1580 - 1587, 1589 - 1600 REFURBISHED. ADDED SUBWAY TYPE 3RD RAIL SHOES AND WERE ASSIGNED TO DYRE AVENUE LINE (1941).
NEW HEADLIGHTS INSTALLED (1941).
THIS SERIES OF MOTOR CARS HAD MARKER LIGHT HOUSING WITH CONDUIT INSTALLED BEHIND THE LIGHT INTO SLOPE OF ROOF AND FOR FOLLOWING ELEVATED MOTOR CARS.
ALL CARS WERE SOLD OR SCRAPPED (1940's - 1950's).

CAR NUMBERS: 1613 - 1672
TOTAL: 60
BUILT BY: BARNEY & SMITH, JEWETT (SEE NOTE 2)

DATE: 1910
AVERAGE COST PER CAR: N/A

# 1910 (IRT)

MANHATTAN ELEVATED, MOTOR

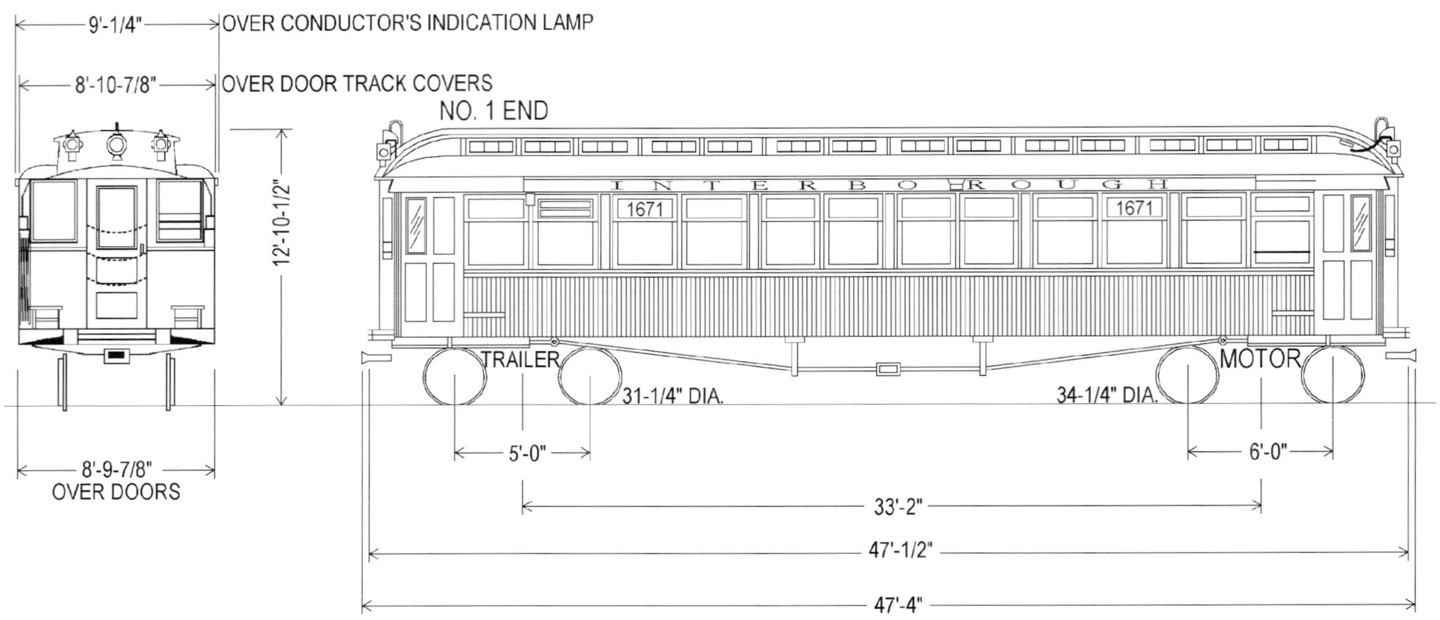

9'-1/4" OVER CONDUCTOR'S INDICATION LAMP
8'-10-7/8" OVER DOOR TRACK COVERS
NO. 1 END
12'-10-1/2"
8'-9-7/8" OVER DOORS
INTERBOROUGH
1671
1671
TRAILER
MOTOR
31-1/4" DIA.
34-1/4" DIA.
5'-0"
6'-0"
33'-2"
47'-1/2"
47'-4"

Copyright © 1997 by NYC Transit

## CAR BODY - EQUIPMENT

| | | | |
|---|---|---|---|
| COUPLER | VAN DORN | SEAT MATERIAL | RATTAN |
| END DOOR | SLIDING | SEAT TYPE | INDIVIDUAL CUSHION |
| HEAD LIGHT | ELECTRIC LAMP | STRUCTURE MATERIAL | WOOD |
| MARKER LIGHT | ELECTRIC LAMP | TAIL LIGHT | KEROSENE SINGLE LENS RED LAMP |
| SEAT ARRANGEMENT & CAPACITY | MANHATTAN [44], SEE NOTE 1 | | |

## TRUCK

| | | | |
|---|---|---|---|
| JOURNAL BEARING | FRICTION | WEIGHT (2 TRUCKS) | 20,800 LBS |

## GENERAL

| | | | |
|---|---|---|---|
| CAR BODY & EQUIPMENT WEIGHT | 42,360 LBS | CAR LOADED WEIGHT | 80,660 LBS |
| CAR LIGHT WEIGHT | 63,160 LBS | STANDING CAPACITY | [77] |

NOTE 1: LONGITUDINAL SEATING EXCEPT FOUR CROSS SEATS ON EACH SIDE OF AISLE TWO OF WHICH FACED EACH OTHER AND WERE LOCATED AT CENTER OF CAR LENGTH.

NOTE 2:

| CARS | CAR BUILDER |
|---|---|
| 1613 - 1652 | BARNEY & SMITH |
| 1653 - 1672 | JEWETT |

STRUCTURE BUILT WITH CORK BETWEEN STEEL POSTS AND PANEL TO REDUCE NOISE.
ALL CARS EXCEPT 1613 - 1633, 1653, 1672 WERE CONVERTED TO MUDC OPERATION (1923 - 1924).
CARS 1660 - 1671 CONVERTED TO LOW VOLTAGE OPERATION (1920's).
ALL CARS WERE SCRAPPED (1940's - 1950's).

CAR NUMBERS: 1673 - 1752
TOTAL: 80 CARS
BUILT BY: ST. LOUIS, WASON (SEE NOTE 2)

DATE: 1909, 1911 (SEE NOTE 2)
AVERAGE COST PER CAR: N/A

# 1909, 1911 (IRT)

MANHATTAN ELEVATED, TRAILER

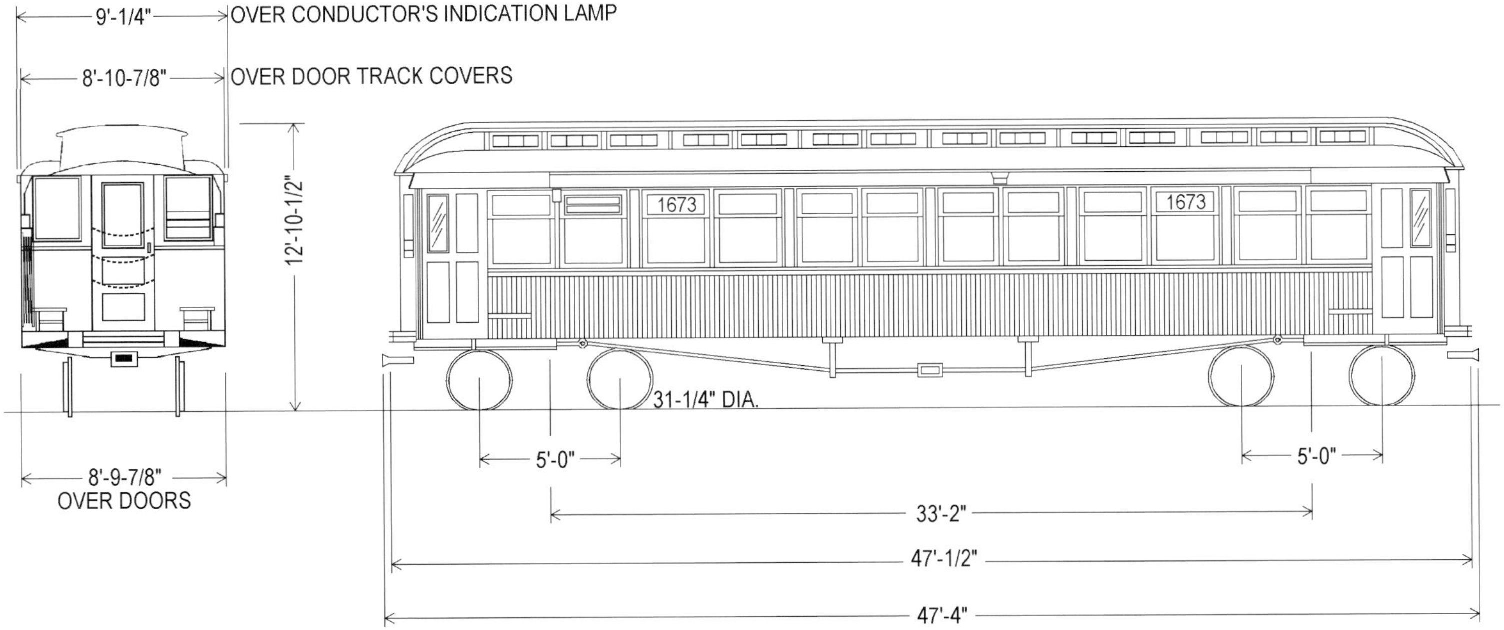

9'-1/4" OVER CONDUCTOR'S INDICATION LAMP
8'-10-7/8" OVER DOOR TRACK COVERS
12'-10-1/2"
8'-9-7/8" OVER DOORS
1673
1673
31-1/4" DIA.
5'-0"
5'-0"
33'-2"
47'-1/2"
47'-4"

Copyright © 1997 by NYC Transit

## CAR BODY - EQUIPMENT

| | | | |
|---|---|---|---|
| COUPLER | VAN DORN | SEAT MATERIAL | RATTAN |
| END DOOR | SLIDING | SEAT TYPE | INDIVIDUAL CUSHION |
| SEAT ARRANGEMENT & CAPACITY | MANHATTAN [48], SEE NOTE 1 | STRUCTURE MATERIAL | WOOD |

## TRUCK

| | | | |
|---|---|---|---|
| AXLE LOAD | 8,575 LBS | WEIGHT (2 TRUCKS) | 16,480 LBS |
| JOURNAL BEARING | FRICTION | | |

## GENERAL

| | | | |
|---|---|---|---|
| CAR BODY & EQUIPMENT WEIGHT | 25,600 LBS | STANDING CAPACITY | [77] |
| CAR LIGHT WEIGHT | 43,000 LBS (AVERAGE) | | |

NOTE 1: LONGITUDINAL SEATING EXCEPT FOUR CROSS SEATS ON EACH SIDE OF AISLE TWO OF WHICH FACED EACH OTHER AND WERE LOCATED AT CENTER OF CAR LENGTH.

NOTE 2:

| CARS | CAR BUILDER | YEAR BUILT |
|---|---|---|
| 1673 - 1692 | WASON | 1909 |
| 1713 - 1752 | | 1911 |
| 1693 - 1712 | ST. LOUIS | 1911 |

STRUCTURE BUILT WITH CORK BETWEEN STEEL POSTS AND PANEL TO REDUCE NOISE.
ALL CARS EXCEPT 1686 AND 1713 CONVERTED TO MUDC OPERATION (1923 - 1924).
CARS 1706 - 1711 AND 1724 - 1752 CONVERTED TO LOW VOLTAGE OPERATION (1920'S). CAR 1718 CONVERTED BACK TO GATE CAR (1927).
ALL CARS WERE SCRAPPED (1940's - 1950's).

CAR NUMBERS: 1753 - 1812
TOTAL: 60 CARS
BUILT BY: CINCINNATI, JEWETT (SEE NOTE 2)

DATE: 1911
AVERAGE COST PER CAR: N/A

# 1911 (IRT)

MANHATTAN ELEVATED, MOTOR

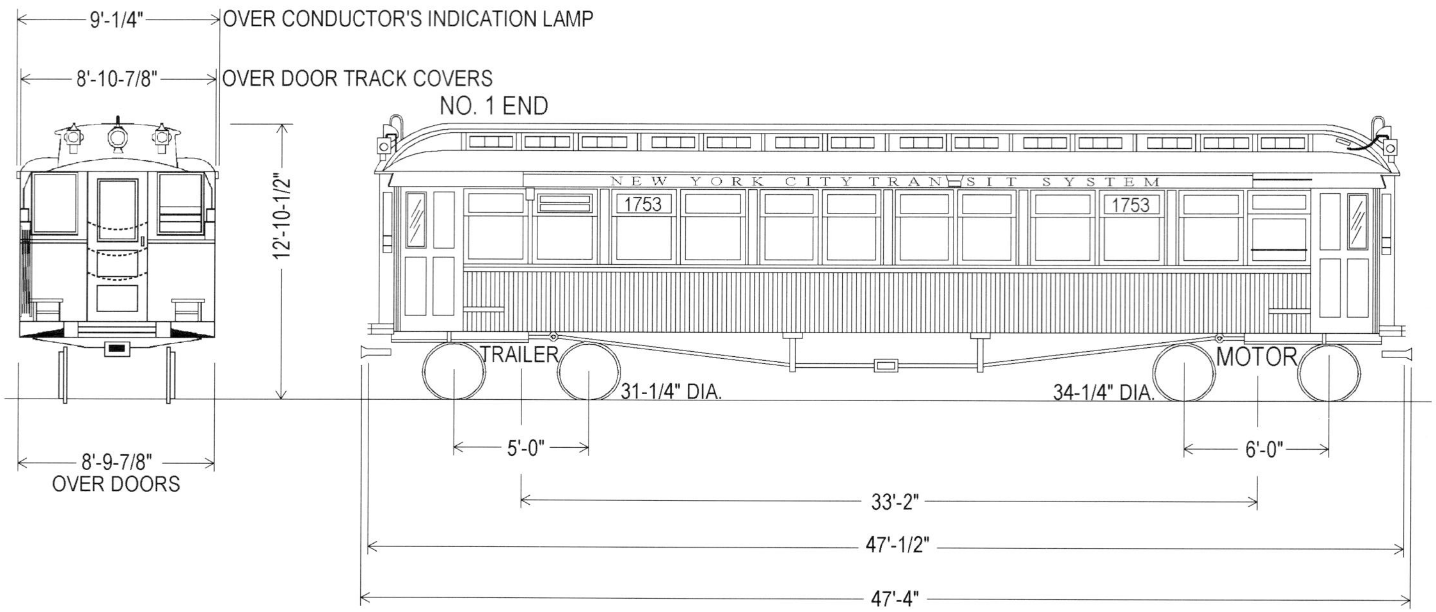

9'-1/4" OVER CONDUCTOR'S INDICATION LAMP
8'-10-7/8" OVER DOOR TRACK COVERS
NO. 1 END
12'-10-1/2"
NEW YORK CITY TRANSIT SYSTEM
1753
TRAILER
MOTOR
31-1/4" DIA.
34-1/4" DIA.
8'-9-7/8" OVER DOORS
5'-0"
6'-0"
33'-2"
47'-1/2"
47'-4"

Copyright © 1997 by NYC Transit

## CAR BODY - EQUIPMENT

| | | | |
|---|---|---|---|
| COUPLER | VAN DORN | SEAT MATERIAL | RATTAN |
| END DOOR | SLIDING | SEAT TYPE | INDIVIDUAL CUSHION |
| HEAD LIGHT | ELECTRIC LAMP | STRUCTURE MATERIAL | WOOD |
| MARKER LIGHT | ELECTRIC LAMP | TAIL LIGHT | KEROSENE SINGLE LENS RED LAMP |
| SEAT ARRANGEMENT & CAPACITY | MANHATTAN [48], SEE NOTE 1 | | |

## TRUCK

| | | | |
|---|---|---|---|
| JOURNAL BEARING | FRICTION | WEIGHT (2 TRUCKS) | 20,800 LBS |

## GENERAL

| | | | |
|---|---|---|---|
| CAR BODY & EQUIPMENT WEIGHT | 42,400 LBS | CAR LOADED WEIGHT | 80,660 LBS |
| CAR LIGHT WEIGHT | 63,200 LBS (AVERAGE) | STANDING CAPACITY | [81] |

NOTE 1: LONGITUDINAL SEATING EXCEPT FOUR CROSS SEATS ON EACH SIDE OF AISLE TWO OF WHICH FACED EACH OTHER AND WERE LOCATED AT CENTER OF CAR LENGTH.

NOTE 2:

| CARS | CAR BUILDERS |
|---|---|
| 1753 - 1792 | JEWETT |
| 1793 - 1812 | CINCINNATI |

STRUCTURE BUILT WITH CORK BETWEEN STEEL POSTS AND PANEL TO REDUCE NOISE.
ALL CARS CONVERTED TO MUDC OPERATION (1923 - 1924).
ALL CARS EXCEPT 1762 AND 1810 WERE CONVERTED TO LOW VOLTAGE OPERATION (1920's).
CAR NUMBERS 1813 - 1999 WERE NOT ASSIGNED.
ALL CARS WERE SCRAPPED (1950's).

CAR NUMBERS. 1600 - 1629 A, B, C
TOTAL: 90 CARS (30 UNITS), SEE NOTE 1
BUILT BY: BRADLEY, BRILL, JEWETT, LACONIA
DATE: 1904 - 1908
AVERAGE COST PER CAR: N/A

REBUILT & OVERHAULED BY: BMT, BOT
REBUILT DATE: 1938 - 1941
OVERHAUL DATE: 1950
AVERAGE COST PER CAR: N/A

# Q (IRT)

9'-1/4"
8'-9-1/2"
5'-9"
8'-8"
NON-OPERATING END
"C" CAR (MOTOR)
SIMILAR TO "A" CAR
"B" CAR (TRAILER)
29'-4-1/2"
DIMENSIONS APPLY TO ALL THREE CARS
19'-10-1/2"
"A" CAR (MOTOR)
DOOR OPENING
3'-7"X6'-2"
32" DIA.
6'-4"
34" DIA.
6'-0"
17"
4'-0"
2"
12'-5"
OPERATING END
33'-2" OR 33'-6" (SEE NOTE)
48'-11" OVER ANTICLIMBERS
49'-3" OVER COUPLERS - COMPLETE UNIT
147'-9" OVER COUPLERS FOR THE CONSIST
56,200 LBS TOTAL (AVG.)
70,450 LBS TOTAL (AVG.)

Copyright © 1997 by NYC Transit

## CAR BODY - EQUIPMENT

| | | | | |
|---|---|---|---|---|
| AIR BRAKE | | WH AML A & C; WH ATL B | HANDHOLD | CANVAS WEBBING |
| AIR COMPRESSOR, A & C | | WH D2F | HEATER, A & C CARS | CONSOLIDATED, 143LL [16], 146X (CAB) [2],32 [2] |
| BRAKE EQUIPMENT | | WABCO L TRIPLE | HEATER, B CAR | CONSOLIDATED, 143LL [12], 146X [2] |
| COUPLER | | VAN DORN | LIGHTING | LAMPS [25], MARKER/VESTIBULE [2] |
| DOOR OPERATOR ENGINE | | CONSOLIDATED, R5082A & R5083A [4] | MASTER CONTROLLER(A & C CARS) | WH NO. 8 |
| DRAFT GEAR | #1 END | SLAB & BARREL | PROPULSION CONTROL, A & C | WH 251-I-3 |
| | #2 END | RAIL & BARREL | SEAT ARRANGEMENT & CAPACITY | CROSS & LONGITUDINAL, A & C [50], B [52] |
| HAND BRAKE | | STAFF, A & C; NONE IN B CAR | SEAT MATERIAL | RATTAN |

## TRUCK

| | | | |
|---|---|---|---|
| MANUFACTURER | INTERBOROUGH R.T. CO. (A & C CARS); PECKHAM (B CAR) | CONTACT SHOE | IRT ELEVATED TYPE |
| TYPE | "I" BEAM SINGLE-MOTOR (A & C CARS); PECKHAM 40 TRAILER (B CAR) | JOURNAL BEARING | 4-3/4X8 PLAIN BEARING (A & C CARS); 4-1/4X8 PLAIN BEARING (B CAR) |
| TRUCK NUMBERS | 1200 - 2173 | TRACTION MOTOR | WH, 333-L3 (120 HP) [1/MOTOR TRUCK] |
| WEIGHT | 13,600 LBS (A & C CARS); 9,200 LBS (B CAR) | | GE, 259 (120 HP) [1/MOTOR TRUCK] |
| BRAKE RIGGING | INSIDE HUNG, SINGLE-SHOE BRAKE | TRIP COCK | WABCO, A-2-B [1/CAR] |

## GENERAL

| | | | |
|---|---|---|---|
| CAR LIGHT WEIGHT | A & C CARS: 70,450 LBS (AVERAGE) | STANDING CAPACITY | [108] |
| | B CAR: 56, 200 LBS (AVERAGE) | | |

NOTE 1: UNIT CONSIST OF THREE CARS (A,B,C). MOTOR CARS ARE A & C, TRAILER CAR IS B.
NOTE 2:

| | | | | | |
|---|---|---|---|---|---|
| BRADLEY | A TYPE | CARS 1604,1621 | BRILL | A TYPE | CARS 1618,1623 |
| DATE: 1904 - 1905 | B TYPE | CARS 1600,1604,1606,1607,1609,1610,1614,1620,1625 | DATE: 1904 - 1905 | B TYPE | CARS 1601,1603,1605,1608,1616-1617,1619,1623,1627,1629 |
| TOTAL: 12 CARS | C TYPE | CAR 1625 | TOTAL: 15 CARS | C TYPE | CARS 1604,1623,1624 |
| JEWETT | A TYPE | CARS 1601-1603,1605-1606,1608-1614,1617,1619-1620,1622 | LACONIA | A TYPE | 1600,1607,1615-1616,1624-1629 |
| DATE: 1908 | C TYPE | CAR 1600-1603,1605-1616,1619-1621 | DATE: 1904, 1905, 1908 | B TYPE | 1602,1611-1613,1615,1618,1621-1622,1624,1626,1628 |
| TOTAL: 35 CARS | | | TOTAL: 28 CARS | C TYPE | 1617-1618,1622,1626-1629 |

ORIGINALLY, THE CARS WERE 'BU' TYPE NUMBERED 1200 AND 1400 SERIES AND WERE OPEN PLATFORM ELEVATED MOTOR CARS (1903 - 1908).
ASSIGNED TO IRT DIVISION (1950 - 1956).
'B' CARS WERE TRAILERS AND RECEIVED POWER FROM 'C' CAR BY JUMPER CABLE.
REPLACED ORIGINAL PECKHAM 40 TRUCKS TO I-BEAM TYPE ON 'A' & 'C' CARS. ADDED SAFETY GATES, MARKER LIGHTS RELOCATED TOWARD CENTER LINE OF CARS FOR THE 3RD AVE ELEVATED SERVICE, IRT DIVISION (1949 - 1956).

CAR NUMBERS: 2000 - 2159, 3000 - 3339
TOTAL: 500 CARS
BUILT BY: JEWETT, J. STEPHENSON, ST. LOUIS, WASON (SEE NOTE 6)

DATE: 1903, 1904 (SEE NOTE 6)
AVERAGE COST PER CAR: N/A

# 1903, 1904 (IRT)

COMPOSITE, LV MOTOR

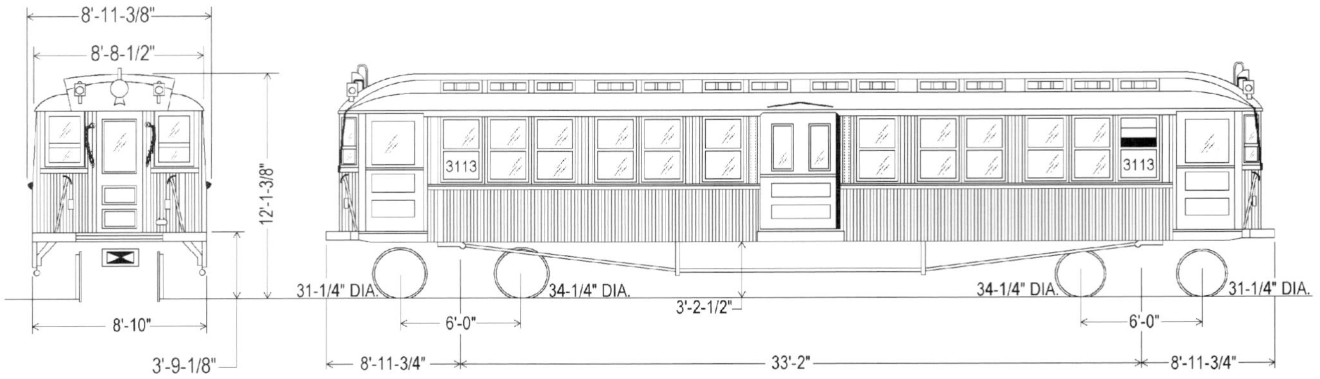

Copyright © 1997 by NYC Transit

## CAR BODY-EQUIPMENT

| | | | |
|---|---|---|---|
| COUPLER | J | SEAT ARRANGEMENT & CAPACITY | MANHATTAN, SEE NOTE 1 |
| END DOOR | SLIDING | SEAT MATERIAL | RATTAN |
| FAN | REMOVED 1915-1916 | SEAT TYPE | INDIVIDUAL CUSHION |
| HEADLIGHT | ELECTRIC LAMP | STRUCTURE MATERIAL | WOOD & STEEL |
| MARKER LIGHT | ELECTRIC LAMP | TAIL LIGHT | KEROSENE LANTERN, SEE NOTE 4 |

## TRUCK

| | | | | |
|---|---|---|---|---|
| AXLE LOAD | (TOWARD CAR CENTER) | 19,890 LBS | JOURNAL BEARING | FRICTION |
| | (TOWARD CAR END) | 17,004 LBS | WEIGHT (2 TRUCKS) | 27,200 LBS |

## GENERAL

| | | | |
|---|---|---|---|
| CAR BODY & EQUIPMENT WEIGHT | 46,588 LBS | CAR LOADED WEIGHT | 96,468 LBS |
| CAR LIGHT WEIGHT | 73,788 LBS, SEE NOTE 5 | STANDING CAPACITY | [118] |

NOTE 1: LONGITUDINAL SEATING EXCEPT FOUR SEATS ACROSS ON EACH SIDE OF AISLE TWO OF WHICH FACED EACH OTHER AND WERE LOCATED AT CENTER OF CAR LENGTH AND WERE CONVERTED TO LONGITUDINAL SEATS WHEN CENTER DOORS WERE ADDED, SEATING [44] (1916). ORIGINAL SEATING [52] (1912).

NOTE 2: ORIGINALLY HIGH VOLTAGE TRAILERS PRIOR TO CONVERSION (1915-1916).

NOTE 3: ORIGINALLY HIGH VOLTAGE MOTORS; SOME WERE CONVERTED TO HIGH VOLTAGE TRAILERS (1905 - 1910), PRIOR TO CONVERSION (1915-1916).

NOTE 4: RED LENS FOR TAIL, WHITE LENS FOR RUNNING.

NOTE 5: ORIGINAL WEIGHT OF CARS BEFORE CONVERSION: MOTOR 81,600 LBS; TRAILER 60,000 LBS (APPROX.).

NOTE 6:

| CARS | CAR BUILDERS | YEAR BUILT |
|---|---|---|
| 2000 - 2059, SEE NOTE 2 | JEWETT | 1903 - 04 |
| 3000 - 3039, SEE NOTE 3 | | |
| 3040 - 3139, SEE NOTE 3 | J. STEPHENSON | 1903 - 04 |
| 2060 - 2119, SEE NOTE 2 | ST. LOUIS | 1903 - 04 |
| 3140 - 3279, SEE NOTE 3 | | |
| 2120 - 2159, SEE NOTE 2 | WASON | 1903 |
| 3280 - 3339, SEE NOTE 3 | | 1903 - 04 |

ORIGINALLY BUILT AS SUBWAY CARS AND WERE CONVERTED FOR ELEVATED OPERATION (1915-1916). SOME EQUIPMENT REMOVED DURING CONVERSION WAS INSTALLED ON FLIVVER CARS AND HV TRAILERS (1915-1916).
CAR 3282 WAS A TEST CAR PRIOR TO PASSENGER SERVICE. TRUCK: ONE AXLE HAD 31-1/4" WHEELS, THE OTHER AXLE HAD 34-1/4" WHEELS. 120 TRUCKS WERE TRANSFERRED TO BMT 'Q' TYPES.
ORIGINAL DOOR CONTROL LEVERS WERE REPLACED WITH THE GIBBS CAR REVISED TYPE AND INSTALLATION WAS COMPLETED 1904/1905.
CARS ORIGINALLY HAD CAR NUMBERS PAINTED ON LETTER BOARD NEAR END DOORS AND WERE LATER PAINTED ON GLASS PANELS (BEFORE 1910), VAN DORN COUPLERS AND NO ANTICLIMBERS.
CARS HAD WOOD LATTICE FLOORS, LEATHER HAND STRAPS. CENTER DOOR PANELS HAD EITHER ONE, TWO, OR THREE PANEL CONFIGURATIONS.
CAR NUMBERS 2160-2999 WERE NOT ASSIGNED. CAR 2135 CONVERTED TO PUMP CAR NO. "03" (1907). CAR 3260 WAS CONVERTED TO AN INSTRUCTION CAR AND LATER RETURNED TO PASSENGER SERVICE (CA. 1917).
VESTIBULE AT CAR ENDS. ALL CARS WERE SOLD OR SCRAPPED (1938 TO EARLY 1950'S).

CAR NUMBERS: 3340, 3341
TOTAL: 2 CARS
BUILT BY: WASON

DATE: 1902
AVERAGE COST PER CAR: N/A

# 1902 (IRT)

COMPOSITE SAMPLES, MOTOR

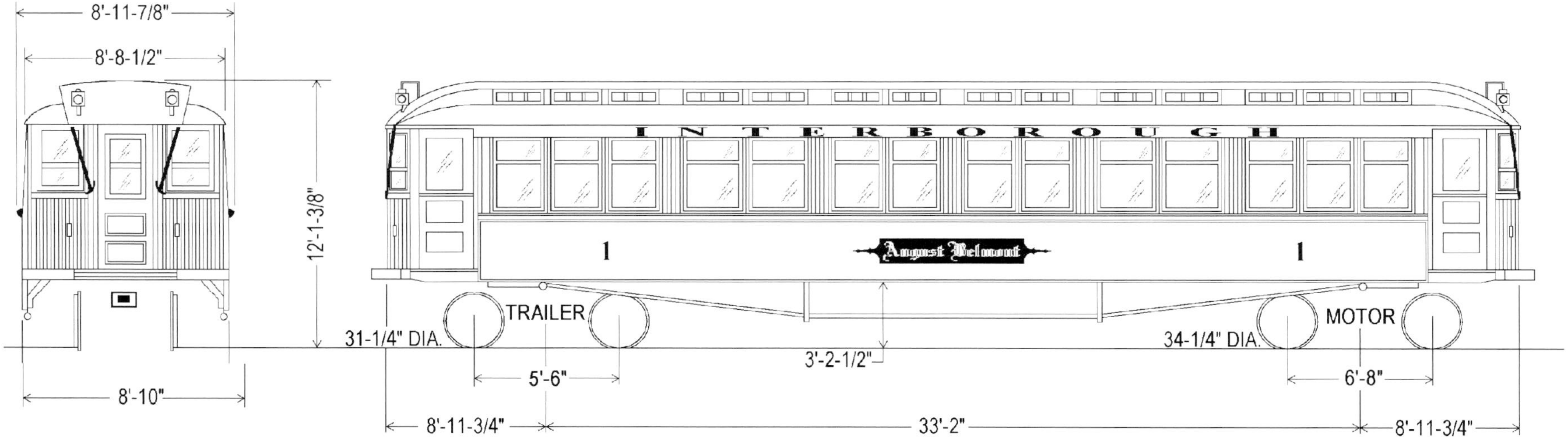

Copyright © 1997 by NYC Transit

| CAR BODY - EQUIPMENT | | | |
|---|---|---|---|
| COUPLER | SEE NOTE 1 | SEAT MATERIAL | RATTAN |
| END DOOR | SWING | SEAT TYPE | INDIVIDUAL CUSHION |
| MARKER LIGHT | ELECTRIC LAMP | STRUCTURE MATERIAL | WOOD & STEEL |
| SEAT ARRANGEMENT & CAPACITY | N/A | TAIL LIGHT | KEROSENE LANTERN, SEE NOTE 2 |
| **TRUCK** | | | |
| JOURNAL BEARING | FRICTION | WEIGHT (2 TRUCKS) | 20,800 LBS |
| **GENERAL** | | | |
| CAR BODY & EQUIPMENT WEIGHT | 53,860 LBS | CAR LIGHT WEIGHT | 74,660 LBS |

NOTE 1: ORIGINALLY VAN DORN. CAR 3340 REMAINED VAN DORN; CAR 3341 CHANGED TO J (APPROX. 1917).
NOTE 2: RED LENS FOR TAIL, WHITE LENS FOR RUNNING.

INITIALLY OPERATED AS TEST PASSENGER CARS.
CARS ORIGINALLY HAD NO ANTICLIMBERS. MARKER LIGHTS REPOSITIONED ON CAR 3341 (1903).
BOTH CARS ORIGINALLY HAD HEADLIGHTS (REMOVED 1903).
CAR 3340 WAS NUMBERED "NO. 1" AND NAMED AUGUST BELMONT. WAS CONVERTED TO INSTRUCTION CAR (BY 1908). HAD NO CENTER DOOR. CAR 3340 WAS SCRAPPED.
CAR 3341 WAS NUMBERED "NO. 2" AND NAMED JOHN B. MCDONALD. WAS CONVERTED TO TRAILER PAY CAR (BY 1908). END DOOR CONVERTED TO SLIDING AT A LATER DATE. USED AS TIME CLOCK CAR AND ALSO CANTEEN CAR (APPROX. 1917). RECONVERTED TO MOTOR PASSENGER CAR WITH CENTER DOOR ADDED. CAR 3341 WAS SCRAPPED.

CAR NUMBER: 3342
TOTAL: 1 CAR
BUILT BY: PENNSYLVANIA RAILROAD (ALTOONA)

DATE: 1903
AVERAGE COST PER CAR: N/A

# 1903 (IRT)

STEEL CAR SAMPLE, MOTOR

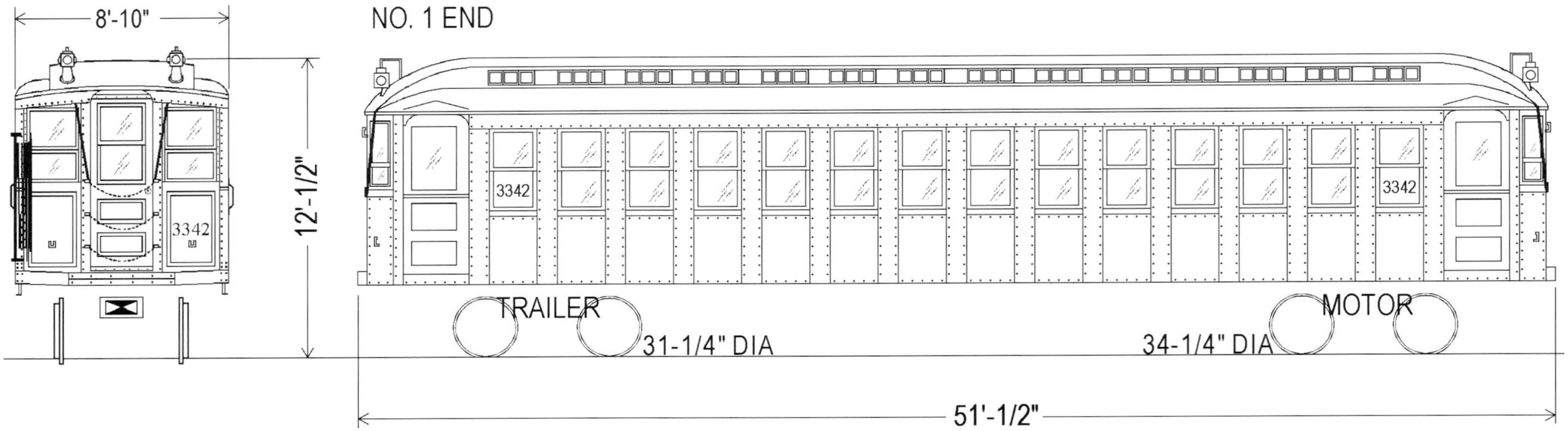

8'-10"

NO. 1 END

12'-1/2"

3342

TRAILER

31-1/4" DIA

MOTOR

34-1/4" DIA

51'-1/2"

Copyright © 1997 by NYC Transit

| CAR BODY - EQUIPMENT | | | |
|---|---|---|---|
| COUPLER | F | SEAT TYPE | INDIVIDUAL CUSHION |
| MARKER LIGHT | ELECTRIC LAMP | STRUCTURE MATERIAL | STEEL |
| SEAT ARRANGEMENT & CAPACITY | MANHATTAN [54], SEE NOTE 1 | TAIL LIGHT | KEROSENE LANTERN, SEE NOTE 2 |
| SEAT MATERIAL | RATTAN | | |
| **TRUCK** | | | |
| JOURNAL BEARING | FRICTION | | |
| **GENERAL** | | | |
| CAR LIGHT WEIGHT | 89,960 LBS | STANDING CAPACITY | [152] |

NOTE 1: LONGITUDINAL SEATING EXCEPT FOUR CROSS SEATS ON EACH SIDE OF AISLE TWO OF WHICH SEATS FACED EACH OTHER AND WERE LOCATED AT CENTER OF CAR LENGTH.
NOTE 2: RED LENS FOR TAIL, WHITE LENS FOR RUNNING.

FIRST ALL STEEL PASSENGER CAR.
CAR ORIGINALLY HAD VAN DORN COUPLERS AND NO ANTICLIMBERS.
CONVERTED TO PAY CAR BY 1908.
CAR SCRAPPED 1956.

CAR NUMBER: 3344
TOTAL: 1 CAR
BUILT BY: WASON

DATE: 1904
AVERAGE COST PER CAR: $11,429

# 1904 (IRT)

MINEOLA (DIRECTOR'S CAR), MOTOR

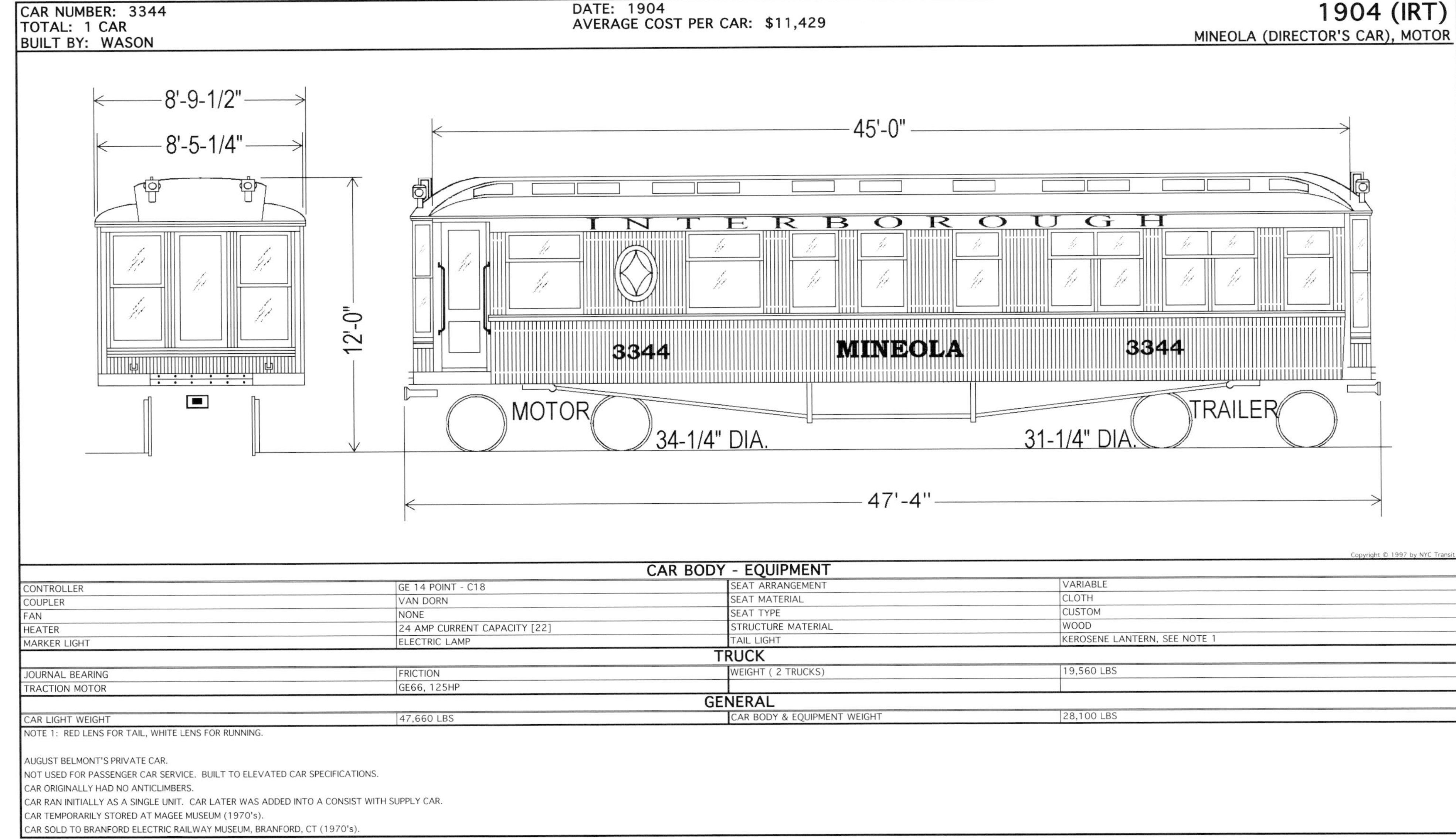


Copyright © 1997 by NYC Transit

| CAR BODY - EQUIPMENT | | | |
|---|---|---|---|
| CONTROLLER | GE 14 POINT - C18 | SEAT ARRANGEMENT | VARIABLE |
| COUPLER | VAN DORN | SEAT MATERIAL | CLOTH |
| FAN | NONE | SEAT TYPE | CUSTOM |
| HEATER | 24 AMP CURRENT CAPACITY [22] | STRUCTURE MATERIAL | WOOD |
| MARKER LIGHT | ELECTRIC LAMP | TAIL LIGHT | KEROSENE LANTERN, SEE NOTE 1 |
| **TRUCK** | | | |
| JOURNAL BEARING | FRICTION | WEIGHT ( 2 TRUCKS) | 19,560 LBS |
| TRACTION MOTOR | GE66, 125HP | | |
| **GENERAL** | | | |
| CAR LIGHT WEIGHT | 47,660 LBS | CAR BODY & EQUIPMENT WEIGHT | 28,100 LBS |

NOTE 1: RED LENS FOR TAIL, WHITE LENS FOR RUNNING.

AUGUST BELMONT'S PRIVATE CAR.
NOT USED FOR PASSENGER CAR SERVICE. BUILT TO ELEVATED CAR SPECIFICATIONS.
CAR ORIGINALLY HAD NO ANTICLIMBERS.
CAR RAN INITIALLY AS A SINGLE UNIT. CAR LATER WAS ADDED INTO A CONSIST WITH SUPPLY CAR.
CAR TEMPORARILY STORED AT MAGEE MUSEUM (1970's).
CAR SOLD TO BRANFORD ELECTRIC RAILWAY MUSEUM, BRANFORD, CT (1970's).

CAR NUMBERS: 3350 - 3513, 3515, 3516
TOTAL: 166 CARS
BUILT BY: AMERICAN CAR FOUNDRY

DATE: 1904
AVERAGE COST PER CAR: N/A

# 1904 (IRT)

MODIFIED GIBBS, HV MOTOR

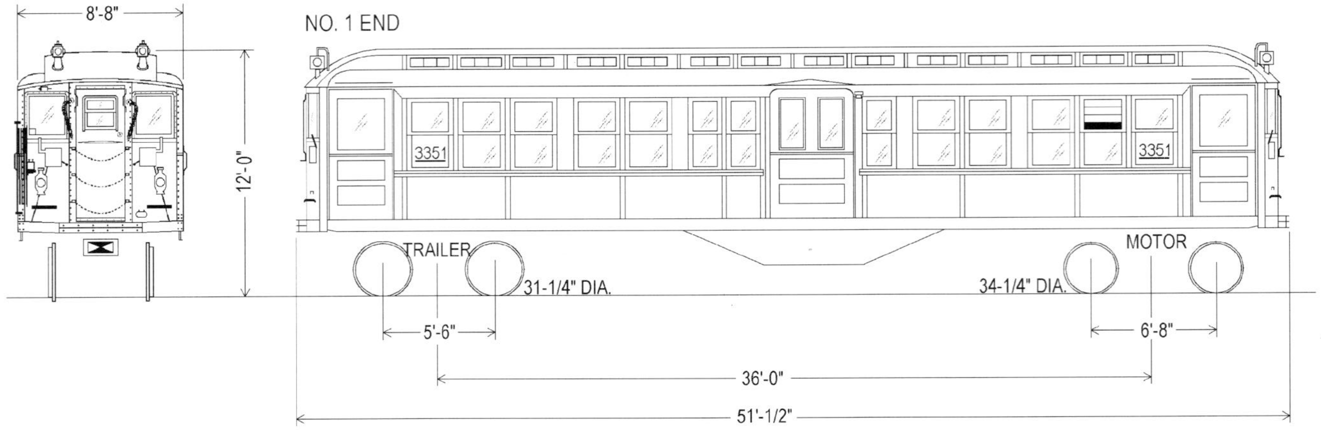

8'-8"
12'-0"
NO. 1 END
3351
3351
TRAILER
31-1/4" DIA.
MOTOR
34-1/4" DIA.
5'-6"
6'-8"
36'-0"
51'-1/2"

Copyright © 1997 by NYC Transit

## CAR BODY - EQUIPMENT

| | | | |
|---|---|---|---|
| AIR COMPRESSOR | WH D2-EG/D2-EY/D2F | PROPULSION CONTROL | GE "K" |
| BRAKE TYPE | WABCO, AMRE/ME21 | SEAT ARRANGEMENT & CAPACITY | LONGITUDINAL [44], SEE NOTE 1 |
| COUPLER | F | SEAT MATERIAL | RATTAN |
| END DOOR | SWING | SEAT TYPE | INDIVIDUAL CUSHION |
| FAN | PADDLE [4] | STRUCTURE MATERIAL | STEEL |
| MARKER LIGHT | ELECTRIC LAMP | TAIL LIGHT | KEROSENE LANTERN, SEE NOTE 2 |
| MASTER CONTROLLER | C18 | | |

## TRUCK

| | | | | |
|---|---|---|---|---|
| MANUFACTURER | MOTOR | BALDWIN | JOURNAL BEARING | FRICTION |
| | TRAILER | BALDWIN | TRACTION MOTOR | GE 69 (200 HP); GE 212 (200 HP) [2/MOTOR TRUCK] |
| AXLE LOAD | # 1 END - TRAILER | 17,485 LBS | | WH 86 (200 HP); WH 300 (200 HP) [2/MOTOR TRUCK] |
| | # 2 END - MOTOR | 25,405 LBS | WEIGHT ( 2 TRUCKS) | 35,920 LBS |

## GENERAL

| | | | |
|---|---|---|---|
| CAR BODY & EQUIPMENT WEIGHT | 49,860 LBS | CAR LOADED WEIGHT | 113,220 LBS |
| CAR LIGHT WEIGHT | 85,780 LBS | STANDING CAPACITY | [152] |

NOTE 1: DROP SEATS AT CENTER DOORS WERE REMOVED (1942 - 1948).
NOTE 2: RED LENS FOR TAIL, WHITE LENS FOR RUNNING.

CARS HAD EXPOSED BEAMED CEILING.
CARS HAD CAR BODY BOLSTERS REINFORCED (BY 1916).
CARS ORIGINALLY HAD FOUR CROSS SEATS ON EACH SIDE OF AISLE WHICH WERE REMOVED WHEN CENTER DOORS WERE ADDED (BY 1912). CENTER DOOR PANELS HAD EITHER ONE, TWO OR THREE PANEL CONFIGURATIONS AND WERE METAL OR WOOD.
CARS 3376, 3386 & 3447 ASSIGNED TO WORK CARS (1924) AND WERE KNOWN AS UNMODIFIED, REGULAR OR STANDARD. CAR 3350 CONVERTED TO PAY CAR.
BUILT WITH WOOD LATTICE FLOORS, LEATHER HAND STRAPS, VAN DORN COUPLERS, AND NO ANTICLIMBER. CAR NUMBERS WERE PAINTED ON LETTER BOARD NEAR END DOOR AND WERE LATER PAINTED ON GLASS PANELS (BEFORE 1910).
DAMAGED/WORN SEATS WERE REPLACED WITH SARAN (RATTAN SIMULATED) HAVING CHAIN STRIPES OF GREEN OR BLUE ON A BACKGROUND OF YELLOW (STARTING 1942). ORIGINAL SEATING CAPACITY WAS 52.
ORIGINAL DOOR LEVERS ON EARLY CARS REPLACED WITH GIBBS REVISED TYPE (1904).
CARS WERE MODIFIED WITH END DOORS THAT WERE MANUALLY OPERATED & EQUIPPED WITH RUBBER CUSHION.
CENTER DOORS WERE EQUIPPED WITH DOOR OPERATOR ENGINE AND COLLAPSIBLE STEEL SHOE.
ALL GIBBS EXCEPT WORK CARS HAD WHITE LINES UNDER THE NUMBER GLASSES (1936).
CAR 3352 SOLD TO SEASHORE ELECTRIC RAILWAY, (MUSEUM), KENNEBUNKPORT, ME (1958).
VESTIBULE AT ENDS OF CAR.
ALL OTHER CARS WERE SCRAPPED (BY 1959).

CAR NUMBERS: 3407, 3419, 3421, 3425, 3427, 3429, 3435, 3445
TOTAL: 8 CARS
BUILT BY: AMERICAN CAR FOUNDRY
DATE: 1904
AVERAGE COST PER CAR: N/A

REBUILT BY: IRT
DATE: 1908
REBUILT COST PER CAR: N/A

# 1904 (IRT)

MODIFIED GIBBS, HV MOTOR

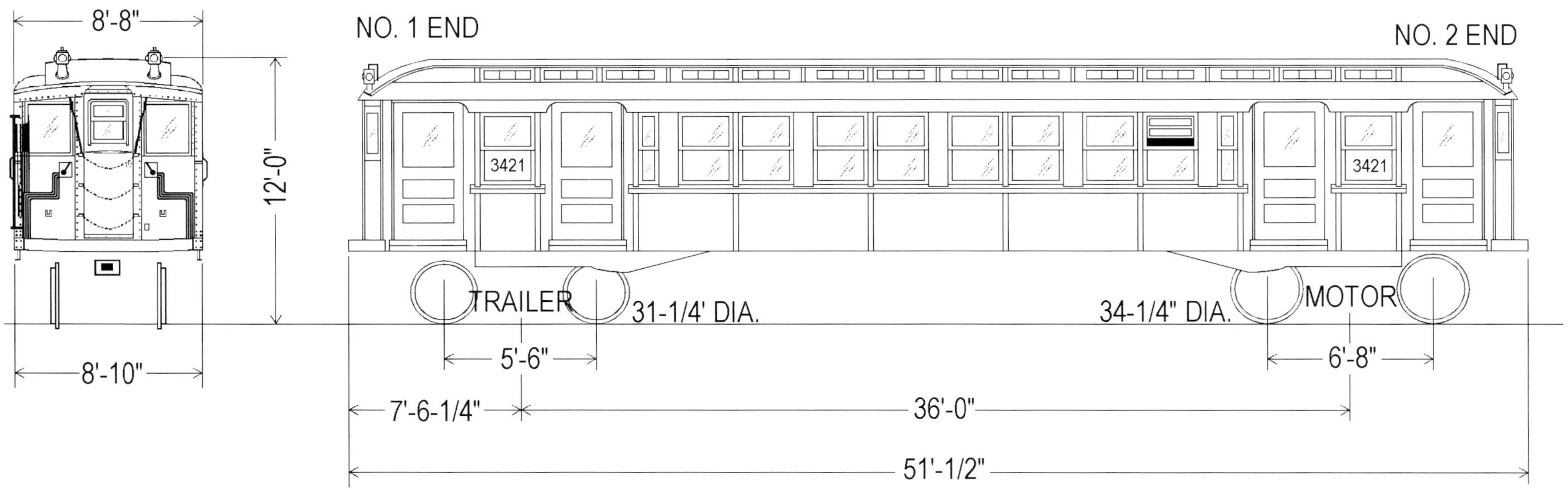

8'-8"
12'-0"
8'-10"
NO. 1 END
NO. 2 END
3421
3421
TRAILER
31-1/4' DIA.
34-1/4" DIA.
MOTOR
5'-6"
6'-8"
7'-6-1/4"
36'-0"
51'-1/2"

Copyright © 1997 by NYC Transit

| CAR BODY - EQUIPMENT | | | | |
|---|---|---|---|---|
| COUPLER | | VAN DORN | SEAT MATERIAL | RATTAN |
| END DOOR | | SLIDING | SEAT TYPE | INDIVIDUAL CUSHION |
| MARKER LIGHT | | ELECTRIC LAMP | STRUCTURE MATERIAL | STEEL |
| SEAT ARRANGEMENT & CAPACITY | | MANHATTAN [44], SEE NOTE 1 | TAIL LIGHT | KEROSENE LANTERN, SEE NOTE 2 |
| **TRUCK** | | | | |
| AXLE LOAD | # 1 END - TRAILER | 17,485 LBS | JOURNAL BEARING | FRICTION |
| | # 2 END - MOTOR | 25,405 LBS | WEIGHT ( 2 TRUCKS) | 35,920 LBS |
| **GENERAL** | | | | |
| CAR BODY & EQUIPMENT WEIGHT | | 49,860 LBS (APPROX.) | CAR LOADED WEIGHT (APPROX.) | 113,220 LBS |
| CAR LIGHT WEIGHT | | 85,780 LBS (APPROX.) | STANDING CAPACITY | [152] |

NOTE 1: LONGITUDINAL EXCEPT FOUR CROSS SEATS ON EACH SIDE OF AISLE AT CENTER OF CAR.
NOTE 2: RED LENS FOR TAIL, WHITE LENS FOR RUNNING.

CARS HAD EXPOSED BEAMED CEILING.
1908 GIBBS MODIFICATION. DOOR CONTROLS BY TWO COMPANIES (4 CARS EACH).
CONVERTED TO THE APPEARANCE OF 1904 MODIFIED GIBBS (BY 1910 - SEE PAGE 366).
ALL CARS WERE SCRAPPED

CAR NUMBERS: 3514, 3517 - 3649
TOTAL: 134 CARS
BUILT BY: AMERICAN CAR FOUNDRY

DATE: 1904, 1905
AVERAGE COST PER CAR: N/A

# 1904, 1905 (IRT)

MUDC GIBBS, HV MOTOR

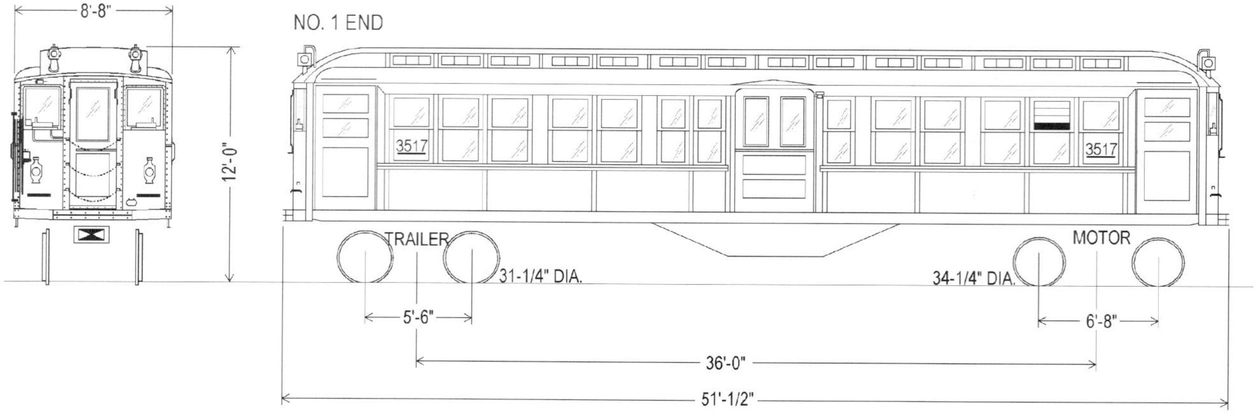

8'-8"
12'-0"
NO. 1 END
3517
TRAILER
MOTOR
31-1/4" DIA.
34-1/4" DIA.
5'-6"
6'-8"
36'-0"
51'-1/2"

Copyright © 1997 by NYC Transit

## CAR BODY - EQUIPMENT

| | | | |
|---|---|---|---|
| AIR COMPRESSOR | WH D2-EG/D2-EY/D2F | PROPULSION CONTROL | GE "K" |
| BRAKE TYPE | WABCO, AMRE/ME21 | SEAT ARRANGEMENT & CAPACITY | LONGITUDINAL [44], SEE NOTE 2 |
| COUPLER | F | SEAT MATERIAL | RATTAN |
| END DOOR | SLIDING | SEAT TYPE | INDIVIDUAL CUSHION |
| FAN | PADDLE [4] | STRUCTURE MATERIAL | STEEL |
| MARKER LIGHT | ELECTRIC LAMP | TAIL LIGHT | KEROSENE LANTERN, SEE NOTE 3 |
| MASTER CONTROLLER | C18 | | |

## TRUCK

| | | | | |
|---|---|---|---|---|
| MANUFACTURER | MOTOR | BALDWIN | JOURNAL BEARING | FRICTION |
| | TRAILER | BALDWIN | TRACTION MOTOR | GE 69 (200 HP); GE 212 (200 HP) [2/MOTOR TRUCK] |
| AXLE LOAD | # 1 END - TRAILER | 18,325 LBS | | WH 86 (200 HP); WH 300 (200 HP) [2/MOTOR TRUCK] |
| | # 2 END - MOTOR | 26,245 LBS | WEIGHT ( 2 TRUCKS) | 35,920 LBS |

## GENERAL

| | | | |
|---|---|---|---|
| CAR BODY & EQUIPMENT WEIGHT | 53,220 LBS | CAR LOADED WEIGHT | 116,580 LBS |
| CAR LIGHT WEIGHT | 89,140 LBS, SEE NOTE 1 | STANDING CAPACITY | [152] |

NOTE 1: ORIGINAL WEIGHT 85,780 LBS.
NOTE 2: DROP SEATS AT CENTER DOORS WERE REMOVED (1942 - 1948).
NOTE 3: RED LENS FOR TAIL, WHITE LENS FOR RUNNING.

CARS ORIGINALLY HAD FOUR CROSS SEATS, ON EACH SIDE OF AISLE AT CENTER OF CAR WHICH WERE REMOVED (BY 1912).
CARS ORIGINALLY HAD CAR NUMBERS PAINTED ON LETTER BOARD NEAR END DOORS WHICH WERE LATER PAINTED ON GLASS PANELS (BEFORE 1910).
CARS HAD EXPOSED BEAMED CEILING.
CARS HAD CAR BODY BOLSTERS REINFORCED (BY 1916).
ADDED CENTER DOORS (BY 7/19/1912) AND NEW END (STORM) DOORS WHEN CONVERTED TO MUDC OPERATION (1936). CENTER DOOR PANELS HAD EITHER ONE, TWO OR THREE PANEL CONFIGURATIONS.
CENTER DOORS WERE EITHER METAL OR WOOD.
BUILT WITH WOOD LATTICE FLOORS AND LEATHER HAND STRAPS, VAN DORN COUPLERS AND NO ANTICLIMBER. ORIGINAL SEATING CAPACITY WAS 52.
CARS 3514, 3524, 3567, 3591, & 3638 ASSIGNED TO WORK MOTORS (1924) AND WERE KNOWN AS UNMODIFIED, REGULAR OR STANDARD.
ALL GIBBS EXCEPT WORK CARS HAD WHITE LINES UNDER THE NUMBER GLASSES (1936).
CAR 3514 WAS RETURNED FROM WORK TO PASSENGER SERVICE AND CONVERTED TO MUDC OPERATION USING THE EQUIPMENT FROM CAR 3583 WHICH WAS INVOLVED IN A COLLISION (1942).
DAMAGED/WORN SEATS WERE REPLACED WITH SARAN (RATTAN SIMULATED) HAVING CHAIN STRIPE OF GREEN OR BLUE ON A BACKGROUND OF YELLOW (INSTALLATION STARTED IN 1942).
VESTIBULE AT CAR ENDS.
ALL CARS WERE SCRAPPED (BY 1959).

CAR NUMBERS: 3650 - 3699
TOTAL: 50 CARS
BUILT BY: AMERICAN CAR FOUNDRY

DATE: 1907, 1908
AVERAGE COST PER CAR: N/A

# 1907, 1908 (IRT)

MODIFIED DECK ROOF, HV MOTOR

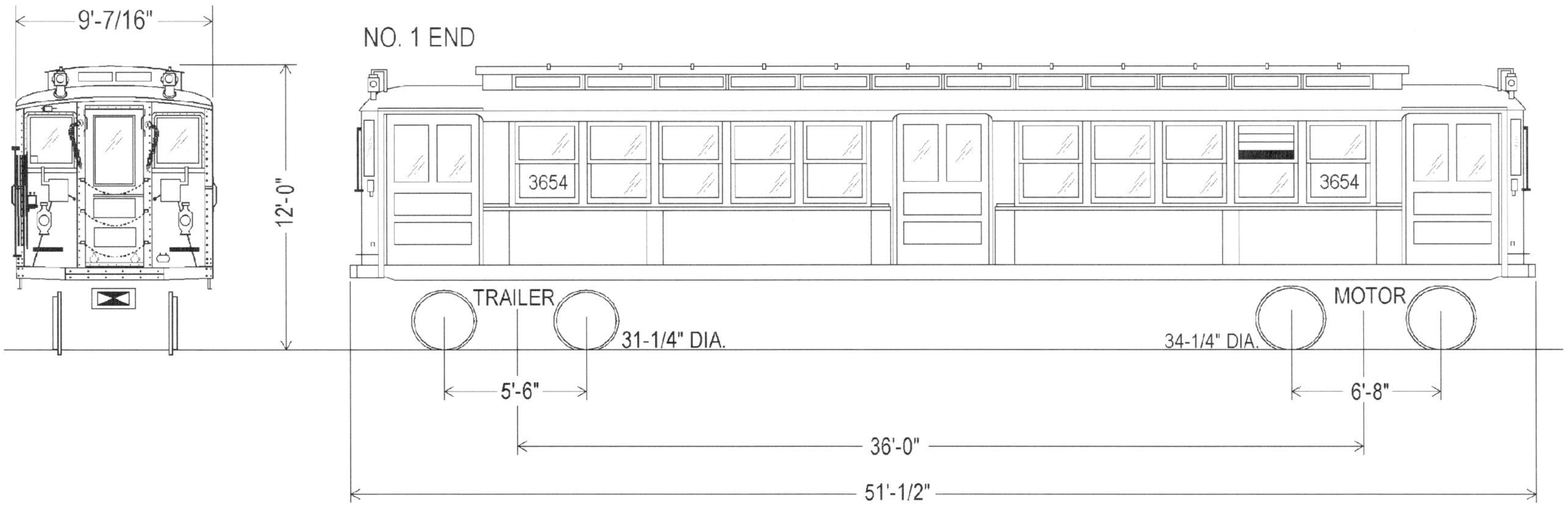

9'-7/16"
12'-0"
NO. 1 END
3654
3654
TRAILER
MOTOR
31-1/4" DIA.
34-1/4" DIA.
5'-6"
6'-8"
36'-0"
51'-1/2"

Copyright © 1997 by NYC Transit

## CAR BODY - EQUIPMENT

| | | | |
|---|---|---|---|
| AIR COMPRESSOR | WH D2-EG/D2-EY/D2F | PROPULSION CONTROL | GE "B" |
| BRAKE TYPE | WABCO, AMRE/ME21 | SEAT ARRANGEMENT & CAPACITY | LONGITUDINAL [44], SEE NOTE 1 |
| COUPLER | F | SEAT MATERIAL | RATTAN |
| FAN | PADDLE [4] | SEAT TYPE | INDIVIDUAL CUSHION, SEE NOTE 2 |
| MARKER LIGHT | ELECTRIC LAMP | STUCTURE MATERIAL | STEEL |
| MASTER CONTROLLER | C18 | TAIL LIGHT | KEROSENE LANTERN, SEE NOTE 3 |

## TRUCK

| | | | | |
|---|---|---|---|---|
| MANUFACTURER | MOTOR | BALDWIN | JOURNAL BEARING | FRICTION |
| | TRAILER | BALDWIN | TRACTION MOTOR | GE 69 (200 HP); GE 212 (200 HP) [2/MOTOR TRUCK] |
| AXLE LOAD | #1 END - TRAILER | 16,985 LBS | | WH 86 (200 HP); WH 300 (200 HP) [2/MOTOR TRUCK] |
| | #2 END - MOTOR | 24,905 LBS | WEIGHT ( 2 TRUCKS) | 35,920 LBS |

## GENERAL

| | | | |
|---|---|---|---|
| CAR BODY & EQUIPMENT WEIGHT | 47,860 LBS | CAR LOADED WEIGHT | 111,220 LBS |
| CAR LIGHT WEIGHT | 83,780 LBS | STANDING CAPACITY | [152] |

NOTE 1: DROP SEATS AT CENTER DOORS WERE REMOVED (1942 - 1948).
NOTE 2: LONGITUDINAL SEATING EXCEPT TWO SEATS ACROSS ON EACH SIDE OF AISLE TWO OF WHICH FACED EACH OTHER AND WERE LOCATED AT CENTER OF CAR LENGTH. CONVERTED WHEN CENTER DOORS WERE ADDED (1910).
NOTE 3: RED LENS FOR TAIL, WHITE LENS FOR RUNNING.

BUILT WITH WOOD LATTICE FLOORS, LEATHER HAND STRAPS. FIRST CARS DELIVERED WITH CAR NUMBERS PAINTED ON GLASS PANELS. CAR WAS DESIGNED FOR ADDITION OF CENTER DOOR WITHOUT FISHBELLY.
CARS HAD EXPOSED BEAMED CEILING.
CARS (3650 - 3699) ORIGINALLY HAD VAN DORN COUPLERS AND ANTICLIMBERS. SEVERAL CARS WERE CONVERTED TO WORK MOTORS. ORIGINAL SEATING CAPACITY WAS 48.
CAR 3662 SOLD TO BRANFORD ELECTRIC RAILWAY MUSEUM (1958).
CARS (3650 - 3653) WERE EQUIPPED WITH BATTERY OPERATED TAIL LIGHT/RUNNING LIGHT FOR THE TIMES SQUARE SHUTTLE AND HAD WHITE LINES UNDER THE NUMBER GLASSES (1940's).
CENTER DOORS WERE EQUIPPED WITH DOOR OPERATOR ENGINE AND COLLAPSIBLE STEEL SHOE. END DOORS MANUALLY OPERATED & EQUIPPED WITH RUBBER CUSHION.
VESTIBULE AT CAR ENDS.
ALL OTHER CARS WERE SCRAPPED (BY 1961).

CAR NUMBERS: 3700 - 3756, 3815, 3915
TOTAL: 59 CARS
BUILT BY: AMERICAN CAR FOUNDRY, PRESSED STEEL, STANDARD STEEL (SEE NOTE 3)

DATE: 1909
AVERAGE COST PER CAR: N/A

# 1909 (IRT)

MODIFIED HEDLEY, HV MOTOR

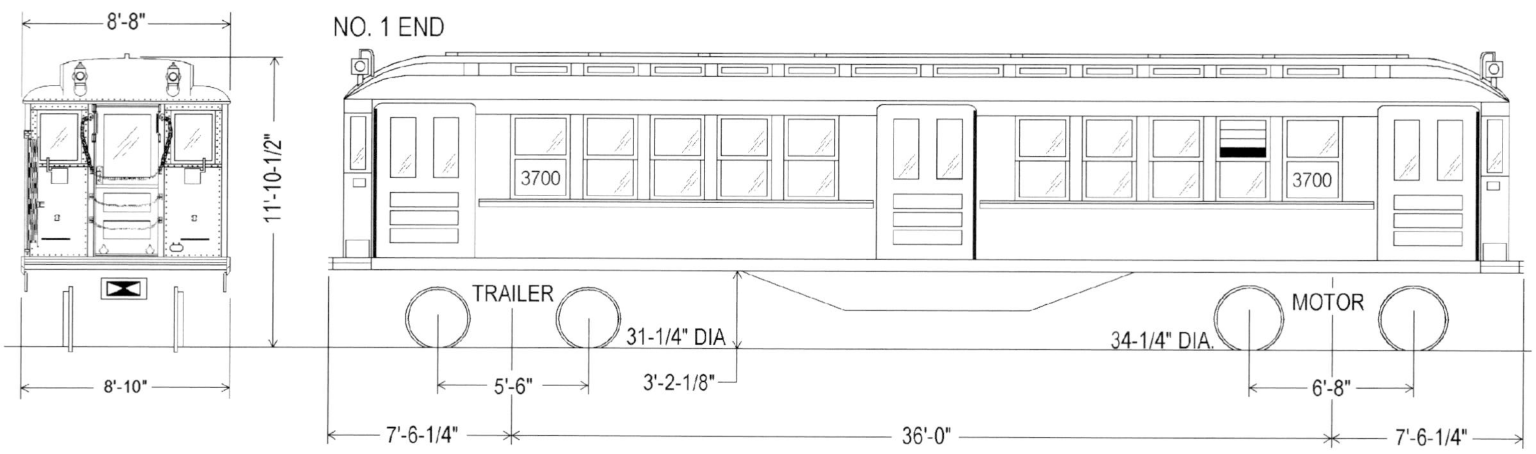

8'-8"
11'-10-1/2"
8'-10"
NO. 1 END
3700
3700
TRAILER
MOTOR
31-1/4" DIA
34-1/4" DIA.
5'-6"
3'-2-1/8"
6'-8"
7'-6-1/4"
36'-0"
7'-6-1/4"

Copyright © 1997 by NYC Transit

## CAR BODY-EQUIPMENT

| | | | |
|---|---|---|---|
| AIR COMPRESSOR | WH D2-EG/D2-EY/D2F | PROPULSION CONTROL | GE "B" |
| BRAKE TYPE | WABCO, AMRE/ME21 | SEAT ARRANGEMENT & CAPACITY | LONGITUDINAL [44], SEE NOTE 1 |
| COUPLER | F | SEAT MATERIAL | RATTAN |
| END DOOR | SLIDING | SEAT TYPE | INDIVIDUAL CUSHION |
| FAN | PADDLE [4] | STRUCTURE MATERIAL | STEEL |
| MARKER LIGHT | ELECTRIC LAMP | TAIL LIGHT | KEROSENE LANTERN, SEE NOTE 2 |
| MASTER CONTROLLER | C18 | | |

## TRUCK

| | | | | |
|---|---|---|---|---|
| MANUFACTURER | MOTOR | BALDWIN/ HEDLEY | JOURNAL BEARING | FRICTION |
| | TRAILER | BALDWIN/ HEDLEY | TRACTION MOTOR | GE 69 (200 HP); GE 212 (200 HP) [2/MOTOR TRUCK] |
| AXLE LOAD | # 1 END - TRAILER | 15,637 LBS | | WH 86 (200 HP); WH 300 (200 HP) [2/MOTOR TRUCK] |
| | # 2 END - MOTOR | 23,113 LBS | WEIGHT (2 TRUCKS) | 36,450 LBS |
| GEAR UNIT | | 62:21 | | |

## GENERAL

| | | | |
|---|---|---|---|
| CAR BODY & EQUIPMENT WEIGHT | 41,050 LBS | CAR LOADED WEIGHT | 104,940 LBS |
| CAR LIGHT WEIGHT | 77,500 LBS | STANDING CAPACITY | [152] |

NOTE 1: DROP SEATS AT CENTER DOORS WERE REMOVED (1942 - 1948).
NOTE 2: RED LENS FOR TAIL, WHITE LENS FOR RUNNING.
NOTE 3:

| CARS | CAR BUILDER |
|---|---|
| 3700 - 3756 | AMERICAN CAR FOUNDRY |
| 3815 | STANDARD STEEL |
| 3915 | PRESSED STEEL |

CARS HAD EXPOSED BEAMED CEILING.
CARS 3730 AND 3748 WERE THE FIRST CARS TO BE EQUIPPED WITH FANS (1910). FIRST SERIES OF CARS EQUIPPED WITH STANDARD IRT COUPLER (F TYPE).
CARS 3712 AND 3737 CONVERTED AS WELDING CARS. SEVERAL CARS WERE LATER ASSIGNED AS WORK MOTORS ( LATE 1950's).
ON MODIFIED CARS, END DOORS WERE MANUALLY OPERATED & EQUIPPED WITH RUBBER CUSHION. CENTER DOORS HAD DOOR OPERATOR ENGINE AND EQUIPPED WITH COLLAPSIBLE STEEL SHOES.
VESTIBULE AT CAR ENDS.
ALL CARS WERE SCRAPPED BY 1961.

CAR NUMBERS: 3757 - 3814, 3816 - 3914, 3916 - 4024
TOTAL: 266 CARS
BUILT BY: AMERICAN CAR FOUNDRY, PRESSED STEEL, STANDARD STEEL (SEE NOTE 3)

DATE: 1909
AVERAGE COST PER CAR: N/A

# 1909 (IRT)

MUDC HEDLEY, HV MOTOR

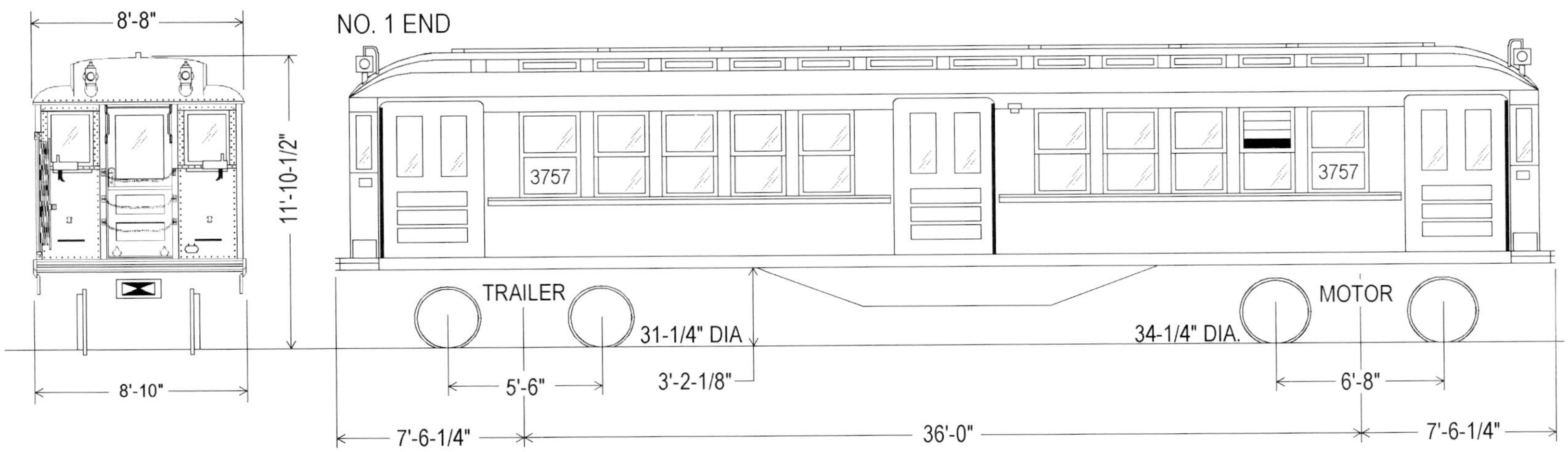

8'-8"
11'-10-1/2"
8'-10"
NO. 1 END
3757
3757
TRAILER
MOTOR
31-1/4" DIA
34-1/4" DIA.
5'-6"
3'-2-1/8"
6'-8"
7'-6-1/4"
36'-0"
7'-6-1/4"

Copyright © 1997 by NYC Transit

## CAR BODY-EQUIPMENT

| | | | |
|---|---|---|---|
| AIR COMPRESSOR | WH D2-EG/D2-EY/D2F | PROPULSION CONTROL | GE "B" |
| BRAKE TYPE | WABCO, AMRE/ME21 | SEAT ARRANGEMENT & CAPACITY | LONGITUDINAL [44], SEE NOTE 1 |
| COUPLER | F | SEAT MATERIAL | RATTAN |
| END DOOR | SLIDING | SEAT TYPE | INDIVIDUAL CUSHION |
| FAN | PADDLE [4] | STRUCTURE MATERIAL | STEEL |
| MARKER LIGHT | ELECTRIC LAMP | TAIL LIGHT | KEROSENE LANTERN, SEE NOTE 2 |
| MASTER CONTROLLER | C18 | | |

## TRUCK

| | | | | |
|---|---|---|---|---|
| MANUFACTURER | MOTOR | BALDWIN/ HEDLEY | JOURNAL BEARING | FRICTION |
| | TRAILER | BALDWIN/ HEDLEY | TRACTION MOTOR | GE 69 (200 HP); GE 212 (200 HP) [2/MOTOR TRUCK] |
| AXLE LOAD | # 1 END - TRAILER | 16,625 LBS | | WH 86 (200 HP); WH 300 (200 HP) [2/MOTOR TRUCK] |
| | # 2 END - MOTOR | 24,000 LBS | WEIGHT (2 TRUCKS) | 36,450 LBS |
| GEAR UNIT | | 62:21 | | |

## GENERAL

| | | | |
|---|---|---|---|
| CAR BODY & EQUIPMENT WEIGHT | 45,000 LBS | CAR LOADED WEIGHT | 108,890 LBS |
| CAR LIGHT WEIGHT | 81,450 LBS | STANDING CAPACITY | [152] |

NOTE 1: DROP SEATS AT CENTER DOORS WERE REMOVED (1942 - 1948).
NOTE 2: RED LENS FOR TAIL, WHITE LENS FOR RUNNING.
NOTE 3:

| CARS | CAR BUILDER |
|---|---|
| 3757 - 3809. | AMERICAN CAR FOUNDRY |
| 3810 - 3814, 3816 - 3849. | STANDARD STEEL |
| 3850 - 3914, 3916 - 4024. | PRESSED STEEL |

CARS HAD EXPOSED BEAMED CEILING.
CARS EQUIPPED WITH MUDC (EARLY 1920's).
CARS 4015-4024 OPERATED ON TIMES SQUARE SHUTTLE LINE. FOR CARS 3815 AND 3915 (SEE PAGE 370).
VESTIBULE AT CAR ENDS.
ALL CARS WERE SCRAPPED (1950's - 1960).

CAR NUMBERS: 4025 - 4036
TOTAL: 12 CARS
BUILT BY: PRESSED STEEL

DATE: 1915
AVERAGE COST PER CAR: N/A

## 1915 (IRT)
STEINWAY, LV MOTOR

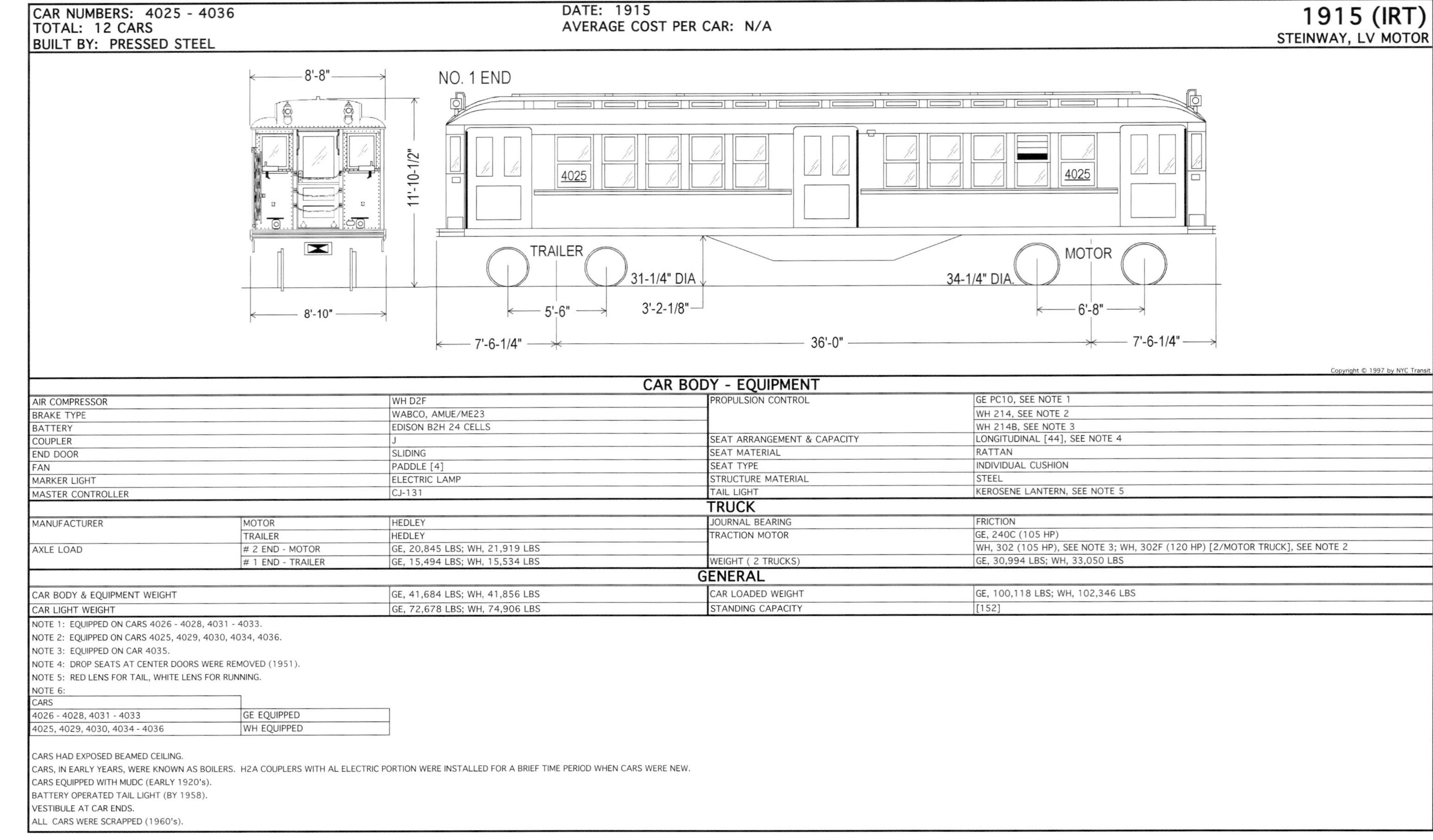


Copyright © 1997 by NYC Transit

### CAR BODY - EQUIPMENT

| | | | |
|---|---|---|---|
| AIR COMPRESSOR | WH D2F | PROPULSION CONTROL | GE PC10, SEE NOTE 1 |
| BRAKE TYPE | WABCO, AMUE/ME23 | | WH 214, SEE NOTE 2 |
| BATTERY | EDISON B2H 24 CELLS | | WH 214B, SEE NOTE 3 |
| COUPLER | J | SEAT ARRANGEMENT & CAPACITY | LONGITUDINAL [44], SEE NOTE 4 |
| END DOOR | SLIDING | SEAT MATERIAL | RATTAN |
| FAN | PADDLE [4] | SEAT TYPE | INDIVIDUAL CUSHION |
| MARKER LIGHT | ELECTRIC LAMP | STRUCTURE MATERIAL | STEEL |
| MASTER CONTROLLER | CJ-131 | TAIL LIGHT | KEROSENE LANTERN, SEE NOTE 5 |

### TRUCK

| | | | | |
|---|---|---|---|---|
| MANUFACTURER | MOTOR | HEDLEY | JOURNAL BEARING | FRICTION |
| | TRAILER | HEDLEY | TRACTION MOTOR | GE, 240C (105 HP) |
| AXLE LOAD | # 2 END - MOTOR | GE, 20,845 LBS; WH, 21,919 LBS | | WH, 302 (105 HP), SEE NOTE 3; WH, 302F (120 HP) [2/MOTOR TRUCK], SEE NOTE 2 |
| | # 1 END - TRAILER | GE, 15,494 LBS; WH, 15,534 LBS | WEIGHT ( 2 TRUCKS) | GE, 30,994 LBS; WH, 33,050 LBS |

### GENERAL

| | | | |
|---|---|---|---|
| CAR BODY & EQUIPMENT WEIGHT | GE, 41,684 LBS; WH, 41,856 LBS | CAR LOADED WEIGHT | GE, 100,118 LBS; WH, 102,346 LBS |
| CAR LIGHT WEIGHT | GE, 72,678 LBS; WH, 74,906 LBS | STANDING CAPACITY | [152] |

NOTE 1: EQUIPPED ON CARS 4026 - 4028, 4031 - 4033.
NOTE 2: EQUIPPED ON CARS 4025, 4029, 4030, 4034, 4036.
NOTE 3: EQUIPPED ON CAR 4035.
NOTE 4: DROP SEATS AT CENTER DOORS WERE REMOVED (1951).
NOTE 5: RED LENS FOR TAIL, WHITE LENS FOR RUNNING.
NOTE 6:

| CARS | |
|---|---|
| 4026 - 4028, 4031 - 4033 | GE EQUIPPED |
| 4025, 4029, 4030, 4034 - 4036 | WH EQUIPPED |

CARS HAD EXPOSED BEAMED CEILING.
CARS, IN EARLY YEARS, WERE KNOWN AS BOILERS. H2A COUPLERS WITH AL ELECTRIC PORTION WERE INSTALLED FOR A BRIEF TIME PERIOD WHEN CARS WERE NEW.
CARS EQUIPPED WITH MUDC (EARLY 1920's).
BATTERY OPERATED TAIL LIGHT (BY 1958).
VESTIBULE AT CAR ENDS.
ALL CARS WERE SCRAPPED (1960's).

CAR NUMBERS: 4037 - 4160
TOTAL: 124 CARS
BUILT BY: PULLMAN

DATE: 1915
AVERAGE COST PER CAR: N/A

# 1915 (IRT)
## FLIVVER, LV MOTOR

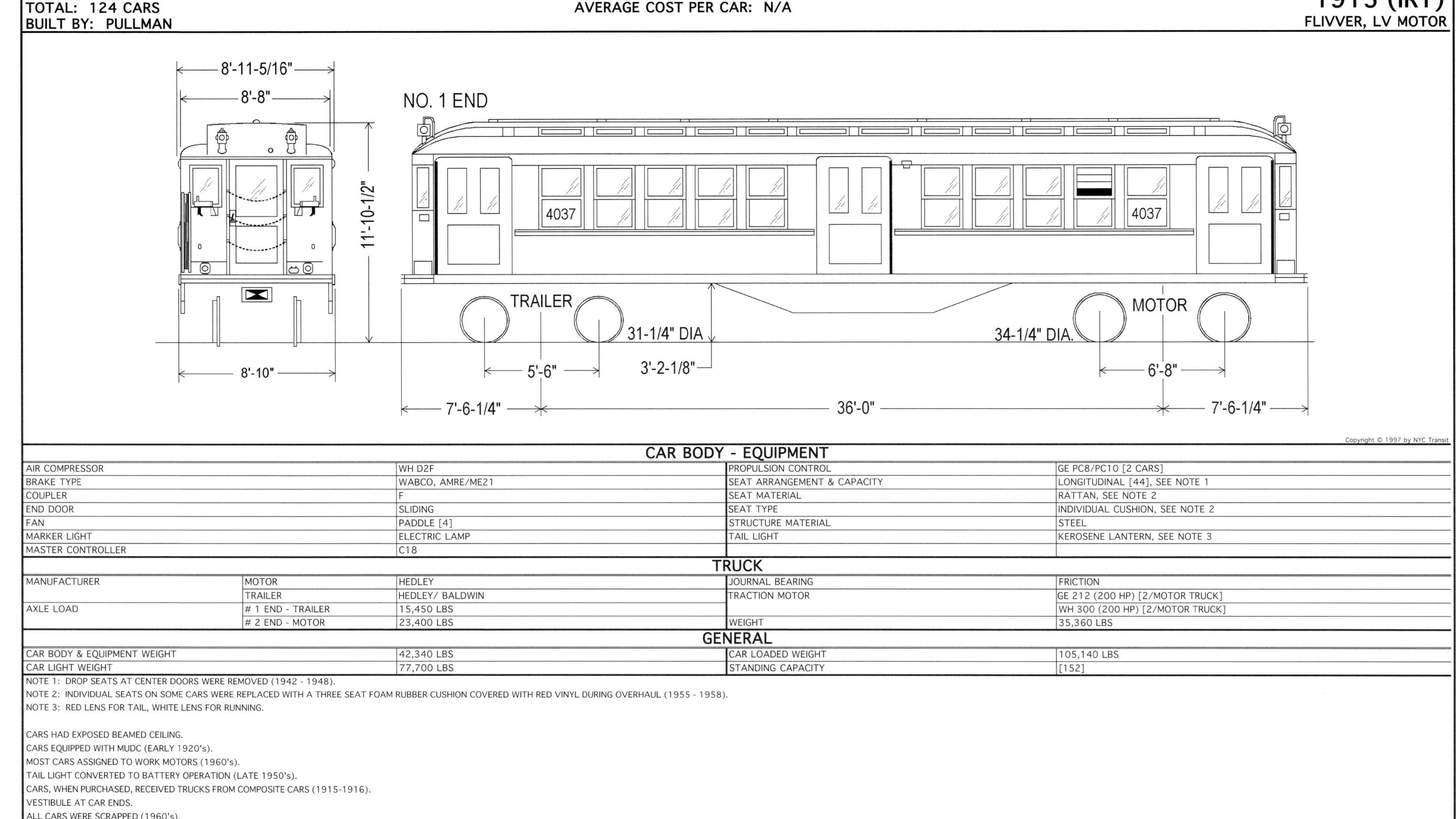


Copyright © 1997 by NYC Transit

| CAR BODY - EQUIPMENT | | | | |
|---|---|---|---|---|
| AIR COMPRESSOR | | WH D2F | PROPULSION CONTROL | GE PC8/PC10 [2 CARS] |
| BRAKE TYPE | | WABCO, AMRE/ME21 | SEAT ARRANGEMENT & CAPACITY | LONGITUDINAL [44], SEE NOTE 1 |
| COUPLER | | F | SEAT MATERIAL | RATTAN, SEE NOTE 2 |
| END DOOR | | SLIDING | SEAT TYPE | INDIVIDUAL CUSHION, SEE NOTE 2 |
| FAN | | PADDLE [4] | STRUCTURE MATERIAL | STEEL |
| MARKER LIGHT | | ELECTRIC LAMP | TAIL LIGHT | KEROSENE LANTERN, SEE NOTE 3 |
| MASTER CONTROLLER | | C18 | | |
| TRUCK | | | | |
| MANUFACTURER | MOTOR | HEDLEY | JOURNAL BEARING | FRICTION |
| | TRAILER | HEDLEY/ BALDWIN | TRACTION MOTOR | GE 212 (200 HP) [2/MOTOR TRUCK] |
| AXLE LOAD | # 1 END - TRAILER | 15,450 LBS | | WH 300 (200 HP) [2/MOTOR TRUCK] |
| | # 2 END - MOTOR | 23,400 LBS | WEIGHT | 35,360 LBS |
| GENERAL | | | | |
| CAR BODY & EQUIPMENT WEIGHT | | 42,340 LBS | CAR LOADED WEIGHT | 105,140 LBS |
| CAR LIGHT WEIGHT | | 77,700 LBS | STANDING CAPACITY | [152] |

NOTE 1: DROP SEATS AT CENTER DOORS WERE REMOVED (1942 - 1948).
NOTE 2: INDIVIDUAL SEATS ON SOME CARS WERE REPLACED WITH A THREE SEAT FOAM RUBBER CUSHION COVERED WITH RED VINYL DURING OVERHAUL (1955 - 1958).
NOTE 3: RED LENS FOR TAIL, WHITE LENS FOR RUNNING.

CARS HAD EXPOSED BEAMED CEILING.
CARS EQUIPPED WITH MUDC (EARLY 1920's).
MOST CARS ASSIGNED TO WORK MOTORS (1960's).
TAIL LIGHT CONVERTED TO BATTERY OPERATION (LATE 1950's).
CARS, WHEN PURCHASED, RECEIVED TRUCKS FROM COMPOSITE CARS (1915-1916).
VESTIBULE AT CAR ENDS.
ALL CARS WERE SCRAPPED (1960's).

CAR NUMBERS: 4161 - 4214
TOTAL: 54 CARS
BUILT BY: PULLMAN

DATE: 1915
AVERAGE COST PER CAR: N/A

# 1915 (IRT)

FLIVVER, LV TRAILER

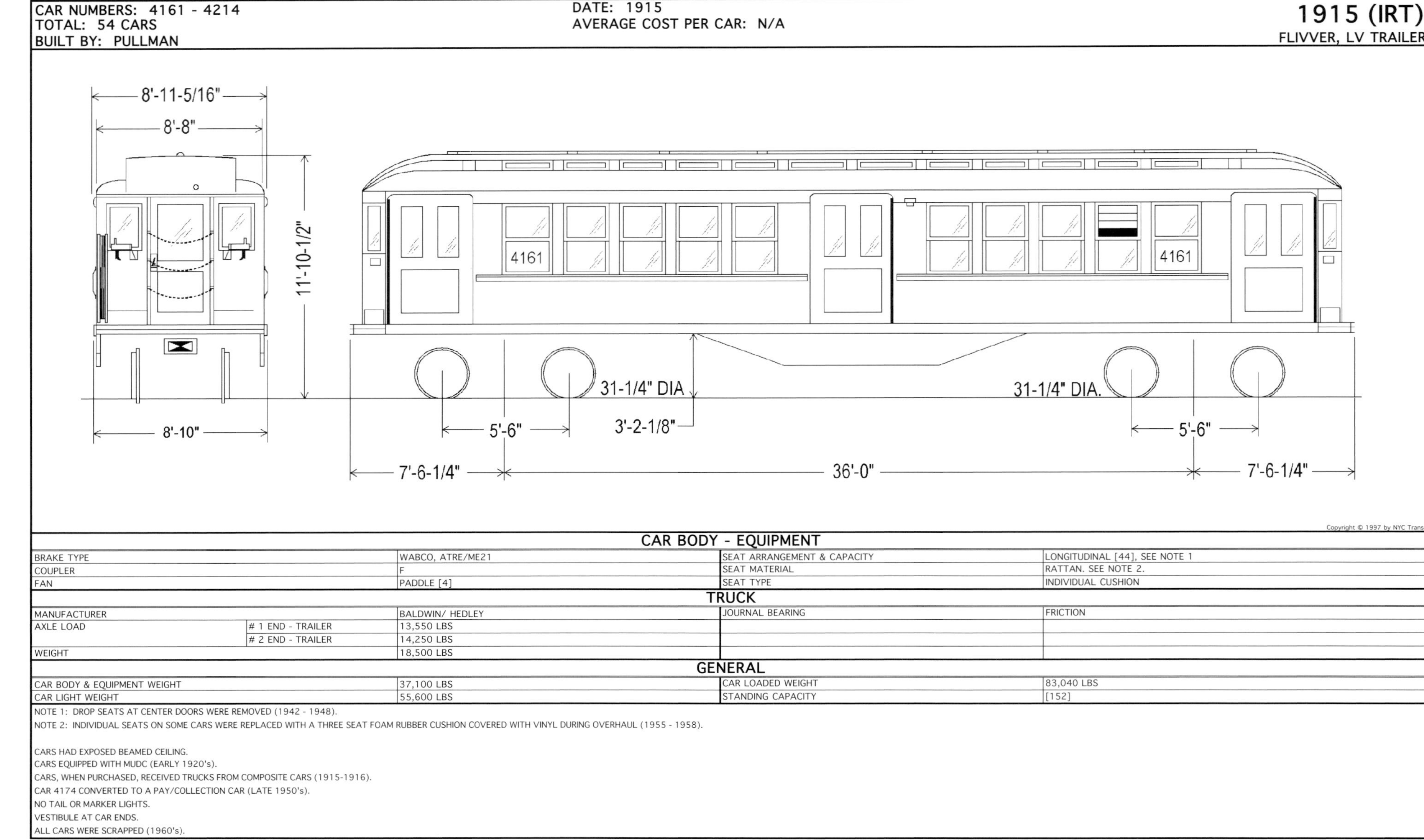


Copyright © 1997 by NYC Transit

## CAR BODY - EQUIPMENT

| | | | | |
|---|---|---|---|---|
| BRAKE TYPE | | WABCO, ATRE/ME21 | SEAT ARRANGEMENT & CAPACITY | LONGITUDINAL [44], SEE NOTE 1 |
| COUPLER | | F | SEAT MATERIAL | RATTAN. SEE NOTE 2. |
| FAN | | PADDLE [4] | SEAT TYPE | INDIVIDUAL CUSHION |

## TRUCK

| | | | | |
|---|---|---|---|---|
| MANUFACTURER | | BALDWIN/ HEDLEY | JOURNAL BEARING | FRICTION |
| AXLE LOAD | # 1 END - TRAILER | 13,550 LBS | | |
| | # 2 END - TRAILER | 14,250 LBS | | |
| WEIGHT | | 18,500 LBS | | |

## GENERAL

| | | | |
|---|---|---|---|
| CAR BODY & EQUIPMENT WEIGHT | 37,100 LBS | CAR LOADED WEIGHT | 83,040 LBS |
| CAR LIGHT WEIGHT | 55,600 LBS | STANDING CAPACITY | [152] |

NOTE 1: DROP SEATS AT CENTER DOORS WERE REMOVED (1942 - 1948).
NOTE 2: INDIVIDUAL SEATS ON SOME CARS WERE REPLACED WITH A THREE SEAT FOAM RUBBER CUSHION COVERED WITH VINYL DURING OVERHAUL (1955 - 1958).

CARS HAD EXPOSED BEAMED CEILING.
CARS EQUIPPED WITH MUDC (EARLY 1920's).
CARS, WHEN PURCHASED, RECEIVED TRUCKS FROM COMPOSITE CARS (1915-1916).
CAR 4174 CONVERTED TO A PAY/COLLECTION CAR (LATE 1950's).
NO TAIL OR MARKER LIGHTS.
VESTIBULE AT CAR ENDS.
ALL CARS WERE SCRAPPED (1960's).

CAR NUMBERS: 4215 - 4222
TOTAL: 8 CARS
BUILT BY: PULLMAN

DATE: 1915
AVERAGE COST PER CAR: N/A

# 1915 (IRT)

STEINWAY, LV MOTOR

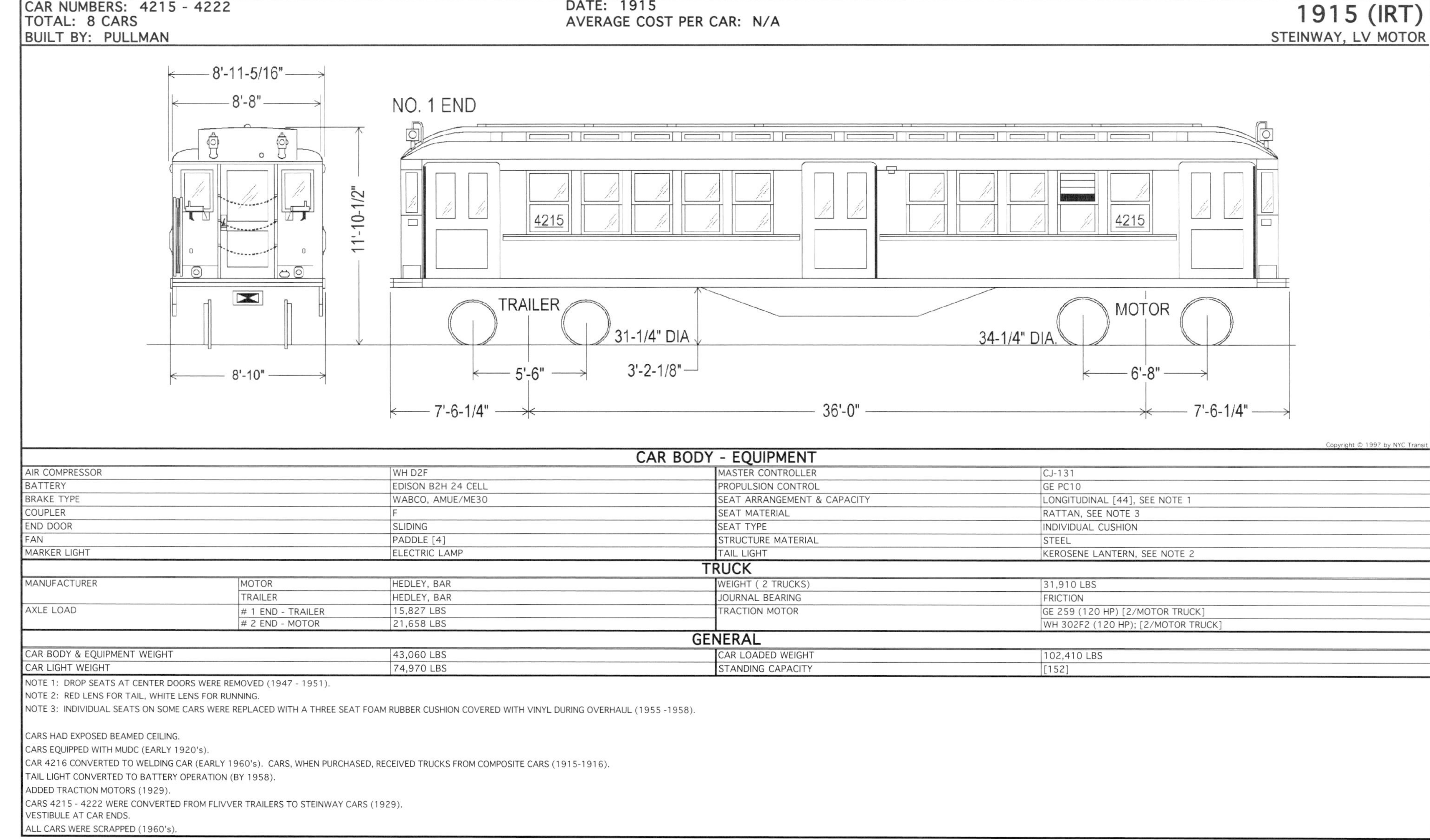


Copyright © 1997 by NYC Transit

## CAR BODY - EQUIPMENT

| | | | |
|---|---|---|---|
| AIR COMPRESSOR | WH D2F | MASTER CONTROLLER | CJ-131 |
| BATTERY | EDISON B2H 24 CELL | PROPULSION CONTROL | GE PC10 |
| BRAKE TYPE | WABCO, AMUE/ME30 | SEAT ARRANGEMENT & CAPACITY | LONGITUDINAL [44], SEE NOTE 1 |
| COUPLER | F | SEAT MATERIAL | RATTAN, SEE NOTE 3 |
| END DOOR | SLIDING | SEAT TYPE | INDIVIDUAL CUSHION |
| FAN | PADDLE [4] | STRUCTURE MATERIAL | STEEL |
| MARKER LIGHT | ELECTRIC LAMP | TAIL LIGHT | KEROSENE LANTERN, SEE NOTE 2 |

## TRUCK

| | | | | |
|---|---|---|---|---|
| MANUFACTURER | MOTOR | HEDLEY, BAR | WEIGHT ( 2 TRUCKS) | 31,910 LBS |
| | TRAILER | HEDLEY, BAR | JOURNAL BEARING | FRICTION |
| AXLE LOAD | # 1 END - TRAILER | 15,827 LBS | TRACTION MOTOR | GE 259 (120 HP) [2/MOTOR TRUCK] |
| | # 2 END - MOTOR | 21,658 LBS | | WH 302F2 (120 HP); [2/MOTOR TRUCK] |

## GENERAL

| | | | |
|---|---|---|---|
| CAR BODY & EQUIPMENT WEIGHT | 43,060 LBS | CAR LOADED WEIGHT | 102,410 LBS |
| CAR LIGHT WEIGHT | 74,970 LBS | STANDING CAPACITY | [152] |

NOTE 1: DROP SEATS AT CENTER DOORS WERE REMOVED (1947 - 1951).
NOTE 2: RED LENS FOR TAIL, WHITE LENS FOR RUNNING.
NOTE 3: INDIVIDUAL SEATS ON SOME CARS WERE REPLACED WITH A THREE SEAT FOAM RUBBER CUSHION COVERED WITH VINYL DURING OVERHAUL (1955 -1958).

CARS HAD EXPOSED BEAMED CEILING.
CARS EQUIPPED WITH MUDC (EARLY 1920's).
CAR 4216 CONVERTED TO WELDING CAR (EARLY 1960's). CARS, WHEN PURCHASED, RECEIVED TRUCKS FROM COMPOSITE CARS (1915-1916).
TAIL LIGHT CONVERTED TO BATTERY OPERATION (BY 1958).
ADDED TRACTION MOTORS (1929).
CARS 4215 - 4222 WERE CONVERTED FROM FLIVVER TRAILERS TO STEINWAY CARS (1929).
VESTIBULE AT CAR ENDS.
ALL CARS WERE SCRAPPED (1960's).

CAR NUMBERS: 4223 - 4514
TOTAL: 292 CARS
BUILT BY: PULLMAN

DATE: 1915
AVERAGE COST PER CAR: N/A

# 1915 (IRT)
HV TRAILER

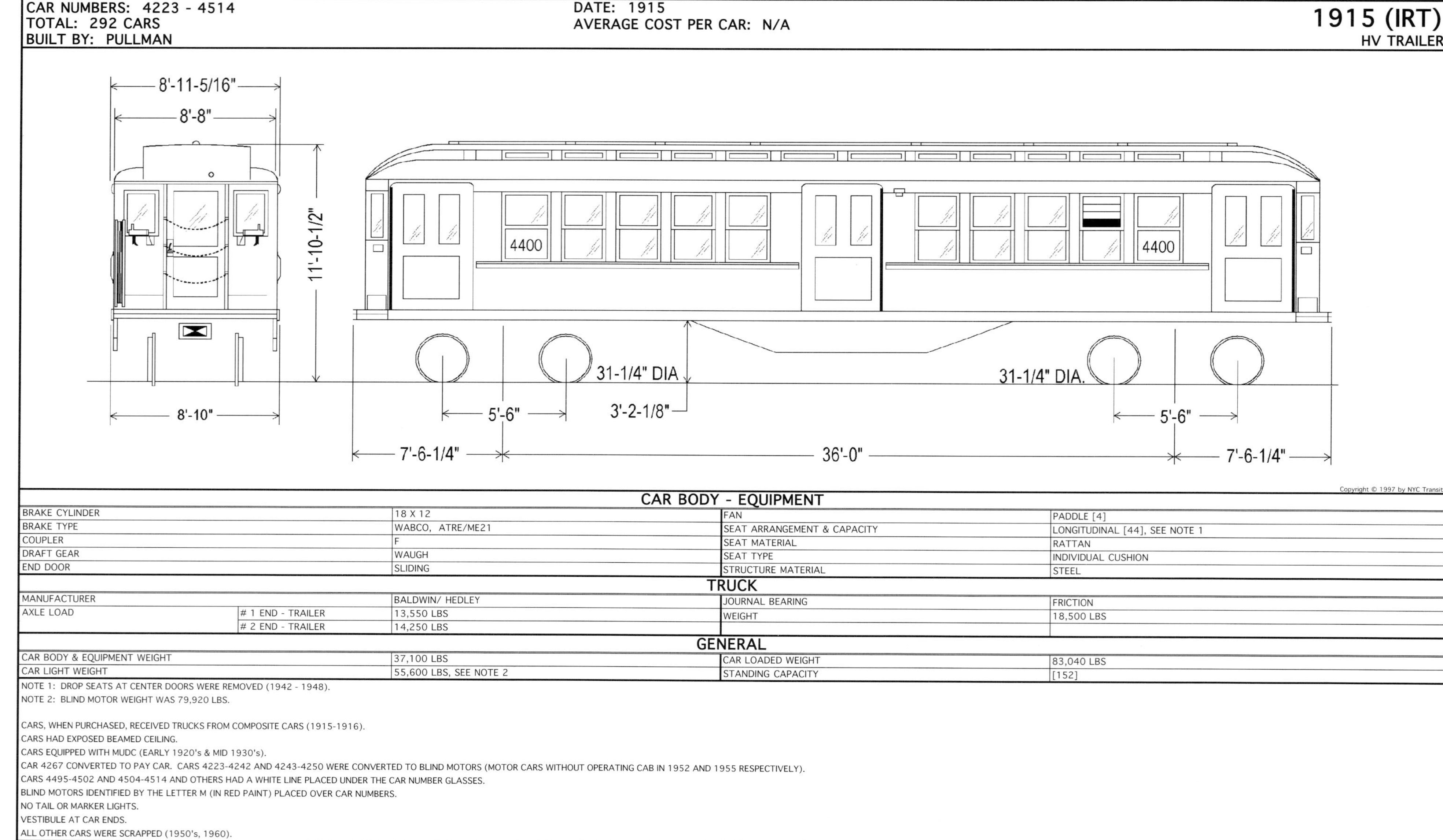


Copyright © 1997 by NYC Transit

## CAR BODY - EQUIPMENT

| | | | |
|---|---|---|---|
| BRAKE CYLINDER | 18 X 12 | FAN | PADDLE [4] |
| BRAKE TYPE | WABCO, ATRE/ME21 | SEAT ARRANGEMENT & CAPACITY | LONGITUDINAL [44], SEE NOTE 1 |
| COUPLER | F | SEAT MATERIAL | RATTAN |
| DRAFT GEAR | WAUGH | SEAT TYPE | INDIVIDUAL CUSHION |
| END DOOR | SLIDING | STRUCTURE MATERIAL | STEEL |

## TRUCK

| | | | | |
|---|---|---|---|---|
| MANUFACTURER | | BALDWIN/ HEDLEY | JOURNAL BEARING | FRICTION |
| AXLE LOAD | # 1 END - TRAILER | 13,550 LBS | WEIGHT | 18,500 LBS |
| | # 2 END - TRAILER | 14,250 LBS | | |

## GENERAL

| | | | |
|---|---|---|---|
| CAR BODY & EQUIPMENT WEIGHT | 37,100 LBS | CAR LOADED WEIGHT | 83,040 LBS |
| CAR LIGHT WEIGHT | 55,600 LBS, SEE NOTE 2 | STANDING CAPACITY | [152] |

NOTE 1: DROP SEATS AT CENTER DOORS WERE REMOVED (1942 - 1948).
NOTE 2: BLIND MOTOR WEIGHT WAS 79,920 LBS.

CARS, WHEN PURCHASED, RECEIVED TRUCKS FROM COMPOSITE CARS (1915-1916).
CARS HAD EXPOSED BEAMED CEILING.
CARS EQUIPPED WITH MUDC (EARLY 1920's & MID 1930's).
CAR 4267 CONVERTED TO PAY CAR. CARS 4223-4242 AND 4243-4250 WERE CONVERTED TO BLIND MOTORS (MOTOR CARS WITHOUT OPERATING CAB IN 1952 AND 1955 RESPECTIVELY).
CARS 4495-4502 AND 4504-4514 AND OTHERS HAD A WHITE LINE PLACED UNDER THE CAR NUMBER GLASSES.
BLIND MOTORS IDENTIFIED BY THE LETTER M (IN RED PAINT) PLACED OVER CAR NUMBERS.
NO TAIL OR MARKER LIGHTS.
VESTIBULE AT CAR ENDS.
ALL OTHER CARS WERE SCRAPPED (1950's, 1960).

CAR NUMBERS: 4515 - 4554
TOTAL: 40 CARS
BUILT BY: PULLMAN

DATE: 1916
AVERAGE COST PER CAR: N/A

# 1916 (IRT)
LV TRAILER

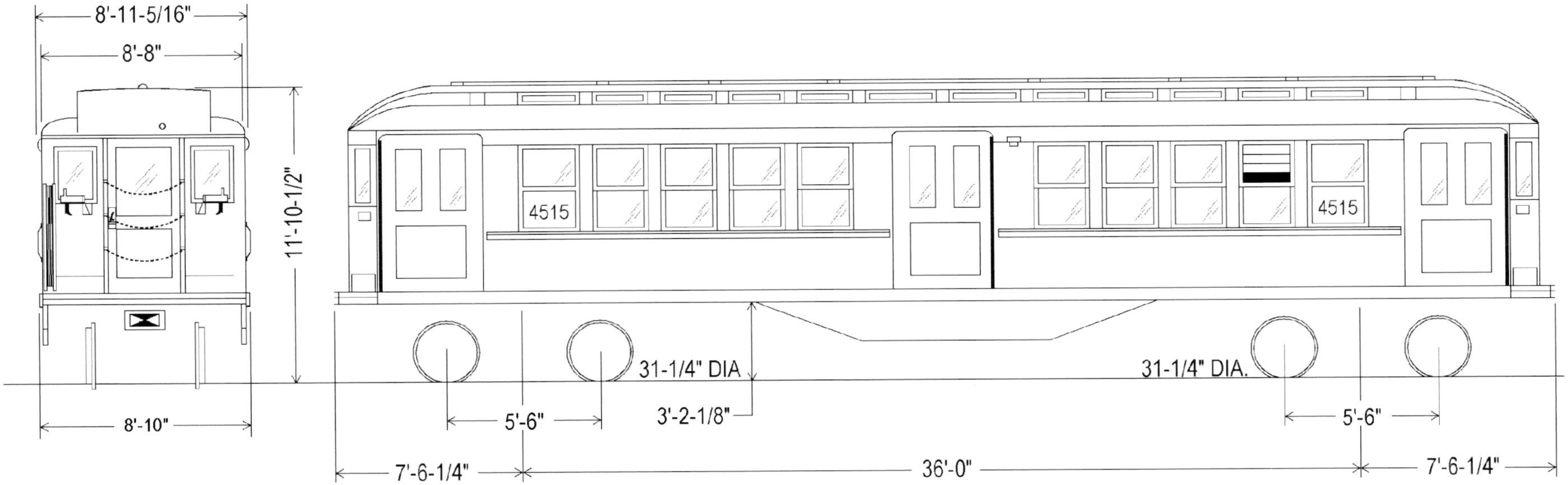

8'-11-5/16"
8'-8"
11'-10-1/2"
8'-10"
4515
4515
31-1/4" DIA
31-1/4" DIA.
3'-2-1/8"
5'-6"
5'-6"
7'-6-1/4"
36'-0"
7'-6-1/4"

Copyright © 1997 by NYC Transit

## CAR BODY - EQUIPMENT

| | | | | |
|---|---|---|---|---|
| BATTERY | | LEAD ACID | FAN | PADDLE [4] |
| BRAKE CYLINDER | | 18 X 12 | SEAT ARRANGEMENT & CAPACITY | LONGITUDINAL [44], SEE NOTE 1 |
| BRAKE TYPE | | WABCO, ATUE/ME23 | SEAT MATERIAL | RATTAN, SEE NOTE 2 |
| COUPLER | | J | SEAT TYPE | INDIVIDUAL CUSHION |
| DRAFT GEAR | | WAUGH | STRUCTURE MATERIAL | STEEL |
| END DOOR | | SLIDING | | |

## TRUCK

| | | | | |
|---|---|---|---|---|
| MANUFACTURER | | HEDLEY/ COMMONWEALTH | JOURNAL BEARING | FRICTION |
| AXLE LOAD | # 1 END - TRAILER | 13,700 LBS | WEIGHT | 18,500 LBS |
| | # 2 END - TRAILER | 14,400 LBS | | |

## GENERAL

| | | | |
|---|---|---|---|
| CAR BODY & EQUIPMENT WEIGHT | 37,700 LBS | CAR LOADED WEIGHT | 83,640 LBS |
| CAR LIGHT WEIGHT | 56,200 LBS | STANDING CAPACITY | [152] |

NOTE 1: DROP SEATS AT CENTER DOORS WERE REMOVED (1942 - 1948).
NOTE 2: INDIVIDUAL SEATS ON SOME CARS WERE REPLACED WITH A THREE SEAT FOAM RUBBER CUSHION COVERED WITH VINYL DURING OVERHAUL (1955 - 1958).

CARS HAD EXPOSED BEAMED CEILING.
CARS EQUIPPED WITH MUDC (EARLY 1920's).
CAR 4517 CONVERTED TO PAY/COLLECTION CAR AND OPERATED ON 3RD AVE. ELEVATED LINE IN THE BRONX.
NO TAIL OR MARKER LIGHTS.
VESTIBULE AT CAR ENDS.
ALL CARS WERE SCRAPPED (1960's, 1970's, 1980's).

CAR NUMBERS: 4555 - 4576
TOTAL: 22 CARS
BUILT BY: PULLMAN

DATE: 1916
AVERAGE COST PER CAR: N/A

## 1916 (IRT)

STEINWAY, LV MOTOR

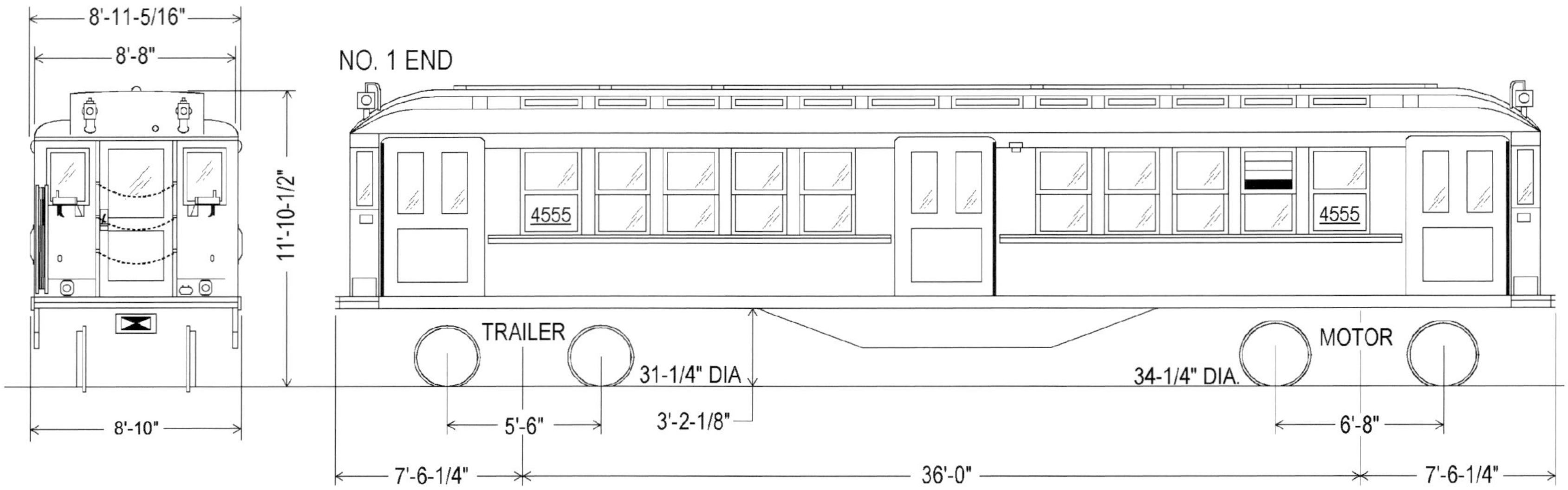

8'-11-5/16"
8'-8"
11'-10-1/2"
8'-10"
NO. 1 END
4555
4555
TRAILER
31-1/4" DIA
34-1/4" DIA.
MOTOR
5'-6"
3'-2-1/8"
6'-8"
7'-6-1/4"
36'-0"
7'-6-1/4"

Copyright © 1997 by NYC Transit

| CAR BODY - EQUIPMENT | | | | |
|---|---|---|---|---|
| AIR COMPRESSOR | | WH D2F | MASTER CONTROLLER | CJ-131 |
| BATTERY | | EDISON, B2H 24 CELLS | PROPULSION CONTROL | GE PC10 |
| BRAKE TYPE | | WABCO, AMUE/ME30 | SEAT ARRANGEMENT & CAPACITY | LONGITUDINAL [44], SEE NOTE 1 |
| COUPLER | | J | SEAT MATERIAL | RATTAN, SEE NOTE 2 |
| END DOOR | | SLIDING | SEAT TYPE | INDIVIDUAL CUSHION |
| FAN | | PADDLE [4] | STRUCTURE MATERIAL | STEEL |
| MARKER LIGHT | | ELECTRIC LAMP | TAIL LIGHT | KEROSENE LANTERN, SEE NOTE 3 |
| **TRUCK** | | | | |
| MANUFACTURER | MOTOR | HEDLEY/ BAR | JOURNAL BEARING | FRICTION |
| | TRAILER | HEDLEY/ BAR | TRACTION MOTOR | GE 259 (120 HP) [2/MOTOR TRUCK] |
| AXLE LOAD | # 1 END - TRAILER | 15,827 LBS | | WH 302F2 (120 HP) [2/MOTOR TRUCK] |
| | # 2 END - MOTOR | 21,658 LBS | WEIGHT ( 2 TRUCKS) | 31,910 LBS |
| **GENERAL** | | | | |
| CAR BODY & EQUIPMENT WEIGHT | | 43,060 LBS | CAR LOADED WEIGHT | 102,410 LBS |
| CAR LIGHT WEIGHT | | 74,970 LBS | STANDING CAPACITY | [152] |

NOTE 1: DROP SEATS AT CENTER DOORS WERE REMOVED (1947 - 1951).
NOTE 2: INDIVIDUAL SEATS ON SOME CARS WERE REPLACED WITH A THREE SEAT FOAM RUBBER CUSHION COVERED WITH VINYL DURING OVERHAUL (1955 -1958).
NOTE 3: RED LENS FOR TAIL, WHITE LENS FOR RUNNING.

CARS HAD EXPOSED BEAMED CEILING.
CARS EQUIPPED WITH MUDC (EARLY 1920's).
CARS WERE CONVERTED FROM L.V. TRAILERS TO STEINWAY CARS (1929).
ADDED TRACTION MOTORS (1929).
TAIL LIGHT CONVERTED TO BATTERY OPERATION (BY 1958).
VESTIBULE AT CAR ENDS.
ALL CARS WERE SCRAPPED (1960's).

CAR NUMBERS: 4577 - 4699 & 4719
TOTAL: 124 CARS
BUILT BY: PULLMAN

DATE: 1916
AVERAGE COST PER CAR: N/A

# 1916 (IRT)

LV MOTOR

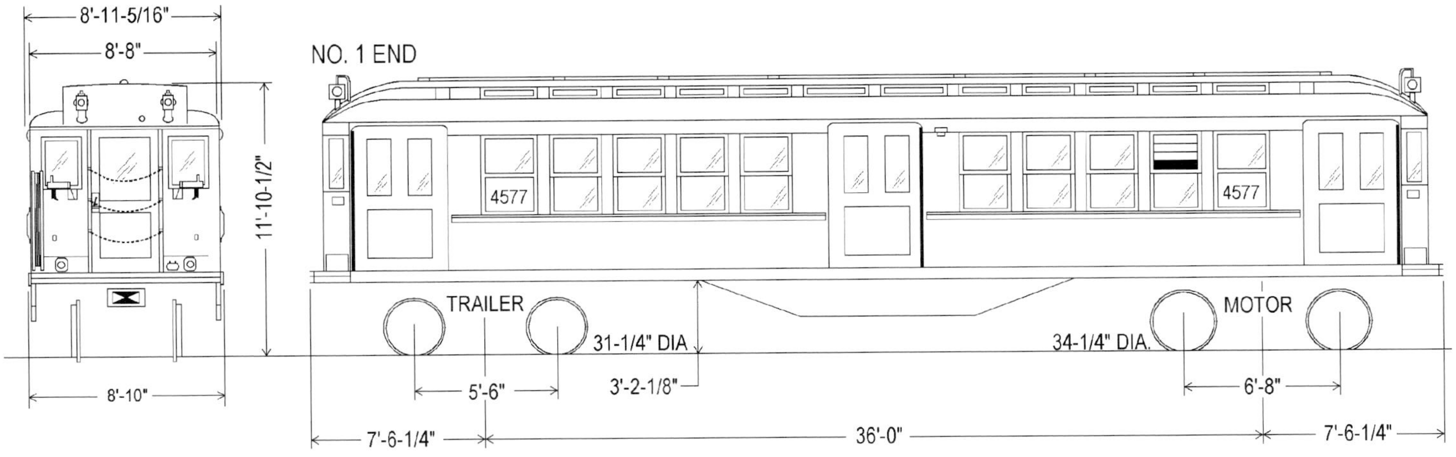

8'-11-5/16"
8'-8"
11'-10-1/2"
8'-10"
NO. 1 END
4577
4577
TRAILER
MOTOR
31-1/4" DIA
34-1/4" DIA.
5'-6"
3'-2-1/8"
6'-8"
7'-6-1/4"
36'-0"
7'-6-1/4"

Copyright © 1997 by NYC Transit

## CAR BODY - EQUIPMENT

| | | | |
|---|---|---|---|
| AIR COMPRESSOR | WH D2F | MASTER CONTROLLER | C27 |
| BATTERY | EDISON, B2H 24 CELL | PROPULSION CONTROL | WH 214B2 |
| BRAKE TYPE | WABCO, AMUE/ME23 | SEAT ARRANGEMENT & CAPACITY | LONGITUDINAL [44], SEE NOTE 1 |
| COUPLER | J | SEAT MATERIAL | RATTAN, SEE NOTE 2 |
| END DOOR | SLIDING | SEAT TYPE | INDIVIDUAL CUSHION, SEE NOTE 2 |
| FAN | PADDLE [4] | STRUCTURE MATERIAL | STEEL |
| MARKER LIGHT | ELECTRIC LAMP | TAIL LIGHT | KEROSENE LANTERN, SEE NOTE 3 |

## TRUCK

| | | | | |
|---|---|---|---|---|
| MANUFACTURER | MOTOR | MCB/ HEDLEY/ BAR | JOURNAL BEARING | FRICTION |
| | TRAILER | COMMONWEALTH/ HEDLEY/ BAR | TRACTION MOTOR | GE 260 (195 HP) [2/MOTOR TRUCK] |
| AXLE LOAD | # 1 END - TRAILER | 16,000 LBS | | WH 577 (200 HP) [2/MOTOR TRUCK] |
| | # 2 END - MOTOR | 22,650 LBS | WEIGHT | 35,000 LBS |

## GENERAL

| | | | |
|---|---|---|---|
| CAR BODY & EQUIPMENT WEIGHT | 42,300 LBS | CAR LOADED WEIGHT | 104,740 LBS |
| CAR LIGHT WEIGHT | 77,300 LBS | STANDING CAPACITY | [152] |

NOTE 1: DROP SEATS AT CENTER DOORS WERE REMOVED (1942 - 1948).
NOTE 2: INDIVIDUAL SEATS ON SOME CARS WERE REPLACED WITH A THREE SEAT FOAM RUBBER CUSHION COVERED WITH RED VINYL DURING OVERHAUL (1955 - 1958).
NOTE3: RED LENS FOR TAIL, WHITE LENS FOR RUNNING.

CARS HAD EXPOSED BEAMED CEILING.
CARS EQUIPPED WITH MUDC (EARLY 1920's).
CARS WITH KEROSENE TAIL LIGHT WERE CONVERTED TO BATTERY OPERATION (BY 1958).
CARS 4577 - 4580 WERE ASSIGNED TO THE BOWLING GREEN SHUTTLE (1958 - 1964).
CARS 4581, 4583 - 4591 WERE PILOT CARS. CARS 4581, 4583 - 4605 TRANSFERRED TO BMT SERVICE - CULVER & FRANKLIN SHUTTLES (1959 - 1960).
CAR 4617 ASSIGNED AS INSTRUCTION CAR AND REASSIGNED TO PASSENGER SERVICE IN 1921. CAR 4642 CONVERTED TO SNOW PLOW OPERATION (EARLY 1960's).
CAR 4719 ORIGINALLY A STEINWAY CAR, CONVERTED TO L.V. OPERATION (FISHBELLY REMOVED ON BOTH SIDES).
VESTIBULE AT CAR ENDS.
ALL CARS WERE SCRAPPED (1960's, 1970's, 1980's).

CAR NUMBERS: 4700 - 4718, 4720 - 4771
TOTAL: 71 CARS
BUILT BY: PULLMAN

DATE: 1916
AVERAGE COST PER CAR: N/A

# 1916 (IRT)

STEINWAY, LV MOTOR

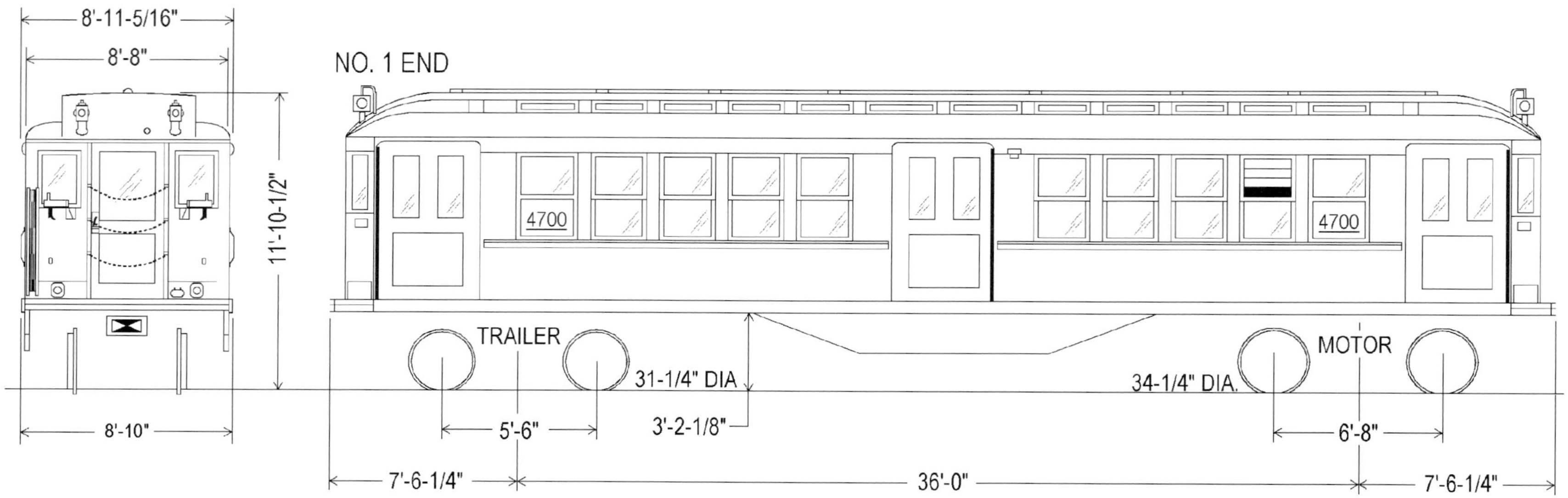

8'-11-5/16"
8'-8"
8'-10"
11'-10-1/2"
NO. 1 END
4700
4700
TRAILER
MOTOR
31-1/4" DIA
34-1/4" DIA.
5'-6"
3'-2-1/8"
6'-8"
7'-6-1/4"
36'-0"
7'-6-1/4"

Copyright © 1997 by NYC Transit

## CAR BODY - EQUIPMENT

| | | | |
|---|---|---|---|
| AIR COMPRESSOR | WH D2F | MASTER CONTROLLER | C27 |
| BATTERY | EDISON, B2H 24 CELLS | PROPULSION CONTROL | WH 214B |
| BRAKE TYPE | WABCO, AMUE/ME23 | SEAT ARRANGEMENT & CAPACITY | LONGITUDINAL[44], SEE NOTE 1 |
| COUPLER | J | SEAT MATERIAL | RATTAN, SEE NOTE 3 |
| END DOOR | SLIDING | SEAT TYPE | INDIVIDUAL CUSHION |
| FAN | PADDLE [4] | STRUCTURE MATERIAL | STEEL |
| MARKER LIGHT | ELECTRIC LAMP | TAIL LIGHT | KEROSENE LANTERN, SEE NOTE 2 |

## TRUCK

| | | | | |
|---|---|---|---|---|
| MANUFACTURER | MOTOR | HEDLEY/ BAR | JOURNAL BEARING | FRICTION |
| | TRAILER | HEDLEY/ BAR | TRACTION MOTOR | GE 259 (120 HP) [2/MOTOR TRUCK] |
| AXLE LOAD | # 1 END - TRAILER | 15,827 LBS | | WH 302F2 (120 HP) [2/MOTOR TRUCK] |
| | # 2 END - MOTOR | 21,658 LBS | WEIGHT ( 2 TRUCKS) | 31,700 LBS |
| GEAR UNIT | | 61:16/57:15 | | |

## GENERAL

| | | | |
|---|---|---|---|
| CAR BODY & EQUIPMENT WEIGHT | 42,300 LBS | CAR LOADED WEIGHT | 101,440 LBS |
| CAR LIGHT WEIGHT | 74,000 LBS | STANDING CAPACITY | [152] |

NOTE 1: DROP SEATS AT CENTER DOORS WERE REMOVED (1947 - 1951).
NOTE 2: RED LENS FOR TAIL, WHITE LENS FOR RUNNING.
NOTE 3: INDIVIDUAL SEATS ON SOME CARS WERE REPLACED WITH A THREE SEAT FOAM RUBBER CUSHION COVERED WITH VINYL DURING OVERHAUL (1955 - 1958).

CARS HAD EXPOSED BEAMED CEILING.
CARS WERE EQUIPPED WITH MUDC (1920's).
TAIL LIGHT CONVERTED TO BATTERY OPERATION (BY 1958).
CARS 4700-4718 AND 4720-4725 WERE PILOT CARS. CAR 4702 CONVERTED TO COLLECTION CAR (1950'S). CAR 4771 WAS ORIGINALLY A LV CAR.
VESTIBULE AT CAR ENDS.
ALL CARS WERE SCRAPPED (1960's).

CAR NUMBERS: 4772 - 4810
TOTAL: 39 CARS
BUILT BY: PULLMAN

DATE: 1916
AVERAGE COST PER CAR: N/A

# 1916 (IRT)

LV MOTOR

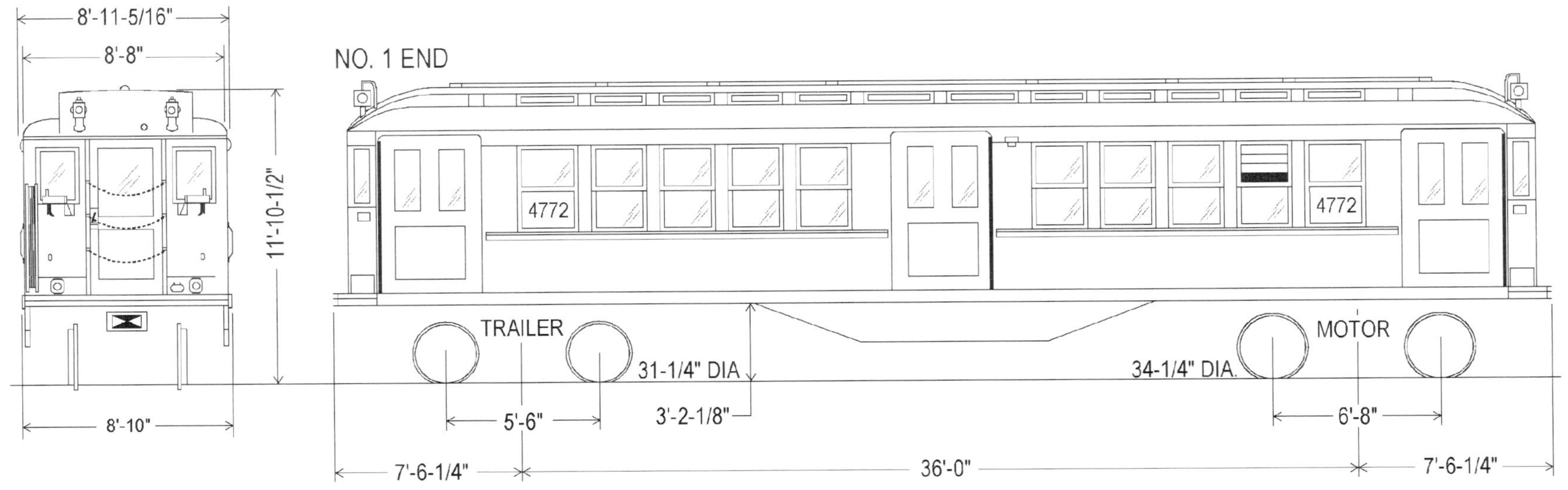

8'-11-5/16"
8'-8"
11'-10-1/2"
8'-10"
NO. 1 END
4772
4772
TRAILER
MOTOR
31-1/4" DIA
34-1/4" DIA.
5'-6"
3'-2-1/8"
6'-8"
7'-6-1/4"
36'-0"
7'-6-1/4"

Copyright © 1997 by NYC Transit

## CAR BODY - EQUIPMENT

| | | | |
|---|---|---|---|
| AIR COMPRESSOR | WH D2F | MASTER CONTROLLER | C27 |
| BATTERY | EDISON, B2H 24 CELLS | PROPULSION CONTROL | WH 214A |
| BRAKE TYPE | WABCO, AMUE/ME23 | SEAT ARRANGEMENT & CAPACITY | LONGITUDINAL[44], SEE NOTE 1 |
| COUPLER | J | SEAT MATERIAL | RATTAN, SEE NOTE 2 |
| END DOOR | SLIDING | SEAT TYPE | INDIVIDUAL CUSHION, SEE NOTE 2 |
| FAN | PADDLE [4] | STRUCTURE MATERIAL | STEEL |
| MARKER LIGHT | ELECTRIC LAMP | TAIL LIGHT | KEROSENE LANTERN, SEE NOTE 3 |

## TRUCK

| | | | | |
|---|---|---|---|---|
| MANUFACTURER | MOTOR | MCB/ HEDLEY/ BAR | TRACTION MOTOR | GE, 260, 195 HP |
| | TRAILER | COMMONWEALTH/ HEDLEY/ BAR | | WH, 577, 200HP |
| AXLE LOAD | # 1 END - TRAILER | 16,000 LBS | TRIP COCK | FISH TAIL |
| | # 2 END - MOTOR | 22,650 LBS | WEIGHT ( 2 TRUCKS) | 35,000 LBS |
| JOURNAL BEARING | | FRICTION | | |

## GENERAL

| | | | |
|---|---|---|---|
| CAR BODY & EQUIPMENT WEIGHT | 42,300 LBS | CAR LOADED WEIGHT | 104,740 LBS |
| CAR LIGHT WEIGHT | 77,300 LBS | STANDING CAPACITY | [152] |

NOTE 1: DROP SEATS AT CENTER DOORS WERE REMOVED (1942 - 1948).
NOTE 2: INDIVIDUAL SEATS ON SOME CARS WERE REPLACED WITH A THREE SEAT FOAM RUBBER CUSHION COVERED WITH RED VINYL DURING OVERHAUL (1955 - 1958).
NOTE 3: RED LENS FOR TAIL, WHITE LENS FOR RUNNING.

CARS HAD EXPOSED BEAMED CEILING.
CARS EQUIPPED WITH MUDC (EARLY 1920's)
CARS HAD KEROSENE TAIL LIGHT CONVERTED TO BATTERY OPERATION (BY 1958).
CARS 4785 CONVERTED TO SNOW PLOW (EARLY 1960's).
VESTIBULE AT CAR ENDS.
ALL CARS WERE SCRAPPED (1960's, 1980's).

CAR NUMBERS: 4811 - 4825
TOTAL: 15 CARS
BUILT BY: PULLMAN

DATE: 1916
AVERAGE COST PER CAR: N/A

# 1916 (IRT)

LV TRAILER

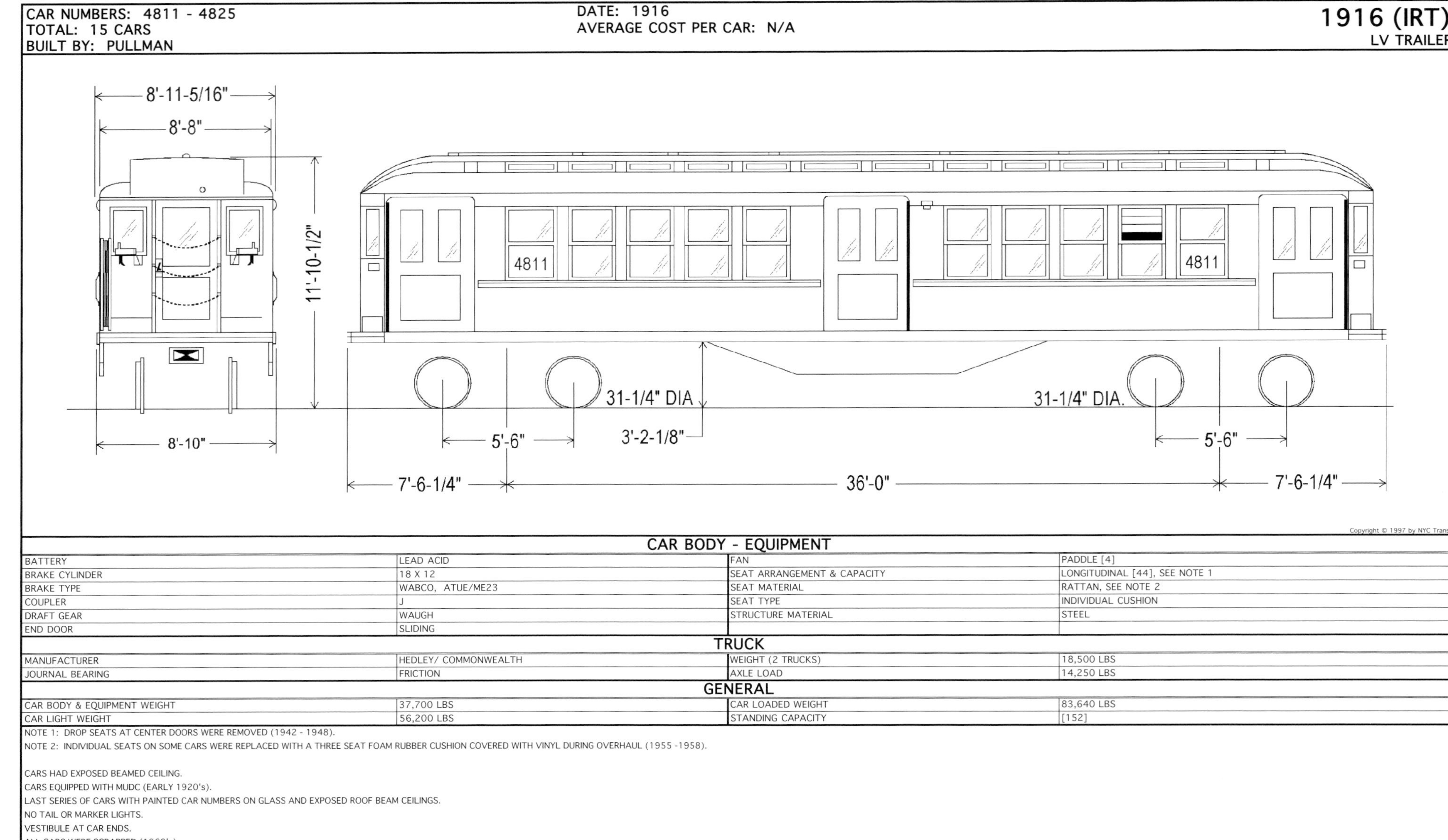


Copyright © 1997 by NYC Transit

## CAR BODY - EQUIPMENT

| | | | |
|---|---|---|---|
| BATTERY | LEAD ACID | FAN | PADDLE [4] |
| BRAKE CYLINDER | 18 X 12 | SEAT ARRANGEMENT & CAPACITY | LONGITUDINAL [44], SEE NOTE 1 |
| BRAKE TYPE | WABCO, ATUE/ME23 | SEAT MATERIAL | RATTAN, SEE NOTE 2 |
| COUPLER | J | SEAT TYPE | INDIVIDUAL CUSHION |
| DRAFT GEAR | WAUGH | STRUCTURE MATERIAL | STEEL |
| END DOOR | SLIDING | | |

## TRUCK

| | | | |
|---|---|---|---|
| MANUFACTURER | HEDLEY/ COMMONWEALTH | WEIGHT (2 TRUCKS) | 18,500 LBS |
| JOURNAL BEARING | FRICTION | AXLE LOAD | 14,250 LBS |

## GENERAL

| | | | |
|---|---|---|---|
| CAR BODY & EQUIPMENT WEIGHT | 37,700 LBS | CAR LOADED WEIGHT | 83,640 LBS |
| CAR LIGHT WEIGHT | 56,200 LBS | STANDING CAPACITY | [152] |

NOTE 1: DROP SEATS AT CENTER DOORS WERE REMOVED (1942 - 1948).
NOTE 2: INDIVIDUAL SEATS ON SOME CARS WERE REPLACED WITH A THREE SEAT FOAM RUBBER CUSHION COVERED WITH VINYL DURING OVERHAUL (1955 -1958).

CARS HAD EXPOSED BEAMED CEILING.
CARS EQUIPPED WITH MUDC (EARLY 1920's).
LAST SERIES OF CARS WITH PAINTED CAR NUMBERS ON GLASS AND EXPOSED ROOF BEAM CEILINGS.
NO TAIL OR MARKER LIGHTS.
VESTIBULE AT CAR ENDS.
ALL CARS WERE SCRAPPED (1960's).

CAR NUMBERS: 4826 - 4965
TOTAL: 140 CARS
BUILT BY: PULLMAN

DATE: 1917
AVERAGE COST PER CAR: N/A

# 1917 (IRT)

LV TRAILER

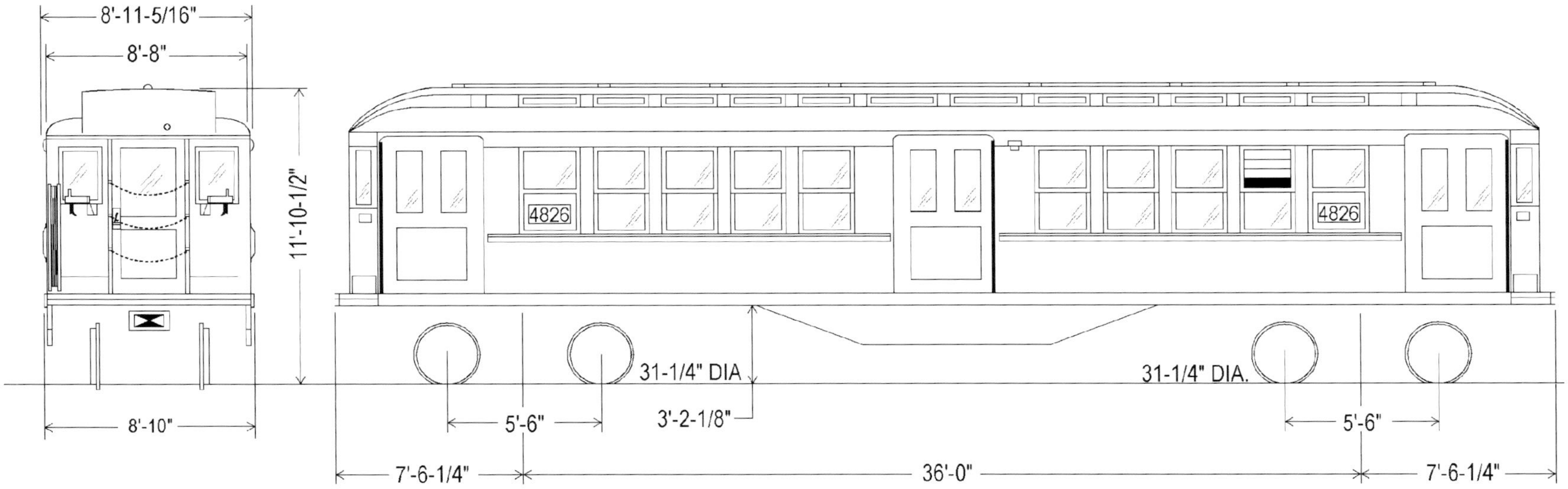

8'-11-5/16"
8'-8"
11'-10-1/2"
8'-10"
4826
31-1/4" DIA
31-1/4" DIA.
3'-2-1/8"
5'-6"
5'-6"
7'-6-1/4"
36'-0"
7'-6-1/4"

Copyright © 1997 by NYC TRANSIT

## CAR BODY - EQUIPMENT

| | | | |
|---|---|---|---|
| BATTERY | LEAD ACID | FAN | PADDLE [4] |
| BRAKE CYLINDER | 18 X 12 | SEAT ARRANGEMENT & CAPACITY | LONGITUDINAL [44], SEE NOTE 1 |
| BRAKE TYPE | WABCO, ATUE/ME23 | SEAT MATERIAL | RATTAN |
| COUPLER | J | SEAT TYPE | INDIVIDUAL CUSHION |
| DRAFT GEAR | WAUGH | STRUCTURE MATERIAL | STEEL |
| END DOOR | SLIDING | | |

## TRUCK

| | | | | |
|---|---|---|---|---|
| MANUFACTURER | | HEDLEY/ COMMONWEALTH | JOURNAL BEARING | FRICTION |
| AXLE LOAD | # 1 END - TRAILER | 13,700 LBS | WEIGHT | 18,500 LBS |
| | # 2 END - TRAILER | 14,400 LBS | | |

## GENERAL

| | | | |
|---|---|---|---|
| CAR BODY & EQUIPMENT WEIGHT | 37,700 LBS | CAR LOADED WEIGHT | 83,640 LBS |
| CAR LIGHT WEIGHT | 56,200 LBS | STANDING CAPACITY | [152] |

NOTE 1: DROP SEATS AT CENTER DOORS WERE REMOVED (1942 - 1948).

FIRST SERIES OF CARS WITH CAR NUMBERS PAINTED ON LOWER METAL WINDOW PANEL ADJACENT TO END DOOR. PORCELAIN CAR NUMBER PLATES WERE POSITIONED OVER THE PAINTED CAR NUMBERS AND ALL FOLLOWING CARS THROUGH CAR 5652 (1940's).
THIS SERIES OF CARS WAS FIRST TO HAVE MASONITE CEILING.
CARS EQUIPPED WITH MUDC (EARLY 1920's).
CARS 4838, 4849, 4851 WERE IN CONSIST WITH BLOWER TRAIN. WOOD WINDOW SASH ON CARS 4937 AND BELOW. BRASS WINDOW SASHES ON CARS 4876, 4938 TO 4965.
CAR 4954 MODIFIED AND USED AS BLOWER CAR IN SPENO RAIL GRINDER TRAIN.
NO TAIL OR MARKER LIGHTS.
VESTIBULE AT CAR ENDS.
CAR 4902 AT NEW YORK TRANSIT MUSEUM (1976).
ALL OTHER CARS WERE SCRAPPED (1960's).

CAR NUMBERS: 4966 - 5302
TOTAL: 337 CARS
BUILT BY: PULLMAN

DATE: 1917
AVERAGE COST PER CAR: N/A

# 1917 (IRT)
LV MOTOR

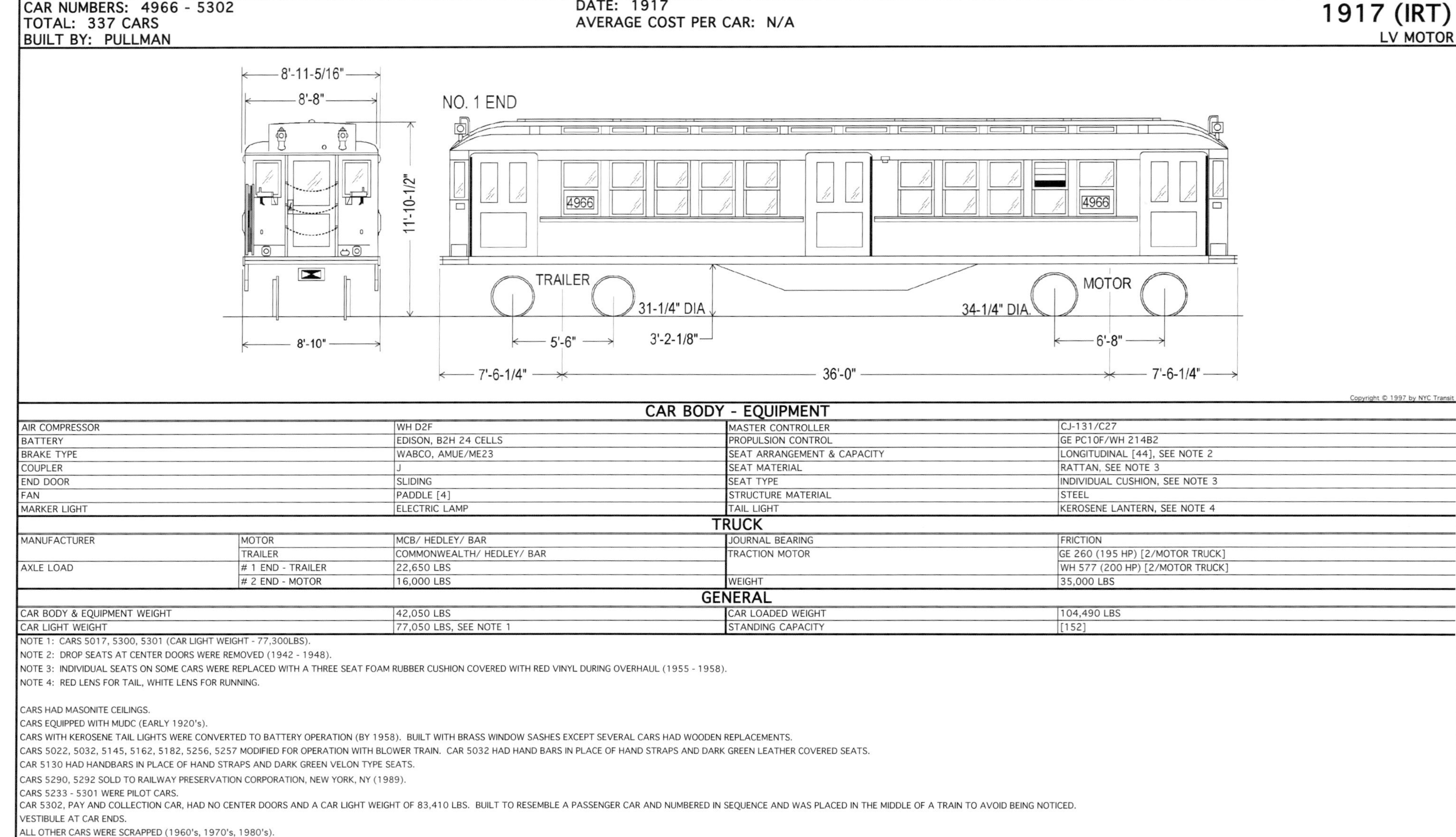


Copyright © 1997 by NYC Transit

## CAR BODY - EQUIPMENT

| | | | |
|---|---|---|---|
| AIR COMPRESSOR | WH D2F | MASTER CONTROLLER | CJ-131/C27 |
| BATTERY | EDISON, B2H 24 CELLS | PROPULSION CONTROL | GE PC10F/WH 214B2 |
| BRAKE TYPE | WABCO, AMUE/ME23 | SEAT ARRANGEMENT & CAPACITY | LONGITUDINAL [44], SEE NOTE 2 |
| COUPLER | J | SEAT MATERIAL | RATTAN, SEE NOTE 3 |
| END DOOR | SLIDING | SEAT TYPE | INDIVIDUAL CUSHION, SEE NOTE 3 |
| FAN | PADDLE [4] | STRUCTURE MATERIAL | STEEL |
| MARKER LIGHT | ELECTRIC LAMP | TAIL LIGHT | KEROSENE LANTERN, SEE NOTE 4 |

## TRUCK

| | | | | |
|---|---|---|---|---|
| MANUFACTURER | MOTOR | MCB/ HEDLEY/ BAR | JOURNAL BEARING | FRICTION |
| | TRAILER | COMMONWEALTH/ HEDLEY/ BAR | TRACTION MOTOR | GE 260 (195 HP) [2/MOTOR TRUCK] |
| AXLE LOAD | # 1 END - TRAILER | 22,650 LBS | | WH 577 (200 HP) [2/MOTOR TRUCK] |
| | # 2 END - MOTOR | 16,000 LBS | WEIGHT | 35,000 LBS |

## GENERAL

| | | | |
|---|---|---|---|
| CAR BODY & EQUIPMENT WEIGHT | 42,050 LBS | CAR LOADED WEIGHT | 104,490 LBS |
| CAR LIGHT WEIGHT | 77,050 LBS, SEE NOTE 1 | STANDING CAPACITY | [152] |

NOTE 1: CARS 5017, 5300, 5301 (CAR LIGHT WEIGHT - 77,300LBS).
NOTE 2: DROP SEATS AT CENTER DOORS WERE REMOVED (1942 - 1948).
NOTE 3: INDIVIDUAL SEATS ON SOME CARS WERE REPLACED WITH A THREE SEAT FOAM RUBBER CUSHION COVERED WITH RED VINYL DURING OVERHAUL (1955 - 1958).
NOTE 4: RED LENS FOR TAIL, WHITE LENS FOR RUNNING.

CARS HAD MASONITE CEILINGS.
CARS EQUIPPED WITH MUDC (EARLY 1920's).
CARS WITH KEROSENE TAIL LIGHTS WERE CONVERTED TO BATTERY OPERATION (BY 1958). BUILT WITH BRASS WINDOW SASHES EXCEPT SEVERAL CARS HAD WOODEN REPLACEMENTS.
CARS 5022, 5032, 5145, 5162, 5182, 5256, 5257 MODIFIED FOR OPERATION WITH BLOWER TRAIN. CAR 5032 HAD HAND BARS IN PLACE OF HAND STRAPS AND DARK GREEN LEATHER COVERED SEATS.
CAR 5130 HAD HANDBARS IN PLACE OF HAND STRAPS AND DARK GREEN VELON TYPE SEATS.
CARS 5290, 5292 SOLD TO RAILWAY PRESERVATION CORPORATION, NEW YORK, NY (1989).
CARS 5233 - 5301 WERE PILOT CARS.
CAR 5302, PAY AND COLLECTION CAR, HAD NO CENTER DOORS AND A CAR LIGHT WEIGHT OF 83,410 LBS. BUILT TO RESEMBLE A PASSENGER CAR AND NUMBERED IN SEQUENCE AND WAS PLACED IN THE MIDDLE OF A TRAIN TO AVOID BEING NOTICED.
VESTIBULE AT CAR ENDS.
ALL OTHER CARS WERE SCRAPPED (1960's, 1970's, 1980's).

CAR NUMBERS: 5303 - 5402
TOTAL: 100 CARS
BUILT BY: PULLMAN

DATE: 1922
AVERAGE COST PER CAR: N/A

# 1922 (IRT)

LV TRAILER

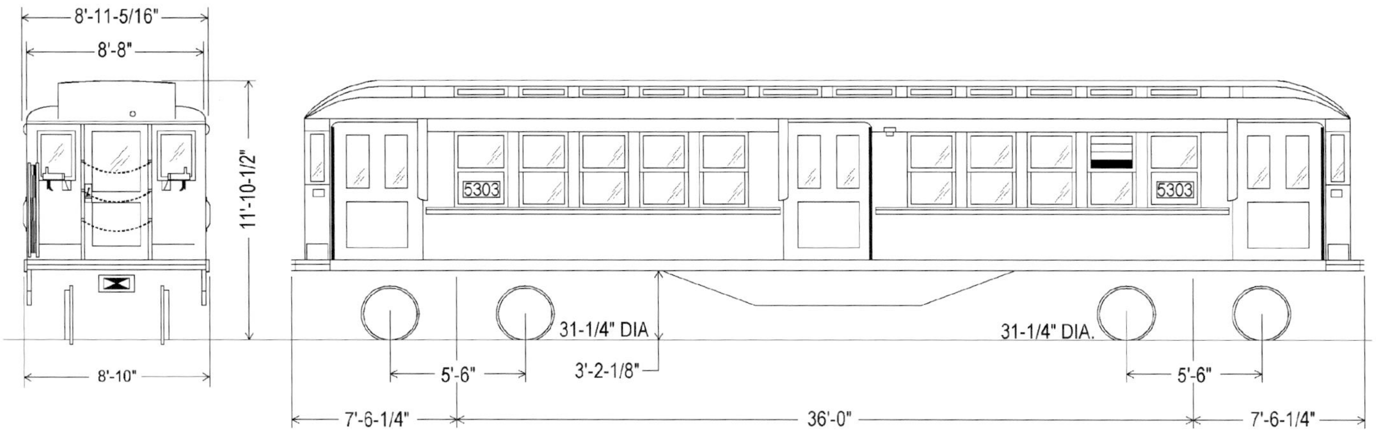

Copyright © 1997 by NYC Transit

## CAR BODY - EQUIPMENT

| | | | |
|---|---|---|---|
| AIR COMPRESSOR | GE CP28 | SEAT ARRANGEMENT & CAPACITY | LONGITUDINAL [44], SEE NOTE 1 |
| BRAKE TYPE | WABCO, ATUE/ME23 | SEAT MATERIAL | RATTAN, SEE NOTE 2 |
| COUPLER | J | SEAT TYPE | INDIVIDUAL CUSHION |
| END DOOR | SLIDING | STRUCTURE MATERIAL | STEEL |
| FAN | PADDLE [4] | | |

## TRUCK

| | | | | |
|---|---|---|---|---|
| MANUFACTURER | COMMONWEALTH | | JOURNAL BEARING | FRICTION |
| AXLE LOAD<br># 1 END - TRAILER | CARS WITH AIR COMPRESSOR | 15,406 LBS | WEIGHT (2 TRUCKS) | 22,320 LBS |
| | CARS WITHOUT AIR COMPRESSOR | 15,273 LBS | | |
| AXLE LOAD<br># 2 END - TRAILER | CARS WITH AIR COMPRESSOR | 15,120 LBS | | |
| | CARS WITHOUT AIR COMPRESSOR | 14,572 LBS | | |

## GENERAL

| | | | | | |
|---|---|---|---|---|---|
| CAR BODY & EQUIPMENT WEIGHT | CARS WITH AIR COMPRESSOR | 38,732 LBS | CAR LOADED WEIGHT | CARS WITH AIR COMPRESSOR | 88,492 LBS |
| | CARS WITHOUT AIR COMPRESSOR | 37,370 LBS | | CARS WITHOUT AIR COMPRESSOR | 87,130 LBS |
| CAR LIGHT WEIGHT | CARS WITH AIR COMPRESSOR | 61,052 LBS | STANDING CAPACITY | [152] | |
| | CARS WITHOUT AIR COMPRESSOR | 59,690 LBS | | | |

NOTE 1: DROP SEATS AT CENTER DOORS WERE REMOVED (1941 - 1948).

NOTE 2: INDIVIDUAL SEATS ON SOME CARS WERE REPLACED WITH A THREE SEAT FOAM CUSHION COVERED WITH RED VINYL DURING OVERHAUL (1955 - 1958).

NOTE 3:

| CARS | |
|---|---|
| 5303 - 5377 | CARS WITH AIR COMPRESSOR |
| 5378 - 5402 | CARS WITHOUT AIR COMPRESSOR |

CARS HAD MASONITE CEILINGS.
CARS EQUIPPED WITH MUDC UPON DELIVERY OR SOON AFTER.
SOME CARS HAD LARGE SINGLE-PIECE METAL ROUTE/DESTINATION SIGN INSTALLED (1968).
CARS HAD WOOD WINDOW SASHES. SOME AIR COMPRESSORS REMOVED FROM CARS 5303-5377.
CAR 5365 CONVERTED TO INSTRUCTION CAR.
NO TAIL OR MARKER LIGHTS.
VESTIBULE AT CAR ENDS.
ALL CARS WERE SCRAPPED (1960's, 1970's, 1980's).

CAR NUMBERS: 5403 - 5502
TOTAL: 100 CARS
BUILT BY: AMERICAN CAR FOUNDRY

DATE: 1924
AVERAGE COST PER CAR: N/A

# 1924 (IRT)

LV MOTOR

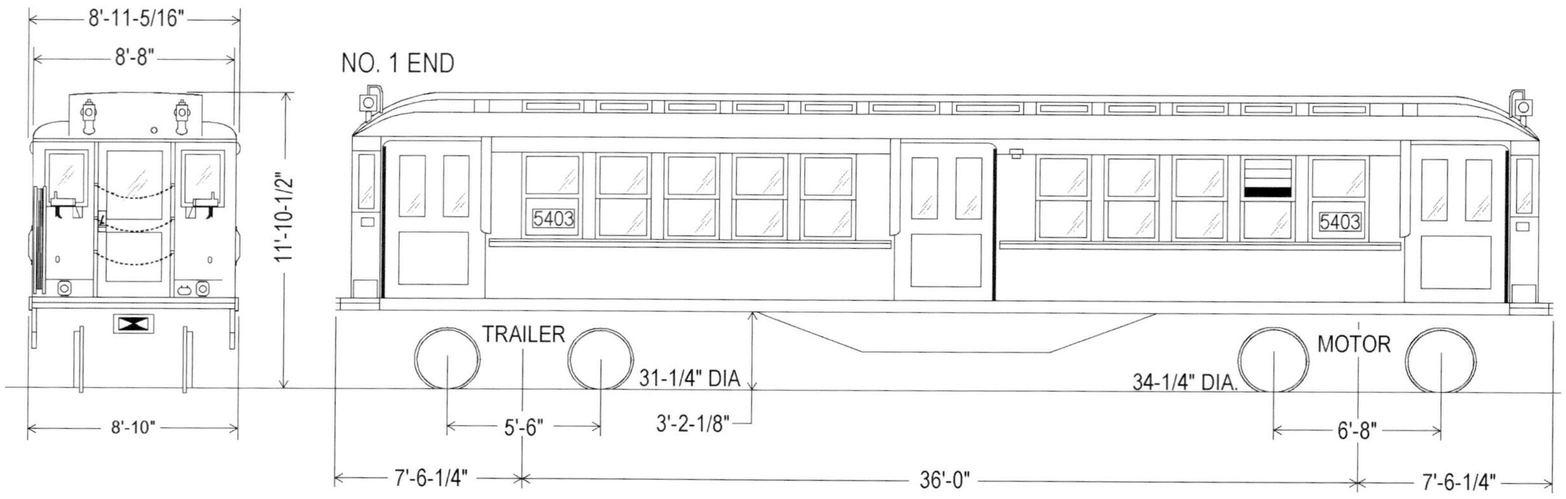

8'-11-5/16"
8'-8"
8'-10"
11'-10-1/2"
NO. 1 END
5403
TRAILER
MOTOR
31-1/4" DIA
34-1/4" DIA.
5'-6"
3'-2-1/8"
6'-8"
7'-6-1/4"
36'-0"
7'-6-1/4"

Copyright © 1997 by NYC Transit

## CAR BODY-EQUIPMENT

| | | | |
|---|---|---|---|
| AIR COMPRESSOR | WH D2F | PROPULSION CONTROL | GE PC10K |
| BATTERY | EDISON, B2H 24 CELLS | SEAT ARRANGEMENT & CAPACITY | LONGITUDINAL [44], SEE NOTE 1 |
| BRAKE TYPE | WABCO, AMUE/ME23 | SEAT MATERIAL | RATTAN |
| COUPLER | J | SEAT TYPE | INDIVIDUAL CUSHION |
| FAN | PADDLE [4] | STRUCTURE MATERIAL | STEEL |
| MARKER LIGHT | ELECTRIC LAMP | TAIL LIGHT | KEROSENE LANTERN, SEE NOTE 2 |
| MASTER CONTROLLER | CJ-131 | | |

## TRUCK

| | | | | |
|---|---|---|---|---|
| MANUFACTURER | MOTOR | MCB/ HEDLEY/ BAR | JOURNAL BEARING | FRICTION |
| | TRAILER | COMMONWEALTH/ HEDLEY/ BAR | TRACTION MOTOR | GE 260 (195 HP) [2/MOTOR TRUCK] |
| AXLE LOAD | # 1 END - TRAILER | 16,151 LBS | | WH 577 (200 HP) [2/MOTOR TRUCK] |
| | # 2 END - MOTOR | 23,214 LBS | WEIGHT ( 2 TRUCKS) | 36,224 LBS |

## GENERAL

| | | | |
|---|---|---|---|
| CAR BODY & EQUIPMENT WEIGHT | 42,506 LBS | CAR LOADED WEIGHT | 106,170 LBS |
| CAR LIGHT WEIGHT | 78,730 LBS | STANDING CAPACITY | [152] |

NOTE 1: DROP SEAT AT CENTER DOORS REMOVED (1941 - 1948).
NOTE 2: RED LENS FOR TAIL, WHITE LENS FOR RUNNING.

CARS HAD MASONITE CEILINGS.
MUDC INSTALLED UPON DELIVERY.
MOST CARS CONVERTED TO WORK MOTORS (EARLY 1960's). CARS HAD WOOD WINDOW SASHES.
BATTERY OPERATED TAIL LIGHTS (BY 1958).
CARS 5425, 5439, 5456, 5457, 5485 CONVERTED TO WORK MOTORS HAVING TWO TRUCKS WITH TWO MOTORS PER TRUCK.
CAR 5470 CONVERTED TO SNOW PLOW (EARLY 1960's). CARS 5403-5502 WERE PILOT CARS. CARS 5432, 5462, 5488 CONVERTED TO REVENUE COLLECTION CARS.
INDIVIDUAL SEATS ON SOME CARS WERE REPLACED WITH A THREE SEAT FOAM RUBBER CUSHION COVERED WITH RED VINYL DURING OVERHAUL (1955 - 1958).
CAR 5466 SOLD TO BRANFORD ELECTRIC RAILWAY ASSOCIATION (MUSEUM), EAST HAVEN, CT (1989). CARS 5443, 5483 SOLD TO RAILWAY PRESERVATION CORPORATION, NEW YORK, NY (1989).
VESTIBULE AT CAR ENDS.
ALL OTHER CARS WERE SCRAPPED (1960's, 1970's, 1980's).

CAR NUMBERS: 5503 - 5627
TOTAL: 125 CARS
BUILT BY: AMERICAN CAR FOUNDRY

DATE: 1925
AVERAGE COST PER CAR: N/A

# 1925 (IRT)

LV MOTOR

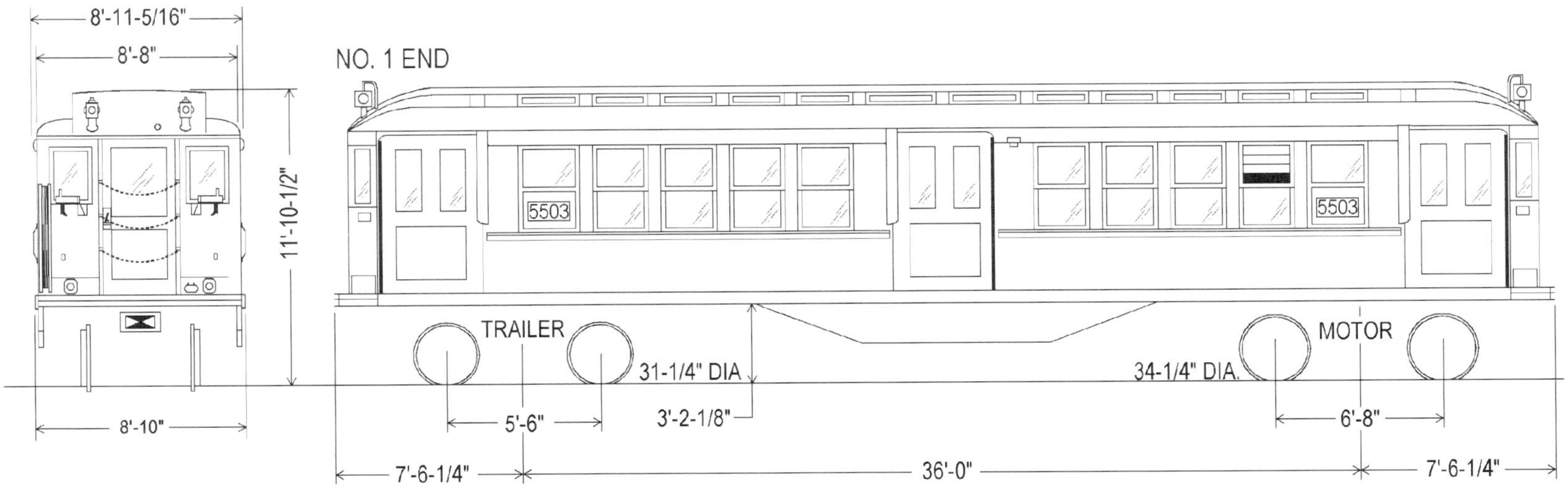

8'-11-5/16"
8'-8"
11'-10-1/2"
8'-10"
NO. 1 END
5503
5503
TRAILER
MOTOR
31-1/4" DIA
34-1/4" DIA.
5'-6"
3'-2-1/8"
6'-8"
7'-6-1/4"
36'-0"
7'-6-1/4"

Copyright © 1997 by NYC Transit

## CAR BODY-EQUIPMENT

| | | | |
|---|---|---|---|
| AIR COMPRESSOR | WH D3F | PROPULSION CONTROL | GE PC10K1 |
| BATTERY | EDISON, B2H 24 CELLS | SEAT ARRANGEMENT & CAPACITY | LONGITUDINAL [44], SEE NOTE 2 |
| BRAKE TYPE | WABCO, AMUE/ME30 | SEAT MATERIAL | RATTAN |
| COUPLER | J | SEAT TYPE | INDIVIDUAL CUSHION |
| MARKER LIGHT | ELECTRIC LAMP | STRUCTURE MATERIAL | STEEL |
| MASTER CONTROLLER | CJ-131 | TAIL LIGHT | KEROSENE LANTERN, SEE NOTE 3 |

## TRUCK

| | | | | |
|---|---|---|---|---|
| MANUFACTURER | MOTOR | MCB/ HEDLEY/ BAR | JOURNAL BEARING | FRICTION |
| | TRAILER | COMMONWEALTH/ HEDLEY/ BAR | TRACTION MOTOR | GE 260 (195 HP) [2/MOTOR TRUCK] |
| AXLE LOAD | # 1 END - TRAILER | 16,254 LBS | | WH 577 (200 HP) [2/MOTOR TRUCK] |
| | # 2 END - MOTOR | 23,440 LBS | WEIGHT ( 2 TRUCKS) | 36,224 LBS |

## GENERAL

| | | | |
|---|---|---|---|
| CAR BODY & EQUIPMENT WEIGHT | 43,164 LBS | CAR LOADED WEIGHT | 106,828 LBS |
| CAR LIGHT WEIGHT | 79,388 LBS, SEE NOTE 1 | STANDING CAPACITY | [152] |

NOTE 1: CARS 5507, 5542, 5573, 5587 WEIGHT 78,888 LBS.
NOTE 2: DROP SEATS AT CENTER DOORS WERE REMOVED (1942 - 1948).
NOTE 3: RED LENS FOR TAIL, WHITE LENS FOR RUNNING.

CARS HAD MASONITE CEILINGS.
MUDC INSTALLED UPON DELIVERY.
CARS HAD WOODEN SASHES. MOST CARS IN THIS SERIES BECAME WORK CARS (EARLY 1960's).
CARS 5554 CONVERTED TO WORK MOTOR HAVING TWO TRUCKS WITH TWO MOTORS PER TRUCK. CARS 5504, 5606 CONVERTED TO SNOW PLOW (EARLY 1960's).
CARS 5518, 5566 CONVERTED TO WELDING CARS. CAR 5600 CONVERTED TO SCHOOL CAR. CARS 5524, 5559, 5602, 5615, 5622 CONVERTED TO REVENUE COLLECTION CARS.
INDIVIDUAL SEATS ON SOME CARS WERE REPLACED WITH A THREE SEAT FOAM CUSHION COVERED WITH RED VINYL DURING OVERHAUL (1955 - 1958).
CAR 5600 SOLD TO TROLLEY MUSEUM OF NEW YORK, KINGSTON, NY (1980) AND WAS LAST STANDARD BODY IRT CAR REMOVED FROM NYC TRANSIT PROPERTY (APPROXIMATELY 1990).
VESTIBULE AT CAR ENDS.
ALL OTHER CARS WERE SCRAPPED (1960's, 1970's, 1980's).

CAR NUMBERS: 5628 - 5652
TOTAL: 25 CARS
BUILT BY: AMERICAN CAR FOUNDRY

DATE: 1925
AVERAGE COST PER CAR: N/A

# 1925 (IRT)

STEINWAY, LV MOTOR

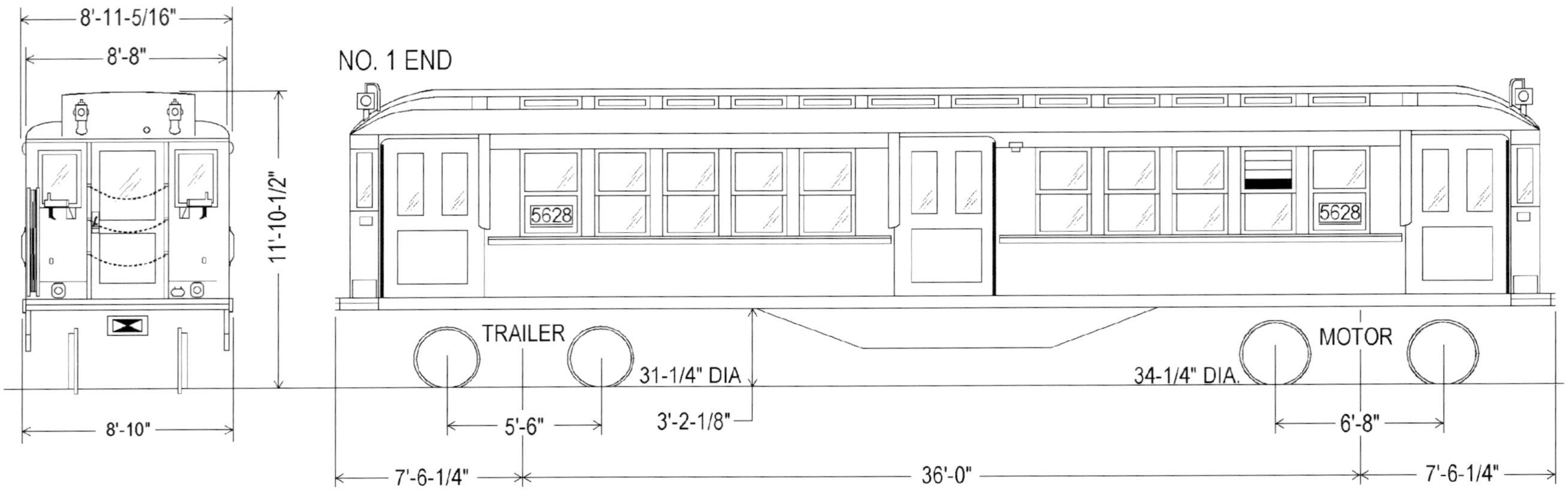

8'-11-5/16"
8'-8"
11'-10-1/2"
8'-10"
NO. 1 END
5628
5628
TRAILER
31-1/4" DIA
MOTOR
34-1/4" DIA.
5'-6"
3'-2-1/8"
6'-8"
7'-6-1/4"
36'-0"
7'-6-1/4"

Copyright © 1997 by NYC Transit

## CAR BODY-EQUIPMENT

| | | | |
|---|---|---|---|
| AIR COMPRESSOR | WH D2F | PROPULSION CONTROL | GE PC10K1 |
| BATTERY | EDISON, B2H 24 CELLS | SEAT ARRANGEMENT & CAPACITY | LONGITUDINAL [44], SEE NOTE 1 |
| BRAKE TYPE | WABCO, AMUE/ME30 | SEAT MATERIAL | RATTAN |
| COUPLER | J | SEAT TYPE | INDIVIDUAL CUSHION |
| END DOOR | SLIDING | STRUCTURE MATERIAL | STEEL |
| MARKER LIGHT | ELECTRIC LAMP | TAIL LIGHT | KEROSENE LANTERN, SEE NOTE 2 |
| MASTER CONTROLLER | CJ-131 | | |

## TRUCK

| | | | | |
|---|---|---|---|---|
| MANUFACTURER | MOTOR | BAR/ HEDLEY | JOURNAL BEARING | FRICTION |
| | TRAILER | BAR/ HEDLEY | TRACTION MOTOR | GE 259 (120 HP) [2/MOTOR TRUCK] |
| AXLE LOAD | # 1 END - TRAILER | 16,125 LBS | WEIGHT ( 2 TRUCKS) | 32,760 LBS |
| | # 2 END - MOTOR | 21,565 LBS | | |

## GENERAL

| | | | |
|---|---|---|---|
| CAR BODY & EQUIPMENT WEIGHT | 42,620 LBS | CAR LOADED WEIGHT | 102,820 LBS |
| CAR LIGHT WEIGHT | 75,380 LBS | STANDING CAPACITY | [152] |

NOTE 1: DROP SEATS AT CENTER DOORS WERE REMOVED (1947 - 1951).
NOTE 2: RED LENS FOR TAIL, WHITE LENS FOR RUNNING.

CARS HAD MASONITE CEILINGS.
MUDC INSTALLED UPON DELIVERY.
TAIL LIGHT CONVERTED TO BATTERY OPERATION (BY 1958).
CARS HAD WOOD WINDOW SASHES.
SOME CARS HAD LARGE SINGLE-PIECE METAL ROUTE/DESTINATION SIGNS INSTALLED (1968).
INDIVIDUAL SEATS ON SOME CARS WERE REPLACED WITH A THREE SEAT FOAM CUSHION COVERED WITH RED VINYL DURING OVERHAUL (1955 - 1958).
VESTIBULE AT CAR ENDS.
ALL CARS WERE SCRAPPED (1960's & 1970's).

CAR NUMBERS: 5653 - 5702
TOTAL: 50 CARS
BUILT BY: ST. LOUIS CAR

DATE: 1938
AVERAGE COST PER CAR: $30,786.60

# 1938 (IRT)

WORLD'S FAIR, STEINWAY, LV MOTOR

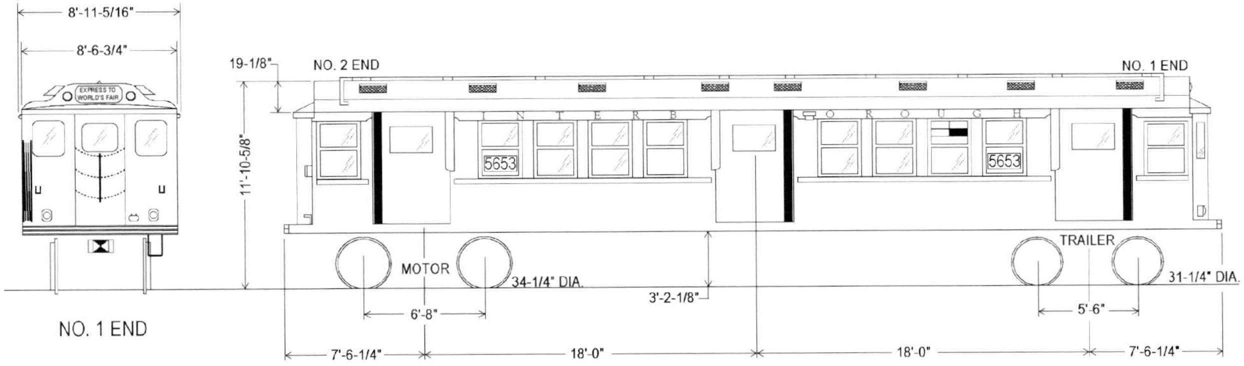

8'-11-5/16"
8'-6-3/4"
EXPRESS TO WORLD'S FAIR
NO. 1 END
19-1/8"
11'-10-5/8"
NO. 2 END
NO. 1 END
5653
5653
MOTOR
34-1/4" DIA.
TRAILER
31-1/4" DIA.
3'-2-1/8"
6'-8"
5'-6"
7'-6-1/4"
18'-0"
18'-0"
7'-6-1/4"

Copyright © 1997 by NYC Transit

## CAR BODY - EQUIPMENT

| | | | |
|---|---|---|---|
| AIR COMPRESSOR | WABCO, D2F | HEATER | CONSOLIDATED CAR HEATING |
| BATTERY | EDISON, B2H 24 CELL | LIGHTING | 30 V [22/CIRCUIT], 600 V CIRCUIT[2] |
| BRAKE EQUIPMENT | WABCO, AMUE | MASTER CONTROLLER | WH, XM129 |
| BRAKE TYPE | ELECTROPNEUMATIC | PROPULSION CONTROL GROUP MODE C | WH, UP231B |
| COUPLER | J | PROPULSION CONTROL TYPE | WH, UNIT SWITCH |
| DOOR ENGINE | CONSOLIDATED CAR HEATING | SEAT ARRANGEMENT & CAPACITY | LONGINTUDINAL [48] |
| ELECTRIC PORTION | JUMPER CABLES | SEAT TYPE | TWO SEAT CUSHION |
| END DOOR | SLIDING | SIDE SIGN | PLATE |
| END SIGN | HUNTER CAR SIGN | STRUCTURE MATERIAL | STEEL |
| FAN | WH [4] | SWITCH PANEL | CONSOLIDATED CAR HEATING PS 57A, #1 END |
| FIRE EXTINGUISHER | 2 LBS DRY POWDER | TAIL LIGHT | KEROSENE LANTERN, SEE NOTE 1 |
| HAND BRAKE | BLACKALL | | |

## TRUCK

| | | | | | |
|---|---|---|---|---|---|
| MANUFACTURER | | ST. LOUIS CAR | TRIP COCK | | WABCO, A2B |
| AXLE LOAD | #1 END - TRAILER | 16,120 LBS | TRUCK NUMBERS | MOTOR | 3873 - 3922 |
| BRAKE RIGGING | | SINGLE SHOE | | TRAILER | 1150 - 1199 |
| JOURNAL BEARING | | FRICTION, 5 X 9 | TYPE | | BUILT UP FRAME EQUALIZED BAR |
| QUANTITY | | [100] | WEIGHT | MOTOR | 22,160 LBS |
| TRACTION MOTOR | | WH, 336A1 [2/MOTOR TRUCK] 125 HP | | TRAILER | 11,510 LBS |

## GENERAL

| | | | |
|---|---|---|---|
| CAR BODY & EQUIPMENT WEIGHT | 41,460 LBS | CAR LOADED WEIGHT | 103,410 LBS |
| CAR LIGHT WEIGHT | 75,130 LBS | STANDING CAPACITY | [154] |

NOTE 1: RED LENS FOR TAIL, WHITE LENS FOR RUNNING.

ALL CARS HAD ONE CAB AT NO.1 END AND DOOR CONTROLS AT OPPOSITE END.
ALL CARS HAD LARGE SINGLE-PIECE METAL ROUTE/DESTINATION SIGNS INSTALLED AND ADDED TO MOST CARS LARGE CAR NUMBER STICKERS (1968).
FIRST SERIES OF CARS DELIVERED WITH PORCELAIN CAR NUMBER PLATES AND SMALL SIDE SIGN PLATES. ROOF TYPE KNOWN AS OGEE.
ONLY SERIES OF IRT SUBWAY CARS WITHOUT WINDOW SHADES.
CARS HAD BRASS WINDOW SASHES. ADDED THE FOLLOWING: END STEP (1951), TRICKLE CHARGE (1954), ELECTRIC RUNNING LIGHTS (BY 1958), AND HEADLIGHTS ADDED TO MOST CARS (EARLY 1960's).
MUDC INSTALLED ON DELIVERY.
CARS 5660 & 5689 CONVERTED TO LOW SPEED WORK MOTORS (1963) FOR OPERATION WITH VACUUM CLEANING TRAIN (R31 CONTRACT - 1963). CARS 5660 & 5689 WERE RENUMBERED TO 20501 & 20502 AND THEN WERE RENUMBERED TO V286 & V287.
CARS 5693 - 5702 HAD INSULATED ROOFS.
ALL CARS CONVERTED TO WORK MOTORS (AFTER 1969).
CAR 5655 AT NYC TRANSIT (CONEY ISLAND SHOP) UNDER RESTORATION. ALL OTHER CARS WERE SCRAPPED (1970's & EARLY 1980's).

999

BROOKLYN RAPID TRANSIT.
999
INSTRUCTION CAR.
999
D150

BRIGH
CONE

1479

CAR NUMBERS: 1 - 51
TOTAL: 51 CARS
BUILT BY: PULLMAN

DATE: 1884
AVERAGE COST PER CAR: $4,500

# BU CLASS (BMT)

TRAILER

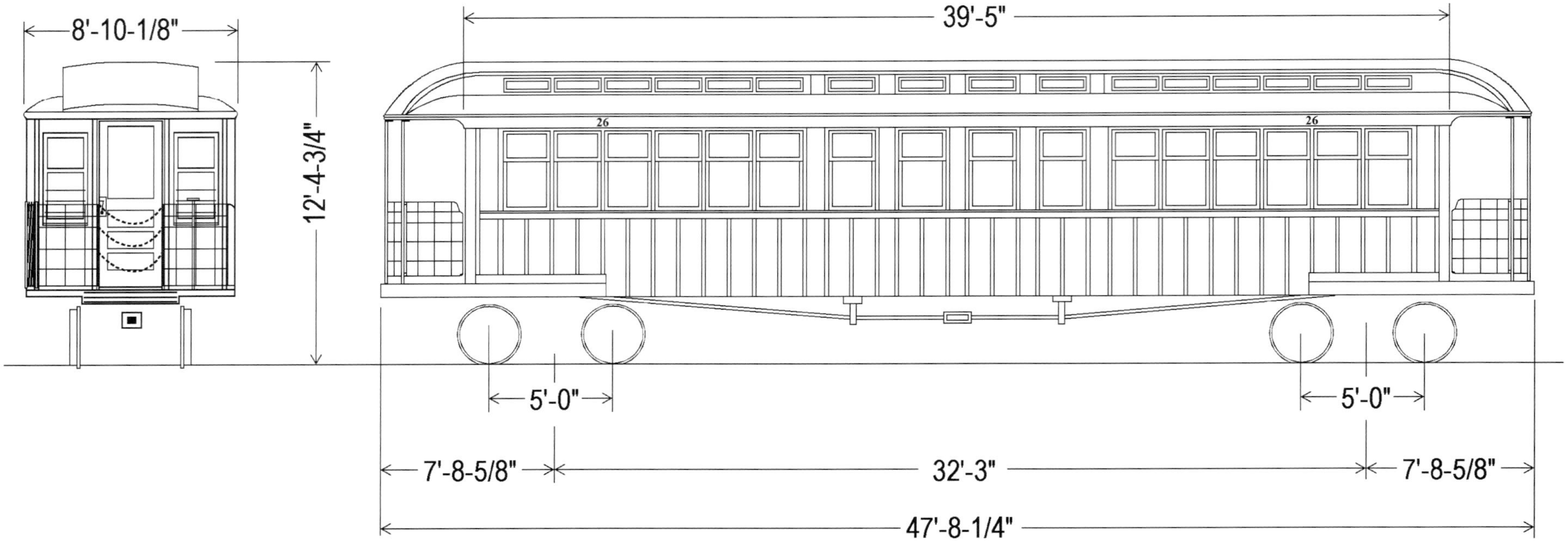

8'-10-1/8"
12'-4-3/4"
39'-5"
26
26
5'-0"
5'-0"
7'-8-5/8"
32'-3"
7'-8-5/8"
47'-8-1/4"

Copyright © 1997 by NYC Transit

## CAR BODY - EQUIPMENT

| | | | |
|---|---|---|---|
| BRAKE EQUIPMENT TYPE | WH, ATL | SEAT ARRANGEMENT & CAPACITY | MANHATTAN [48]: CARS 1-29, 31-34, 45-51 , SEE NOTE 1 |
| HAND BRAKE | STAFF | | CROSS [60]: CARS 30, 35-44 |
| HAND HOLD | LEATHER STRAP | SEAT MATERIAL | RATTAN |
| HEATER | CONSOLIDATED CAR HEATING | STRUCTURE MATERIAL | WOOD |
| LIGHTING | LAMPS [25] | WINDOW CURTAIN | BURROWS, PANTASOTE 83 |

## TRUCK

| | | | |
|---|---|---|---|
| MANUFACTURER | PULLMAN | | |

## GENERAL

| | | | |
|---|---|---|---|
| CAR LIGHT WEIGHT | 33,200 LBS - CARS 1-29, 31-34, 45-51 | STANDING CAPACITY | CARS 1-29, 31-34, 45-51 [99]; CARS 30, 35-44 [70] |
| | 34,300 LBS - CARS 30, 35-44 | | |

NOTE 1: LONGITUDINAL EXCEPT FOUR CROSS SEATS ON EACH SIDE OF AISLE OF WHICH TWO SEATS FACED EACH OTHER AND WERE LOCATED AT CENTER OF CAR LENGTH.

ORIGINALLY HAD DECK ROOF MODIFIED TO EXISTING ROOF.
WINDOW CURTAINS REMOVED (1920's).
PAINTED VENTILATOR GLASS IN CLERESTORY OR REPLACED WITH SHEET IRON (BEGINNING 1921).
CAR 33 DESTROYED BY FIRE (12/7/1922).
CARS ORIGINALLY BUILT AS STEAM COACHES FOR BROOKLYN ELEVATED RAILROAD AND NUMBERED IN THE 100-189 SERIES.
CARS REBUILT FOR ELECTRIC OPERATION AT A COST OF $3000 PER CAR (1904 - 1906).
ALL CARS SCRAPPED (1940's).

CAR NUMBERS: 52 - 190, SEE NOTE 1
TOTAL: 139 CARS
BUILT BY: GILBERT, BRADLEY, PULLMAN, SEE NOTE 1

DATE: 1887, 1891, 1893, SEE NOTE 1
AVERAGE COST PER CAR: N/A

# BU CLASS (BMT)

TRAILER

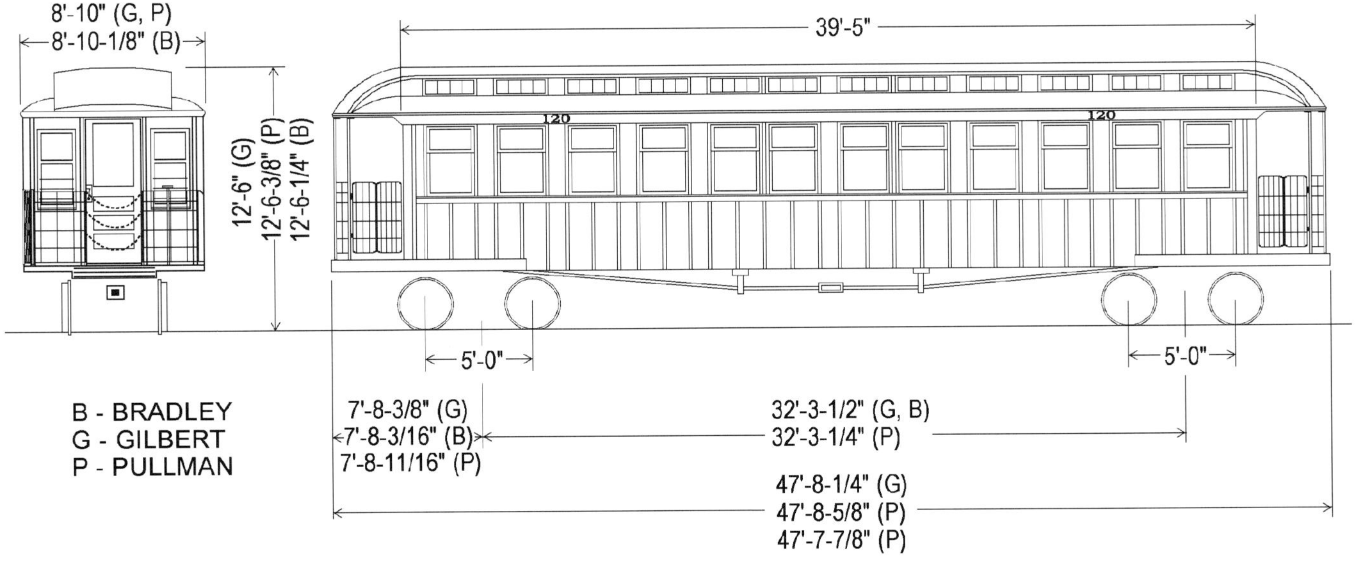

8'-10" (G, P)
8'-10-1/8" (B)
12'-6" (G)
12'-6-3/8" (P)
12'-6-1/4" (B)
39'-5"
5'-0"
5'-0"
7'-8-3/8" (G)
7'-8-3/16" (B)
7'-8-11/16" (P)
32'-3-1/2" (G, B)
32'-3-1/4" (P)
47'-8-1/4" (G)
47'-8-5/8" (P)
47'-7-7/8" (P)

B - BRADLEY
G - GILBERT
P - PULLMAN

Copyright © 1997 by NYC Transit

## CAR BODY - EQUIPMENT

| | | | |
|---|---|---|---|
| BRAKE EQUIPMENT TYPE | WH, ATL | SEAT ARRANGEMENT & CAPACITY | CROSS [60], CARS 53, 54, 58-60, 63-68 |
| HAND BRAKE | STAFF | | LONGITUDINAL [52], CARS 57,62,70-79, 81, 83-95 |
| HAND HOLD | LEATHER STRAP | | MANHATTAN [48], CARS 133 -190, SEE NOTE 2 |
| HEATER | CONSOLIDATED CAR HEATING | WINDOW CURTAIN | BURROWS, PANTASOTE 83 |
| LIGHTING | LAMPS [25] | | |

## TRUCK

| | | | |
|---|---|---|---|
| MANUFACTURER | GILBERT AND PULLMAN | | |

## GENERAL

| | | | |
|---|---|---|---|
| CAR LIGHT WEIGHT | 33,720 LBS: CARS 52, 54, 58-60, 63-68 | STANDING CAPACITY | [70] CARS 52, 54, 58-60, 63-68 |
| | 33,350 LBS: CARS 57, 62, 70-79, 81, 83-95 | | [101] CARS 57, 62, 70-79, 81, 83-95 |
| | 34,150 LBS: CARS 133-190 | | [99] CARS 133-190 |

NOTE 1: GILBERT, CAR NUMBERS 52 - 132 (1887); PULLMAN, CAR NUMBERS 133 - 137 (1891); BRADLEY, 138 - 190 (1893).
NOTE 2: LONGITUDINAL EXCEPT FOUR CROSS SEATS ON EACH SIDE OF AISLE OF WHICH TWO SEATS FACED EACH OTHER AND WERE LOCATED AT CENTER OF CAR LENGTH.

CARS WERE ORIGINALLY BUILT AS STEAM COACHES FOR THE BROOKLYN ELEVATED RAILROAD AND NUMBERED IN THE 190 - 309 SERIES.
CAR 82 DAMAGED AND REBUILT INTO MOTOR CAR 684 (1910) AND RENUMBERED 1261 (1930).
CARS REBUILT FOR ELECTRIC OPERATION AT A COST OF $3000 PER CAR (1904 - 1906).
WINDOW CURTAINS REMOVED (1920's).
PAINTED VENTILATOR GLASS IN CLERESTORY OR REPLACED WITH SHEET IRON (BEGINNING 1921).
SOME CARS WERE CONVERTED TO 'B' CARS OF THE 'C' TYPE CAR UNITS (1923 - 1925).
CARS 52, 55, 61, 69 SOLD TO SAND SPRINGS RAILWAY, TULSA, OK. (3/5/1918).
CAR 56 DEMOLISHED IN WRECK (1/12/1917).
CARS 80 & 100 DESTROYED IN MALBONE STREET DERAILMENT, BRIGHTON LINE, BROOKLYN (11/1/1918).
ALL CARS SCRAPPED (1940's).

CAR NUMBERS: 191 - 271
TOTAL: 81 CARS
BUILT BY: PULLMAN, HARLAN, SEE NOTE 1

DATE: 1888-1889, 1893, SEE NOTE 2
AVERAGE COST PER CAR: N/A

# BU CLASS (BMT)

KINGS COUNTY TRAILER

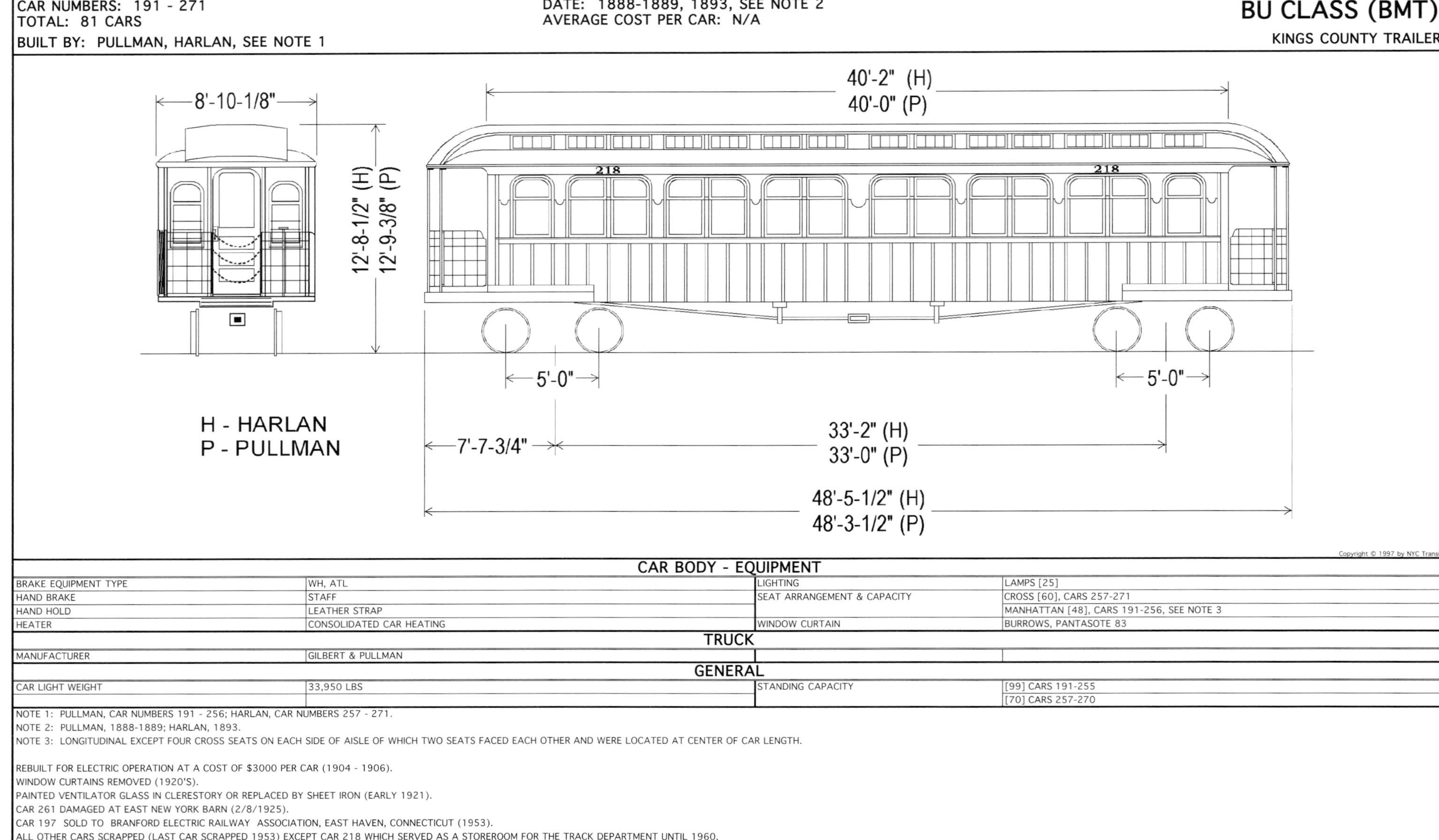


Copyright © 1997 by NYC Transit

| CAR BODY - EQUIPMENT | | | |
|---|---|---|---|
| BRAKE EQUIPMENT TYPE | WH, ATL | LIGHTING | LAMPS [25] |
| HAND BRAKE | STAFF | SEAT ARRANGEMENT & CAPACITY | CROSS [60], CARS 257-271 |
| HAND HOLD | LEATHER STRAP | | MANHATTAN [48], CARS 191-256, SEE NOTE 3 |
| HEATER | CONSOLIDATED CAR HEATING | WINDOW CURTAIN | BURROWS, PANTASOTE 83 |
| **TRUCK** | | | |
| MANUFACTURER | GILBERT & PULLMAN | | |
| **GENERAL** | | | |
| CAR LIGHT WEIGHT | 33,950 LBS | STANDING CAPACITY | [99] CARS 191-255 |
| | | | [70] CARS 257-270 |

NOTE 1: PULLMAN, CAR NUMBERS 191 - 256; HARLAN, CAR NUMBERS 257 - 271.
NOTE 2: PULLMAN, 1888-1889; HARLAN, 1893.
NOTE 3: LONGITUDINAL EXCEPT FOUR CROSS SEATS ON EACH SIDE OF AISLE OF WHICH TWO SEATS FACED EACH OTHER AND WERE LOCATED AT CENTER OF CAR LENGTH.

REBUILT FOR ELECTRIC OPERATION AT A COST OF $3000 PER CAR (1904 - 1906).
WINDOW CURTAINS REMOVED (1920'S).
PAINTED VENTILATOR GLASS IN CLERESTORY OR REPLACED BY SHEET IRON (EARLY 1921).
CAR 261 DAMAGED AT EAST NEW YORK BARN (2/8/1925).
CAR 197 SOLD TO BRANFORD ELECTRIC RAILWAY ASSOCIATION, EAST HAVEN, CONNECTICUT (1953).
ALL OTHER CARS SCRAPPED (LAST CAR SCRAPPED 1953) EXCEPT CAR 218 WHICH SERVED AS A STOREROOM FOR THE TRACK DEPARTMENT UNTIL 1960.

CAR NUMBERS: 600, 601, 683
TOTAL: 3 CARS
BUILT BY: GILBERT

DATE: 1887
AVERAGE COST PER CAR: N/A

# BU CLASS (BMT)

600 SERIES MOTOR

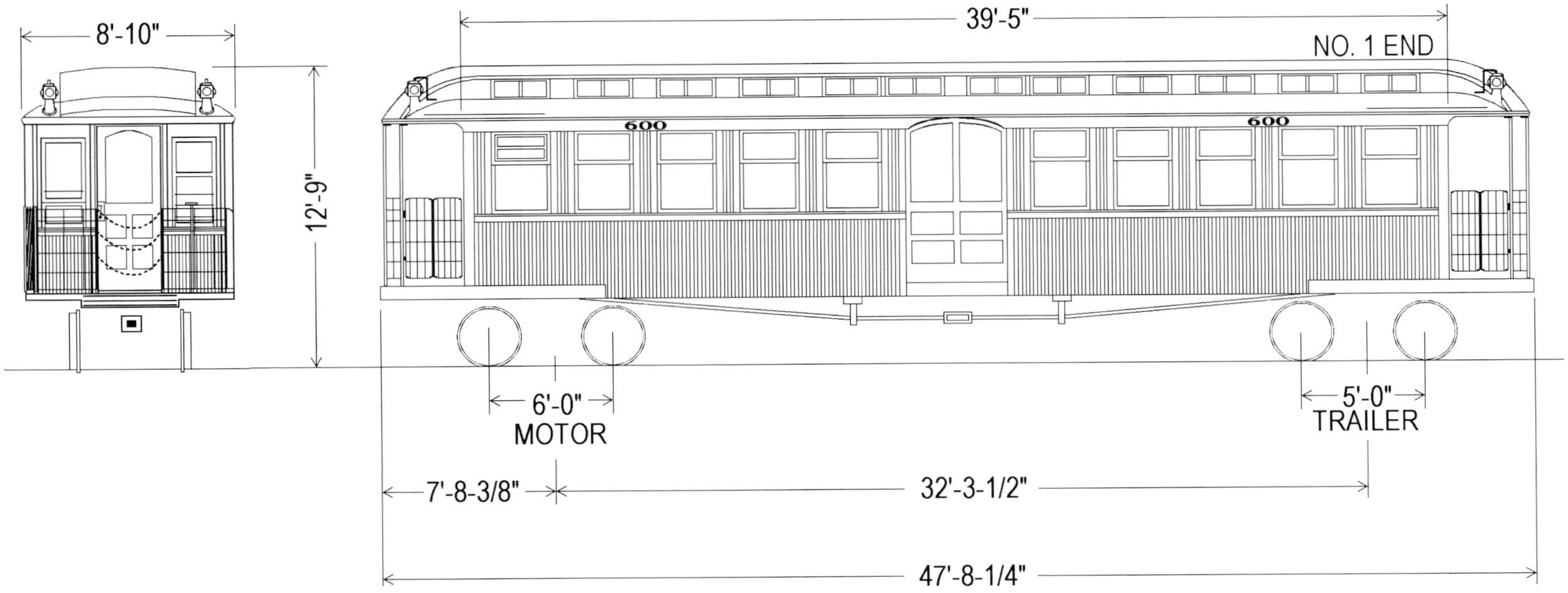

8'-10"
12'-9"
39'-5"
NO. 1 END
600
6'-0"
MOTOR
5'-0"
TRAILER
7'-8-3/8"
32'-3-1/2"
47'-8-1/4"

Copyright © 1997 by NYC Transit

## CAR BODY - EQUIPMENT

| | | | |
|---|---|---|---|
| AIR COMPRESSOR | CHRIS, B2 | LIGHTING | LAMPS [29] |
| BRAKE EQUIPMENT TYPE | WH, AML | PROPULSION CONTROL TYPE | WH, 131 |
| COUPLER | VAN DORN | SEAT ARRANGEMENT & CAPACITY | LONGITUDINAL [50] |
| HAND BRAKE | STAFF | SIGN, SIDE | METAL PLATES |
| HEATER | GOLD CAR HEATING | STRUCTURE MATERIAL | WOOD |
| | CONSOLIDATED CAR HEATING | WINDOW CURTAIN | BURROWS, PANTASOTE 83 |
| HAND HOLD | LEATHER STRAP | | |

## TRUCK

| | | | | |
|---|---|---|---|---|
| MANUFACTURER | MOTOR | BRILL, 27E | GEAR RATIO | 19:64 |
| | TRAILER | BRILL 27, CARS 600, 601 | TRACTION MOTOR | WH, 50E (150 HP) [2 /MOTOR TRUCK] |
| | | PECKHAM 40, CAR 683 | | |

## GENERAL

| | | | |
|---|---|---|---|
| CAR LIGHT WEIGHT | 66,130 LBS, CAR 601-602; 65,290 LBS, CAR 683 | STANDING CAPACITY | [96] |

ORIGINALLY BUILT AS STEAM COACHES FOR BROOKLYN ELEVATED RAILROAD.
CARS REBUILT INTO MOTOR CARS AT A COST OF $5000 PER CAR BY LENGTHENING PLATFORM, HOOD AND SEATS (6/1906).
WINDOW CURTAINS REMOVED (1920's).
PAINTED VENTILATOR GLASS IN CLERESTORY OR REPLACED WITH SHEET IRON (BEGINNING 1921).
CAR 601 PLACED OUT OF SERVICE DUE TO CONSTANT EQUIPMENT FAILURE (7/12/1924).
CARS 600 & 601 ORIGINALLY NUMBERED 145 & 190 RESPECTIVELY.
CAR 683 ORIGINALLY NUMBERED 188.
CENTER DOOR ADDED AT A LATER DATE.
ALL CARS SCRAPPED (1946).

CAR NUMBERS: 602 - 619, SEE NOTE 1
TOTAL: 18 CARS
BUILT BY: BRADLEY, PULLMAN

DATE: 1891-1893
AVERAGE COST PER CAR: N/A

# BU CLASS (BMT)

600 SERIES MOTOR

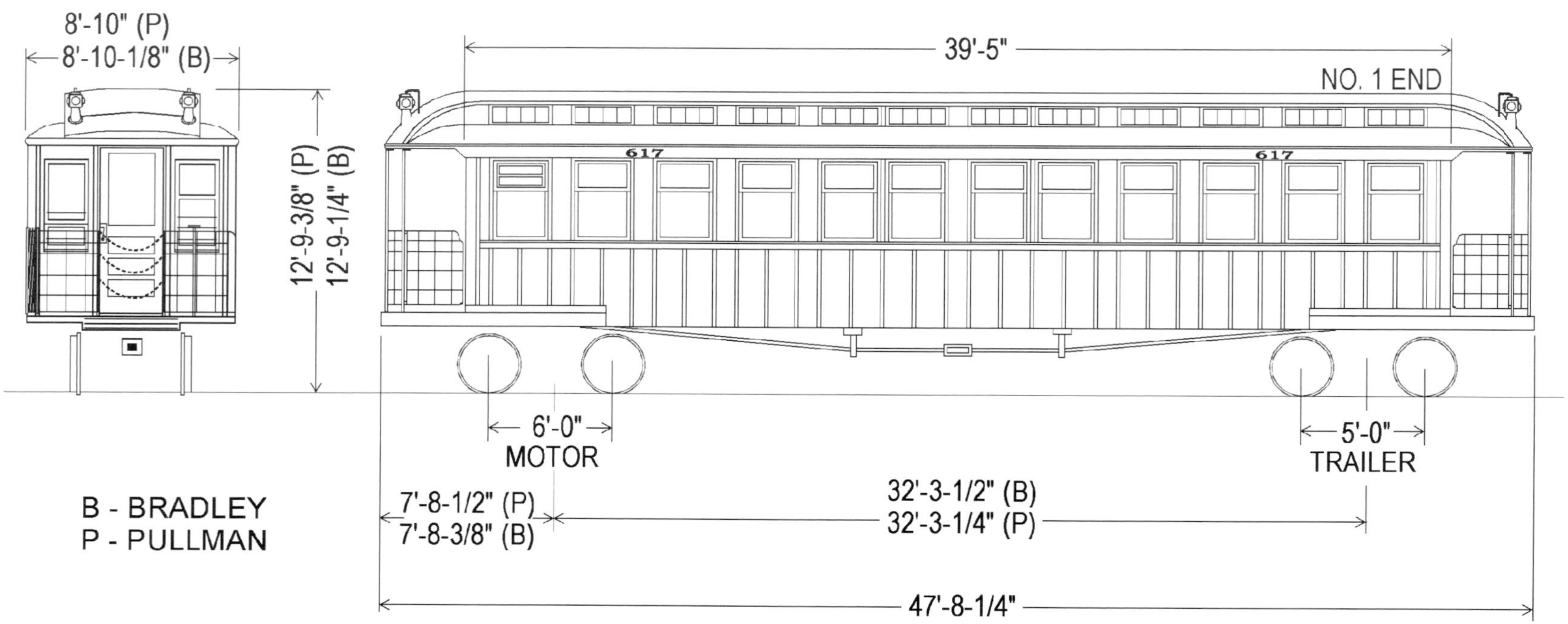

8'-10" (P)
8'-10-1/8" (B)
12'-9-3/8" (P)
12'-9-1/4" (B)
39'-5"
NO. 1 END
617
6'-0"
MOTOR
5'-0"
TRAILER
7'-8-1/2" (P)
7'-8-3/8" (B)
32'-3-1/2" (B)
32'-3-1/4" (P)
47'-8-1/4"

B - BRADLEY
P - PULLMAN

Copyright © 1997 by NYC Transit

## CAR BODY - EQUIPMENT

| | | | |
|---|---|---|---|
| AIR COMPRESSOR | CHRIS, B2 | LIGHTING | LAMPS [29] |
| BRAKE EQUIPMENT TYPE | WH, AML | PROPULSION CONTROL TYPE | WH, 131 |
| COUPLER | VAN DORN | SEAT ARRANGEMENT & CAPACITY | MANHATTAN [48], SEE NOTE 2 |
| HAND BRAKE | STAFF | WINDOW CURTAIN | BURROWS, PANTASOTE 83 |
| HEATER | CONSOLIDATED CAR HEATING, 143LL | | |

## TRUCK

| | | | | |
|---|---|---|---|---|
| MANUFACTURER | MOTOR | BRILL, 27E | GEAR RATIO | 19:64 |
| | TRAILER | PULLMAN | TRACTION MOTOR | WH, 50E (150 HP) [2 /MOTOR TRUCK] |

## GENERAL

| | | | |
|---|---|---|---|
| CAR LIGHT WEIGHT | 64, 900 LBS | STANDING CAPACITY | [88] |

NOTE 1: CAR NUMBERS - PULLMAN, 602 - 614; BRADLEY,615 - 619.
NOTE 2: LONGITUDINAL EXCEPT FOUR CROSS SEATS ON EACH SIDE OF AISLE OF WHICH TWO SEATS FACED EACH OTHER AND WERE LOCATED AT CENTER OF CAR LENGTH.

ORIGINALLY BUILT AS STEAM COACHES FOR BROOKLYN ELEVATED RAILROAD AND NUMBERED 231 - 301.
REBUILT AS MOTOR CARS AND RENUMBERED 412, 413, 415 - 431 AT A COST OF $5000 PER CAR (2/21/1905 TO 5/11/1905).
WINDOW CURTAINS REMOVED (1920's).
PAINTED VENTILATOR GLASS IN CLERESTORY OR REPLACED WITH SHEET IRON (BEGINNING 1921).
CAR 609 USED AS RUBBISH COLLECTION CAR (1940 - 1950).
ALL CARS SCRAPPED (1940's).

CAR NUMBERS: 620 - 627
TOTAL: 8 CARS
BUILT BY: PULLMAN

DATE: 1898
AVERAGE COST PER CAR: N/A

# BU CLASS (BMT)

600 SERIES MOTOR

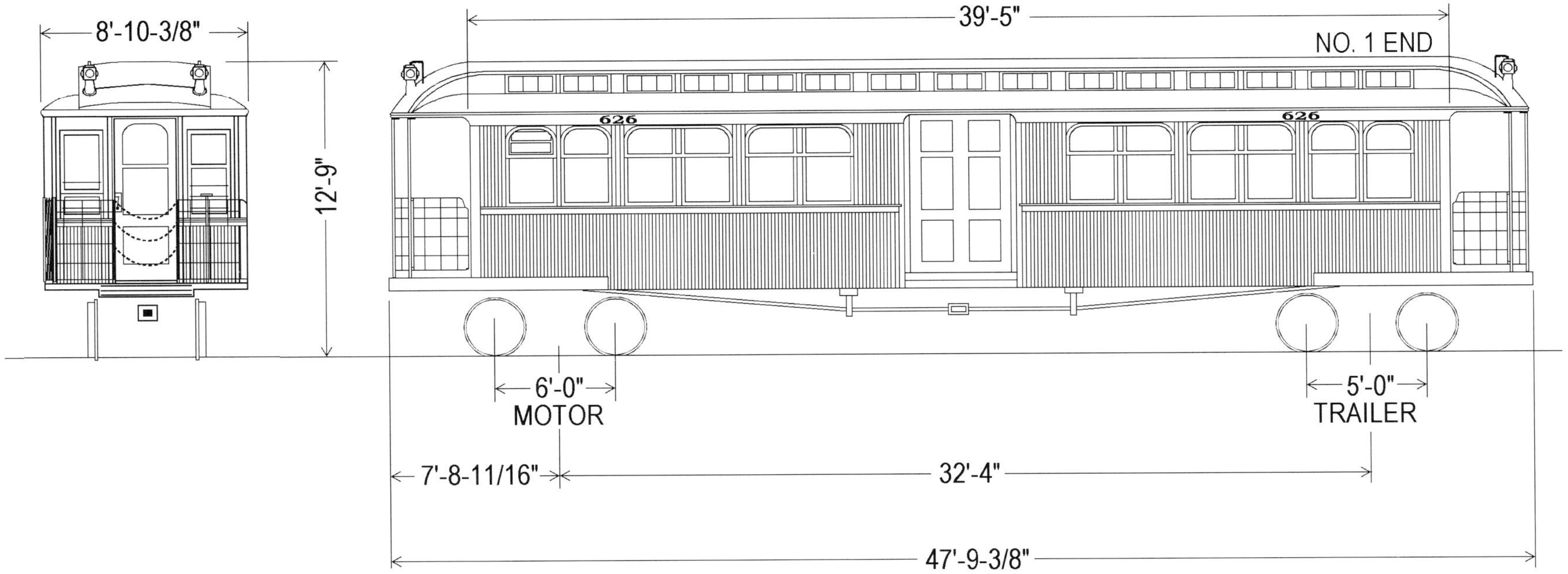

8'-10-3/8"
12'-9"
39'-5"
NO. 1 END
626
626
6'-0"
MOTOR
5'-0"
TRAILER
7'-8-11/16"
32'-4"
47'-9-3/8"

Copyright © 1997 by NYC Transit

## CAR BODY - EQUIPMENT

| | | | |
|---|---|---|---|
| AIR COMPRESSOR | CHRIS, B2 | LIGHTING | LAMPS [29] |
| BRAKE EQUIPMENT TYPE | WH, AML | PROPULSION CONTROL TYPE | WH, 131 |
| COUPLER | VAN DORN | SEAT ARRANGEMENT & CAPACITY | LONGITUDINAL [52], SEE NOTE 1 |
| HAND BRAKE | STAFF | WINDOW CURTAIN | FORSYTHE, PANTASOTE 86 |
| HEATER | CONSOLIDATED CAR HEATING, 143LL | | |

## TRUCK

| | | | | |
|---|---|---|---|---|
| MANUFACTURER | MOTOR | BRILL, 27E | GEAR RATIO | 19:64 |
| | TRAILER | PULLMAN | TRACTION MOTOR | WH 50E (150 HP) [2/MOTOR TRUCK] |

## GENERAL

| | | | |
|---|---|---|---|
| CAR LIGHT WEIGHT | 64,374 LBS | STANDING CAPACITY | [98] |

NOTE 1: LONGITUDINAL EXCEPT FOUR CROSS SEATS ON EACH SIDE OF AISLE OF WHICH TWO SEATS FACE EACH OTHER AND IS LOCATED AT CENTER OF CAR LENGTH.

REBUILT CARS BY: RELOCATING MOTORMAN'S CAB INSIDE, OIL LAMPS REPLACED BY ELECTRIC, SPRAGUE CONTROL REPLACED BY WESTINGHOUSE UNIT, LENGTHENED PLATFORMS AND HOODS, AND ADDING NEW FOLDING GATES (1904 - 1905).
WINDOW CURTAINS REMOVED (1920's).
PAINTED VENTILATOR GLASS IN CLERESTORY OR REPLACED WITH SHEET IRON (EARLY 1921).
GLASS REINSTALLED (1941).
LETTERBOARD ABOVE WINDOWS EXTENDED THROUGH UPPER SECTION OF CENTER DOOR (1941).
CARS 620 - 627 WERE ORIGINALLY INCLUDED IN THE GROUP OF CARS 400 - 411.
ALL CARS SCRAPPED (1950's).

CAR NUMBERS: 633 - 682
TOTAL: 50 CARS
BUILT BY: JEWETT

DATE: 1901
AVERAGE COST PER CAR: N/A

# BU CLASS (BMT)

## 600 SERIES MOTOR

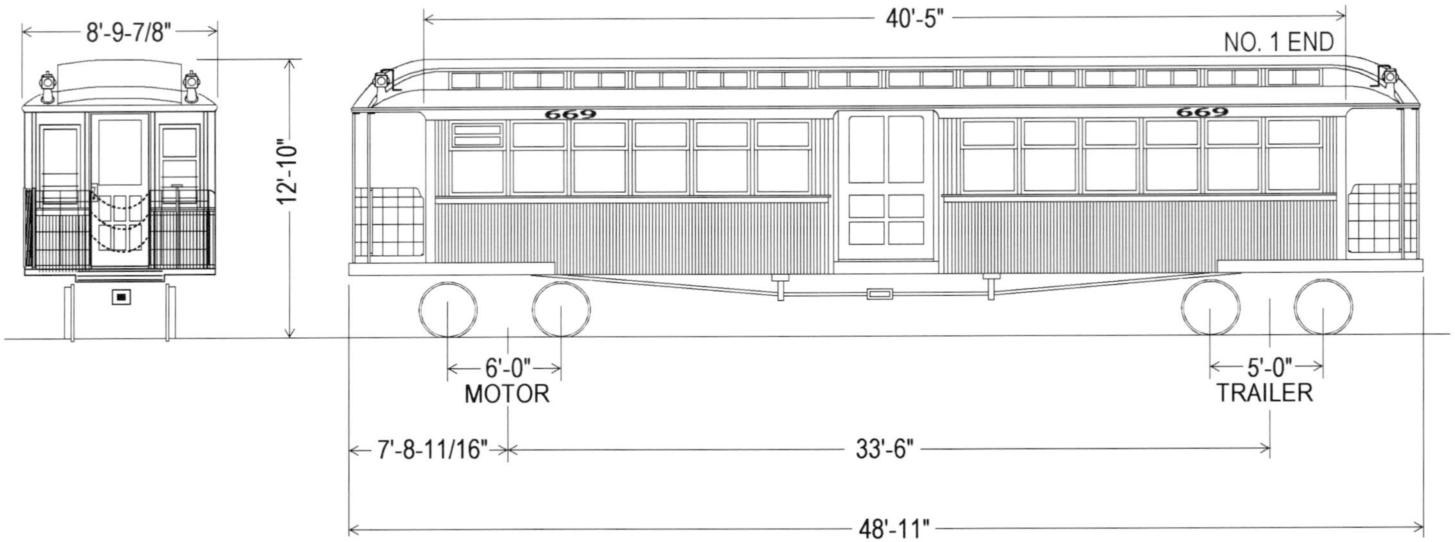

8'-9-7/8"
12'-10"
40'-5"
NO. 1 END
669
669
6'-0" MOTOR
5'-0" TRAILER
7'-8-11/16"
33'-6"
48'-11"

Copyright © 1997 by NYC Transit

### CAR BODY - EQUIPMENT

| | | | |
|---|---|---|---|
| AIR COMPRESSOR | CHRIS, B2 | LIGHTING | LAMPS [29] |
| BRAKE EQUIPMENT TYPE | WH, AML | PROPULSION CONTROL TYPE | WH 132 |
| COUPLER | VAN DORN | SEAT ARRANGEMENT & CAPACITY | LONGITUDINAL [52] |
| HAND BRAKE | STAFF | WINDOW CURTAIN | BURROWS, PANTASOTE #83 |
| HEATER | GOLD CAR HEATING | | |

### TRUCK

| | | | | |
|---|---|---|---|---|
| MANUFACTURER | MOTOR | BRILL, 27E | GEAR RATIO | 19:64 |
| | TRAILER | BRILL, 27 | TRACTION MOTOR | WH, 50E (150 HP) [2/MOTOR TRUCK] |

### GENERAL

| | | | |
|---|---|---|---|
| CAR LIGHT WEIGHT | 64,374 LBS | STANDING CAPACITY | [91] |

ORIGINAL SEATING ARRANGEMENT WAS SEMI-MANHATTAN (TWO SETS OF FACING SEATS TO ALLOW ACCESS TO CENTER DOOR).
CENTER DOOR WAS HAND OPERATED.
CAR NUMBERS 685 - 694 WERE NOT ASSIGNED.
CAR 695 (FORMERLY CAR 26) WAS A WORK CAR.
CARS 696 - 699 (FORMERLY CARS 432-435) WERE SEA VIEW RAILROAD ELECTRIC CARS BUILT BY PULLMAN IN 1898 AND LATER CONVERTED TO WORK CARS.
WINDOW CURTAINS REMOVED (1920'S).
PAINTED VENTILATOR GLASS IN CLERESTORY OR REPLACED WITH SHEET IRON (BEGINNING 1921). GLASS REINSTALLED (1941 - 1943).
CENTER DOOR WAS PERMANENTLY SEALED AND SEATS ARRANGED LONGITUDINALLY (1930's).
CARS 641 & 643 CONVERTED TO WORK MOTORS 11/25/1924 & 3/31/1923 AND RENUMBERED 992 & 993 RESPECTIVELY (1930).
CAR 684 WAS RENUMBERED 1261(1930).
CAR 648 CONVERTED TO WORK MOTOR AND RENUMBERED 996 (7/24/1940).
LETTER BOARD ABOVE WINDOWS EXTENDED OVER TOP OF CENTER DOOR (1941 - 1943).
CAR 659 SOLD TO BRANFORD ELECTRIC RAILWAY, EAST HAVEN, CT. (1961).
CARS 633 - 682 WERE ORIGINALLY NUMBERED 450 - 499.
ALL OTHER CARS SCRAPPED (1950's).

CAR NUMBERS: 700 - 760
TOTAL: 61 CARS
BUILT BY: PULLMAN

DATE: 1888
AVERAGE COST PER CAR: N/A

# BU CLASS (BMT)

700 SERIES MOTOR

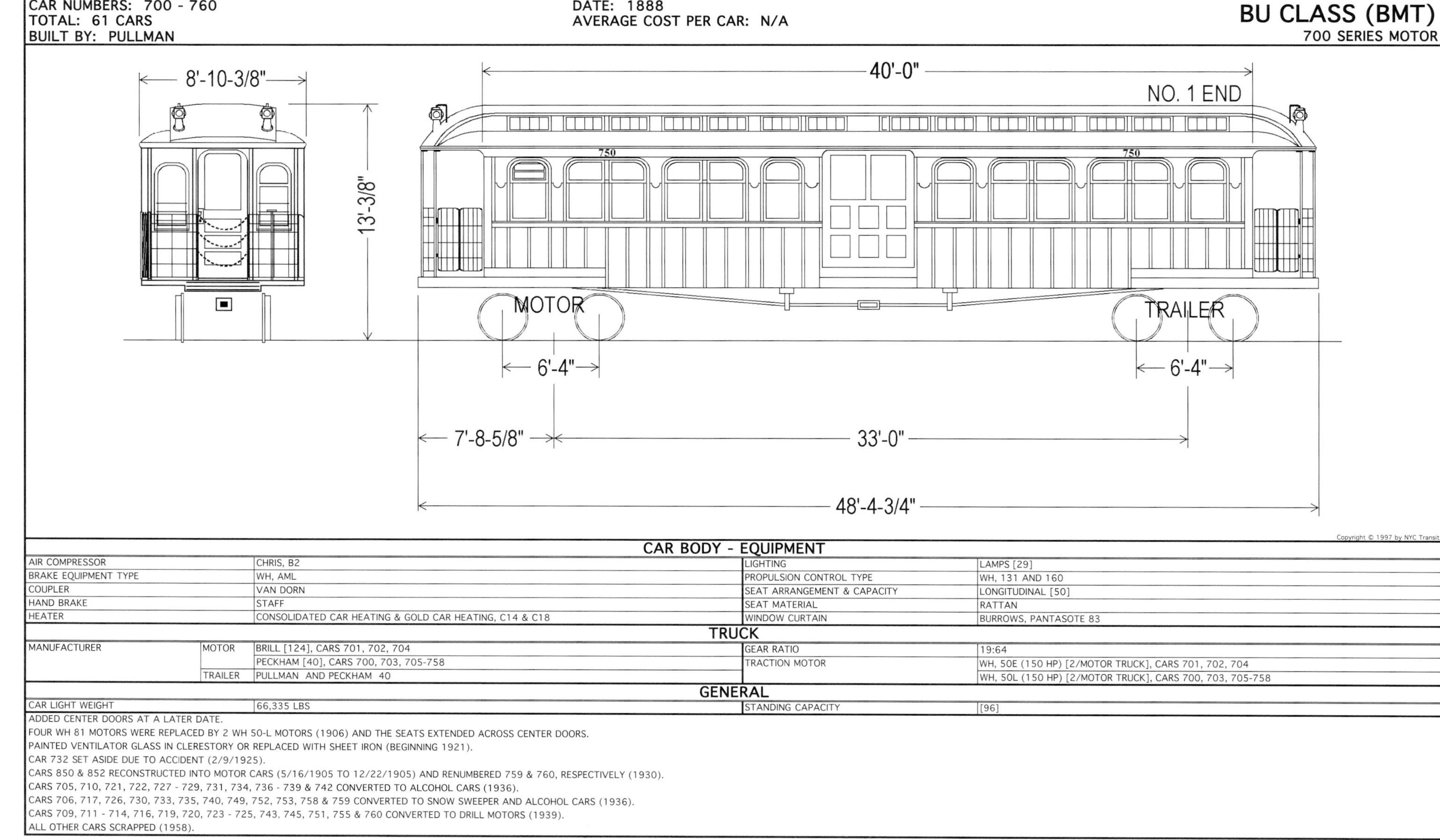


Copyright © 1997 by NYC Transit

## CAR BODY - EQUIPMENT

| | | | |
|---|---|---|---|
| AIR COMPRESSOR | CHRIS, B2 | LIGHTING | LAMPS [29] |
| BRAKE EQUIPMENT TYPE | WH, AML | PROPULSION CONTROL TYPE | WH, 131 AND 160 |
| COUPLER | VAN DORN | SEAT ARRANGEMENT & CAPACITY | LONGITUDINAL [50] |
| HAND BRAKE | STAFF | SEAT MATERIAL | RATTAN |
| HEATER | CONSOLIDATED CAR HEATING & GOLD CAR HEATING, C14 & C18 | WINDOW CURTAIN | BURROWS, PANTASOTE 83 |

## TRUCK

| | | | | |
|---|---|---|---|---|
| MANUFACTURER | MOTOR | BRILL [124], CARS 701, 702, 704 | GEAR RATIO | 19:64 |
| | | PECKHAM [40], CARS 700, 703, 705-758 | TRACTION MOTOR | WH, 50E (150 HP) [2/MOTOR TRUCK], CARS 701, 702, 704 |
| | TRAILER | PULLMAN AND PECKHAM 40 | | WH, 50L (150 HP) [2/MOTOR TRUCK], CARS 700, 703, 705-758 |

## GENERAL

| | | | |
|---|---|---|---|
| CAR LIGHT WEIGHT | 66,335 LBS | STANDING CAPACITY | [96] |

ADDED CENTER DOORS AT A LATER DATE.
FOUR WH 81 MOTORS WERE REPLACED BY 2 WH 50-L MOTORS (1906) AND THE SEATS EXTENDED ACROSS CENTER DOORS.
PAINTED VENTILATOR GLASS IN CLERESTORY OR REPLACED WITH SHEET IRON (BEGINNING 1921).
CAR 732 SET ASIDE DUE TO ACCIDENT (2/9/1925).
CARS 850 & 852 RECONSTRUCTED INTO MOTOR CARS (5/16/1905 TO 12/22/1905) AND RENUMBERED 759 & 760, RESPECTIVELY (1930).
CARS 705, 710, 721, 722, 727 - 729, 731, 734, 736 - 739 & 742 CONVERTED TO ALCOHOL CARS (1936).
CARS 706, 717, 726, 730, 733, 735, 740, 749, 752, 753, 758 & 759 CONVERTED TO SNOW SWEEPER AND ALCOHOL CARS (1936).
CARS 709, 711 - 714, 716, 719, 720, 723 - 725, 743, 745, 751, 755 & 760 CONVERTED TO DRILL MOTORS (1939).
ALL OTHER CARS SCRAPPED (1958).

CAR NUMBERS: 800 - 832
TOTAL: 33 CARS
BUILT BY: PULLMAN

DATE: 1884
AVERAGE COST PER CAR: N/A

# BU CLASS (BMT)

## 800 SERIES MOTOR

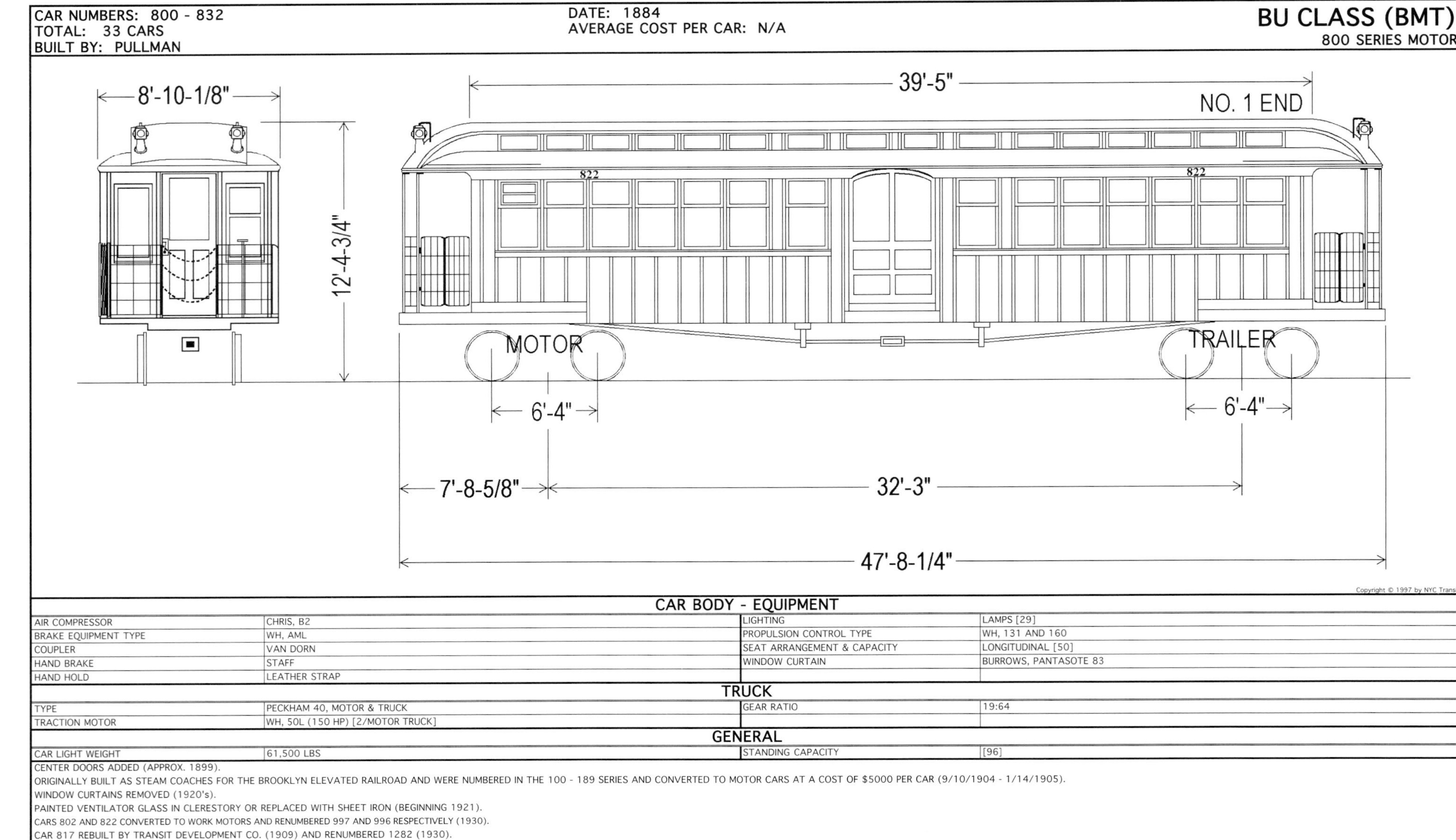


Copyright © 1997 by NYC Transit

| CAR BODY - EQUIPMENT | | | |
|---|---|---|---|
| AIR COMPRESSOR | CHRIS, B2 | LIGHTING | LAMPS [29] |
| BRAKE EQUIPMENT TYPE | WH, AML | PROPULSION CONTROL TYPE | WH, 131 AND 160 |
| COUPLER | VAN DORN | SEAT ARRANGEMENT & CAPACITY | LONGITUDINAL [50] |
| HAND BRAKE | STAFF | WINDOW CURTAIN | BURROWS, PANTASOTE 83 |
| HAND HOLD | LEATHER STRAP | | |
| **TRUCK** | | | |
| TYPE | PECKHAM 40, MOTOR & TRUCK | GEAR RATIO | 19:64 |
| TRACTION MOTOR | WH, 50L (150 HP) [2/MOTOR TRUCK] | | |
| **GENERAL** | | | |
| CAR LIGHT WEIGHT | 61,500 LBS | STANDING CAPACITY | [96] |

CENTER DOORS ADDED (APPROX. 1899).
ORIGINALLY BUILT AS STEAM COACHES FOR THE BROOKLYN ELEVATED RAILROAD AND WERE NUMBERED IN THE 100 - 189 SERIES AND CONVERTED TO MOTOR CARS AT A COST OF $5000 PER CAR (9/10/1904 - 1/14/1905).
WINDOW CURTAINS REMOVED (1920's).
PAINTED VENTILATOR GLASS IN CLERESTORY OR REPLACED WITH SHEET IRON (BEGINNING 1921).
CARS 802 AND 822 CONVERTED TO WORK MOTORS AND RENUMBERED 997 AND 996 RESPECTIVELY (1930).
CAR 817 REBUILT BY TRANSIT DEVELOPMENT CO. (1909) AND RENUMBERED 1282 (1930).
ALL OTHER CARS SCRAPPED (1924 - 1930).

CAR NUMBERS: 833 - 858
TOTAL: 26 CARS
BUILT BY: GILBERT

DATE: 1887
AVERAGE COST PER CAR: N/A

# BU CLASS (BMT)

800 SERIES MOTOR

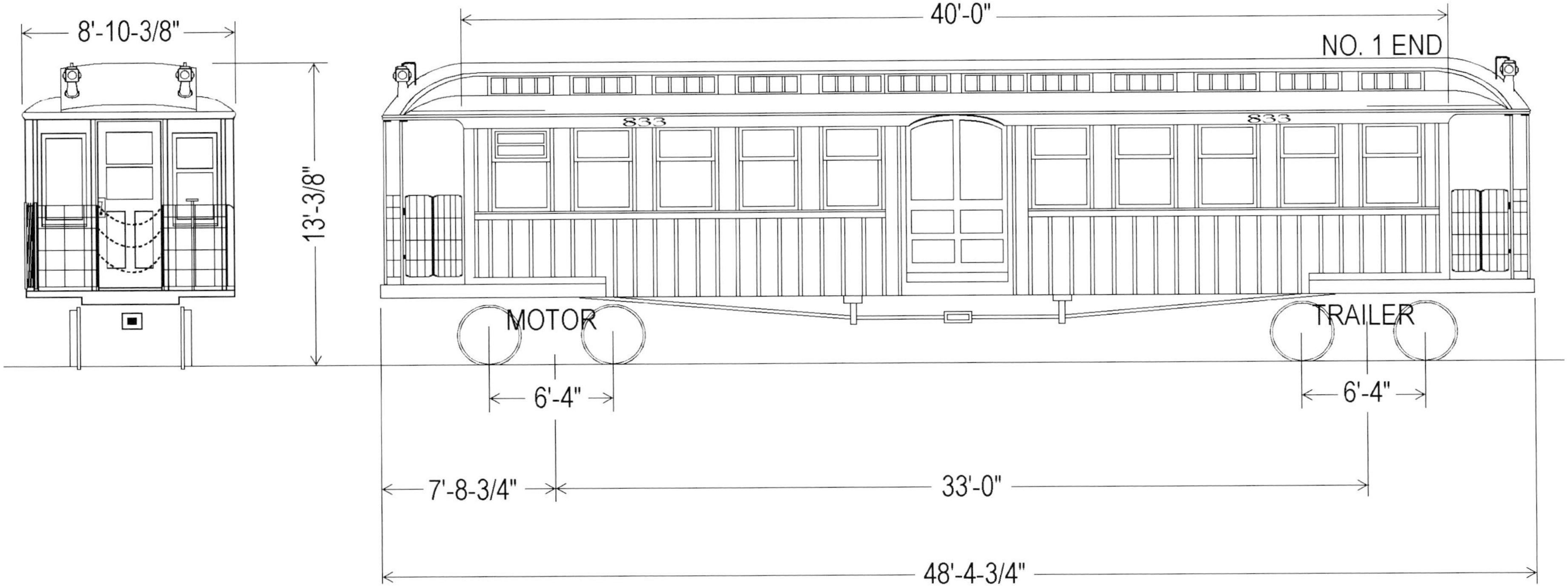

8'-10-3/8"
13'-3/8"
40'-0"
NO. 1 END
833
833
MOTOR
TRAILER
6'-4"
6'-4"
7'-8-3/4"
33'-0"
48'-4-3/4"

Copyright © 1997 by NYC Transit

## CAR BODY - EQUIPMENT

| | | | |
|---|---|---|---|
| AIR COMPRESSOR | CHRIS | LIGHTING | LAMPS [29] |
| BRAKE EQUIPMENT TYPE | WH, AML | PROPULSION CONTROL TYPE | WH, 160 |
| COUPLER | VAN DORN | SEAT ARRANGEMENT & CAPACITY | LONGITUDINAL [50] |
| HAND BRAKE | STAFF | WINDOW CURTAIN | BURROWS, PANTASOTE 83 |
| HAND HOLD | LEATHER STRAP | | |

## TRUCK

| | | | |
|---|---|---|---|
| MANUFACTURER | PECKHAM | GEAR RATIO | 19:64 |
| TYPE | PECKHAM 40 | TRACTION MOTOR | WH 50L (150 HP) [2/MOTOR TRUCK] |

## GENERAL

| | | | |
|---|---|---|---|
| CAR LIGHT WEIGHT | 61,620 LBS | STANDING CAPACITY | [96] |

CENTER DOORS ADDED (APPROX. 1899).
ORIGINALLY BUILT AS STEAM COACHES FOR THE BROOKLYN ELEVATED RAILROAD AND NUMBERED IN THE 190 - 309 SERIES AND CONVERTED TO MOTOR CARS (12/6/1904 - 4/11/1905).
WINDOW CURTAINS REMOVED (1920's).
PAINTED VENTILATOR GLASS IN CLERESTORY OR REPLACED WITH SHEET STEEL (BEGINNING 1921).
CARS 850 AND 852 WERE RENUMBERED 759 AND 760 RESPECTIVELY (1930).
CAR 843 RENUMBERED TO WORK MOTOR 995 (1930) AND DESTROYED BY FIRE (7/7/1942).
ALL OTHER CARS SCRAPPED (1924 - 1930).

CAR NUMBERS: 900 - 936
TOTAL: 37 CARS
BUILT BY: WASON

DATE: 1898
AVERAGE COST PER CAR: N/A

# BU CLASS (BMT)

900 SERIES MOTOR

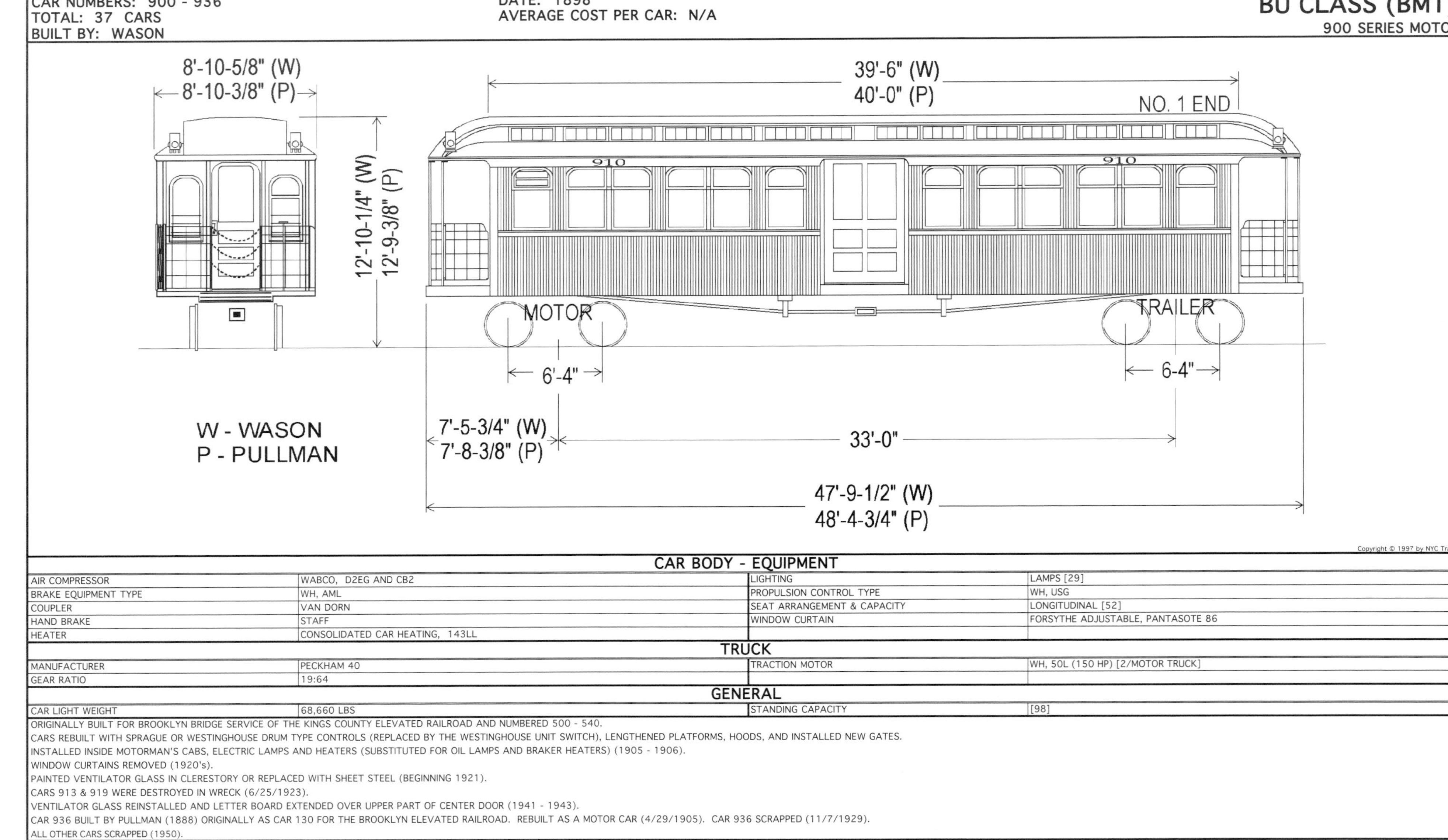


Copyright © 1997 by NYC Transit

| CAR BODY - EQUIPMENT | | | |
|---|---|---|---|
| AIR COMPRESSOR | WABCO, D2EG AND CB2 | LIGHTING | LAMPS [29] |
| BRAKE EQUIPMENT TYPE | WH, AML | PROPULSION CONTROL TYPE | WH, USG |
| COUPLER | VAN DORN | SEAT ARRANGEMENT & CAPACITY | LONGITUDINAL [52] |
| HAND BRAKE | STAFF | WINDOW CURTAIN | FORSYTHE ADJUSTABLE, PANTASOTE 86 |
| HEATER | CONSOLIDATED CAR HEATING, 143LL | | |
| **TRUCK** | | | |
| MANUFACTURER | PECKHAM 40 | TRACTION MOTOR | WH, 50L (150 HP) [2/MOTOR TRUCK] |
| GEAR RATIO | 19:64 | | |
| **GENERAL** | | | |
| CAR LIGHT WEIGHT | 68,660 LBS | STANDING CAPACITY | [98] |

ORIGINALLY BUILT FOR BROOKLYN BRIDGE SERVICE OF THE KINGS COUNTY ELEVATED RAILROAD AND NUMBERED 500 - 540.
CARS REBUILT WITH SPRAGUE OR WESTINGHOUSE DRUM TYPE CONTROLS (REPLACED BY THE WESTINGHOUSE UNIT SWITCH), LENGTHENED PLATFORMS, HOODS, AND INSTALLED NEW GATES.
INSTALLED INSIDE MOTORMAN'S CABS, ELECTRIC LAMPS AND HEATERS (SUBSTITUTED FOR OIL LAMPS AND BRAKER HEATERS) (1905 - 1906).
WINDOW CURTAINS REMOVED (1920's).
PAINTED VENTILATOR GLASS IN CLERESTORY OR REPLACED WITH SHEET STEEL (BEGINNING 1921).
CARS 913 & 919 WERE DESTROYED IN WRECK (6/25/1923).
VENTILATOR GLASS REINSTALLED AND LETTER BOARD EXTENDED OVER UPPER PART OF CENTER DOOR (1941 - 1943).
CAR 936 BUILT BY PULLMAN (1888) ORIGINALLY AS CAR 130 FOR THE BROOKLYN ELEVATED RAILROAD. REBUILT AS A MOTOR CAR (4/29/1905). CAR 936 SCRAPPED (11/7/1929).
ALL OTHER CARS SCRAPPED (1950).

CAR NUMBERS: 937 - 940
TOTAL: 4 CARS
BUILT BY: BRILL

DATE: 1900
AVERAGE COST PER CAR: N/A

# BU CLASS (BMT)

BRILL 900 SERIES MOTOR

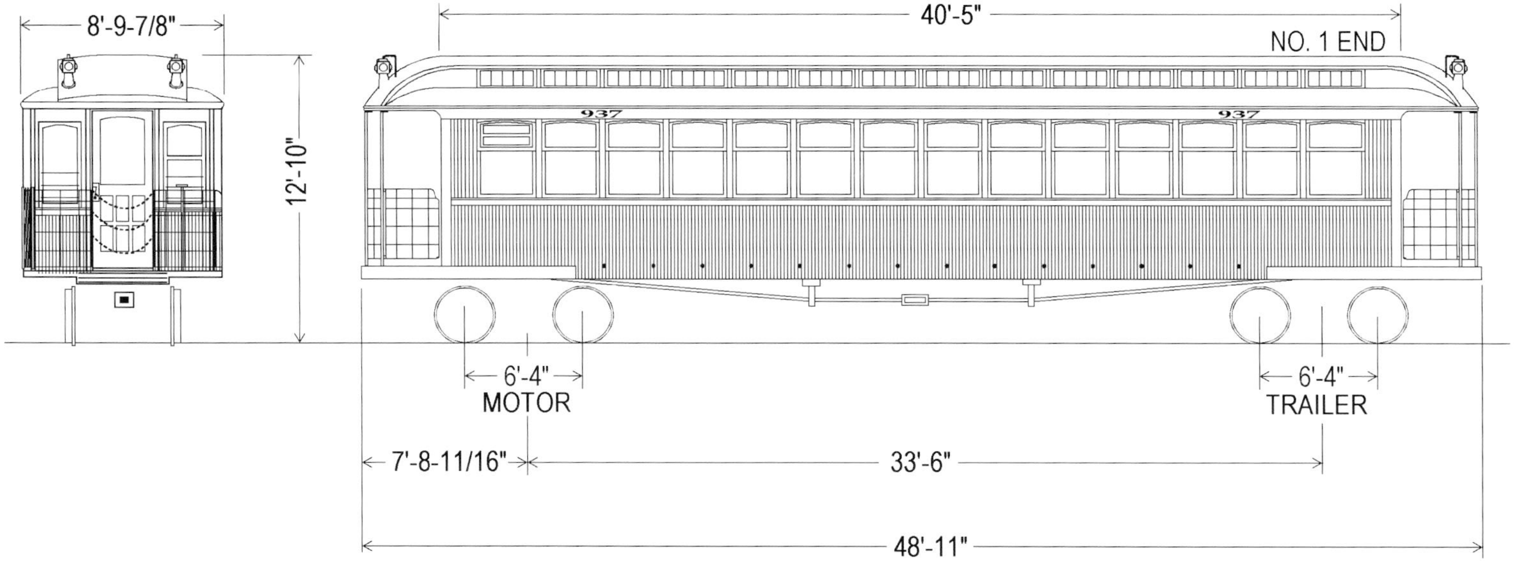

8'-9-7/8"
12'-10"
40'-5"
NO. 1 END
937
6'-4" MOTOR
6'-4" TRAILER
7'-8-11/16"
33'-6"
48'-11"

Copyright © 1997 by NYC Transit

## CAR BODY - EQUIPMENT

| | | | |
|---|---|---|---|
| AIR COMPRESSOR | CHRIS, B2 | PROPULSION CONTROL TYPE | WH, USG |
| BRAKE EQUIPMENT TYPE | WH, AML | SEAT ARRANGEMENT & CAPACITY | CROSS [58] |
| COUPLER | VAN DORN | WINDOW CURTAIN | FRONT: PANTASOTE 'J', 2; BACK: PANTASOTE 86 |
| HAND BRAKE | STAFF | WINDOW FIXTURE | ACME, CLOSED CAR CABLE |
| LIGHTING | LAMPS [29] | | |

## TRUCK

| | | | |
|---|---|---|---|
| MANUFACTURER & TYPE | PECKHAM 40 | TRACTION MOTOR | WH 50L (150 HP) [2/MOTOR TRUCK] |
| GEAR RATIO | 19:64 | | |

## GENERAL

| | | | |
|---|---|---|---|
| CAR LIGHT WEIGHT | 71,640 LBS | STANDING CAPACITY | [72] |

CAR NUMBERS 941 - 987 WERE NOT ASSIGNED.
CARS REBUILT BY LENGTHENING PLATFORMS AND HOODS, INSTALLED NEW HEATERS, OIL LAMPS REPLACED BY ELECTRIC, SPRAGUE CONTROL (REPLACED BY WESTINGHOUSE UNIT SWITCH GROUP) AND INSTALLED INSIDE MOTORMAN'S CABS (1904 -1905).
WINDOW CURTAINS REMOVED (1920's).
PAINTED VENTILATOR GLASS IN CLERESTORY OR REPLACED WITH SHEET IRON (BEGINNING 1921).
ONE CAR SCRAPPED AND REMAINING CARS WERE RENUMBERED 937 - 940 (1930).
VENTILATOR GLASS REINSTALLED (1941).
CAR 999 BUILT AS AN INSTRUCTION CAR AND LATER SOLD TO BRANFORD ELECTRIC RAILWAY, EAST HAVEN, CT. (1960).
CARS 988 - 997 WERE CONVERTED FROM OTHER SERIES TO WORK CARS.
CAR NO. 998 HAD STEEL FABRICATED STRUCTURE AND WAS LATER CONVERTED TO A PAY/REVENUE COLLECTION CAR.
ORIGINALLY, CARS WERE NUMBERED 436 - 440 AND RENUMBERED 628 -632 (1905). CAR 632 SCRAPPED (3/11/1926).
ALL OTHER CARS SCRAPPED (1950's).

CAR NUMBER: 998
TOTAL: 1 CAR
BUILT BY: PRESSED STEEL

DATE: 2/12/1909
AVERAGE COST PER CAR: N/A

## BU CLASS (BMT)

900 SERIES MOTOR

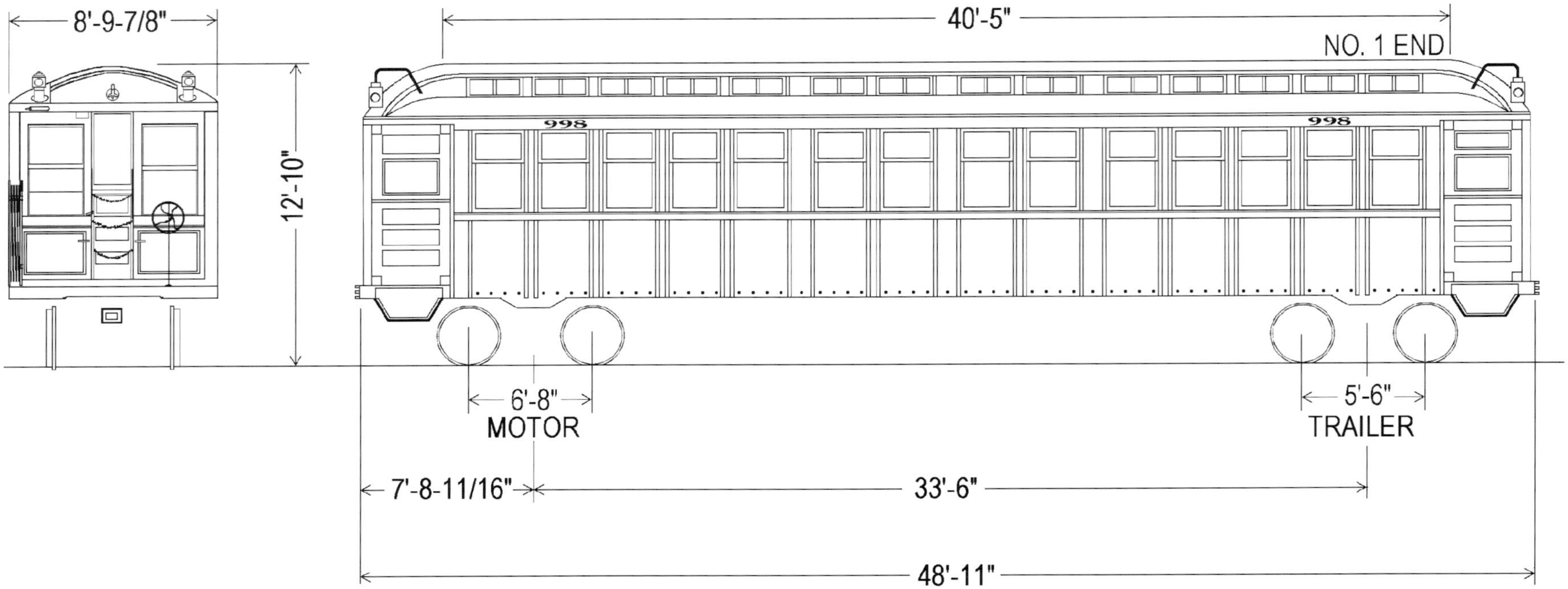

Copyright © 1997 by NYC Transit

### CAR BODY - EQUIPMENT

| | | | |
|---|---|---|---|
| AIR COMPRESSOR | WABCO, D2EG | LIGHTING | LAMPS [25] |
| BRAKE EQUIPMENT TYPE | WH, AML | PROPULSION CONTROL TYPE | WH, USG 251-1-3 |
| COUPLER | VAN DORN | SEAT ARRANGEMENT & CAPACITY | MANHATTAN [56], SEE NOTE 1 |
| HAND HOLD | LEATHER STRAP | WINDOW CURTAIN | PANTASOTE 'J', 72 |
| HEATER | CONSOLIDATED CAR HEATING | | |

### TRUCK

| | | | | |
|---|---|---|---|---|
| MANUFACTURER | MOTOR | AMERICAN LOCOMOTIVE | TRACTION MOTOR | WH, 300 (200 HP) [2/MOTOR TRUCK] |
| | TRAILER | ST. LOUIS | | |

### GENERAL

| | | | |
|---|---|---|---|
| CAR LIGHT WEIGHT | 74,400 LBS | STANDING CAPACITY | [85] |

NOTE 1: LONGITUDINAL EXCEPT FOUR CROSS SEATS ON EACH SIDE OF AISLE OF WHICH TWO SEATS FACED EACH OTHER AND WERE LOCATED AT CENTER OF CAR LENGTH.
ANTICLIMBER ADDED LATER.
BUILT AS AN EXPERIMENTAL ALL STEEL STRUCTURE SUBWAY CAR. CONVERTED TO PAY/REVENUE COLLECTION CAR (9/9/1924).
CONVERTED TO PAPER BALING CAR WITH ADDED CENTER DOOR(1946).
CAR SCRAPPED (1960).

CAR NUMBERS: 1000 - 1119
TOTAL: 120 CARS
BUILT BY: STEPHENSON

DATE: 1903
AVERAGE COST PER CAR: N/A

# BU CLASS (BMT)

CONVERTIBLE 1000 SERIES MOTOR

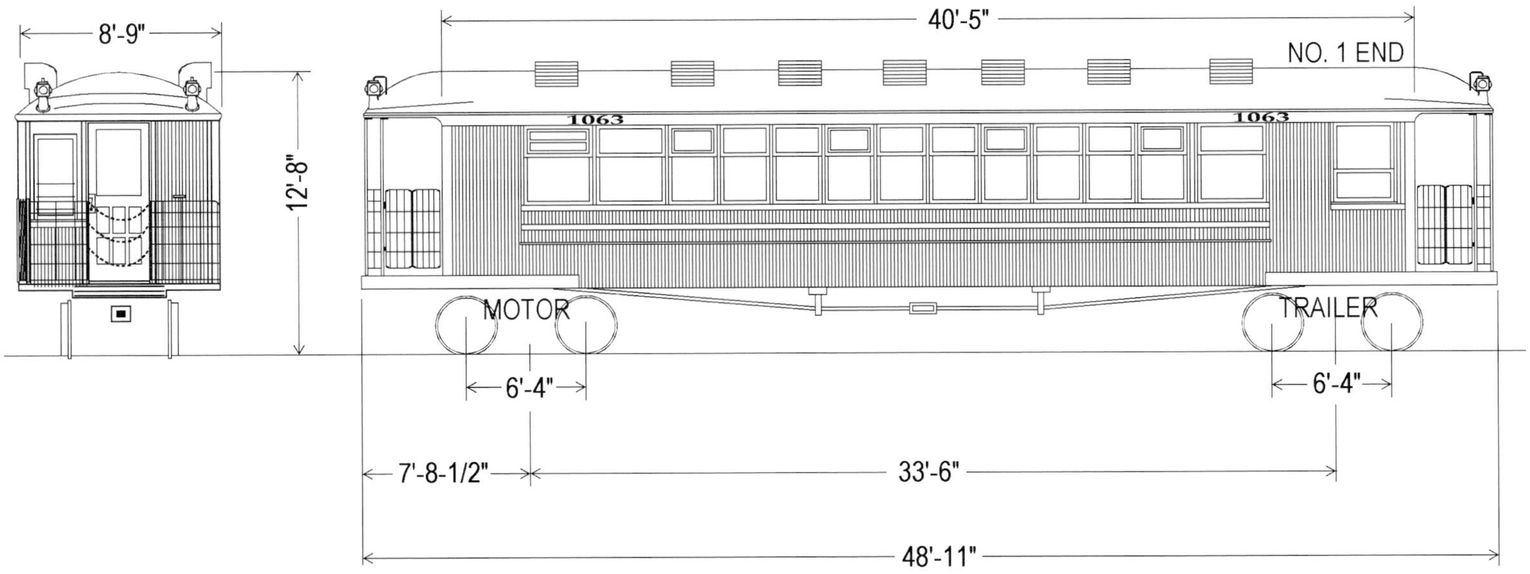

8'-9"
12'-8"
40'-5"
NO. 1 END
1063
MOTOR
TRAILER
6'-4"
6'-4"
7'-8-1/2"
33'-6"
48'-11"

Copyright © 1997 by NYC Transit

## CAR BODY - EQUIPMENT

| | | | |
|---|---|---|---|
| AIR COMPRESSOR | CHRIS B2 | LIGHTING | LAMP [24] |
| BRAKE EQUIPMENT TYPE | WH, AML | PROPULSION CONTROL TYPE | WH, 160 |
| COUPLER | VAN DORN | SEAT ARRANGEMENT & CAPACITY | CROSS [60] |
| HAND BRAKE | STAFF | STRUCTURE MATERIAL | WOOD |
| HEATER | CONSOLIDATED CAR HEATING & GOLD CAR HEATING | WINDOW CURTAIN | FRONT:PANTASOTE 'B'; BACK:PANTASOTE 82 |

## TRUCK

| | | | |
|---|---|---|---|
| MANUFACTURER & TYPE | PECKHAM 40 | TRACTION MOTOR | WH 50L (150 HP) [2/MOTOR TRUCK] |
| GEAR RATIO | 19:64 | | |

## GENERAL

| | | | |
|---|---|---|---|
| CAR LIGHT WEIGHT | 72,088 LBS | STANDING CAPACITY | [70] |

CAR NUMBERS 1120 - 1199 WERE NOT ASSIGNED.
WOOD SIDING EXTENDED TO REPLACE ONE REMOVABLE PANEL AT EACH END ON BOTH SIDES OF CAR (BEHIND MOTORMAN'S CAB ON ONE SIDE AND AT A WINDOW OPENING AT VERTICLE CONTROLLER ON OTHER SIDE), ALSO END WINDOW OPPOSITE MOTORMAN'S VISION WINDOW ON EACH END; ADDITIONAL ROOF VENTILATORS ADDED AND ORIGINAL VENTS RELOCATED DURING 1920's.
REMOVABLE SIDE PANELS (SASH) REPLACED WITH WOOD SAFETY BARS IN SUMMER.
CARS 1079 & 1081 WERE REBUILT BY TRANSIT DEVELOPMENT CO. (1908) AND RENUMBERED 1286 & 1287 (1930).
CARS 1016, 1049 & 1076 DESTROYED BY FIRES (2/13/1918, 4/6/1918 & 12/7/1922 RESPECTIVELY).
CAR 1045 DAMAGED BY FIRE (11/1/1920) AND CONVERTED INTO GONDOLA 3063.
CAR 1087 DAMAGED BY FIRE (4/19/1921) AND CONVERTED INTO FLAT CAR 3062.
CAR 1020 DAMAGED BY FIRE (4/23/1924) AND CONVERTED INTO WHEEL FLAT 3064.
CAR 1080 DEMOLISHED DUE TO COLLISION AT OCEAN PARKWAY (8/5/1924).
CAR 1030 WAS REBUILT BY TRANSIT DEVELOPMENT CO. (1910) AND RENUMBERED 1283 (1930).
CARS 1093, 1096, 1097, 1099 CONVERTED TO WORK MOTORS AND RENUMBERED 988 (4/16/1933), 989 (3/5/1934), 990 (1/10/1933) AND 991 (5/5/1932) RESPECTIVELY.
ALL OTHER CARS SCRAPPED (1946 - 1947).

CAR NUMBERS: 1200 - 1299
TOTAL: 100 CARS
BUILT BY: SEE NOTE 1

DATE: 1904 - 1905
AVERAGE COST PER CAR: $8,800

# BU CLASS (BMT)

1200 SERIES MOTOR

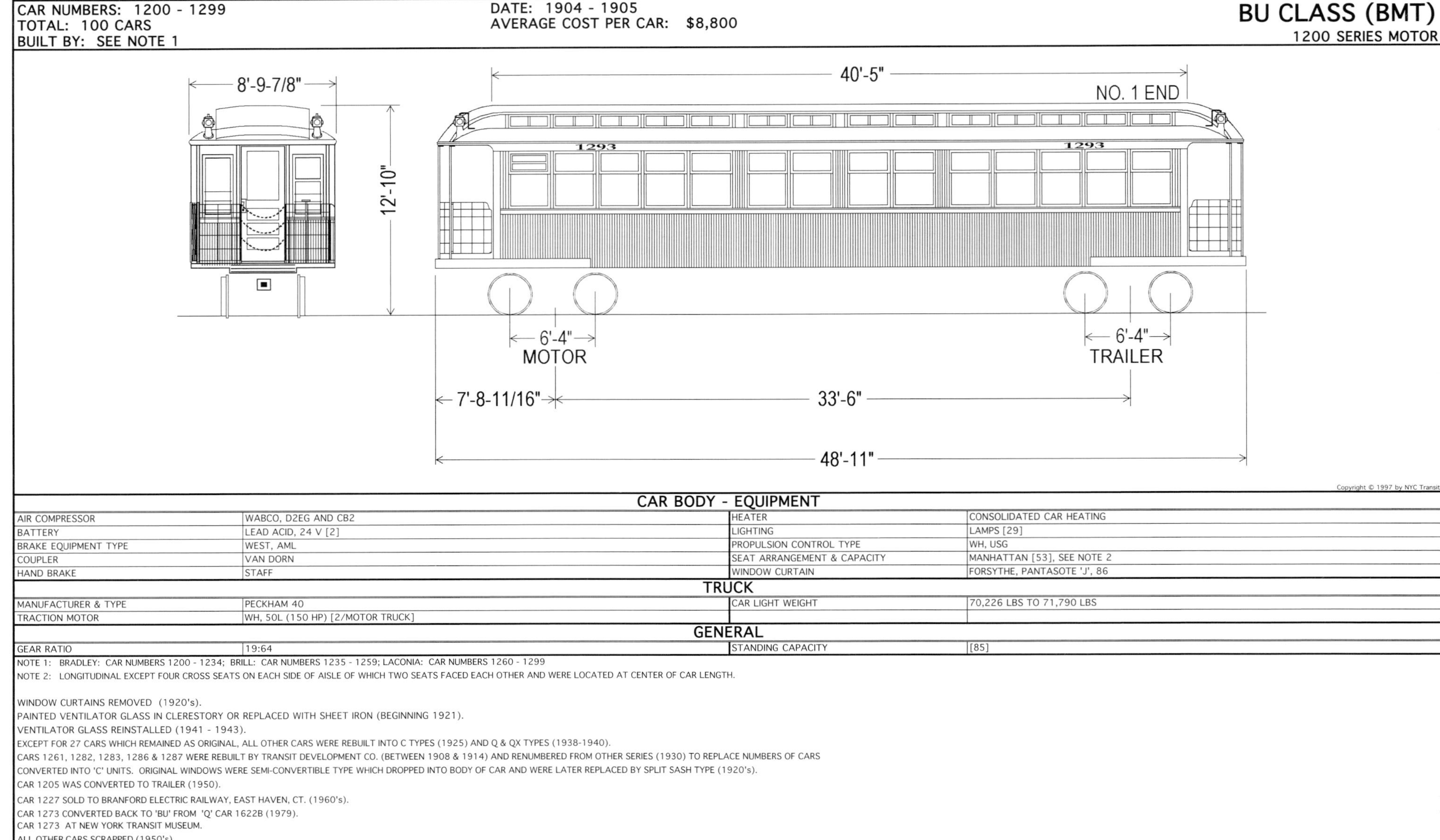


Copyright © 1997 by NYC Transit

## CAR BODY - EQUIPMENT

| | | | |
|---|---|---|---|
| AIR COMPRESSOR | WABCO, D2EG AND CB2 | HEATER | CONSOLIDATED CAR HEATING |
| BATTERY | LEAD ACID, 24 V [2] | LIGHTING | LAMPS [29] |
| BRAKE EQUIPMENT TYPE | WEST, AML | PROPULSION CONTROL TYPE | WH, USG |
| COUPLER | VAN DORN | SEAT ARRANGEMENT & CAPACITY | MANHATTAN [53], SEE NOTE 2 |
| HAND BRAKE | STAFF | WINDOW CURTAIN | FORSYTHE, PANTASOTE 'J', 86 |

## TRUCK

| | | | |
|---|---|---|---|
| MANUFACTURER & TYPE | PECKHAM 40 | CAR LIGHT WEIGHT | 70,226 LBS TO 71,790 LBS |
| TRACTION MOTOR | WH, 50L (150 HP) [2/MOTOR TRUCK] | | |

## GENERAL

| | | | |
|---|---|---|---|
| GEAR RATIO | 19:64 | STANDING CAPACITY | [85] |

NOTE 1: BRADLEY: CAR NUMBERS 1200 - 1234; BRILL: CAR NUMBERS 1235 - 1259; LACONIA: CAR NUMBERS 1260 - 1299
NOTE 2: LONGITUDINAL EXCEPT FOUR CROSS SEATS ON EACH SIDE OF AISLE OF WHICH TWO SEATS FACED EACH OTHER AND WERE LOCATED AT CENTER OF CAR LENGTH.

WINDOW CURTAINS REMOVED (1920's).
PAINTED VENTILATOR GLASS IN CLERESTORY OR REPLACED WITH SHEET IRON (BEGINNING 1921).
VENTILATOR GLASS REINSTALLED (1941 - 1943).
EXCEPT FOR 27 CARS WHICH REMAINED AS ORIGINAL, ALL OTHER CARS WERE REBUILT INTO C TYPES (1925) AND Q & QX TYPES (1938-1940).
CARS 1261, 1282, 1283, 1286 & 1287 WERE REBUILT BY TRANSIT DEVELOPMENT CO. (BETWEEN 1908 & 1914) AND RENUMBERED FROM OTHER SERIES (1930) TO REPLACE NUMBERS OF CARS CONVERTED INTO 'C' UNITS. ORIGINAL WINDOWS WERE SEMI-CONVERTIBLE TYPE WHICH DROPPED INTO BODY OF CAR AND WERE LATER REPLACED BY SPLIT SASH TYPE (1920's).
CAR 1205 WAS CONVERTED TO TRAILER (1950).
CAR 1227 SOLD TO BRANFORD ELECTRIC RAILWAY, EAST HAVEN, CT. (1960's).
CAR 1273 CONVERTED BACK TO 'BU' FROM 'Q' CAR 1622B (1979).
CAR 1273 AT NEW YORK TRANSIT MUSEUM.
ALL OTHER CARS SCRAPPED (1950's).

CAR NUMBERS: 1261, 1282, 1283, 1286, 1287
TOTAL: 5 CARS
REBUILT BY: TRANSIT DEVELOPMENT CO.

DATE: 1908 - 1914
AVERAGE COST PER CAR: N/A

# BU CLASS (BMT)

1200 SERIES MOTOR

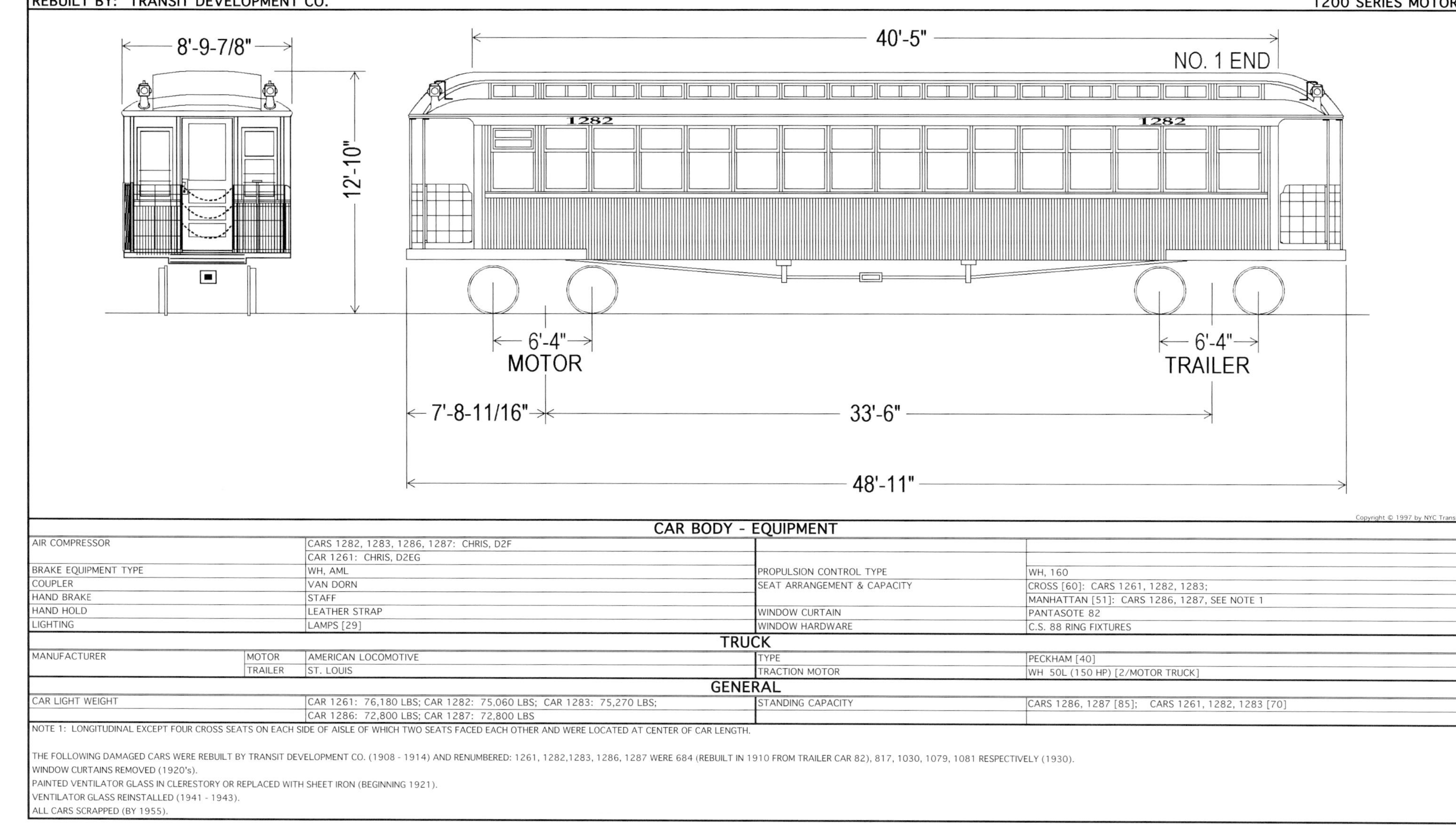


Copyright © 1997 by NYC Transit

## CAR BODY - EQUIPMENT

| | | | |
|---|---|---|---|
| AIR COMPRESSOR | CARS 1282, 1283, 1286, 1287: CHRIS, D2F<br>CAR 1261: CHRIS, D2EG | | |
| BRAKE EQUIPMENT TYPE | WH, AML | PROPULSION CONTROL TYPE | WH, 160 |
| COUPLER | VAN DORN | SEAT ARRANGEMENT & CAPACITY | CROSS [60]: CARS 1261, 1282, 1283;<br>MANHATTAN [51]: CARS 1286, 1287, SEE NOTE 1 |
| HAND BRAKE | STAFF | | |
| HAND HOLD | LEATHER STRAP | WINDOW CURTAIN | PANTASOTE 82 |
| LIGHTING | LAMPS [29] | WINDOW HARDWARE | C.S. 88 RING FIXTURES |

## TRUCK

| | | | | |
|---|---|---|---|---|
| MANUFACTURER | MOTOR | AMERICAN LOCOMOTIVE | TYPE | PECKHAM [40] |
| | TRAILER | ST. LOUIS | TRACTION MOTOR | WH 50L (150 HP) [2/MOTOR TRUCK] |

## GENERAL

| | | | |
|---|---|---|---|
| CAR LIGHT WEIGHT | CAR 1261: 76,180 LBS; CAR 1282: 75,060 LBS; CAR 1283: 75,270 LBS;<br>CAR 1286: 72,800 LBS; CAR 1287: 72,800 LBS | STANDING CAPACITY | CARS 1286, 1287 [85]; CARS 1261, 1282, 1283 [70] |

NOTE 1: LONGITUDINAL EXCEPT FOUR CROSS SEATS ON EACH SIDE OF AISLE OF WHICH TWO SEATS FACED EACH OTHER AND WERE LOCATED AT CENTER OF CAR LENGTH.

THE FOLLOWING DAMAGED CARS WERE REBUILT BY TRANSIT DEVELOPMENT CO. (1908 - 1914) AND RENUMBERED: 1261, 1282,1283, 1286, 1287 WERE 684 (REBUILT IN 1910 FROM TRAILER CAR 82), 817, 1030, 1079, 1081 RESPECTIVELY (1930).
WINDOW CURTAINS REMOVED (1920's).
PAINTED VENTILATOR GLASS IN CLERESTORY OR REPLACED WITH SHEET IRON (BEGINNING 1921).
VENTILATOR GLASS REINSTALLED (1941 - 1943).
ALL CARS SCRAPPED (BY 1955).

CAR NUMBERS: 1300 - 1399
TOTAL: 100 CARS
BUILT BY: SEE NOTE 1

DATE: 1905 - 1906
AVERAGE COST PER CAR: $4,200

# BU CLASS (BMT)

CONVERTIBLE 1300 SERIES MOTOR

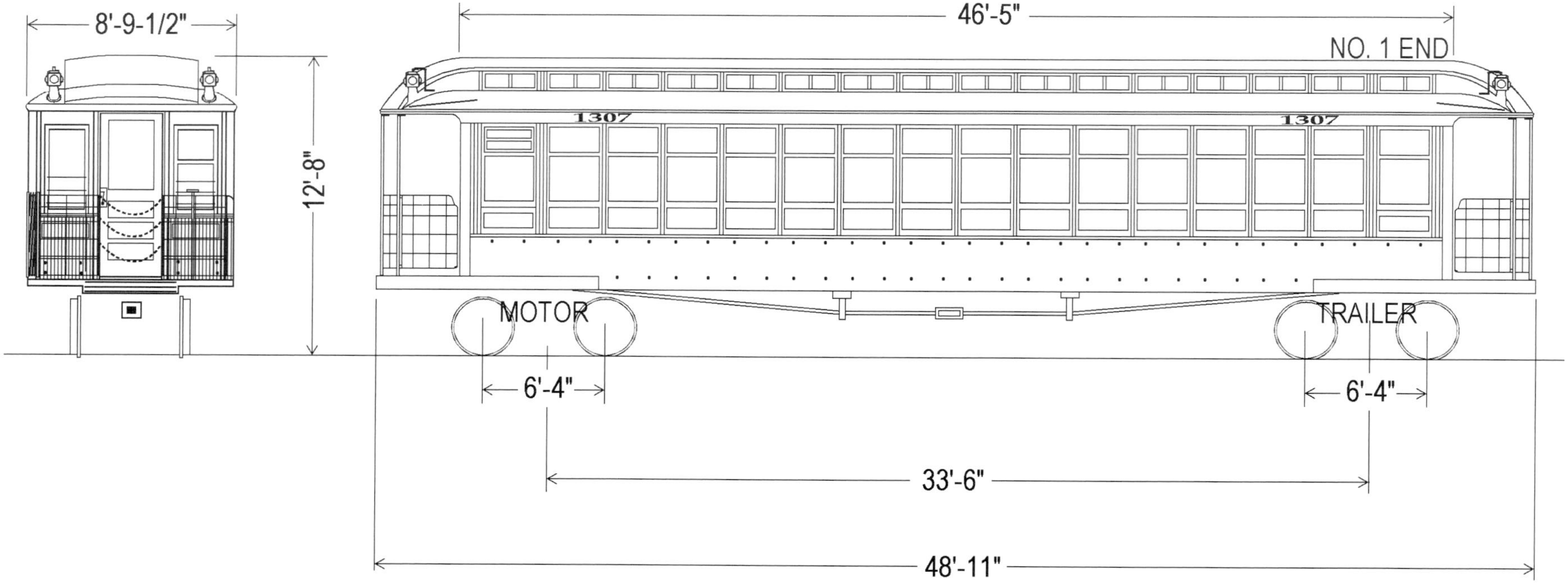

Copyright © 1997 by NYC Transit

## CAR BODY - EQUIPMENT

| | | | |
|---|---|---|---|
| AIR COMPRESSOR | WABCO, D2EG | LIGHTING | LAMPS [29] |
| BATTERY | LEAD ACID, 24 V [2] | PROPULSION CONTROL TYPE | WABCO, UNIT SWITCH GROUP (USG) |
| BRAKE EQUIPMENT TYPE | WABCO, TRACTION | SEAT ARRANGEMENT & CAPACITY | CROSS [62] |
| COUPLER | VAN DORN | WINDOW CURTAIN | FORSYTHE, ADJUSTABLE, PANTASOTE 'J', 86 |
| HEATING | CONSOLIDATED CAR HEATING | | |

## TRUCK

| | | | |
|---|---|---|---|
| MANUFACTURER | PECKHAM | GEAR RATIO | 17:54 |
| TYPE | PECKHAM 40 | TRACTION MOTOR | WH 50L (150 HP) [2/MOTOR TRUCK] |

## GENERAL

| | | | |
|---|---|---|---|
| STANDING CAPACITY | [70] | | |

NOTE 1: CARS 1300 - 1349 BY CINCINNATI CAR; CARS 1350 - 1374 BY JEWETT CAR; CARS 1375 - 1399 BY LACONIA.

PAINTED VENTILATOR GLASS IN CLERESTORY OR REPLACED WITH SHEET IRON (BEGINNING 1921).
VENTILATOR GLASS REINSTALLED (1942 - 1944).
REMOVABLE SIDE PANELS REPLACED WITH IRON SAFETY BARS IN SUMMER. NUMBER OF PANELS REPLACED ON EACH SIDE WAS REDUCED TO EIGHT (1934), SIX (1942) AND FOUR (1950).
CARS 1349, 1362 SOLD TO BRANFORD ELECTRIC RAILWAY, EAST HAVEN, CT. (1962 & 1964 RESPECTIVELY).
CAR 1365 SOLD TO NATIONAL MUSEUM OF TRANSPORT, ST. LOUIS, MO. (1964).
ALL OTHER CARS SCRAPPED (1962).

CAR NUMBERS: 1400 - 1499
TOTAL: 100 CARS
BUILT BY: JEWETT, LACONIA, SEE NOTE 1

DATE: 1908
AVERAGE COST PER CAR: $10,500

# BU CLASS (BMT)

1400 SERIES MOTOR

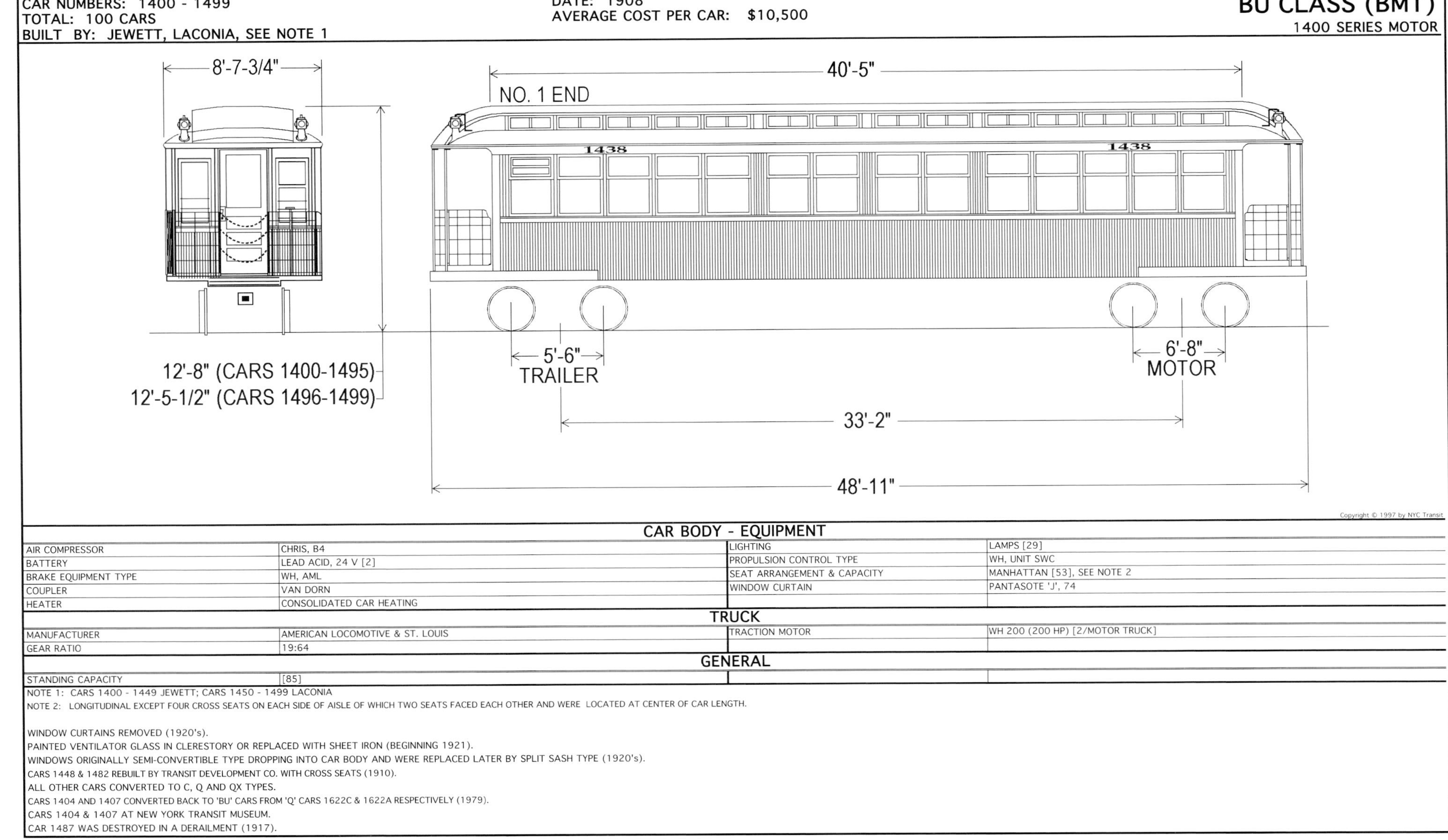


Copyright © 1997 by NYC Transit

## CAR BODY - EQUIPMENT

| | | | |
|---|---|---|---|
| AIR COMPRESSOR | CHRIS, B4 | LIGHTING | LAMPS [29] |
| BATTERY | LEAD ACID, 24 V [2] | PROPULSION CONTROL TYPE | WH, UNIT SWC |
| BRAKE EQUIPMENT TYPE | WH, AML | SEAT ARRANGEMENT & CAPACITY | MANHATTAN [53], SEE NOTE 2 |
| COUPLER | VAN DORN | WINDOW CURTAIN | PANTASOTE 'J', 74 |
| HEATER | CONSOLIDATED CAR HEATING | | |

## TRUCK

| | | | |
|---|---|---|---|
| MANUFACTURER | AMERICAN LOCOMOTIVE & ST. LOUIS | TRACTION MOTOR | WH 200 (200 HP) [2/MOTOR TRUCK] |
| GEAR RATIO | 19:64 | | |

## GENERAL

| | | | |
|---|---|---|---|
| STANDING CAPACITY | [85] | | |

NOTE 1: CARS 1400 - 1449 JEWETT; CARS 1450 - 1499 LACONIA
NOTE 2: LONGITUDINAL EXCEPT FOUR CROSS SEATS ON EACH SIDE OF AISLE OF WHICH TWO SEATS FACED EACH OTHER AND WERE LOCATED AT CENTER OF CAR LENGTH.

WINDOW CURTAINS REMOVED (1920's).
PAINTED VENTILATOR GLASS IN CLERESTORY OR REPLACED WITH SHEET IRON (BEGINNING 1921).
WINDOWS ORIGINALLY SEMI-CONVERTIBLE TYPE DROPPING INTO CAR BODY AND WERE REPLACED LATER BY SPLIT SASH TYPE (1920's).
CARS 1448 & 1482 REBUILT BY TRANSIT DEVELOPMENT CO. WITH CROSS SEATS (1910).
ALL OTHER CARS CONVERTED TO C, Q AND QX TYPES.
CARS 1404 AND 1407 CONVERTED BACK TO 'BU' CARS FROM 'Q' CARS 1622C & 1622A RESPECTIVELY (1979).
CARS 1404 & 1407 AT NEW YORK TRANSIT MUSEUM.
CAR 1487 WAS DESTROYED IN A DERAILMENT (1917).

CAR NUMBERS: 1448, 1482
TOTAL: 2 CARS
BUILT BY: JEWETT, LACONIA
DATE: 1908
AVERAGE COST PER CAR: $2,900

OVERHAULED BY: TRANSIT DEVELOPMENT CO.
DATE: 11/10/1909

# BU CLASS (BMT)

1400 SERIES MOTOR

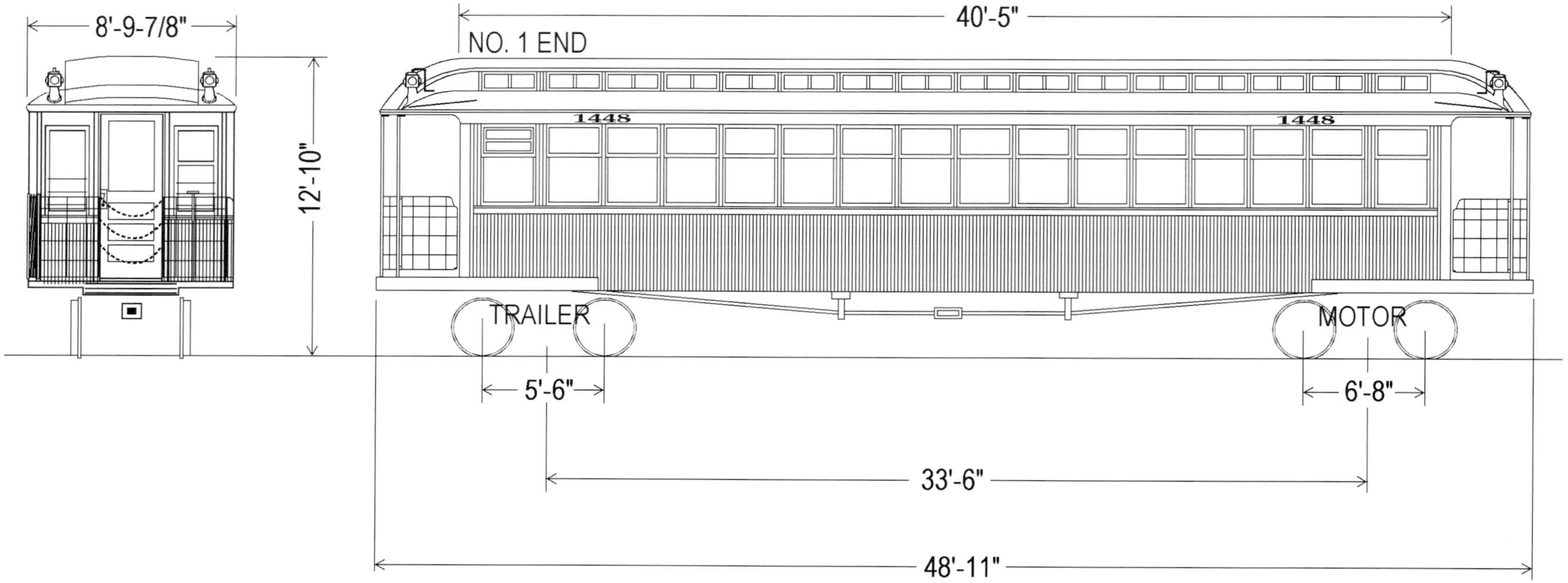

Copyright © 1997 by NYC Transit

## CAR BODY - EQUIPMENT

| | | | |
|---|---|---|---|
| AIR COMPRESSOR | CHRIS, B4 | HEATER | CONSOLIDATED CAR HEATING |
| BATTERY | LEAD ACID, 24 V [2] | LIGHTING | LAMPS [29] |
| BRAKE TYPE | WH, AML | PROPULSION CONTROL TYPE | WH, USG 251 |
| COUPLER | VAN DORN | SEAT ARRANGEMENT & CAPACITY | CROSS [62] |
| HAND BRAKE | STAFF | WINDOW CURTAIN | PANTASOTE 74 |
| HAND HOLD | LEATHER STRAP | WINDOW | EDWARDS |

## TRUCK

| | | | | |
|---|---|---|---|---|
| MANUFACTURER | MOTOR | AMERICAN LOCOMOTIVE | TYPE | PECKHAM 40 |
| | TRAILER | ST. LOUIS | TRACTION MOTOR | WH, 300 (150HP) [2/MOTOR TRUCK] |

## GENERAL

| | | | |
|---|---|---|---|
| CAR LIGHT WEIGHT | CAR 1448, 74,910 LBS | STANDING CAPACITY | [70] |
| | CAR 1482, 75,400 LBS | | |

WINDOW CURTAINS REMOVED (1920's).
PAINTED VENTILATOR GLASS IN CLERESTORY OR REPLACED WITH SHEET IRON (BEGINNING 1921).
VENTILATOR GLASS REINSTALLED (1941).
OVERHAUL INCLUDED REBUILDING OF STRUCTURE.
CAR 1482 SCRAPPED (1955).

CAR NUMBERS: 1500 - 1501 A, B, C
TOTAL: 6 CARS (2 UNITS), SEE NOTE 1
BUILT BY: LACONIA, BRADLEY SEE NOTE 2
DATE: 1893 & 1908 SEE NOTE 2
AVERAGE COST PER CAR: N/A

REBUILT BY: BMT
REBUILT DATE: 1923 - 1925
AVERAGE COST PER CAR: N/A

# C (BMT)

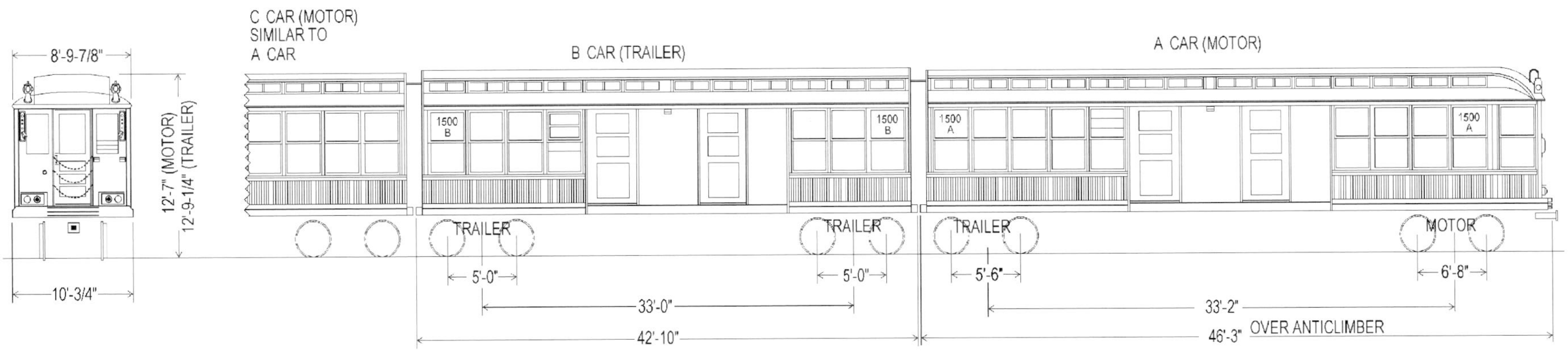

8'-9-7/8"
12'-7" (MOTOR)
12'-9-1/4" (TRAILER)
10'-3/4"
C CAR (MOTOR) SIMILAR TO A CAR
B CAR (TRAILER)
A CAR (MOTOR)
1500 B
1500 A
TRAILER
TRAILER
TRAILER
MOTOR
5'-0"
5'-0"
5'-6"
6'-8"
33'-0"
33'-2"
42'-10"
46'-3" OVER ANTICLIMBER

Copyright © 1997 by NYC Transit

## CAR BODY - EQUIPMENT

| | | | |
|---|---|---|---|
| AIR COMPRESSOR | WABCO, D2F | LIGHTING | LAMPS [20] |
| BATTERY | LEAD ACID, 24 V [2] | MASTER CONTROLLER | A & C CARS, WH, NO. 8 |
| BRAKE EQUIPMENT TYPE | WH, AML | PROPULSION CONTROL TYPE | A & C CARS, WH, USG 251-I-3 |
| COUPLER | VAN DORN | PUBLIC ADDRESS SYSTEM | WHITELAND, LOUD SPEAKING TELEPHONE W/725 RELAYS |
| DOOR OPERATOR ENGINE | CONSOLIDATED, 5020 | SEAT ARRANGEMENT & CAPACITY | LONGITUDINAL, A & C CARS [51]; B CAR [44] |
| HAND BRAKE | STAFF | SEAT MATERIAL | RATTAN |
| HEATER | CONSOLIDATED CAR HEATING | | |

## TRUCK

| | | | |
|---|---|---|---|
| MANUFACTURER | A & C CARS: AMERICAN LOCOMOTIVE<br>B CAR : PULLMAN | TRACTION MOTOR | WH, 300 (200 HP) [2/MOTOR TRUCK] |

## GENERAL

| | | | |
|---|---|---|---|
| CAR LIGHT WEIGHT | A & C CARS, 76, 970 LBS<br>B CAR, 43,170 LBS | | |

NOTE 1: CONSIST OF THREE CARS (A, B, & C). A & C CARS WERE CONVERTED FROM CARS 1496 - 1499 (1923 - 1925). ADDED CONNECTING PASSAGE BETWEEN CARS (1925).
NOTE 2: LACONIA - CARS 1500A, 1500C, 1501A, 1501C (1908); BRADLEY - CARS 1500B (ex-188), 1501B (ex-189) (1893).

ADDED DESTINATION AND LINE SIGNS TO 'B' CARS (1935).
REMOVED PUBLIC ADDRESS SYSTEM (1925). CARS RETAINED ORIGINAL NUMBERS UNTIL 1926 WHEN THEY WERE RENUMBERED INTO 1500 SERIES.
CARS 1500 (A, B & C) SCRAPPED (1955) AND 1501 (A, B, & C) SCRAPPED (1957).

# C (BMT)

CAR NUMBERS: 1502 - 1526 A, B, C
TOTAL: 75 CARS (25 UNITS), SEE NOTE 1
BUILT BY: LACONIA, JEWETT, BRADLEY
DATE: 1893 - 1908
AVERAGE COST PER CAR: N/A

REBUILT BY: BMT
REBUILT DATE: 1925
AVERAGE COST PER CAR: N/A

8'-9-7/8"
12'-7" (MOTOR)
12'-9-1/4" (TRAILER)
9'-9"
C CAR (MOTOR) SIMILAR TO A CAR
B CAR (TRAILER)
A CAR (MOTOR)
1503 C
1503 B
1503 B
1503 A
1503 A
TRAILER
TRAILER
TRAILER
MOTOR
5'-0"
5'-0"
5'-6"
6'-8"
32'-4-1/2"
42'-11-1/2"
33'-2"
46'-4" OVER ANTICLIMBER

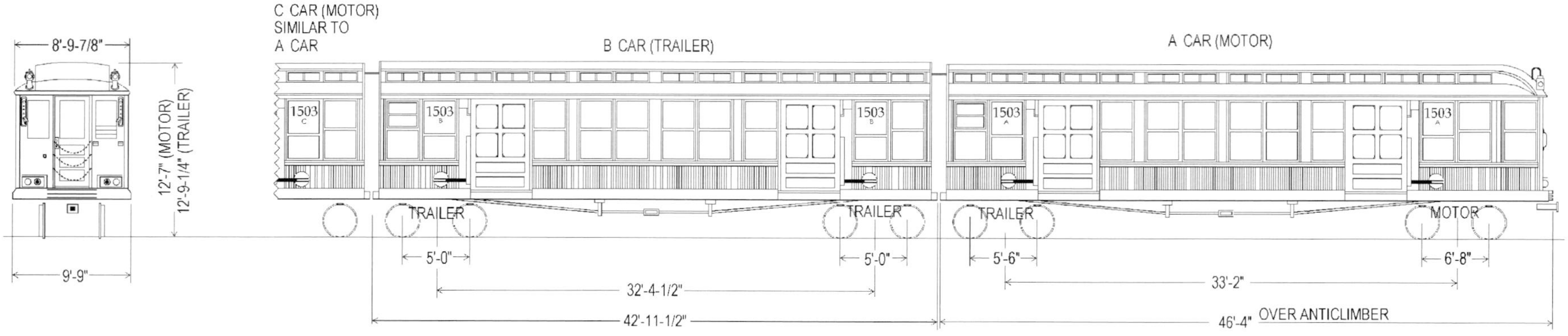

Copyright © 1997 by NYC Transit

## CAR BODY - EQUIPMENT

| | | | |
|---|---|---|---|
| AIR COMPRESSOR | WABCO, D2F | HEATER | CONSOLIDATED CAR HEATING, 146 &148 [20] |
| BATTERY | LEAD ACID, 24 V [2] | LIGHTING | LAMPS [20] |
| BRAKE EQUIPMENT TYPE | WH, AML | MASTER CONTROLLER | WH 27B, SO-776008 |
| COUPLER | VAN DORN | PROPULSION CONTROL TYPE | WH, USG 251-I-3 |
| DOOR OPERATOR ENGINE | CONSOLIDATED CAR HEATING, 5020 | SEAT ARRANGEMENT & CAPACITY | LONGITUDINAL, A & C CARS [46]; B CAR [44] |
| HAND BRAKE | STAFF | SEAT MATERIAL | RATTAN |
| HAND HOLD | LEATHER STRAP | | |

## TRUCK

| | | | | |
|---|---|---|---|---|
| MANUFACTURER | MOTOR | CAR UNITS 1502 - 1512 (A & C CARS), AMERICAN LOCOMOTIVE | TRACTION MOTOR | WH, 300 (200 HP) [1]/MOTOR, [1]/TRAILER |
| | | CAR UNITS 1513 - 1526 (A & C CARS), PECKHAM 40 | | WH, 50L (150 HP) [1]/MOTOR, [1]/TRAILER |
| | TRAILER | CAR UNITS 1504 - 1523 (B CAR), ST. LOUIS | | |
| | | CAR UNITS 1524 - 1526 (B CAR), PECKHAM 40 | | |

## GENERAL

| | | | |
|---|---|---|---|
| CAR LIGHT WEIGHT | A & C CARS, 76,970 LBS | STANDING CAPACITY | A & C CARS [108]; B CAR [110] |
| | B CAR, 43,170 LBS | | |

NOTE 1: UNIT CONSISTS OF THREE CARS (A, B, & C).

CARS WERE CONVERTED FROM 100 SERIES TRAILER CARS AND 1200/1400 SERIES MOTOR (1925).
ADD DESTINATION AND LINE SIGNS TO 'B' CARS (1935).
CARS 1524 - 1526 HAD TURRET 13-SW CONTROL.
WH 333 MOTORS REPLACED WITH WH 50L MOTORS ON SOME CARS (1950's).
CARS 1502 & 1503 WERE EQUIPPED FOR ONE MOTORMAN WITH NO CONDUCTOR OR TRAINMAN OPERATION AND WAS LATER REMOVED (1933).
ALL CARS WERE SCRAPPED (1955 - 1957).

CAR NUMBERS: 1600 - 1629 A, B, C
TOTAL: 90 CARS (30 UNITS), SEE NOTE 1
BUILT BY: BRADLEY, BRILL, JEWETT, LACONIA, SEE NOTE 3
DATE: 1904 - 1908
AVERAGE COST PER CAR: N/A

REBUILT BY: BMT
REBUILT DATE: 1938 - 1941
AVERAGE COST PER CAR: N/A

# Q (BMT)

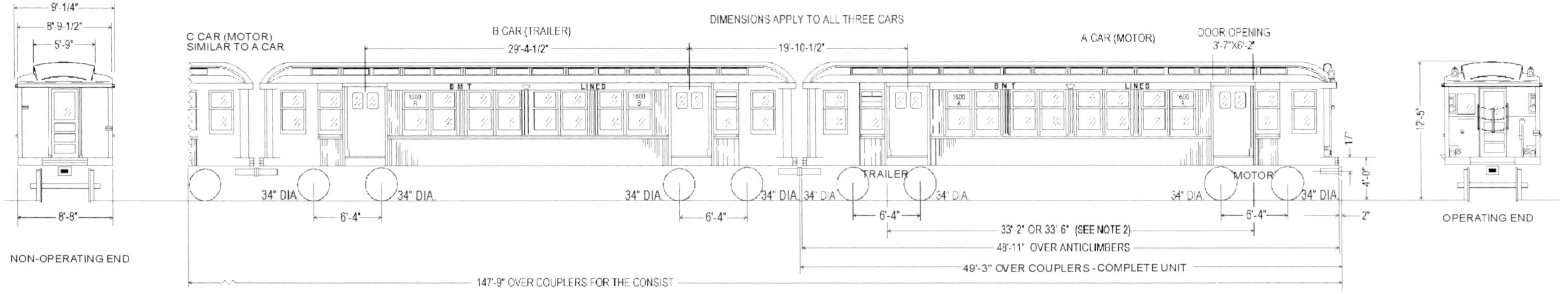

9'-1/4"
8'-9-1/2"
5'-9"
8'-8"
NON-OPERATING END
C CAR (MOTOR) SIMILAR TO A CAR
B CAR (TRAILER)
29'-4-1/2"
DIMENSIONS APPLY TO ALL THREE CARS
19'-10-1/2"
A CAR (MOTOR)
DOOR OPENING 3'-7"X6'-2"
34" DIA
6'-4"
TRAILER
MOTOR
1'7"
4'-0"
2"
33'-2" OR 33'-6" (SEE NOTE 2)
48'-11" OVER ANTICLIMBERS
49'-3" OVER COUPLERS - COMPLETE UNIT
147'-9" OVER COUPLERS FOR THE CONSIST
12'-5"
OPERATING END

Copyright © 1997 by NYC Transit

## CAR BODY - EQUIPMENT

| | | | | |
|---|---|---|---|---|
| AIR COMPRESSOR (A & C CARS) | | WH, D2F | HAND HOLD | LEATHER STRAP |
| BRAKE EQUIPMENT TYPE | | WH, AML | HEATER | A & C CARS: CONSOLIDATED, 143LL PASSENGER [16], 146X (CAB) [2] 32 (VESTIBULE) [2] |
| BRAKE VALVE | | WABCO, L TRIPLE | | B CAR: CONSOLIDATED, 143LL [12], 146X [2] |
| COUPLER | | VAN DORN | LIGHTING | LAMPS [25], MARKER/VESTIBULE [2] |
| DOOR OPERATOR ENGINE | | CONSOLIDATED, R5082A & R5083A [4] | MASTER CONTROLLER | A & C CARS: WH, NO. 8 |
| DRAFT GEAR | #1 END | SLAB & BARREL | PROPULSION CONTROL TYPE | A & C CARS: WH, USG251-I-3 |
| | #2 END | RAIL & BARREL | SEAT ARRANGEMENT & CAPACITY | CROSS & LONGITUDINAL, A & C CARS [50], B CAR [52] |
| HAND BRAKE | | A & C CARS: STAFF; B CAR: NONE | SEAT MATERIAL | RATTAN |

## TRUCK

| | | | | |
|---|---|---|---|---|
| MANUFACTURER | | AMERICAN LOCOMOTIVE, ST. LOUIS | TRACTION MOTOR | WH, 300 (200 HP) [2/MOTOR TRUCK] |
| TYPE | #1 END | MOTOR: ST. LOUIS; TRAILER: PECKHAM 40 | | WH, 50L (150 HP) [2/MOTOR TRUCK] |
| | #2 END | MOTOR: AMERICAN LOCOMOTIVE; TRAILER: PECKHAM 41 | | |

## GENERAL

| | | | |
|---|---|---|---|
| STANDING CAPACITY | [108] | | |

NOTE 1: UNIT CONSIST OF THREE CARS (A,B,C). MOTOR CARS WERE A & C, TRAILER CAR WAS B.

NOTE 2: 33'-6" FOR CARS BUILT IN 1904 - 1905 AND 33'-2" FOR CARS BUILT IN 1908.

NOTE 3:

| | | | | | |
|---|---|---|---|---|---|
| BRADLEY | A CAR | CARS 1604, 1621 | BRILL | A CAR | CARS 1618, 1623 |
| DATE: 1904 - 1905 | B CAR | CARS 1600,1604,1606,1607,1609,1610,1614,1620,1625 | DATE: 1904 - 1905 | B CAR | CARS 1601,1603,1605,1608,1616,1617,1619,1623,1627,1629 |
| TOTAL: 11 CARS | C CAR | CARS 1625 | TOTAL: 16 CARS | C CAR | CARS 1604,1623,1624 |
| JEWETT | A CAR | CARS 1601-1603, 1605-1606, 1608-1614, 1617, 1619-1620, 1622 | LACONIA | A CAR | 1600, 1607, 1615, 1616, 1624-1629 |
| DATE: 1908 | C CAR | CARS 1600-1603, 1605-1616, 1619-1621 | DATE: 1904, 1905, 1908 | B CAR | 1602,1611-1613,1615,1618, 1621,1622,1624,1626,1628 |
| TOTAL: 35 CARS | | | TOTAL: 28 CARS | C CAR | 1617,1618,1622,1624,1626-1629 |

CAR NO. 1621A, BUILT BY JEWETT, BURNED 2/1943 AND WAS REPLACED BY A BRADLEY CAR.
GLASS IN UPPER DOOR POCKET REPLACED WITH SHEET STEEL AND RELOCATED CAR NUMBERS TO UPPER SASH (1941 - 1942)
ORIGINALLY, THE CARS WERE 'BU' TYPE NUMBERED 1200 AND 1400 SERIES AND WERE OPEN PLATFORM ELEVATED MOTOR CARS BUILT (1903 - 1908) AND WERE CONVERTED TO Q TYPE (1938 - 1941) USED ON ASTORIA AND FLUSHING LINES (1/1939 - 10/1949).

CAR NUMBERS: 1600 - 1629 A, B, C
TOTAL: 72 CARS (24 UNITS), SEE NOTE 1
BUILT BY: BRADLEY, BRILL, JEWETT, LACONIA
DATE: 1904 - 1908
AVERAGE COST PER CAR: N/A

REBUILT & OVERHAULED BY: BMT, BOT
REBUILT DATE: 1938 - 1941
OVERHAUL BY: NYC TRANSIT AUTHORITY
DATE: 1958 & 1962
OVERHAUL COST PER CAR: N/A

OVERHAUL DATE: 1950

# Q (BMT)

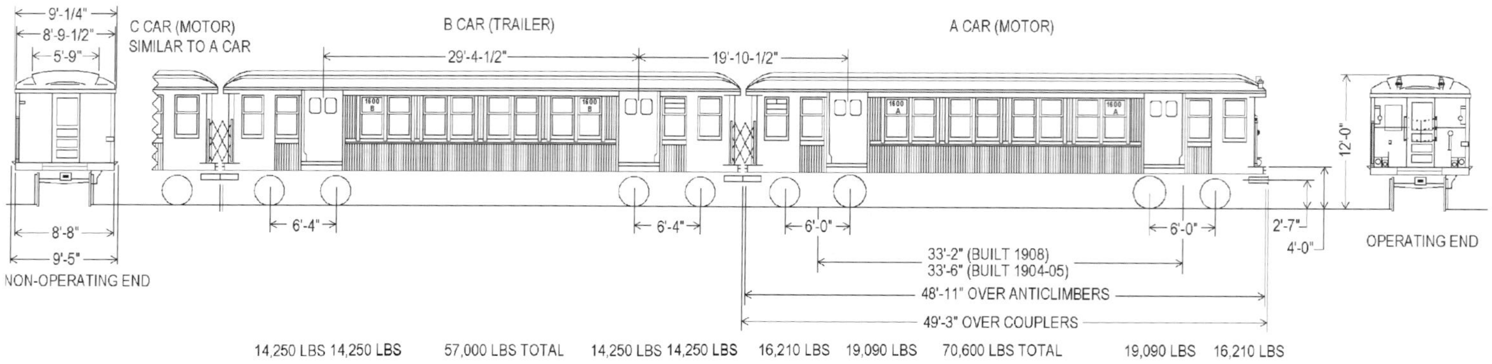

NON-OPERATING END: 9'-1/4", 8'-9-1/2", 5'-9", 8'-8", 9'-5"

C CAR (MOTOR) SIMILAR TO A CAR

B CAR (TRAILER): 29'-4-1/2"; 6'-4"; 6'-4"

A CAR (MOTOR): 19'-10-1/2"; 6'-0"; 6'-0"; 33'-2" (BUILT 1908); 33'-6" (BUILT 1904-05); 48'-11" OVER ANTICLIMBERS; 49'-3" OVER COUPLERS

OPERATING END: 12'-0"; 2'-7"; 4'-0"

14,250 LBS 14,250 LBS 57,000 LBS TOTAL 14,250 LBS 14,250 LBS 16,210 LBS 19,090 LBS 70,600 LBS TOTAL 19,090 LBS 16,210 LBS

Copyright © 1997 by NYC Transit

## CAR BODY - EQUIPMENT

|  | |  |  | |  |
|---|---|---|---|---|---|
| AIR COMPRESSOR (A & C CARS) | | WH, D2F | HAND HOLD | | CANVAS WEBBING |
| BATTERY | | B CAR: B2H [2] | HEAD LIGHT | | LOVELL DRESSEL |
| BRAKE CYCLINDER | | WABCO 12X12 | HEATER | | A & C CARS: CONSOLIDATED, 143LL [16], 146X (CAB) [2] |
| BRAKE EQUIPMENT TYPE | | A & C CARS: WH, AML; B CAR: WH, ATL | | | B CAR: CONSOLIDATED, 143LL [12], 146X [2] |
| BRAKE VALVE | | WABCO, M19A | LIGHTING | | 600 V SERIES [4], 120 V LAMP/CIRCUIT [5] |
| COUPLER | | VAN DORN | PROPULSION CONTROL TYPE | | A & C CARS: WH UNIT SWITCH |
| DOOR HANGER | | OTIS ELEVATOR | SEAT ARRANGEMENT & CAPACITY | | CROSS & LONGITUDINAL, A & C CARS [50], B CAR [52] |
| DOOR OPERATOR ENGINE | | CONSOLIDATED, R5082A & R5083A [4] | SEAT MATERIAL | | RATTAN |
| DRAFT GEAR | #1 END | SLAB & BARREL | SIGN, SIDE | | METAL PLATE |
| | #2 END | RAIL & BARREL | SWITCH GROUP | | A & C CARS: WH 225113 |
| FAN | #1 END | GE, 4X700, BRACKET 10" [4] | SWITCH PANEL | NO. 1 END | A CAR: CONSOLIDATED, PS56A; C CAR: CONSOLIDATED, PS58A |
| | #2 END | SQUARE D, BRACKET 10" [4] | | NO. 2 END | A & B CARS: KNIFE SWITCH |
| FIRE EXTINGUISHER | | 2 LBS DRY POWDER [1/CAR] | TRICKLE CHARGER | | B CAR: WH, UM281A |
| HAND BRAKE | | A & C CARS: STAFF; B CAR: NONE | TRIPLE VALVE | | WABCO, L TYPE |

## TRUCK

| A & C CARS | | B CAR | |
|---|---|---|---|
| MANUFACTURER | INTERBOROUGH RAPID TRANSIT | MANUFACTURER | PECKHAM |
| TYPE | 'I' BEAM, SINGLE MOTOR | TYPE | PECKHAM 40 |
| TRUCK NUMBERS | 1200 - 2173 | WEIGHT | 9,200 LBS |
| WEIGHT | 13,600 LBS | **ALL CARS** | |
| CONTACT SHOE | LO-V TYPE IRT | BRAKE RIGGING | INSIDE HUNG SINGLE SHOE BRAKE |
| TRACTION MOTOR | WH, 336 (120 HP) [1/MOTOR TRUCK]; GE, 259 (120 HP) [1/MOTOR TRUCK] | JOURNAL BEARING | PLAIN, 4-3/4X8 |
| TRIP COCK | WABCO, A2B [2] | | |

## GENERAL

| | | | |
|---|---|---|---|
| CAR LIGHT WEIGHT | A & C CARS: 70,450 LBS (AVG.); B CAR: 56,200 LBS (AVG.) | STANDING CAPACITY | [108] |

NOTE 1: UNIT CONSIST OF THREE CARS (A,B,C). MOTOR CARS WERE A & C, TRAILER CAR WAS B AND RECEIVED POWER FROM C CAR BY JUMPER CABLE.
NOTE 2: 33'-6" TRUCK CENTER FOR CARS BUILT 1904 - 1905 AND 33'-2" TRUCK CENTER FOR CARS BUILT 1908.
1622ABC (EX-1603C) CONVERTED BACK TO BUs 1407, 1273, 1404 (1979).
OVERHAULED CARS BY EXTENDING DOOR SILLS (1958) AND LOWERING ROOF FIVE INCHES (1958 - 1962) FOR MYRTLE AVE. SERVICE, BMT DIVISION. ELEVATED TYPE CONTACT SHOES REPLACED BY SUBWAY TYPE SHOES (1963).
INSTALLED 12" FANS (1958 - 1962).
CAR 1602A SOLD TO TROLLEY MUSEUM OF NEW YORK, KINGSTON, NY (1980). CAR 1612B SCRAPPED 2/2003 (ORIGINAL CAR WAS 1292). CAR 1612C AT NEW YORK TRANSIT MUSEUM.
UNITS 1604, 1614, 1615, 1617, 1623 & 1626 WERE SCRAPPED (1958). ALL OTHER CARS WERE SCRAPPED (1970).

CAR NUMBERS: 1630 - 1642 A & B
TOTAL: 26 CARS (13 UNITS)
BUILT BY: BRADLEY, BRILL, JEWETT, LACONIA, SEE NOTE 1
DATE: 1904 - 1908
AVERAGE COST PER CAR: N/A

REBUILT BY: BMT
REBUILT DATE: 1939 - 1940
AVERAGE COST PER CAR: N/A

# QX (BMT)

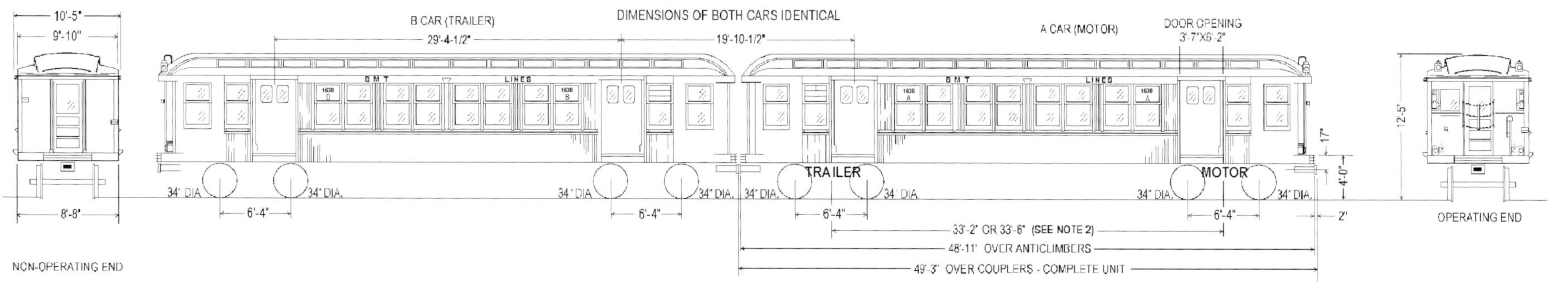

NON-OPERATING END
10'-5"
9'-10"
8'-8"
B CAR (TRAILER)
29'-4-1/2"
DIMENSIONS OF BOTH CARS IDENTICAL
19'-10-1/2"
A CAR (MOTOR)
DOOR OPENING 3'-7"X6'-2"
34" DIA.
6'-4"
TRAILER
MOTOR
17"
4'-0"
2"
33'-2" OR 33'-6" (SEE NOTE 2)
48'-11" OVER ANTICLIMBERS
49'-3" OVER COUPLERS - COMPLETE UNIT
12'-5"
OPERATING END

Copyright © 1997 by NYC Transit

## CAR BODY - EQUIPMENT

| | | | | |
|---|---|---|---|---|
| AIR COMPRESSOR | | WH, D2F (A CAR) | LIGHTING | LAMPS [25] |
| BRAKE EQUIPMENT TYPE | | WH, AML | | [2] MARKER/VESTIBULE LIGHTS |
| COUPLER | | VAN DORN | PROPULSION CONTROL TYPE | WH 251-13 & #8 MASTER (A CAR) |
| DOOR OPERATOR ENGINE | | CONSOLIDATED CAR HEATING, R5082A & R5083A [4] | SEAT ARRANGEMENT & CAPACITY | CROSS & LONGITUDINAL [50] |
| DRAFT GEAR | NO. 1 END | SLAB & BARREL | SIGN, SIDE | METAL PLATE |
| | NO. 2 END | RAIL & BARREL | SWITCH GROUP | WH 225112 |
| HANDBRAKE | | STAFF (A CAR) | SWITCH PANEL | NO.2 END: KNIFE SWITCH (A & B CARS) |
| HAND HOLD | | LEATHER STRAP | | NO. 1 END: CONSOLIDATED PS56A ( A CAR) |
| HEATER | | CONSOLIDATED CAR HEATING, [16] 143LL [2] 146X (CAB), [2] TYPE 32 (A CAR) | TRICKLE CHARGER | WH, UM281A |
| | | CONSOLIDATED CAR HEATING, [12] 143LL, [2] 146X (B CAR) | TRIPLE VALVE | WABCO, L TYPE |

## TRUCK

| | | | | |
|---|---|---|---|---|
| MANUFACTURER | MOTOR | PECKHAM | JOURNAL BEARING | 4 1/4 X 8, PLAIN |
| | TRAILER | AMERICAN LOCOMOTIVE | TRACTION MOTOR | WH, 300 (200 HP) [2/MOTOR TRUCK] |
| TYPE | | PECKHAM 40 | TRIP COCK | WABCO, A2B [1/UNIT] |

## GENERAL

STANDING CAPACITY [108]

NOTE 1:

| A CAR NUMBER | B CAR NUMBER | YEAR BUILT | BUILDER |
|---|---|---|---|
| N/A | 1630-33, 1636, 1638-41 | 1904-05 | BRADLEY |
| N/A | 1635, 1637, 1642 | 1904-05 | BRILL |
| 1630 - 40 | N/A | 1908 | JEWETT |
| 1641 - 42 | 1634 | 1904-05 | LACONIA |

NOTE 2: 33'-6" FOR CARS BUILT IN 1904 - 1905 AND 33'-2" FOR CARS BUILT IN 1908.

CONSIST OF TWO CARS (A,B).
CARS WERE CONVERTED FROM BU TO QX TYPE (1939 - 40) BY MODIFYING THE FOLLOWING: REMOVING GATES AND ENCLOSING END, ADDING SIDE DOORS AND ADDING MUDC FEATURE.
MODIFIED CARS 1630B, 1632B & 1636B INTO "REACH" CARS.
CARS 1630B AND 1636B SCRAPPED 2/2003. ORIGINAL CARS WERE 1231 AND 1209.
CAR 1634A TRANSFERRED TO IRT AND WAS USED AS A REVENUE COLLECTION CAR.
ADDED TROLLEY POLE ON SEVERAL CARS THAT WERE CONVERTED TO WORK CARS.
ALL OTHER CARS SCRAPPED (1951 - 1959).

CAR NUMBERS: 2000 - 2399
TOTAL: 400 CARS
BUILT BY: AMERICAN CAR FOUNDRY

DATE: 1914 - 1917
AVERAGE COST PER CAR: N/A

# B (BMT)

## 2000 SERIES MOTOR

Copyright © 1997 by NYC Transit

## CAR BODY-EQUIPMENT

| | | | |
|---|---|---|---|
| AIR BRAKE PIPE MATERIAL | STEEL | HEADLIGHT | SINGLE SOCKET W/ LENS, 40W |
| AIR COMPRESSOR | WABCO, D2F/D3F | HEATER | GOLD CAR HEATING [30/CAR] |
| BATTERY | EDISON, B4H, 32V, 24 CELL | LIGHTING | MAZDA, 115V 56W S21 FROSTED BULB [15] |
| BRAKE CYLINDER | WABCO, 12" X 18" | MARKER LIGHT | SINGLE SOCKET W/LENS |
| BRAKE EQUIPMENT TYPE | WABCO, AMUE | MASTER CONTROLLER | WH, 214 & 480 |
| CONDUIT MATERIAL | STEEL | PROPULSION CONTROL TYPE | WH, ABF |
| COUPLER | WABCO, H2A | SEAT ARRANGEMENT & CAPACITY | CROSS & LONGITUDINAL [78] |
| DOOR HANGER | J.L. HOWARD | SEAT MANUFACTURER & TYPE | HEYWOOD WAKEFIELD, RATTAN CUSHION & BACK |
| DOOR OPERATOR ENGINE | CONSOLIDATED CAR HEATING, [14] | SIGN | ELECTRIC SERVICE SUPPLY |
| DRAFT GEAR | WAUGH, 'WAUGHMAT' | SWITCH PANEL | CONSOLIDATED CAR HEATING |
| FAN | GE & WH, 38"DIAMETER | TAIL LIGHT | TWIN SOCKET W/LENS , 40W |
| FIRE EXTINGUISHER | PYRENE | WINDOW CURTAIN | C.S. CO., REX FRICTION ROLLER |
| HAND BRAKE | PEACOCK | | |

## TRUCK

| | | | |
|---|---|---|---|
| MANUFACTURER | N/A | JOURNAL BEARING | FRICTION 5X9 |
| TYPE | A55 | TRACTION MOTOR | GE, 248A (140 HP) [1/MOTOR TRUCK] |
| WEIGHT | 18,420 LBS | TRIP COCK | ELECTRIC |
| BRAKE | SIMPLEX, CLASP BRAKE | | |

## GENERAL

| | | | |
|---|---|---|---|
| CAR LIGHT WEIGHT | CARS 2000 - 2099: 89,606 LBS; CARS 2100 - 2199: 89,988 LBS; | STANDING CAPACITY | [182] |
| | CARS 2200 - 2249: 90,028 LBS; CARS 2250 - 2299: 90,843 LBS; | | |
| | CARS 2300 - 2398: 89,641 LBS; CAR 2399: 91,690 LBS | | |

LIGHTING REPLACED BY 20 LAMPS CONNECTED IN SERIES OF 5 CLEAR BULBS EACH CIRCUIT (1920).
WINDOW CURTAINS REMOVED (1928 - 1938). HAND HOLDS INSTALLED ON ALL CARS (1920) AND REPLACED BY PORCELAIN ENAMELED RAILINGS ON MOST CARS (1929 - 1939).
CAR NUMBERS LOCATED ORIGINALLY IN CENTER OF WINDOW IN OUTER DOOR POCKETS AND WERE LATER PAINTED ON CARBODY ABOVE THE WINDOW (1920 -1921). RELOCATED CAR NUMBER WHEN NUMBER PLATES WERE INSTALLED ON MANY CARS (1958 - 1959).
CARS 2000, 2001, 2010, 2021, 2042, 2050, 2051, 2054, 2060 & 2071 EQUIPPED WITH TROLLEY POLES WHICH WERE REMOVED BY 1917.
CAR 2369 EQUIPPED WITH WHITELAND LOUD SPEAKER TELEPHONE SYSTEM WITH NO. 725 RELAY (1923). REMOVED (1928).
CEILING FANS INSTALLED (1920).
ORIGINALLY SINGLE MOTOR CARS. CARS 2000 - 2398 COUPLED SEMI-PERMANENTLY INTO THREE CAR CONSIST AND WERE CLASSIFIED AS 'B' UNITS (1922-23). MOTORMAN'S OPERATING EQUIPMENT REMOVED EXCEPT FOR OPERATING ENDS (1922-23).
CAR 2399 REMAINED SINGLE AND WAS CLASSIFIED AS AN 'A' UNIT AND LATER RENUMBERED TO 2330 AND BECAME A 'B' CAR (CA. 1953).
ADDED 9 POINT MUDC WITH CENTER CAR ACTIVATED INCLUDING CAR 2399 (1927-30).
CONDUCTOR CONTROLS INACTIVATED EXCEPT FOR CENTER CAR.
CAR 2148 EQUIPPED WITH CAB SIGNAL AND SPEED CONTROL BY GENERAL RAILWAY SIGNAL (1916) AND REMOVED (1918).
END DOORS EQUIPPED WITH ELECTRO PNEUMATIC DOOR OPERATOR.
CAR 2204 AT NEW YORK TRANSIT MUSEUM.
CARS 2208 AND 2274 WERE WRECKED AND REPLACED WITH NEW 2500 SERIES STYLE BODIES.
CARS 2390 - 2392 SOLD TO RAILWAY PRESERVATION CORP. (1987). CAR 2006 RENUMBERED TO CAR 2500 (MONEY COLLECTION CAR) AND SCRAPPED (CA. 1980). CAR 2189 RENUMBERED TO CAR 2791. CAR 2330 SCRAPPED IN 1953.
CAR 2351 RENUMBERED TO CAR 2576.
ALMOST ALL OTHER CARS SCRAPPED (1961 - 1965).

CAR NUMBERS: 2400 - 2499
TOTAL: 100 CARS
BUILT BY: AMERICAN CAR FOUNDRY

DATE: 1918
AVERAGE COST PER CAR: N/A

# B, BT, BX (BMT)

## 2400 SERIES MOTOR

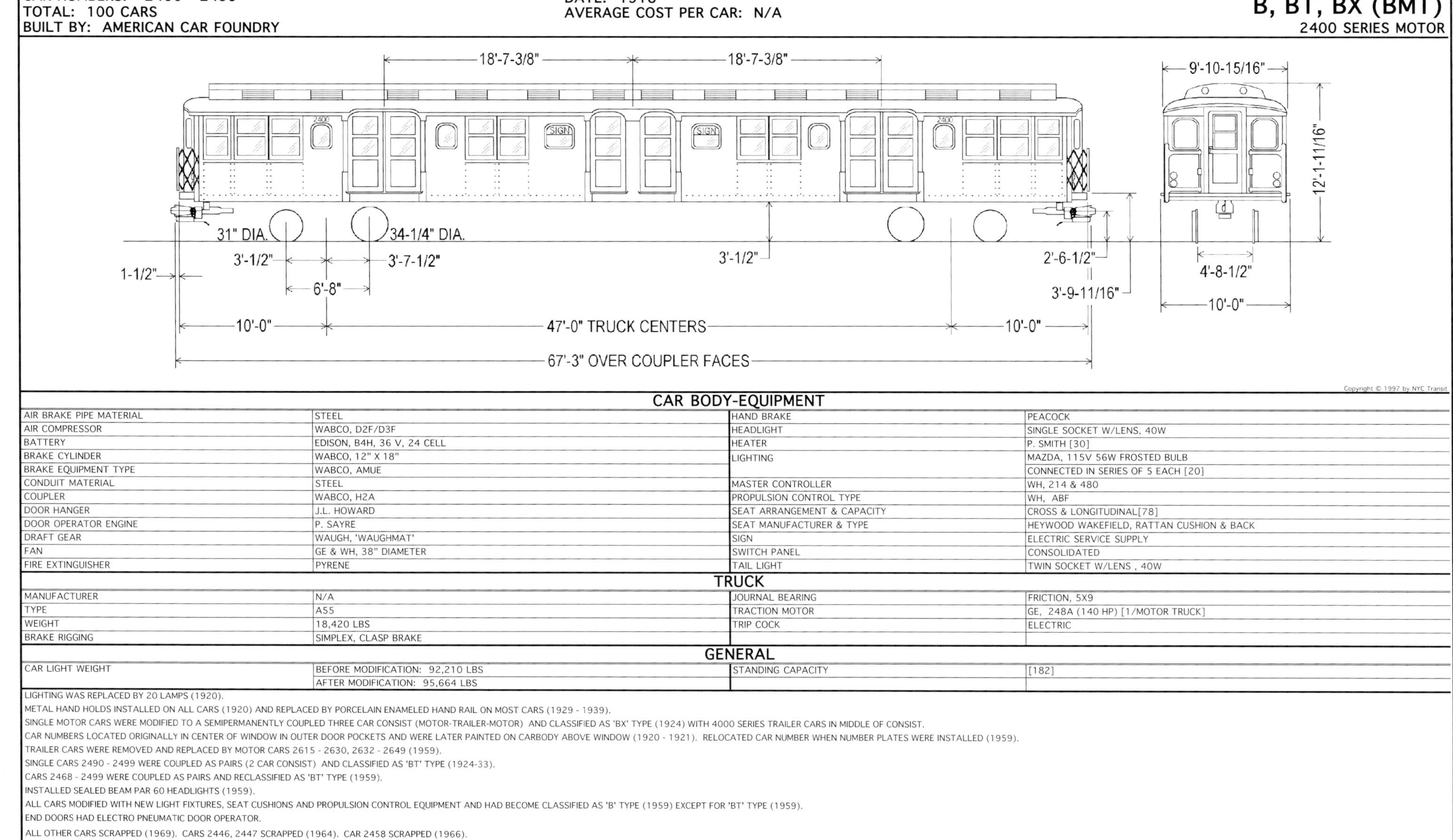


Copyright © 1997 by NYC Transit

## CAR BODY-EQUIPMENT

| | | | |
|---|---|---|---|
| AIR BRAKE PIPE MATERIAL | STEEL | HAND BRAKE | PEACOCK |
| AIR COMPRESSOR | WABCO, D2F/D3F | HEADLIGHT | SINGLE SOCKET W/LENS, 40W |
| BATTERY | EDISON, B4H, 36 V, 24 CELL | HEATER | P. SMITH [30] |
| BRAKE CYLINDER | WABCO, 12" X 18" | LIGHTING | MAZDA, 115V 56W FROSTED BULB |
| BRAKE EQUIPMENT TYPE | WABCO, AMUE | | CONNECTED IN SERIES OF 5 EACH [20] |
| CONDUIT MATERIAL | STEEL | MASTER CONTROLLER | WH, 214 & 480 |
| COUPLER | WABCO, H2A | PROPULSION CONTROL TYPE | WH, ABF |
| DOOR HANGER | J.L. HOWARD | SEAT ARRANGEMENT & CAPACITY | CROSS & LONGITUDINAL[78] |
| DOOR OPERATOR ENGINE | P. SAYRE | SEAT MANUFACTURER & TYPE | HEYWOOD WAKEFIELD, RATTAN CUSHION & BACK |
| DRAFT GEAR | WAUGH, 'WAUGHMAT' | SIGN | ELECTRIC SERVICE SUPPLY |
| FAN | GE & WH, 38" DIAMETER | SWITCH PANEL | CONSOLIDATED |
| FIRE EXTINGUISHER | PYRENE | TAIL LIGHT | TWIN SOCKET W/LENS , 40W |

## TRUCK

| | | | |
|---|---|---|---|
| MANUFACTURER | N/A | JOURNAL BEARING | FRICTION, 5X9 |
| TYPE | A55 | TRACTION MOTOR | GE, 248A (140 HP) [1/MOTOR TRUCK] |
| WEIGHT | 18,420 LBS | TRIP COCK | ELECTRIC |
| BRAKE RIGGING | SIMPLEX, CLASP BRAKE | | |

## GENERAL

| | | | |
|---|---|---|---|
| CAR LIGHT WEIGHT | BEFORE MODIFICATION: 92,210 LBS | STANDING CAPACITY | [182] |
| | AFTER MODIFICATION: 95,664 LBS | | |

LIGHTING WAS REPLACED BY 20 LAMPS (1920).
METAL HAND HOLDS INSTALLED ON ALL CARS (1920) AND REPLACED BY PORCELAIN ENAMELED HAND RAIL ON MOST CARS (1929 - 1939).
SINGLE MOTOR CARS WERE MODIFIED TO A SEMIPERMANENTLY COUPLED THREE CAR CONSIST (MOTOR-TRAILER-MOTOR) AND CLASSIFIED AS 'BX' TYPE (1924) WITH 4000 SERIES TRAILER CARS IN MIDDLE OF CONSIST.
CAR NUMBERS LOCATED ORIGINALLY IN CENTER OF WINDOW IN OUTER DOOR POCKETS AND WERE LATER PAINTED ON CARBODY ABOVE WINDOW (1920 - 1921). RELOCATED CAR NUMBER WHEN NUMBER PLATES WERE INSTALLED (1959).
TRAILER CARS WERE REMOVED AND REPLACED BY MOTOR CARS 2615 - 2630, 2632 - 2649 (1959).
SINGLE CARS 2490 - 2499 WERE COUPLED AS PAIRS (2 CAR CONSIST) AND CLASSIFIED AS 'BT' TYPE (1924-33).
CARS 2468 - 2499 WERE COUPLED AS PAIRS AND RECLASSIFIED AS 'BT' TYPE (1959).
INSTALLED SEALED BEAM PAR 60 HEADLIGHTS (1959).
ALL CARS MODIFIED WITH NEW LIGHT FIXTURES, SEAT CUSHIONS AND PROPULSION CONTROL EQUIPMENT AND HAD BECOME CLASSIFIED AS 'B' TYPE (1959) EXCEPT FOR 'BT' TYPE (1959).
END DOORS HAD ELECTRO PNEUMATIC DOOR OPERATOR.
ALL OTHER CARS SCRAPPED (1969). CARS 2446, 2447 SCRAPPED (1964). CAR 2458 SCRAPPED (1966).

CAR NUMBERS: 2500 - 2599
TOTAL: 100 CARS
BUILT BY: AMERICAN CAR FOUNDRY

DATE: 1919
AVERAGE COST PER CAR: N/A

# B (BMT)

## 2500 SERIES MOTOR

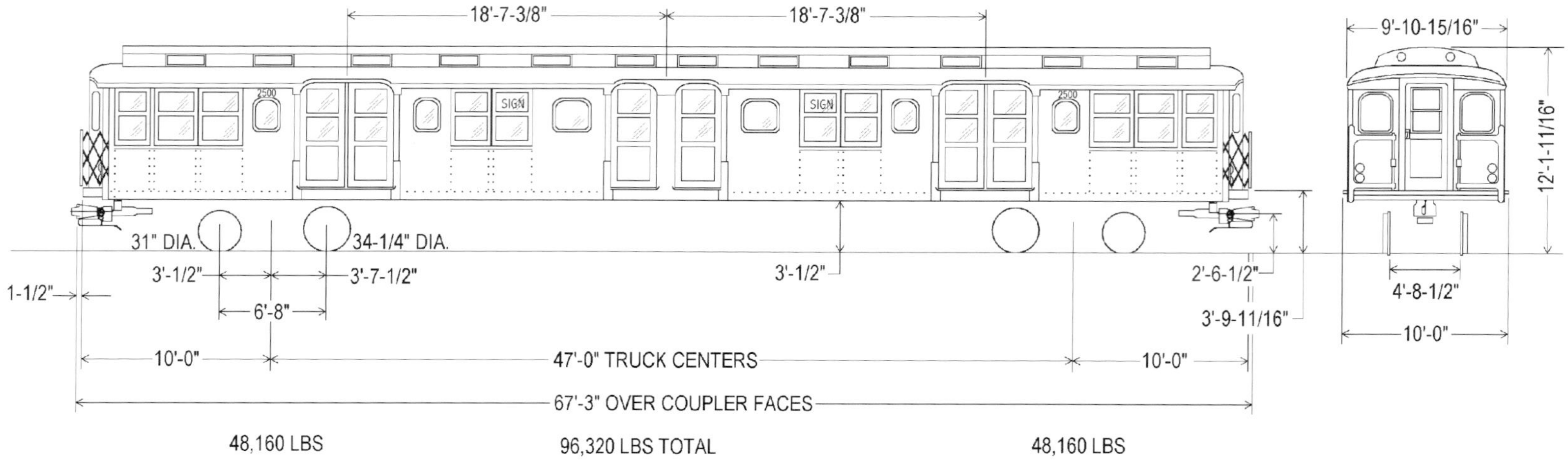

18'-7-3/8" 18'-7-3/8" 9'-10-15/16" 12'-1-11/16"

31" DIA. 34-1/4" DIA. 3'-1/2" 3'-7-1/2" 6'-8" 1-1/2" 3'-1/2" 2'-6-1/2" 3'-9-11/16" 4'-8-1/2" 10'-0"

10'-0" 47'-0" TRUCK CENTERS 10'-0"

67'-3" OVER COUPLER FACES

48,160 LBS 96,320 LBS TOTAL 48,160 LBS

Copyright © 1997 by NYC Transit

### CAR BODY-EQUIPMENT

| | | | |
|---|---|---|---|
| AIR BRAKE PIPE MATERIAL | STEEL | HAND BRAKE | PEACOCK |
| AIR COMPRESSOR | WABCO, D2F/ D3F | HEADLIGHT | SINGLE SOCKET W/ LENS, 40W |
| BATTERY | EDISON, B4H, 36V, 24 CELL | HEATER | P. SMITH [30] |
| BRAKE CYLINDER | WABCO, 12" X 18" | LIGHTING | CONNECTED IN SERIES OF 5 EACH [20] |
| BRAKE EQUIPMENT TYPE | WABCO, AMUE | MARKER LIGHT | SINGLE SOCKET, W/LENS |
| CONDUIT MATERIAL | STEEL | MASTER CONTROLLER | WH, 214 & 480 |
| COUPLER | WABCO, H2A | PROPULSION CONTROL TYPE | WH, ABF |
| DOOR HANGER | J.L. HOWARD | SEAT ARRANGEMENT & CAPACITY | CROSS & LONGITUDINAL [78] |
| DOOR OPERATOR ENGINE | P. SAYRE | SEAT MANUFACTURER AND TYPE | HEYWOOD WAKEFIELD, RATTAN CUSHION & BACK |
| DRAFT GEAR | WAUGH, 'WAUGHMAT' | SIGN | ELECTRIC SERVICE SUPPLY |
| FAN | GE & WH, 38" DIAMETER | SWITCH PANEL | CONSOLIDATED CAR HEATING |
| FIRE EXTINGUISHER | PYRENE | TAIL LIGHT | TWIN SOCKET W/LENS, 40W |

### TRUCK

| | | | |
|---|---|---|---|
| MANUFACTURER | ACF | JOURNAL BEARING | FRICTION, 5X9 |
| TYPE | A55 | TRACTION MOTOR | GE, 248A (140 HP) [1/MOTOR TRUCK] |
| WEIGHT | 18,420 LBS | TRIP COCK | ELECTRIC |
| BRAKE RIGGING | SIMPLEX, CLASP BRAKE | | |

### GENERAL

| | | | |
|---|---|---|---|
| CAR LIGHT WEIGHT | BEFORE MODIFICATION: 91,890 LBS | STANDING CAPACITY | [182] |
| | AFTER MODIFICATION: 95,336 LBS | | |

METAL HAND HOLDS REPLACED BY PORCELAIN ENAMELED HAND RAIL ON MOST CARS (1929 - 1939).
SINGLE MOTOR CARS WERE MODIFIED TO A SEMIPERMANENTLY COUPLED THREE CAR CONSIST AND CLASSIFIED AS A 'B' TYPE (1922) EXCEPT FOR CAR 2500 WHICH REMAINED AS A SINGLE CAR OF THE 'A' TYPE.
CARS WERE MODIFIED WITH NEW LIGHT FIXTURES, SEAT CUSHIONS, PROPULSION CONTROL EQUIPMENT EXCEPT FOR CAR 2500 (1959).
CARS WERE BUILT WITH MODIFIED BOX TYPE VENTILATORS ON ROOF.
END DOORS HAD ELECTRO PNEUMATIC DOOR OPERATOR.
INSTALLED SEALED BEAM PAR 60 HEADLIGHTS (1959).
RELOCATED CAR NUMBER WHEN NUMBER PLATES WERE INSTALLED (1959).
CAR 2500 WAS CONVERTED TO REVENUE COLLECTION CAR (1959).
ALL CARS SCRAPPED (1967 - 1969).

CAR NUMBERS: 2600 - 2899
TOTAL: 300 CARS
BUILT BY: PRESSED STEEL

DATE: 1920 - 1922
AVERAGE COST PER CAR: N/A

# A, B (BMT)

MOTOR

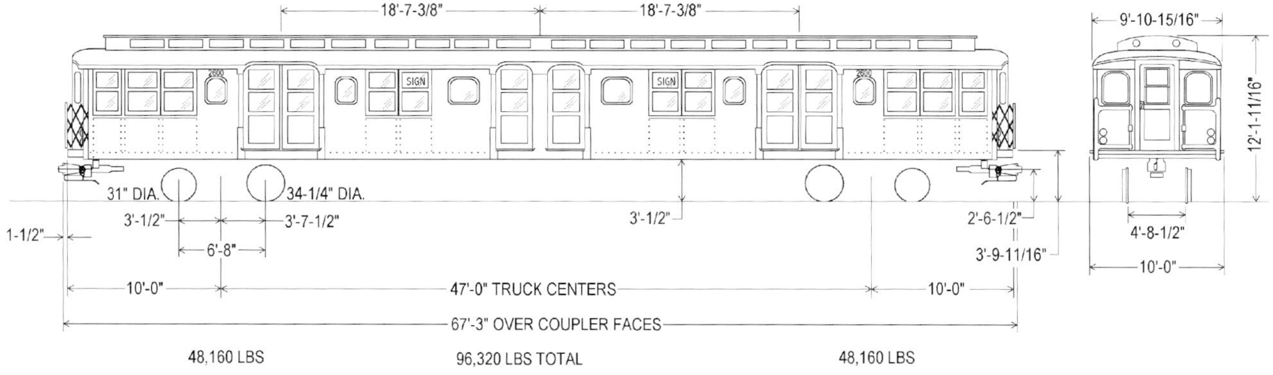

18'-7-3/8" 18'-7-3/8"
9'-10-15/16"
12'-1-11/16"
31" DIA. 34-1/4" DIA.
3'-1/2" 3'-7-1/2" 3'-1/2" 2'-6-1/2"
1-1/2" 6'-8" 3'-9-11/16" 4'-8-1/2"
10'-0" 47'-0" TRUCK CENTERS 10'-0" 10'-0"
67'-3" OVER COUPLER FACES
48,160 LBS 96,320 LBS TOTAL 48,160 LBS

Copyright © 1997 by NYC Transit

## CAR BODY-EQUIPMENT

| | | | |
|---|---|---|---|
| AIR BRAKE PIPE MATERIAL | STEEL | HAND HOLD | METAL |
| AIR COMPRESSOR | WABCO, D2F/ D3F | HEADLIGHT | SINGLE SOCKET W/ LENS, 40W |
| BATTERY | EDISON, B4H, 36V, 24 CELL | HEATER | GOLD CAR HEATING [3] |
| BRAKE CYLINDER | WABCO, 12" X 18" | LIGHTING | CONNECTED IN SERIES OF 5 EACH [20] |
| BRAKE EQUIPMENT TYPE | WABCO, AMUE | MARKER LIGHT | SINGLE SOCKET W/LENS |
| CONDUIT MATERIAL | STEEL | MASTER CONTROLLER | WH, 214 & 480 |
| COUPLER | WABCO, H2A | PROPULSION CONTROL TYPE | WH, ABF |
| DOOR HANGER | J.L. HOWARD | SEAT ARRANGEMENT & CAPACITY | CROSS & LONGITUDINAL [78] |
| DOOR OPERATOR ENGINE | CONSOLIDATED CAR HEATING [14] | SEAT MANUFACTURER AND TYPE | HEYWOOD WAKEFIELD, RATTAN CUSHION & BACK |
| DRAFT GEAR | WAUGH, 'WAUGHMAT' | SIGN | ELECTRIC SERVICE SUPPLY |
| FAN | GE & WH, 38" DIAMETER | SWITCH PANEL | CONSOLIDATED CAR HEATING |
| FIRE EXTINGUISHER | PYRENE | TAIL LIGHT | TWIN SOCKET W/LENS , 40W |
| HAND BRAKE | PEACOCK | WINDOW HARDWARE | BRASS |

## TRUCK

| | | | |
|---|---|---|---|
| MANUFACTURER | PRESSED STEEL | JOURNAL BEARING | FRICTION, 5X9 |
| TYPE | A55 | TRACTION MOTOR | GE, 248A (140 HP) |
| WEIGHT | 18,420 LBS | TRIP COCK | ELECTRIC |
| BRAKE RIGGING | SIMPLEX, CLASP BRAKE | | |

## GENERAL

| | | | |
|---|---|---|---|
| CAR LIGHT WEIGHT | BEFORE MODIFICATION: CARS 2700 - 2749 (95,975 LBS); CARS 2750 - 2799 (95,425 LBS); CARS 2800 - 2899 (95,616 LBS)<br>AFTER MODIFICATION: CARS 2700 - 2749 (99,198 LBS); CARS 2750 - 2799 (99,198 LBS); CARS 2800 - 2899 (99,198 LBS) | STANDING CAPACITY | [182] |

CARS 2600 - 2749 OPERATED AS SINGLE CARS ('A' UNITS).
CARS 2750 - 2899 OPERATED AS THREE CAR UNITS ('B' UNITS).
HAND HOLDS REPLACED BY PORCELAIN ENAMELED BAR ON SOME CARS (1936 - 1939).
SINGLE MOTOR CARS 2615 - 2630 & 2632 - 2649 WERE MODIFIED BY REMOVING CAB AND MASTER CONTROLLER AND BECAME A MIDDLE (KNOWN AS MASTER) CAR OF A THREE CAR CONSIST AND RECLASSIFIED AS 'B' TYPE.
RELOCATED CAR NUMBER WHEN NUMBER PLATES WERE INSTALLED (1959).
CARS 2600 - 2799 AND 2899 MODIFIED WITH NEW LIGHT FIXTURES, SEAT CUSHIONS AND PROPULSION CONTROL EQUIPMENT (1959 - 1960).
END DOORS EQUIPPED WITH ELECTRO PNEUMATIC DOOR OPERATOR. ADDED PAR 60 SEALED BEAM HEADLIGHTS TO CARS 2600 - 2799 (1959).
SOME CARS WERE EQUIPPED WITH WHITELAND LOUD SPEAKER TELEPHONE SYSTEM WITH NO. 725 RELAY (1924) AND WAS REMOVED (1928).
CAR 2639 SCRAPPED (1964). CARS 2794, 2795 & 2796 SCRAPPED (1966).
CAR 2775 WAS SOLD TO BRANFORD ELECTRIC RAILWAY, EAST HAVEN, CT. (1980).
CARS 2774 & 2899 SCRAPPED (1980).
ALL OTHER CARS SCRAPPED (1965 - 1969).

CAR NUMBERS: 4000 - 4049
TOTAL: 50 CARS
BUILT BY: PRESSED STEEL

DATE: 1924
AVERAGE COST PER CAR: N/A

# AX, BX (BMT)

TRAILER

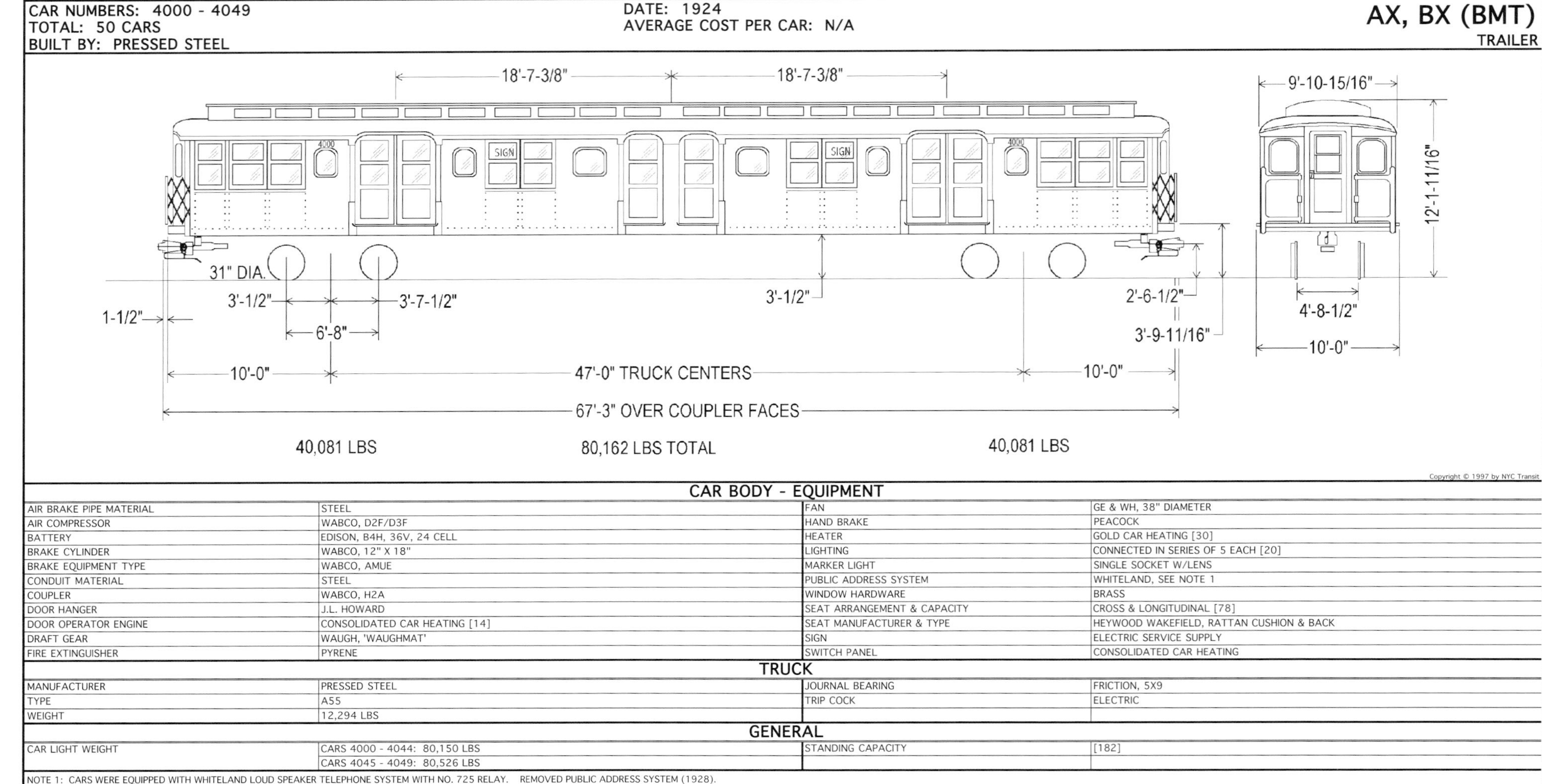


Copyright © 1997 by NYC Transit

## CAR BODY - EQUIPMENT

| | | | |
|---|---|---|---|
| AIR BRAKE PIPE MATERIAL | STEEL | FAN | GE & WH, 38" DIAMETER |
| AIR COMPRESSOR | WABCO, D2F/D3F | HAND BRAKE | PEACOCK |
| BATTERY | EDISON, B4H, 36V, 24 CELL | HEATER | GOLD CAR HEATING [30] |
| BRAKE CYLINDER | WABCO, 12" X 18" | LIGHTING | CONNECTED IN SERIES OF 5 EACH [20] |
| BRAKE EQUIPMENT TYPE | WABCO, AMUE | MARKER LIGHT | SINGLE SOCKET W/LENS |
| CONDUIT MATERIAL | STEEL | PUBLIC ADDRESS SYSTEM | WHITELAND, SEE NOTE 1 |
| COUPLER | WABCO, H2A | WINDOW HARDWARE | BRASS |
| DOOR HANGER | J.L. HOWARD | SEAT ARRANGEMENT & CAPACITY | CROSS & LONGITUDINAL [78] |
| DOOR OPERATOR ENGINE | CONSOLIDATED CAR HEATING [14] | SEAT MANUFACTURER & TYPE | HEYWOOD WAKEFIELD, RATTAN CUSHION & BACK |
| DRAFT GEAR | WAUGH, 'WAUGHMAT' | SIGN | ELECTRIC SERVICE SUPPLY |
| FIRE EXTINGUISHER | PYRENE | SWITCH PANEL | CONSOLIDATED CAR HEATING |

## TRUCK

| | | | |
|---|---|---|---|
| MANUFACTURER | PRESSED STEEL | JOURNAL BEARING | FRICTION, 5X9 |
| TYPE | A55 | TRIP COCK | ELECTRIC |
| WEIGHT | 12,294 LBS | | |

## GENERAL

| | | | |
|---|---|---|---|
| CAR LIGHT WEIGHT | CARS 4000 - 4044: 80,150 LBS | STANDING CAPACITY | [182] |
| | CARS 4045 - 4049: 80,526 LBS | | |

NOTE 1: CARS WERE EQUIPPED WITH WHITELAND LOUD SPEAKER TELEPHONE SYSTEM WITH NO. 725 RELAY. REMOVED PUBLIC ADDRESS SYSTEM (1928).

CARS 4000 - 4044 WERE SEMI PERMANENTLY COUPLED AS A THREE CAR CONSIST (MOTOR-TRAILER-MOTOR) BETWEEN CARS 2400 - 2489 AND CLASSIFIED AS 'BX' TYPE.
CARS 4045 - 4049 WERE ORIGINALLY SINGLE CONTROLLED TRAILERS CLASSIFIED AS 'AX' TYPE. MOTORMAN'S CABS REMOVED (1933) AND WERE COUPLED BETWEEN CARS 2490 - 2499. RECLASSIFIED AS 'BX' TYPE.
HAND HOLDS REPLACED WITH PORCELAIN ENAMELED RAILINGS ON SOME CARS (1936 - 1939).
END DOORS EQUIPPED WITH ELECTRO PNEUMATIC DOOR OPERATOR.
CAR 4036 REWIRED AND SEAT ARRANGEMENT MODIFIED (1959).
RELOCATED CAR NUMBER WHEN NUMBER PLATES WERE INSTALLED ON SOME CARS (1959).
ALL CARS WERE SCRAPPED (1961); EXCEPT FOR CAR 4036 WHICH WAS THEN USED BY THE SOUTH BROOKLYN RAILWAY AS A YARD OFFICE AND SCRAPPED (12/1964).

CAR NUMBERS: 6000 - 6120 A, B, & C
TOTAL: 121 CAR UNITS
BUILT BY: PRESSED STEEL

DATE: 1925, 1927 - 1928 SEE NOTE 1
AVERAGE COST PER CAR: $30,000

# D (BMT)

TRIPLEX

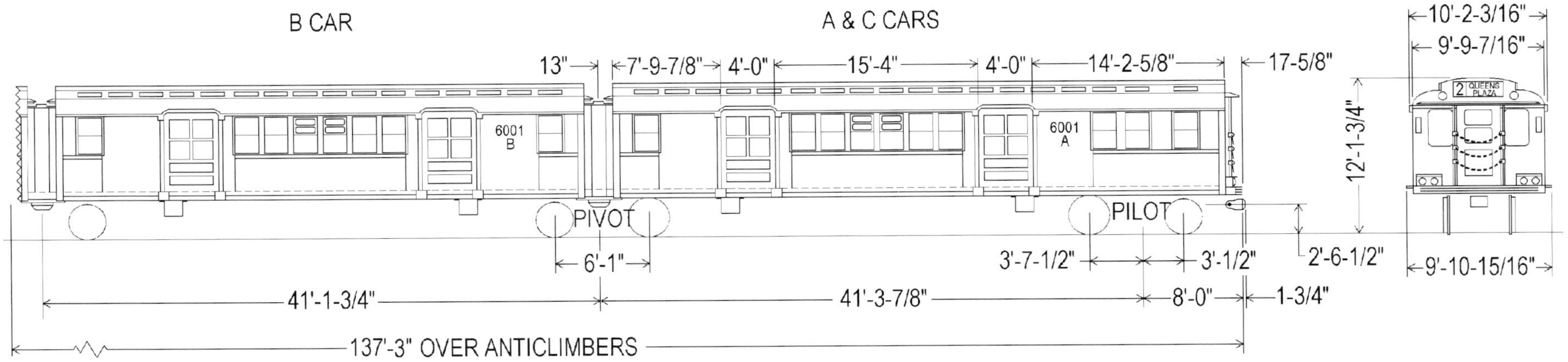

B CAR

A & C CARS

13" 7'-9-7/8" 4'-0" 15'-4" 4'-0" 14'-2-5/8" 17-5/8"

6001 B

6001 A

PIVOT

PILOT

6'-1"

3'-7-1/2" 3'-1/2" 2'-6-1/2"

41'-1-3/4" 41'-3-7/8" 8'-0" 1-3/4"

137'-3" OVER ANTICLIMBERS

10'-2-3/16"

9'-9-7/16"

12'-1-3/4"

2 QUEENS PLAZA

9'-10-15/16"

Copyright © 1997 by NYC Transit

## CAR BODY - EQUIPMENT

| | | | |
|---|---|---|---|
| AIR COMPRESSOR | D3F (2) | HAND BRAKE | PEACOCK |
| BRAKE CYLINDER | WABCO, 12"X18" | HAND HOLD | PORCELAIN ENAMELED RAILING |
| BRAKE EQUIPMENT TYPE | WH, AMUE | HEATER | CONSOLIDATED CAR HEATING [72] |
| COUPLER | WABCO, H2A | | RAILWAY UTILITIES [72] |
| DOOR OPERATOR ENGINE | CONSOLIDATED CAR HEATING, 5021, [12] | LIGHTING | LAMPS [60/CONSIST] |
| | NATIONAL PNEUMATIC, E.U. [12] | PROPULSION CONTROL TYPE | GE, PC15B; WH, 143 |
| DRAFT GEAR | W.T.B. | SEAT ARRANGEMENT & CAPACITY (CONSIST) | CROSS AND LONGITUDINAL [160] |
| FAN | WH, CEILING [12/UNIT] | SIGN | HUNTER ILLUMINATED CAR |

## TRUCK

| | | | |
|---|---|---|---|
| MANUFACTURER | PRESSED STEEL, | TRACTION MOTOR | WH, 584; GE, 282D (200 HP) [1/PILOT TRUCK], [1/PIVOT TRUCK] |
| TYPE | A58 | WHEEL DIAMETER | 34" (MOTOR AXLE), 31"(TRAILER AXLE) |
| JOURNAL BEARING | FRICTION | | |

## GENERAL

| | | | |
|---|---|---|---|
| CONSIST CAR WEIGHT LIGHT RANGE | 207,600 - 213,650 LBS | WEIGHT, LIGHT-A & C CARS | 47,700 - 48,780 LBS |
| STANDING (CONSIST) | [395]/ CONSIST | WEIGHT, LOADED-A & C CARS | 56,280 - 58,200 LBS |

NOTE 1: CARS 6000 - 6003 (1925); CARS 6004 - 6070 (1927); CARS 6071 - 6120 (1928)

UNIT CONSISTS OF THREE CARBODIES (A, B, & C) ON FOUR TRUCKS WITH ONE MOTOR ON EACH TRUCK & CONTACT SHOES ON PIVOT TRUCKS.
CARS 6000 - 6003 REBUILT (1928).
CARS 6000 - 6003 & 6120 HAD TIMKEN ROLLER BEARINGS.
CAR NUMBER PLATES WERE INSTALLED ON SOME CARS (1959).
CARS 6019, 6095 & 6112 AT NY TRANSIT MUSEUM.
CAR 6119B BECAME A CONSTRUCTION SHACK AT WEST HURLEY, N.Y. (1969).
ONE UNIT SCRAPPED (1959). ALL OTHERS SCRAPPED (1964-1965).

CAR NUMBERS: 7003 A, B, C, B1, A1
TOTAL: 1 CAR UNIT, SEE NOTE 1
BUILT BY: PULLMAN

DATE: 1934
AVERAGE COST PER CAR: N/A

# MS (BMT)

GREEN HORNET - MULTI-SECTION

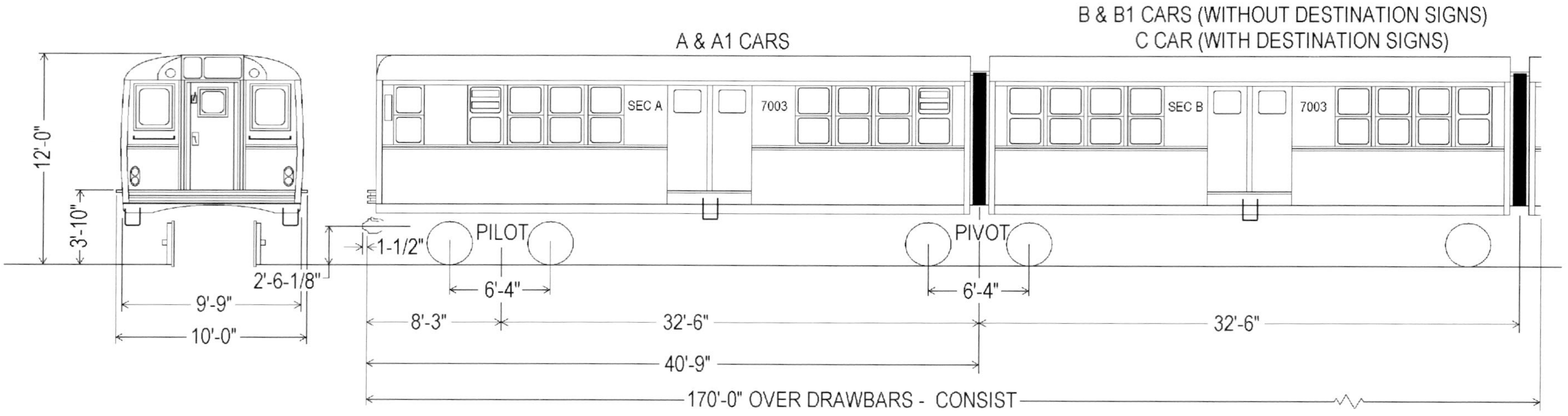

A & A1 CARS

B & B1 CARS (WITHOUT DESTINATION SIGNS)
C CAR (WITH DESTINATION SIGNS)

12'-0"
3'-10"
2'-6-1/8"
9'-9"
10'-0"
1-1/2"
PILOT
PIVOT
SEC A
SEC B
7003
6'-4"
8'-3"
32'-6"
40'-9"
170'-0" OVER DRAWBARS - CONSIST

Copyright © 1997 by NYC Transit

## CAR BODY - EQUIPMENT

| | | | |
|---|---|---|---|
| AIR COMPRESSOR | WH, DH-16 | HAND HOLD | STAINLESS STEEL RAILINGS |
| BRAKE EQUIPMENT TYPE | WH, AMSF/ ME37 | LIGHTING | 80 LAMPS (EACH SECTION) 4 CIRCUITS OF 20 LAMPS EACH MASKED BEHIND A PANEL IN CENTER OF CEILING TO PROVIDE INDIRECT LIGHTING |
| CAR LIGHT WEIGHT | 170,610 LBS | | |
| COUPLER | VAN DORN (EMERGENCY) | PROPULSION CONTROL TYPE | WH, XD29 P.C.C. MULTI NOTCH 47 PTS |
| DOOR OPERATOR MOTOR | CONSOLIDATED | SEAT ARRANGEMENT & CAPACITY | CROSS & LONGITUDINAL [184/CONSIST] |
| FAN | AIR CIRCULATING BLOWERS | SIGN | HUNTER ILLUMINATED CAR AT CAR ENDS |
| HAND BRAKE | DAYTON | | |

## TRUCK

| | | | |
|---|---|---|---|
| MANUFACTURER | PULLMAN | TRACTION MOTOR | WH, M1431 A (70 HP) [2] |

## GENERAL

| | | | |
|---|---|---|---|
| ACCELERATION RATE | 4.00 MPHPS | CAR LIGHT WEIGHT-CONSIST | 170,610 LBS |
| BALANCING SPEED | 53 MPH | STANDING CAPACITY | [490] |
| BRAKING RATE | 4.00 MPHPS | | |

NOTE 1: CONSIST OF FIVE CARS ORIGINALLY NUMBERED: 7000A, 8000B, 9000C, 8001B, 7001A WITH TWO PILOT AND FOUR PIVOT TRUCKS.

SIGNS WERE ELECTRICALLY CONTROLLED WITH PLATES MOUNTED ON ROTATING DRUMS ON SIDES AND INSIDE BULKHEADS BY CONSOLIDATED CAR HEATING.
ADDED 80 LAMPS (EACH SECTION) 4 CIRCUITS OF 20 LAMPS EACH, MASKED BEHIND A PANEL IN CENTER OF CEILING TO PROVIDE INDIRECT LIGHTING.
EQUIPPED FOR ONE MAN OPERATION.
INTERIOR LIGHTS WERE CONTROLLED BY ELECTRIC EYE SENSITIVE TO OUTSIDE LIGHT CONDITIONS.
SPOTLIGHTS OVER OUTSIDE OF DOOR REMAINED LIT UNTIL DOOR CLOSED.
CHIMES WARNED PASSENGERS WHEN DOORS WERE BEING CLOSED.
TRAIN PLACED IN SERVICE ON FULTON ST. LINE (7/23/1934).
REMOVED FROM SERVICE WHEN TWO TRUCK FRAMES WERE FOUND CRACKED (3/1937).
TRAIN WAS PLAGUED WITH MASTER CONTROLLER PROBLEMS (TOTAL MILEAGE - 55,876).
UNIT SCRAPPED (12/1/1942).

CAR NUMBERS: 7004 - 7013 A, B, C, B1, A1
TOTAL: 10 CAR UNITS, SEE NOTE 1
BUILT BY: ST. LOUIS CAR

DATE: 1936
AVERAGE COST PER CAR: N/A

# MS (BMT)

MULTI-SECTION

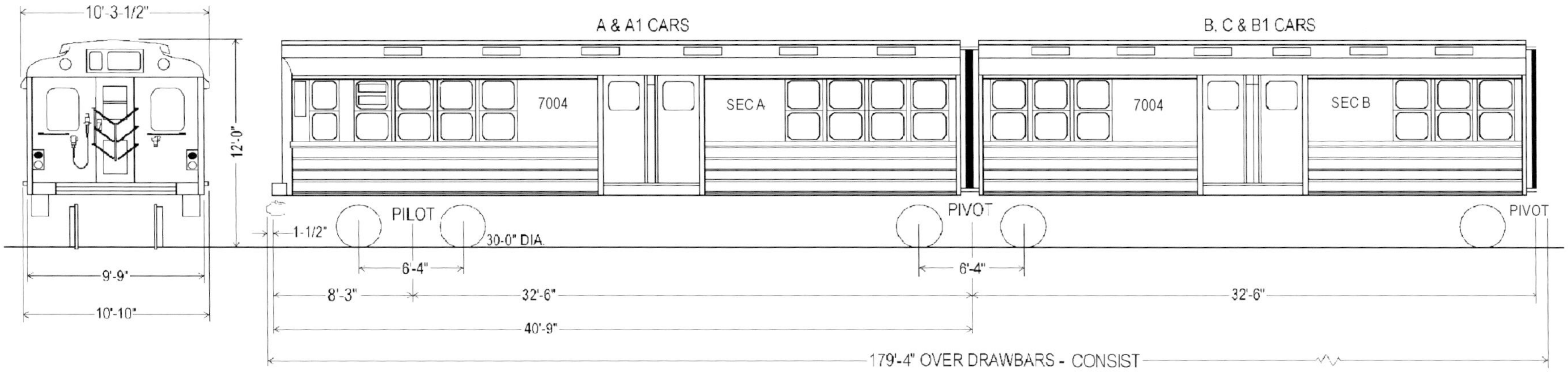

10'-3-1/2"
12'-0"
9'-9"
10'-10"
A & A1 CARS
B, C & B1 CARS
7004
SEC A
SEC B
PILOT
PIVOT
30-0" DIA
1-1/2"
6'-4"
8'-3"
32'-6"
40'-9"
179'-4" OVER DRAWBARS - CONSIST

Copyright © 1997 by NYC Transit

## CAR BODY-EQUIPMENT

| | | | |
|---|---|---|---|
| AIR COMPRESSOR | GE, CP28A (A & A1 CAR) | HEATER | CONSOLIDATED CAR HEATING, R1467 [18] |
| BRAKE EQUIPMENT TYPE | WABCO AMCE ELECTRIC PNEUMATIC/ ME39 | | CONSOLIDATED CAR HEATING, R3050 FOR CAB [4] |
| | | LIGHTING | LAMPS [20], SEE NOTE 2 |
| COUPLER | WABCO, H2A | MASTER CONTROLLER | A & A1 CARS: WH, XM-29-C |
| DOOR OPERATOR ENGINE | CONSOLIDATED, 5037A1 & 75038A1 [4] | PROPULSION CONTROL TYPE | WH, ABF (ACCELERATOR MULTI-NOTCH) |
| DRAFT GEAR | WAUGH | SEAT ARRANGEMENT & CAPACITY | CROSS & LONGITUDINAL [198] |
| ELECTRIC PORTION | WABCO, BL22 | SIGN | CONSOLIDATED CAR HEATING (ENDS AND SIDES) |
| FAN | WH, 10" BRACKET, 150V [20/CONSIST] | VENTILATOR | AUTOMATIC SEMI-EMPIRE ROOF |
| HAND BRAKE | NATIONAL BRAKE, PEACOCK | | |
| | DAYTON, HANDLE & STAFF | | |

## TRUCK

| | | | | |
|---|---|---|---|---|
| MANUFACTURER | | LUKENWELD | JOURNAL BEARING | FRICTION |
| TYPE | PILOT | CPL-P [2] | TRACTION MOTOR | WH, 1433 (70 HP) [2] |
| | PIVOT | CPL-A [4] | | |

## GENERAL

| | | | |
|---|---|---|---|
| ACCELERATION RATE | 4.00 MPHPS | CAR LIGHT WEIGHT | 180,830 LBS |
| BALANCING SPEED | 58 MPH | STANDING CAPACITY | [514]/CONSIST |
| BRAKING RATE | 4.00 MPHPS | | |

NOTE 1: UNIT CONSIST OF FIVE CARS (A-B-C-B1-A1) WITH TWO PILOT & FOUR PIVOT TRUCKS.
NOTE 2: INTERIOR LIGHTING: [20] LAMPS IN SERIES (1.6A @ 30 V); EMERGENCY LIGHTS: [2] LAMPS EACH SECTION (15W @ 40 V).

SIGNS WERE ELECTRICALLY CONTROLLED WITH PLATES MOUNTED ON ROTATING DRUMS ON SIDES AND INSIDE BULKHEADS BY GE & CONSOLIDATED CAR HEATING.
TRUCKS REBUILT BY ST. LOUIS CAR WITH ROLLED STEEL PLATE & MANGANESE STEEL CASTING (1940).
CAR NUMBER PLATES WERE INSTALLED ON SOME CARS (1959).
ALL CARS SCRAPPED (1961).

CAR NUMBERS: 7014 - 7028 A, B, C, B1, A1
TOTAL: 15 CAR UNITS, SEE NOTE 1
BUILT BY: PULLMAN STANDARD

DATE: 1936
AVERAGE COST PER CAR: N/A

# MS (BMT)

MULTI-SECTION

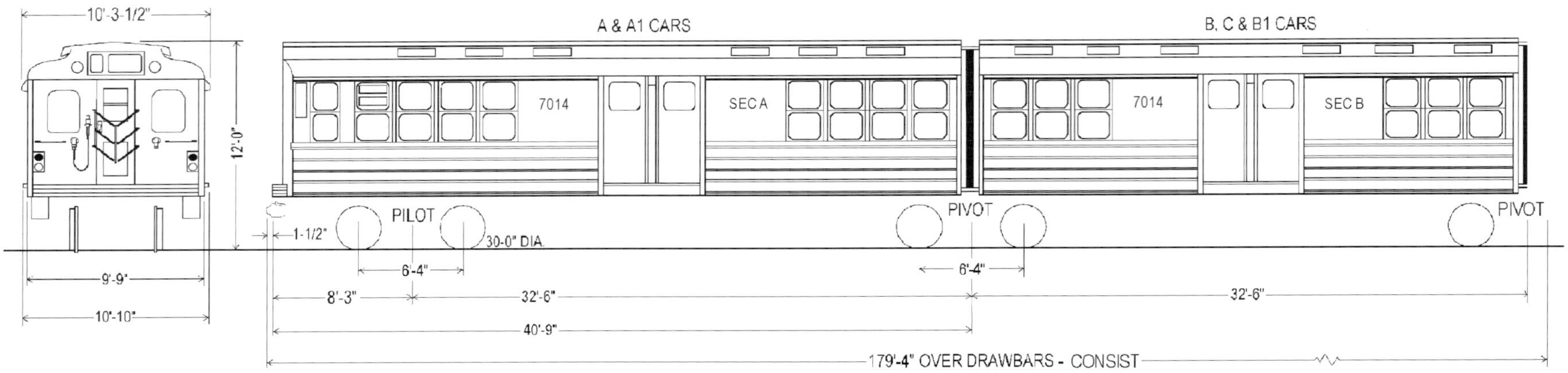

10'-3-1/2"
12'-0"
9'-9"
10'-10"
A & A1 CARS
B, C & B1 CARS
7014
SEC A
7014
SEC B
PILOT
PIVOT
PIVOT
1-1/2"
30-0" DIA
6'-4"
6'-4"
8'-3"
32'-6"
32'-6"
40'-9"
179'-4" OVER DRAWBARS - CONSIST

Copyright © 1997 by NYC Transit

## CAR BODY - EQUIPMENT

| | | | |
|---|---|---|---|
| AIR BRAKE | WH, AMCE ELECTRIC PNEUMATIC | HEATER | CONSOLIDATED, R1467[18] |
| AIR COMPRESSOR | GE, CP28A (A & A1 CARS) | LIGHTING | LAMPS [20], SEE NOTE 2 |
| BRAKE EQUIPMENT TYPE | GE, EDDY CURRENT | | EMERGENCY LIGHTS [2] |
| COUPLER | WABCO, H2A | MASTER CONTROLLER | GE, 17KC 23A1 (A & A1 CARS) |
| DOOR OPERATOR ENGINE | CONSOLIDATED, 5037A1 & 75038A2 [4] | PROPULSION CONTROL TYPE | GE, 17KG39A1 (PCM) |
| | CONSOLIDATED, R3050 FOR CAB [4] | SEAT ARRANGEMENT & CAPACITY | CROSS & LONGITUDINAL [198]/CONSIST |
| DRAFT GEAR | WAUGH | SIGN | ELECTRIC SERVICE SUPPLIES AT CAR ENDS |
| ELECTRIC PORTION | WABCO, BL22 | | CONSOLIDATED CAR HEATING ON SIDES |
| FAN | GE, 10" BRACKET, 150V [20]/CONSIST | VENTILATOR | AUTOMATIC SEMI-EMPIRE ROOF |
| HANDBRAKE | NATIONAL BRAKE, PEACOCK | | |
| | DAYTON, HANDLE & STAFF | | |

## TRUCK

| | | | | |
|---|---|---|---|---|
| MANUFACTURER | | PULLMAN/BUTLER | JOURNAL BEARING | FRICTION |
| TYPE | PILOT | CPL-P [2] | TRACTION MOTOR | GE, 1196A1 [2] |
| | PIVOT | CPL-A [4] | | |

## GENERAL

| | | | |
|---|---|---|---|
| BRAKING RATE | 4.00 MPHPS | CAR LIGHT WEIGHT | 180,830 LBS |
| BALANCING SPEED | 58 MPH | STANDING CAPACITY | [514]/CONSIST |
| ACCELERATION RATE | 4.00 MPHPS | | |

NOTE 1: UNIT CONSISTS OF FIVE CARS (A-B-C-B1-A1) WITH [2] PILOT & [4] PIVOT TRUCKS.
NOTE 2: INTERIOR LIGHTING: [20] LAMPS IN SERIES (1.6A @ 30 V). EMERGENCY LIGHTS: [2] LAMPS EACH SECTION (15 W @ 40 V).

SIGNS WERE ELECTRICALLY CONTROLLED WITH PLATES MOUNTED ON ROTATING DRUMS ON SIDES AND INSIDE BULKHEADS BY GE & CONSOLIDATED CAR HEATING.
TRUCKS REBUILT (ca. 1940).
CAR NUMBER PLATES WERE INSTALLED ON SOME CARS (1958 - 1959).
CARS SCRAPPED (1961).

CAR NUMBERS: 7029 A, B, C, B1, A1
TOTAL: 1 CAR UNIT, SEE NOTE 1
BUILT BY: BUDD

DATE: 1934
AVERAGE COST PER CAR: N/A

# MS (BMT)

BUDD/ZEPHYR - MULTI-SECTION

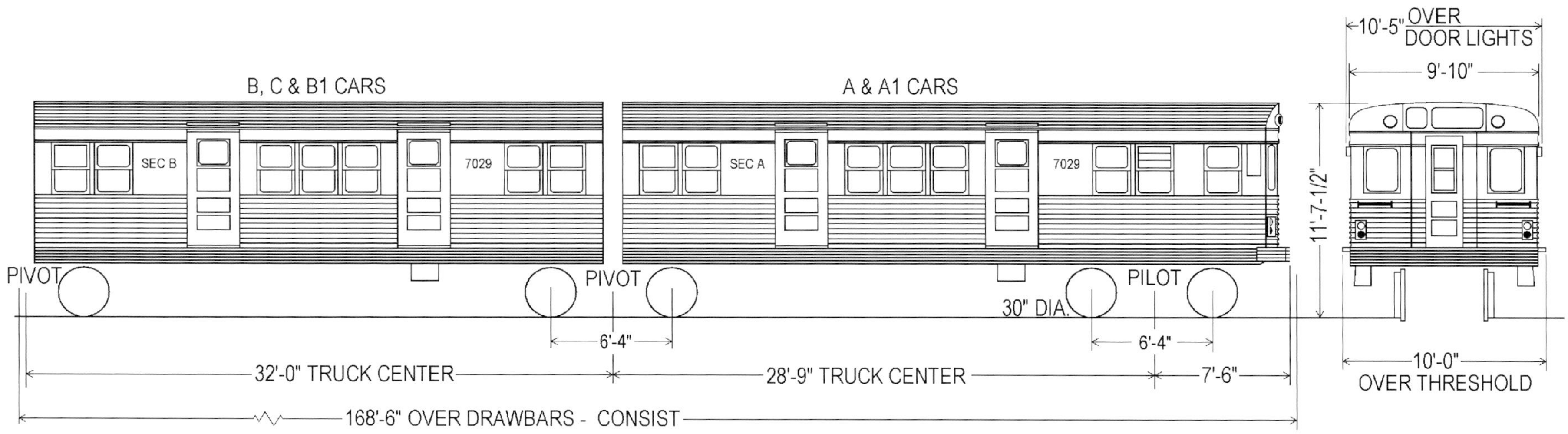

B, C & B1 CARS

A & A1 CARS

SEC B

7029

SEC A

7029

PIVOT

PIVOT

PILOT

30" DIA.

6'-4"

6'-4"

32'-0" TRUCK CENTER

28'-9" TRUCK CENTER

7'-6"

168'-6" OVER DRAWBARS - CONSIST

10'-5" OVER DOOR LIGHTS

9'-10"

11'-7-1/2"

10'-0" OVER THRESHOLD

Copyright © 1997 by NYC Transit

## CAR BODY - EQUIPMENT

| | | | |
|---|---|---|---|
| AIR BRAKE | WH, AMSF | LIGHTING | LAMPS [60], SEE NOTE 2 |
| AIR COMPRESSOR | GE, CP25C11 (A & A1 CARS) | | EMERGENCY LIGHTS[2], SEE NOTE 2 |
| BRAKE EQUIPMENT TYPE | GE, BAND BRAKE - EDDY CURRENT | MASTER CONTROLLER | GE, 17KC16A1 |
| COUPLER | WABCO, H2A, SEE NOTE 3 | PROPULSION CONTROL TYPE | GE, 17KM1C (A & A1 CARS) |
| DOOR OPERATOR ENGINE | CONSOLIDATED, R5033 [4]/CAR | | GE, 17KG21A (C CAR) |
| HAND HOLD | RAILING, ALUMINUM | SEAT ARRANGEMENT & CAPACITY | LONGITUDINAL & CROSS [170]/CONSIST |
| HANDBRAKE | NATIONAL BRAKE, PEACOCK | SEAT MATERIAL | LEATHER |
| HEATER | CONSOLIDATED, R1467 [18] | SIGN | HUNTER (END SIGNS) |
| | CONSOLIDATED, R3050 (CAB) | | |

## TRUCK

| | | | |
|---|---|---|---|
| MANUFACTURER | BUDD/LUKENWELD | TRACTION MOTOR | GE, 1186 [2] 70 HP |
| TYPE | CAST | JOURNAL BEARING | FRICTION |
| | | | |

## GENERAL

| | | | |
|---|---|---|---|
| ACCELERATION RATE | 5.00 MPHPS | CAR LIGHT WEIGHT | 159,250 LBS/CONSIST |
| BALANCING SPEED | 55 MPH | STANDING CAPACITY | [496]/CONSIST |
| BRAKING RATE | 5.00 MPHPS | | |

NOTE 1: UNIT CONSIST OF FIVE CARS (A-B-C-B1-A1) WITH TWO PILOT & FOUR PIVOT TRUCKS.
NOTE 2: INTERIOR LIGHTING: [20] LAMPS IN SERIES (1.0A @ 30 V). EMERGENCY LIGHTING: [ 2] LAMPS EACH SECTION(15W @ 40V).
NOTE 3: VAN DORN COUPLER FOR EMERGENCY USE.

CAR EQUIPPED FOR ONE MAN OPERATION.
SIGNS WERE ELECTRICALLY CONTROLLED WITH PLATES MOUNTED ON ROTATING DRUM ON SIDES AND INSIDE BULKHEADS BY CONSOLIDATED CAR HEATING.
CAR NUMBERS WERE 7002A, 8002B, 9001C, 8003B, 7003A REFERRED TO AS THE "BUDD" & "ZEPHYR".
CAR WAS SCRAPPED (1959).

CAR NUMBERS: 8000 - 8005 A, B, A1
TOTAL: 6 CAR UNITS
BUILT BY: CLARK EQUIPMENT

DATE: 1938, 1940, SEE NOTE 2
AVERAGE COST PER CAR: N/A

# COMPARTMENT (BMT)

BLUEBIRD

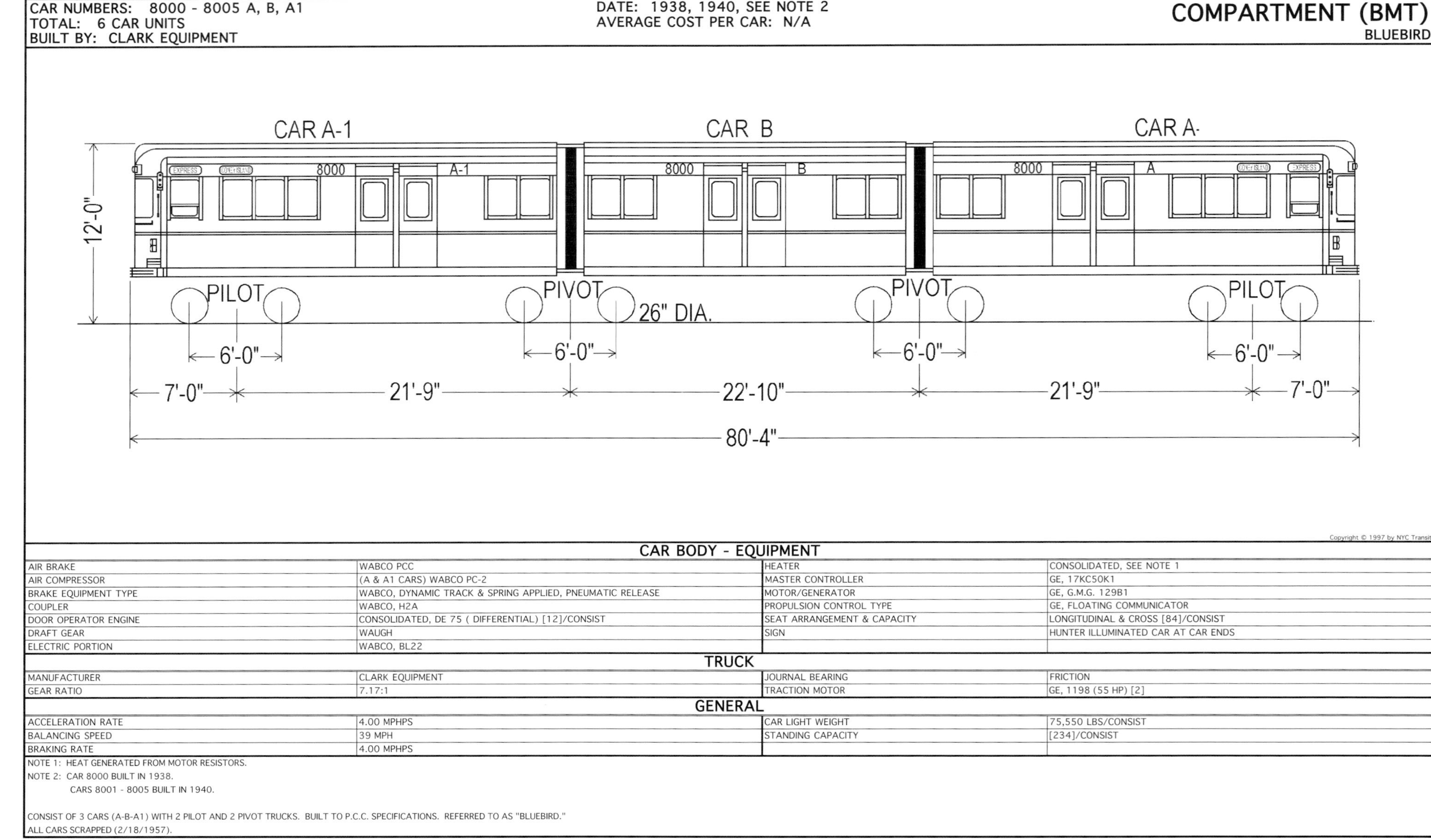


Copyright © 1997 by NYC Transit

| CAR BODY - EQUIPMENT | | | |
|---|---|---|---|
| AIR BRAKE | WABCO PCC | HEATER | CONSOLIDATED, SEE NOTE 1 |
| AIR COMPRESSOR | (A & A1 CARS) WABCO PC-2 | MASTER CONTROLLER | GE, 17KC50K1 |
| BRAKE EQUIPMENT TYPE | WABCO, DYNAMIC TRACK & SPRING APPLIED, PNEUMATIC RELEASE | MOTOR/GENERATOR | GE, G.M.G. 129B1 |
| COUPLER | WABCO, H2A | PROPULSION CONTROL TYPE | GE, FLOATING COMMUNICATOR |
| DOOR OPERATOR ENGINE | CONSOLIDATED, DE 75 ( DIFFERENTIAL) [12]/CONSIST | SEAT ARRANGEMENT & CAPACITY | LONGITUDINAL & CROSS [84]/CONSIST |
| DRAFT GEAR | WAUGH | SIGN | HUNTER ILLUMINATED CAR AT CAR ENDS |
| ELECTRIC PORTION | WABCO, BL22 | | |
| **TRUCK** | | | |
| MANUFACTURER | CLARK EQUIPMENT | JOURNAL BEARING | FRICTION |
| GEAR RATIO | 7.17:1 | TRACTION MOTOR | GE, 1198 (55 HP) [2] |
| **GENERAL** | | | |
| ACCELERATION RATE | 4.00 MPHPS | CAR LIGHT WEIGHT | 75,550 LBS/CONSIST |
| BALANCING SPEED | 39 MPH | STANDING CAPACITY | [234]/CONSIST |
| BRAKING RATE | 4.00 MPHPS | | |

NOTE 1: HEAT GENERATED FROM MOTOR RESISTORS.
NOTE 2: CAR 8000 BUILT IN 1938.
CARS 8001 - 8005 BUILT IN 1940.

CONSIST OF 3 CARS (A-B-A1) WITH 2 PILOT AND 2 PIVOT TRUCKS. BUILT TO P.C.C. SPECIFICATIONS. REFERRED TO AS "BLUEBIRD."
ALL CARS SCRAPPED (2/18/1957).

CAR NUMBERS: 2900 - 2924
TOTAL: 25 CARS
BUILT BY: STANDARD STEEL

DATE: 1925
AVERAGE COST PER CAR: N/A

**SIRT**
MOTOR

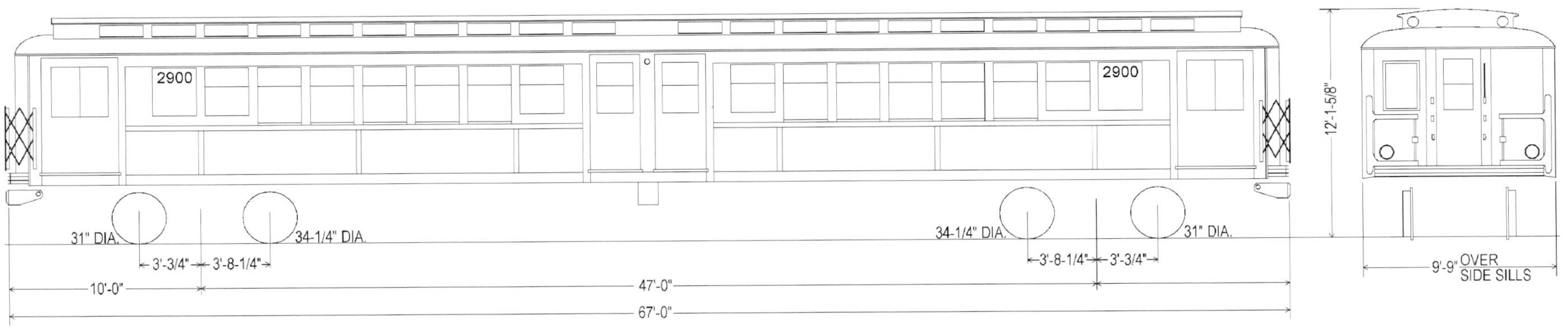

Copyright © 1997 by NYC Transit

## CAR BODY - EQUIPMENT

| | | | |
|---|---|---|---|
| AIR COMPRESSOR | WABCO, D3F | LIGHTING | CONNECTED IN SERIES OF 5 [20] |
| BRAKE EQUIPMENT TYPE | WH, AMUE | MASTER CONTROLLER | GE, C182A |
| BRAKE VALVE | WABCO, ME 30 | PROPULSION CONTROL TYPE | GE, PC10L |
| COUPLER | WABCO, H2A | SEAT ARRANGEMENT & CAPACITY | CROSS [71] |
| DOOR OPERATING EQUIPMENT | CONSOLIDATED CAR HEATING | SEAT MATERIAL | RATTAN |
| HAND BRAKE | NATIONAL BRAKE PEACOCK | SIGN | METAL PLATES |
| HAND HOLD | PORCELAIN METAL STRAP | WINDOW CURTAIN | NONE |
| HEADLIGHT (ELECTRIC) | PYLE INTERNATIONAL | WINDOW MATERIAL | SEAMLESS BRASS |
| HEATER | GOLD CAR HEATING & LIGHTING | | |

## TRUCK

| | | | |
|---|---|---|---|
| MANUFACTURER | COMMONWEALTH | GEAR RATIO | 21:62 |
| TYPE | CAST STEEL EQUALIZED | TRACTION MOTOR | GE, 282A (200 HP) [1/TRUCK] |

## GENERAL

| | | | |
|---|---|---|---|
| CAR LIGHT WEIGHT | 95,750 LBS | STANDING CAPACITY | [169] |

FIVE TRAILERS INCLUDED IN PURCHASE BUT WERE NEVER USED IN PASSENGER SERVICE (ONE WAS ASSIGNED TO FRESH POND YARD AS RCI OFFICE AND RENUMBERED 2925).
CARS WERE PURCHASED BY NYC TRANSIT AUTHORITY FROM SIRT (FORMERLY KNOWN AS STATEN ISLAND RAPID TRANSIT RAILWAY) (1953-1954).
HEADLIGHTS AND WINDOW SHADES REMOVED AND BMT AB TYPE DOOR CONTROLS INSTALLED (1954).
ADDED AXIFLO FANS MOUNTED ON CEILING (1954).
REPLACED GLASS WINDOWS WITH METAL CAR NUMBER PLATE WHEN CONVERTED TO SUBWAY OPERATION (1954).
SCREENS APPLIED TO LOWER SASH ORIGINALLY. REMOVED SCREENS WHEN SASH WAS MODIFIED FOR LIMITED WINDOW OPENING (1957).
CARS ORIGINALLY WERE NUMBERED IN THE 300 SERIES.
ALL CARS SCRAPPED (1961).

CAR NUMBERS: 300 - 394, 500 - 509
TOTAL: 100 CARS
BUILT BY: STANDARD STEEL

DATE: 1925, 1926 (SEE NOTE 1)
AVERAGE COST PER CAR: N/A

# SIRT
MOTOR & TRAILER

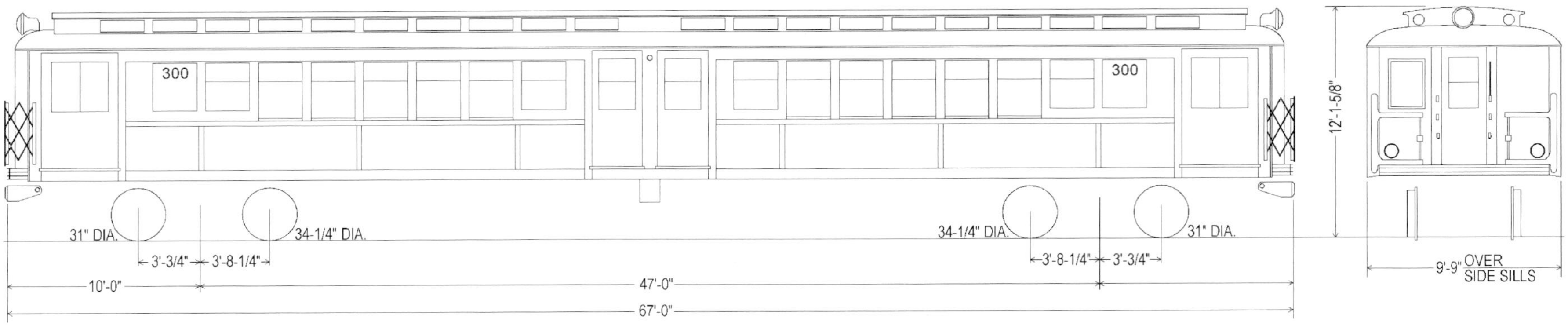

300 300
31" DIA. 34-1/4" DIA. 34-1/4" DIA. 31" DIA.
3'-3/4" 3'-8-1/4" 3'-8-1/4" 3'-3/4"
10'-0" 47'-0" 67'-0"
12'-1-5/8"
9'-9" OVER SIDE SILLS

Copyright © 1997 by NYC Transit

## CAR BODY - EQUIPMENT

| | | | |
|---|---|---|---|
| AIR COMPRESSOR | WABCO, D3F | HEATER | GOLD CAR HEATING & LIGHTING |
| BRAKE EQUIPMENT TYPE | WH, AMUE | LIGHTING | CONNECTED IN SERIES OF 5 [20] |
| BRAKE VALVE | WABCO, ME 30 | MASTER CONTROLLER | GE, C182A |
| COUPLER | WABCO, H2A | PROPULSION CONTROL TYPE | GE, PC10L |
| DOOR OPERATING EQUIPMENT | CONSOLIDATED CAR HEATING | SEAT ARRANGEMENT & CAPACITY | CROSS & LONGITUDINAL [71] |
| HAND BRAKE | NATIONAL BRAKE & PEACOCK | SEAT MATERIAL | RATTAN |
| HAND HOLD | PORCELAIN METAL STRAP | WINDOW CURTAIN | E.I. DUPONT, FABRIKOID |
| HEADLIGHT (ELECTRIC) | PYLE INTERNATIONAL | WINDOW MATERIAL | SEAMLESS BRASS |

## TRUCK

| | | | |
|---|---|---|---|
| MANUFACTURER | COMMONWEALTH | JOURNAL BEARING | FRICTION |
| TYPE | CAST STEEL | TRACTION MOTOR | GE, 282A |
| GEAR RATIO | 21:62 | | |

## GENERAL

| | | | |
|---|---|---|---|
| CAR LIGHT WEIGHT | 95,750 LBS | STANDING CAPACITY | [169] |

NOTE 1:

| CARS | YEAR BUILT |
|---|---|
| 300 - 394 | 1925 |
| 500 - 509 | 1926 |

TRAILER CAR 508 WAS SOLD TO TROLLEY MUSEUM OF NEW YORK, KINGSTON, NY (1980) AND WAS DESTROYED BY FIRE (1992).
CAR 366 SOLD TO SEASHORE TROLLEY MUSEUM, KENNEBUNKPORT, MAINE (1991).
CAR 388 SOLD TO BRANFORD ELECTRIC RAILWAY, EAST HAVEN, CT. (1991).
CAR 353 SOLD TO TROLLEY MUSEUM OF NEW YORK, KINGSTON, NY (1980).
ALL OTHER CARS SCRAPPED (1927 - 1975).

CAR NUMBERS PAINTED ON WINDOWS.
CARS HAD NO SIGNS.
CARS 319, 358, 376, 378 & 382 DESTROYED BY FIRE (TOTTENVILLE) (1927).
CARS 502, 505, 506, 507 & 509 CONVERTED TO MOTOR CARS 390-394 RESPECTIVELY (1928).
CARS 307, 313, 316, 322, 333, 351, 377 & 390 DESTROYED BY FIRE (ST. GEORGE) (1946).
CARS 323 & 363 DESTROYED IN DERAILMENT ON THE SOUTH BEACH BRANCH (1948).
25 MOTOR CARS (RENUMBERED 2900 - 2924) AND 5 TRAILER CARS SOLD TO NYC TRANSIT AUTHORITY (1953-1954) AND ONE TRAILER CAR WAS CONVERTED TO ROAD CAR INSPECTOR OFFICE, FRESH POND YARD AND RENUMBERED 2925 AND SUBSEQUENTLY WAS DESTROYED IN A WRECK (1966).
SCREENS APPLIED TO LOWER SASH ORIGINALLY. REPLACED WITH FULL HEIGHT SCREENS (1960's).

A
8TH AVE.
WASH. HTS
168TH ST.

67
CITY OF NEW YORK
67

A

SPECIAL
WAVE
CREST
ROCKAWAY
HERE WE
COME!

CAR NUMBERS: 100 - 399
TOTAL: 300 CARS
BUILT BY: AMERICAN CAR FOUNDRY

DATE: 1930 - 31
AVERAGE COST PER CAR: $39,201 (SEE NOTE 1)

# R1 (IND)

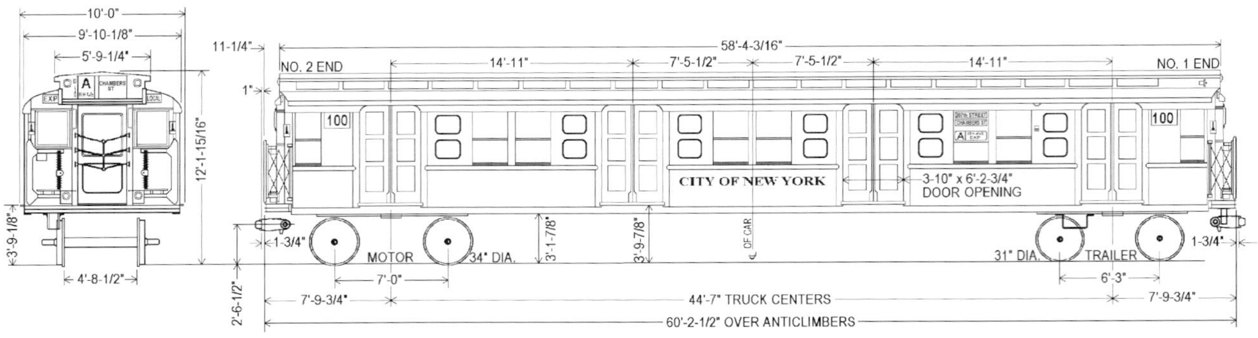

10'-0"
9'-10-1/8"
5'-9-1/4"
12'-1-15/16"
3'-9-1/8"
4'-8-1/2"
11-1/4"
1"
58'-4-3/16"
NO. 2 END
14'-11"
7'-5-1/2"
7'-5-1/2"
14'-11"
NO. 1 END
CITY OF NEW YORK
3-10" x 6'-2-3/4" DOOR OPENING
1-3/4"
MOTOR
34" DIA.
3'-1-7/8"
3'-9-7/8"
℄ OF CAR
31" DIA.
TRAILER
1-3/4"
7'-0"
6'-3"
2'-6-1/2"
7'-9-3/4"
44'-7" TRUCK CENTERS
7'-9-3/4"
60'-2-1/2" OVER ANTICLIMBERS

47,692 LBS
84,081 LBS TOTAL
36,389 LBS

Copyright © 1997 by NYC Transit

## CAR BODY - EQUIPMENT

| | | | |
|---|---|---|---|
| AIR COMPRESSOR | WABCO, D3F | HEATER | CONSOLIDATED CAR HEATING, R1404A |
| BATTERY | EDISON, B4H (24 CELL) | LIGHTING | 30V, 1.0A INCANDESCENT [22] |
| BRAKE CYLINDER | WABCO, Q-18x12 | MARKER LIGHT | ADAMS & WESTLAKE |
| BRAKE EQUIPMENT TYPE | AMUE FRICTION | MASTER CONTROLLER | WH, XM29 |
| COASTING CLOCK | RICO, TYPE H | PROPULSION CONTROL TYPE | WH, ABF, UP143B GROUP |
| COUPLER | WABCO, H2A | SASH | O.M. EDWARDS |
| DOOR HANGER | J.L. HOWARD | SEAT ARRANGEMENT & CAPACITY | LONGITUDINAL & CROSS [56] |
| DOOR OPERATOR ENGINE | NATIONAL PNEUMATIC | SEAT CUSHION & BACK | HALE & KILBURN |
| DRAFT GEAR | WAUGH, E8TV | SEAT MATERIAL | RATTAN |
| ELECTRIC PORTION | WABCO, BL25 | SIGNAL DEVICE | AIR WHISTLE |
| FAN | WH, 800 - 390 [5] | SWITCH PANEL | CONSOLIDATED CAR HEATING: #1 END, R344A; #2 END, R335A |
| FIRE EXTINGUISHER | PYRENE | TAIL LIGHT | SINGLE SOCKET SEMAPHORE |
| HAND BRAKE | BLACKALL | | |

## TRUCK

| BUILT UP | | CAST STEEL FRAME | |
|---|---|---|---|
| MANUFACTURER | AMERICAN CAR FOUNDRY | MANUFACTURER | COMMONWEALTH TRUCK |
| TYPE & QUANTITY | ARCH BAR MOTOR & TRAILER [305 EACH] | TYPE & QUANTITY | MOTOR: EQUALIZER BAR [5]; TRAILER: EQUALIZER BAR [5] |
| TRUCK NUMBERS | MOTOR & TRAILER: 100 - 404 | TRUCK NUMBERS | MOTOR: R1X1 - R1X5; TRAILER: R1X6 - R1X10 |
| WEIGHT | MOTOR: 24,124 LBS; TRAILER: 11,824 LBS | WEIGHT | MOTOR: 25,593 LBS; TRAILER: 13,196 LBS |
| | | BRAKE RIGGING | FOUNDATION SIMPLEX CLASP BRAKE |
| | | JOURNAL BEARING | 5X9 FRICTION |
| | | TRACTION MOTOR | WH, 570D5 (190 HP) [2/MOTOR TRUCK] |
| | | TRIP COCK | WABCO, 90446 |

NOTE 1: R1 WITH $30,483 FOR CARBODY AND $8,718 FOR TRUCKS & MOTORS PURCHASED WITH R2 CONTRACT.

CARS 200-214, 381, 382, 384, 385 & 387 TESTED IN PASSENGER SERVICE ON BMT SEA BEACH EXPRESS (7/8/1931-11/27/1931).
CARS 100 - 211 & 213 - 220 TRANSFERRED TO BMT (1949 - 51).
ALSO, SEVEN SPARE MT 129, 360 (W/MOTORS) & TT 236, 258.
CAR 103 RETURNED FROM BMT & CAR 221 TRANSFERRED TO BMT (7/10/52).
JM-AKOUSTIKOS FELT & PERFORATED HEADLINING INSTALLED ON CAR 295 EQUIPPED AT 207TH ST SHOP (1932).
CARS 100 - 144 EQUIPPED WITH CARBON MANGANESE SIDE SILL STIFFENERS.
INSTALLED NEW TRAIN OPERATOR'S SEATS (1940 - 41).
CHANGED SWING HANGER ANGLE ON ALL TRUCKS (1940 - 41).
COASTING CLOCK REMOVED.
LIGHT BULB WATTAGE INCREASED FROM 30W TO 48W (1950's).
CAR NUMBERS WERE PAINTED ON CARS PRIOR TO ADDING PLATE NUMBERS.

BUILT-UP TRUCK IS FABRICATED STEEL MEMBERS.
AXIFLOW FANS INSTALLED ON CAR 103 (1947).
FIRE EXTINGUISHERS REPLACED (1957).
CARS 212 & 378 DESTROYED IN WRECK (2/17/1936).
MOTOR TRUCK (MT) 186 AND TRAILER TRUCK (TT) 357 DESTROYED IN WRECK (2/17/1936).
CARS 103 & 381 SOLD TO RAILWAY PRESERVATION CORP.
CAR 100 AT NY TRANSIT MUSEUM.
CAR 175 IS AT SEASHORE ELECTRIC MUSEUM, KENNEBUNKPORT, MAINE.
CAR 273 SCRAPPED (6/27/60).
ALL OTHER CARS SCRAPPED (1970's).

CAR NUMBERS: 400 - 899
TOTAL: 500 CARS
BUILT BY: AMERICAN CAR FOUNDRY

DATE: 1932 - 33
AVERAGE COST PER CAR: $30,633 (SEE NOTE 1)

# R4 (IND)

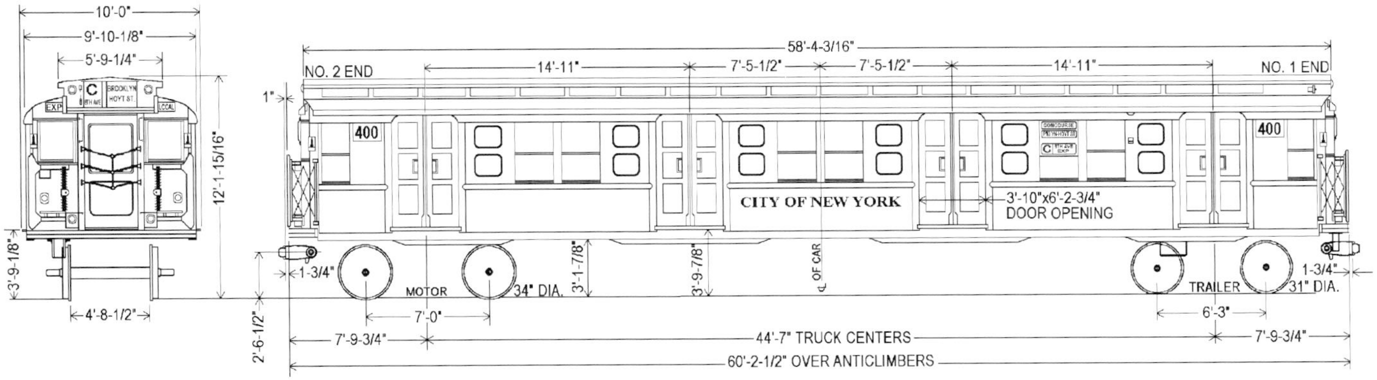

10'-0"
9'-10-1/8"
5'-9-1/4"
12'-1-15/16"
3'-9-1/8"
4'-8-1/2"
58'-4-3/16"
NO. 2 END
14'-11"
7'-5-1/2"
7'-5-1/2"
14'-11"
NO. 1 END
1"
400
CITY OF NEW YORK
3'-10"x6'-2-3/4" DOOR OPENING
1-3/4"
MOTOR
34" DIA.
3'-1-7/8"
3'-9-7/8"
℄ OF CAR
TRAILER
1-3/4"
31" DIA.
7'-0"
6'-3"
2'-6-1/2"
7'-9-3/4"
44'-7" TRUCK CENTERS
7'-9-3/4"
60'-2-1/2" OVER ANTICLIMBERS

47,652 LBS
84,503 LBS TOTAL
36,851 LBS

Copyright © 1997 by NYC Transit

## CAR BODY - EQUIPMENT

| | | | |
|---|---|---|---|
| AIR COMPRESSOR | WABCO, D3F | HEATER | GOLD CAR HEATING , 414E |
| BATTERY | EDISON, B4H, 24 CELL, WOOD BOX | LIGHTING | 30V, 1.0A INCANDESCENT [22] |
| BRAKE CYLINDER | WABCO, Q-18x12 | MARKER LIGHT | ADAMS & WESTLAKE |
| BRAKE EQUIPMENT TYPE | AMUE FRICTION | MASTER CONTROLLER | WH, XM29 |
| COASTING CLOCK | RICOH, TYPE H | PROPULSION CONTROL TYPE | WH, ABF, UP143E GROUP |
| COUPLER | WABCO, H2A | SASH | O.M. EDWARDS |
| DOOR HANGER | J.L. HOWARD | SEAT ARRANGEMENT & CAPACITY | LONGITUDINAL & CROSS [56] |
| DOOR OPERATOR ENGINE | NATIONAL PNEUMATIC | SEAT CUSHION & BACK | HALE & KILBURN |
| DRAFT GEAR | WAUGH, E8TV & WSM 4B | SEAT MATERIAL | RATTAN |
| ELECTRIC PORTION | WABCO, BL25 | SIGNAL DEVICE | AIR WHISTLE |
| FAN | WH, 800-390 [5] | SWITCH PANEL | CONSOLIDATED CAR HEATING: #1 END, R344A; #2 END, R345A |
| FIRE EXTINGUISHER | PYRENE | TAIL LIGHT | SINGLE SOCKET SEMAPHORE |
| HAND BRAKE | BLACKALL | | |

## TRUCK

| | | | |
|---|---|---|---|
| MANUFACTURER | AMERICAN CAR FOUNDRY | BRAKE RIGGING | FOUNDATION SIMPLEX CLASP BRAKE |
| TYPE & QUANTITY | BUILT-UP, ARCH BAR MOTOR & TRAILER [506] | JOURNAL BEARING | 5X9 FRICTION |
| TRUCK NUMBERS | MOTOR & TRAILER: 405-910 | TRACTION MOTOR | GE, 714A1 & 714A2 (190 HP) [2/MOTOR TRUCK] + 2 GE, 717A ROLLER BEARING |
| WEIGHT | MOTOR: 23,978 LBS; TRAILER: 12,096 LBS | TRIP COCK | WABCO, 90446 |

NOTE 1: R4 WITH $21,063 FOR CARBODY AND $9,570 FOR TRUCKS & MOTORS PURCHASED WITH R5 CONTRACT.

DRAFT GEAR EQUIPPED WITH WAUGHMATS - WSM4B FOR CARS 895-899.
TRUCK QUANTITY OF MOTOR TRUCKS [6] & TRAILER TRUCKS [6] PURCHASED ON AGREEMENT, AND MOTOR TRUCK 796 (W/MOTORS) & TRAILER TRUCK 860 TRANSFERRED TO BMT (1/12/1951).
ORIGINAL FIRE EXTINGUISHERS REPLACED (1957).
INSTALLED HEADLIGHTS STARTING WITH CAR 467 (1962).
CARS 600 - 609 ARE EQUIPPED WITH FIN PIPE COOLING COILS.
SWING HANGER ANGLE CHANGED ON TRUCKS (1942, 1943).
CARS 484, 744: 44 BULLSEYE FIXTURES (2 CIRCUITS) AND PUBLIC ADDRESS SYSTEM INSTALLED (1946).
LIGHT BULB WATTAGE INCREASED FROM 30W TO 48W (1950's).
CAR NUMBERS WERE PAINTED ON CARS PRIOR TO ADDING PLATE NUMBERS.

COASTING CLOCK REMOVED.
TRAILER TRUCK 477 DESTROYED IN WRECK (2/17/36).
CAR 401 AT CONEY ISLAND MAINTENANCE SHOP. CAR 484 AT NY TRANSIT MUSEUM.
CAR 800 AT SEASHORE ELECTRIC MUSEUM, KENNEBUNKPORT, MAINE.
CAR 472 DESTROYED IN WRECK (2/17/1936).
CAR 825 AT TROLLEY MUSEUM OF N.Y., KINGSTON, N.Y.
SOME CARS CONVERTED TO WORK CARS.

ALL OTHER CARS SCRAPPED (1970's).

CAR NUMBERS: 900 - 1149
TOTAL: 250 CARS
BUILT BY: AMERICAN CAR FOUNDRY

DATE: 1935
AVERAGE COST PER CAR: $37,683

# R6-3 (IND)

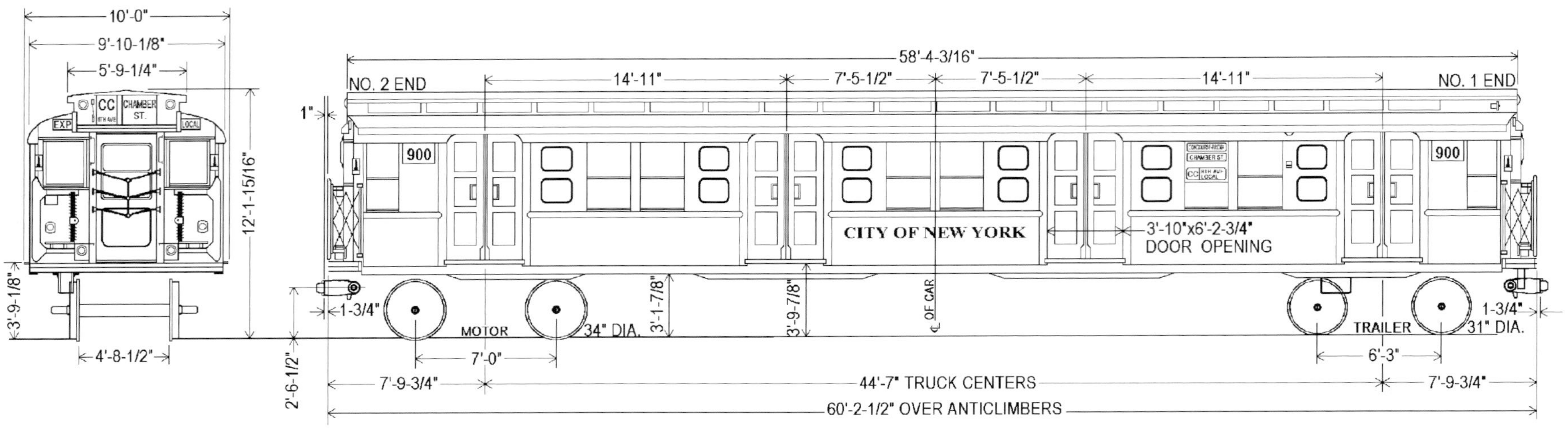

10'-0"
9'-10-1/8"
5'-9-1/4"
12'-1-15/16"
3'-9-1/8"
4'-8-1/2"
2'-6-1/2"
1"
58'-4-3/16"
NO. 2 END
14'-11"
7'-5-1/2"
7'-5-1/2"
14'-11"
NO. 1 END
900
CITY OF NEW YORK
3'-10"x6'-2-3/4" DOOR OPENING
900
1-3/4"
MOTOR
34" DIA.
3'-1-7/8"
3'-9-7/8"
℄ OF CAR
TRAILER
1-3/4"
31" DIA.
7'-0"
6'-3"
7'-9-3/4"
44'-7" TRUCK CENTERS
7'-9-3/4"
60'-2-1/2" OVER ANTICLIMBERS

47,305 LBS
83,963 LBS TOTAL
36,658 LBS

Copyright © 1997 by NYC Transit

## CAR BODY - EQUIPMENT

| | | | |
|---|---|---|---|
| AIR COMPRESSOR | WABCO, D3F | HEAT & VENT CONTROL | ACF |
| BATTERY | EDISON, B4H, 24 CELL, WOOD BOX | HEATER | GOLD CAR HEATING, 414EA |
| BRAKE CYLINDER | WABCO, U-18x12 | LIGHTING | ADAMS & WESTLAKE, 30V, 1.0A INCANDESCENT [22] |
| BRAKE EQUIPMENT TYPE | AMUE FRICTION | MARKER LIGHT | ADAMS & WESTLAKE, SEMAPHORE |
| COUPLER | WABCO, H2A | MASTER CONTROLLER | WH, XM29D |
| DOOR HANGER | J.L. HOWARD | PROPULSION CONTROL TYPE | WH, ABF, UP143 GROUP |
| DOOR OPERATOR ENGINE | NATIONAL PNEUMATIC | SASH | O.M. EDWARDS |
| DOOR PANEL | MORTON | SEAT ARRANGEMENT & CAPACITY | LONGITUDINAL & CROSS[56] |
| DRAFT GEAR | WAUGH, E8TV3 | SEAT CUSHION & BACK | J.G. BRILL |
| ELECTRIC PORTION | WABCO BL25A | SEAT MATERIAL | RATTAN |
| FAN | WH 800 -390 [5] | SIGN | HUNTER ILLUMINATED SIGN |
| FIRE EXTINGUISHER | PYRENE | SIGNAL DEVICE | AIR WHISTLE |
| HAND BRAKE | BLACKALL | TAIL LIGHT | TWIN SOCKET |
| HAND HOLD | WAUGH | WINDSHIELD WIPER | SPRAGUE |

## TRUCK

| | | | |
|---|---|---|---|
| MANUFACTURER | AMERICAN CAR FOUNDRY | GEAR UNIT | GE, 714C1 & C2 (190 HP) |
| TYPE & QUANTITY | BUILT-UP, ARCH BAR MOTOR & TRAILER [255 EACH] | JOURNAL BEARING | 5X9 FRICTION |
| TRUCK NUMBERS | MOTOR & TRAILER: 911 - 1165 | SHOCK CUSHION | FABREEKA & MANHATTAN RUBBER |
| WEIGHT | MOTOR: 23,705 LBS; TRAILER: 12,275 LBS | TRACTION MOTOR | GE, 714C1 & C2 (190 HP) [2/MOTOR TRUCK] |
| BRAKE RIGGING | FOUNDATION SIMPLEX CLASP BRAKE | TRIP COCK | WABCO, 506418 |

FIRE EXTINGUISHER; 2 LB DRY POWDER INSTALLED (1957).
INSTALLED SHOCK INSULATED CUSHIONS FURNISHED BY MANHATTAN RUBBER ON TRUCKS 911 - 1108 & 1159 - 1165.
ALSO INSTALLED SHOCK INSULATED CUSHIONS FURNISHED BY FABREEKA ON TRUCKS 1109-1158.
INSTALLED NEW TRAIN OPERATOR'S SEAT CUSHIONS (1941 - 42).
CHANGED SWING HANGER ANGLE ON ALL TRUCKS.
LIGHT BULB WATTAGE INCREASED FROM 30W TO 48W (1950's).

INSTALLED HEADLIGHTS (1962).
CARS 923, 925, & 1000 SOLD TO RAILWAY PRESERVATION CORP.
CAR 1079 SCRAPPED (7/1958).
CAR 1144 WAS CONVERTED TO A DINER IN BUCKINGHAMSHIRE, ENGLAND (NEAR LONDON).
SOME CARS CONVERTED TO WORK CARS.
ALL OTHER CARS WERE SCRAPPED (1970'S).

PASSENGER CAR NUMBERS: 1150 - 1299
TOTAL: 150 CARS
BUILT BY: PULLMAN STANDARD

DATE: 1936
AVERAGE COST PER CAR: $37,835

# R6-2 (IND)

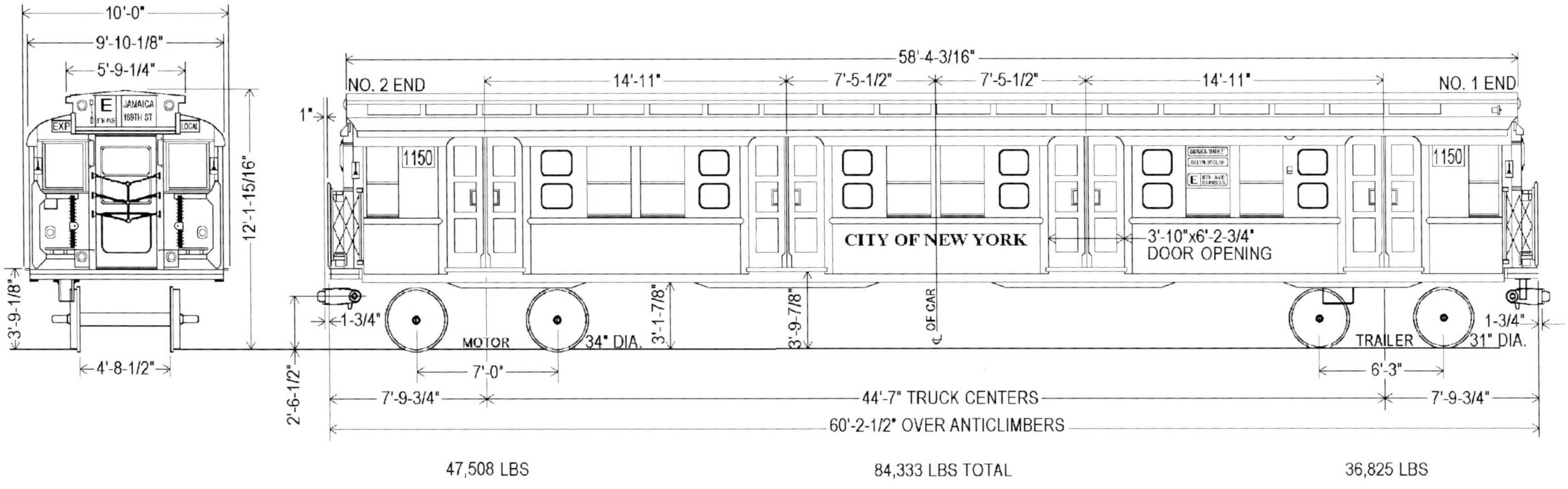

10'-0"
9'-10-1/8"
5'-9-1/4"
12'-1-15/16"
3'-9-1/8"
4'-8-1/2"
58'-4-3/16"
NO. 2 END
14'-11"
7'-5-1/2"
7'-5-1/2"
14'-11"
NO. 1 END
1"
1150
CITY OF NEW YORK
3'-10"x6'-2-3/4" DOOR OPENING
1150
1-3/4"
MOTOR
34" DIA.
3'-1-7/8"
3'-9-7/8"
℄ OF CAR
TRAILER
1-3/4"
31" DIA.
2'-6-1/2"
7'-0"
6'-3"
7'-9-3/4"
44'-7" TRUCK CENTERS
7'-9-3/4"
60'-2-1/2" OVER ANTICLIMBERS

47,508 LBS
84,333 LBS TOTAL
36,825 LBS

Copyright © 1997 by NYC Transit

## CAR BODY - EQUIPMENT

| | | | |
|---|---|---|---|
| AIR COMPRESSOR | WABCO, D3F | LIGHTING | 30V, 1.0A INCANDESCENT [22] |
| BATTERY | EDISON, B4H, 24 CELL, WOOD BOX | MARKER LIGHT | ADAMS & WESTLAKE, SEMAPHORE |
| BRAKE CYLINDER | WABCO, U-18x12 | MASTER CONTROLLER | GE, 17KC15A1 |
| BRAKE EQUIPMENT TYPE | AMUE FRICTION | PROPULSION CONTROL TYPE | GE, PC, PC15C1 GROUP |
| COUPLER | WABCO, H2A | SASH | ADAMS & WESTLAKE |
| DOOR HANGER | J.L. HOWARD | SEAT ARRANGEMENT & CAPACITY | LONGITUDINAL & CROSS [56] |
| DOOR OPERATOR ENGINE | NATIONAL PNEUMATIC | SEAT CUSHION & BACK | HEYWOOD WAKEFIELD |
| DRAFT GEAR | WAUGH, E8TV3 | SEAT MATERIAL | RATTAN |
| FAN | GE, DH [5] | SIGNAL DEVICE | AIR WHISTLE |
| FIRE EXTINGUISHER | PYRENE | SWITCH PANEL | CONSOLIDATED CAR HEATING, R344B/R345B |
| HAND BRAKE | BLACKALL | TAIL LIGHT | TWIN SOCKET |
| HEATER | GOLD CAR HEATING, 414EA | | |

## TRUCK

| | | | |
|---|---|---|---|
| MANUFACTURER | PULLMAN STANDARD | GEAR UNIT | GE |
| TYPE & QUANTITY | BUILT-UP, ARCH BAR MOTOR & TRAILER [155 EACH] | JOURNAL BEARING | 5X9 FRICTION |
| TRUCK NUMBERS | MOTOR & TRAILER: 1166 - 1320 | SHOCK CUSHION | FABREEKA & MANHATTAN RUBBER |
| WEIGHT | MOTOR: 23,696 LBS; TRAILER: 12,288 LBS | TRACTION MOTOR | WH, 570D5 [234]; GE, 714C1 & C2 [76] 190 HP [2/MOTOR TRUCK] |
| BRAKE RIGGING | FOUNDATION SIMPLEX CLASP BRAKE | TRIP COCK | WABCO, 506418 |

INSTALLED SHOCK INSULATED CUSHIONS FURNISHED BY FABREEKA ON TRUCKS 1166, 1198 - 1225 & 1316.
INSTALLED SHOCK INSULATED CUSHIONS FURNISHED BY MANHATTAN RUBBER ON TRUCKS 1167 - 1197, 1226 - 1315 & 1317 - 1320.
CHANGED SWING HANGER ANGLE ON ALL TRUCKS.
LIGHT BULB WATTAGE INCREASED FROM 30W TO 48W (1950's).
ORIGINAL FIRE EXTINGUISHERS REPLACED (1957).
INSTALLED HEADLIGHTS (1962).
CAR 1192 SCRAPPED (8/53). ALL OTHER CARS SCRAPPED (1970's).
CAR 1208 SCRAPPED (1986).

CAR NUMBERS: 1300 - 1399
TOTAL: 100 CARS
BUILT BY: PRESSED STEEL CAR

DATE: 1936
AVERAGE COST PER CAR: $37,485

# R6-1 (IND)

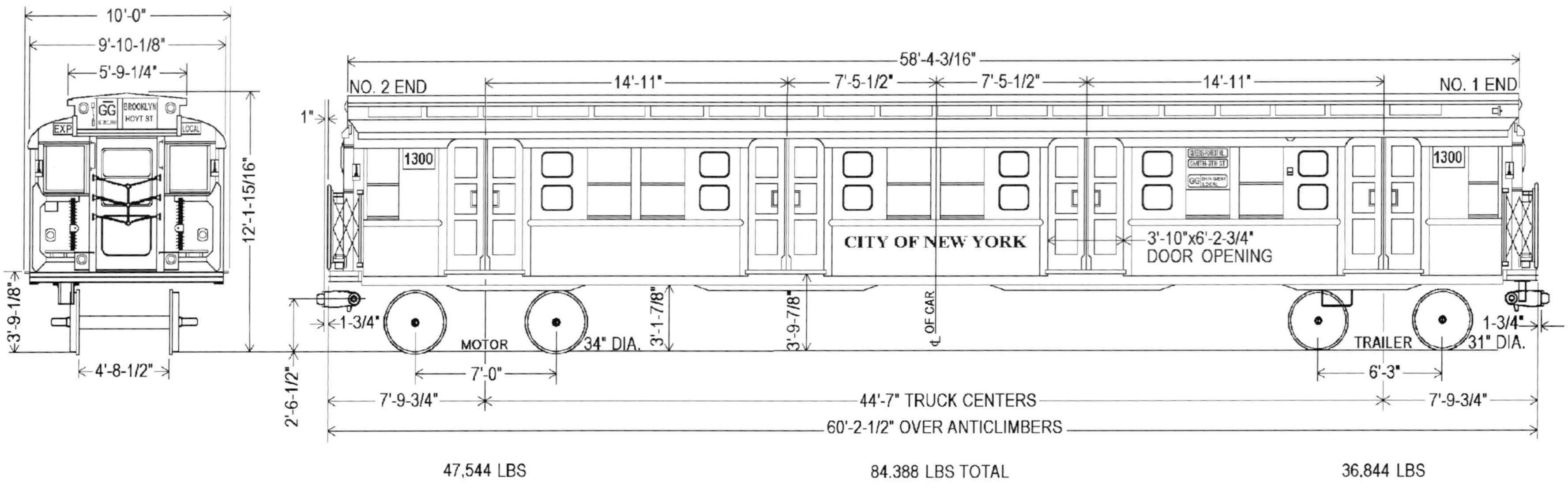

10'-0"
9'-10-1/8"
5'-9-1/4"
12'-1-15/16"
3'-9-1/8"
4'-8-1/2"
58'-4-3/16"
NO. 2 END
14'-11"
7'-5-1/2"
7'-5-1/2"
14'-11"
NO. 1 END
1"
1300
CITY OF NEW YORK
3'-10"x6'-2-3/4" DOOR OPENING
1-3/4"
MOTOR
34" DIA.
3'-1-7/8"
3'-9-7/8"
℄ OF CAR
TRAILER
1-3/4"
31" DIA.
7'-0"
6'-3"
2'-6-1/2"
7'-9-3/4"
44'-7" TRUCK CENTERS
7'-9-3/4"
60'-2-1/2" OVER ANTICLIMBERS

47,544 LBS
84.388 LBS TOTAL
36,844 LBS

Copyright © 1997 by NYC Transit

## CAR BODY - EQUIPMENT

| | | | |
|---|---|---|---|
| AIR COMPRESSOR | WABCO, D3F | HEATER | GOLD CAR HEATING, 414EA |
| BATTERY | EDISON, B4H, 24 CELL, WOOD BOX | LIGHTING | 30V, 1.0A, INCANDESCENT [22] |
| BRAKE CYLINDER | WABCO, U-18x12 | MARKER LIGHT | ADAMS & WESTLAKE, SEMAPHORE |
| BRAKE EQUIPMENT TYPE | AMUE FRICTION | MASTER CONTROLLER | WH, XM29D |
| COUPLER | WABCO, H2A | PROPULSION CONTROL TYPE | WH, ABF, UP143F GROUP |
| DOOR HANGER | J.L. HOWARD | SASH | O.M. EDWARDS |
| DOOR OPERATOR ENGINE | NATIONAL PNEUMATIC | SEAT ARRANGEMENT & CAPACITY | LONGITUDINAL & CROSS [56] |
| DRAFT GEAR | WAUGH, E8TV3 | SEAT CUSHION & BACK | HEYWOOD WAKEFIELD |
| ELECTRIC PORTION | WABCO BL25A | SEAT MATERIAL | RATTAN |
| FAN | WH, 800-390 [5] | SIGNAL DEVICE | AIR WHISTLE |
| FIRE EXTINGUISHER | PYRENE | SWITCH PANEL | CONSOLIDATED CAR HEATING: #1 END, R344B; #2 END, R345B |
| HAND BRAKE | BLACKALL | TAIL LIGHT | TWIN SOCKET |

## TRUCK

| | | | |
|---|---|---|---|
| MANUFACTURER | PRESSED STEEL CAR | GEAR UNIT | GE |
| TYPE & QUANTITY | BUILT-UP, ARCH BAR MOTOR & TRAILER [100 EACH] | JOURNAL BEARING | 5X9 FRICTION |
| TRUCK NUMBERS | MOTOR & TRAILER: 1321 - 1420 | SHOCK CUSHION | FABREEKA & MANHATTAN RUBBER |
| WEIGHT | MOTOR: 23,970 LBS; TRAILER: 12,094 LBS | TRACTION MOTOR | WH, 570D5 (190 HP), NON-OIL SEALED [2/MOTOR TRUCK] |
| BRAKE RIGGING | FOUNDATION SIMPLEX CLASP BRAKE | TRIP COCK | WABCO, 506418 |

SHOCK INSULATED CUSHIONS, FURNISHED BY FABREEKA ON TRUCKS 1321 - 1340.
SHOCK INSULATED CUSHIONS, FURNISHED BY MANHATTAN RUBBER ON TRUCKS 1341 - 1420.
CHANGED SWING HANGER ANGLE ON ALL TRUCKS.
INSTALLED HEADLIGHTS (1962).
INSTALLED NEW TRAIN OPERATOR'S SEAT CUSHIONS (1941-1942).

LIGHT BULB WATTAGE INCREASED FROM 30W TO 48W (1950's).
ORIGINAL FIRE EXTINGUISHERS REPLACED (1957).
CAR 1300 SOLD TO RAILWAY PRESERVATION CORP.
ALL OTHER CARS SCRAPPED (1970's).

CAR NUMBERS: 1400 - 1474
TOTAL: 75 CARS
BUILT BY: AMERICAN CAR FOUNDRY

DATE: 1937
AVERAGE COST PER CAR: $40,375

# R7 (IND)

AMERICAN CAR FOUNDRY

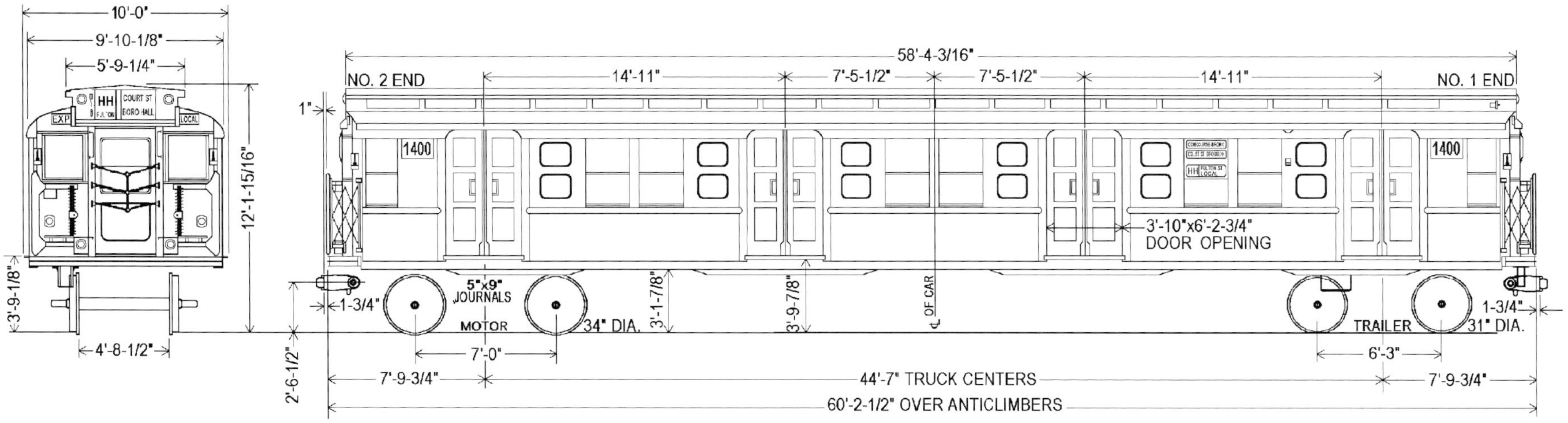

10'-0"
9'-10-1/8"
5'-9-1/4"
12'-1-15/16"
3'-9-1/8"
4'-8-1/2"
58'-4-3/16"
NO. 2 END
14'-11"
7'-5-1/2"
7'-5-1/2"
14'-11"
NO. 1 END
1"
1400
1400
3'-10"x6'-2-3/4" DOOR OPENING
1-3/4"
5"x9" JOURNALS
MOTOR
34" DIA.
3'-1-7/8"
3'-9-7/8"
℄ OF CAR
TRAILER
1-3/4"
31" DIA.
2'-6-1/2"
7'-0"
6'-3"
7'-9-3/4"
44'-7" TRUCK CENTERS
7'-9-3/4"
60'-2-1/2" OVER ANTICLIMBERS

47,774 LBS
84,556 LBS TOTAL
36,782 LBS

Copyright © 1997 by NYC Transit

## CAR BODY - EQUIPMENT

| | | | |
|---|---|---|---|
| AIR COMPRESSOR | WABCO, D3F | HEATER | RAILWAY UTILITY - TYPE 190 |
| BATTERY | EDISON, B4H (24 CELL) | LIGHTING | 30V, 1.0A INCANDESCENT [22] |
| BRAKE CYLINDER | WABCO, U-18x12 | MARKER LIGHT | ADAMS & WESTLAKE, SEMAPHORE |
| BRAKE EQUIPMENT TYPE | AMUE FRICTION | MASTER CONTROLLER | GE, 17KC15A1 |
| COUPLER | WABCO, H2A | PROPULSION CONTROL TYPE | GE, PC, PC15C1 GROUP |
| DOOR HANGER | NATIONAL PNEUMATIC | SASH | O.M. EDWARDS |
| DOOR OPERATOR ENGINE | NATIONAL PNEUMATIC | SEAT ARRANGEMENT & CAPACITY | LONGITUDINAL & CROSS [56] |
| DRAFT GEAR | WAUGH, E8TV3 | SEAT CUSHION & BACK | J.G. BRILL |
| ELECTRIC PORTION | WABCO, BL25A | SEAT MATERIAL | RATTAN |
| FAN | WH, 800-390 [5] | SIGNAL DEVICE | AIR WHISTLE |
| FIRE EXTINGUISHER | PYRENE | SWITCH PANEL | CONSOLIDATED CAR HEATING: #1 END, R344C; #2 END, R345B |
| HAND BRAKE | BLACKALL | TAIL LIGHT | TWIN SOCKET |

## TRUCK

| | | | |
|---|---|---|---|
| MANUFACTURER | AMERICAN CAR FOUNDRY | BRAKE RIGGING | FOUNDATION SIMPLEX CLASP BRAKE |
| TYPE & QUANTITY | BUILT-UP, ARCH BAR MOTOR & TRAILER [75 EACH] | JOURNAL BEARING | 5X9 FRICTION |
| TRUCK NUMBERS | MOTOR & TRAILER : 1421 - 1495 | SHOCK CUSHION | MANHATTAN RUBBER |
| WEIGHT | MOTOR: 23,914 LBS; TRAILER: 12,270 LBS | TRACTION MOTOR | WH, 570D5 (190 HP), NON-OIL SEALED [2/MOTOR TRUCK] |
| | | TRIP COCK | WABCO, 506418 |

CHANGED SWING HANGER ANGLE ON ALL TRUCKS.
ORIGINAL FIRE EXTINGUISHERS REPLACED (1957).
INSTALLED HEADLIGHTS (1962).
LIGHT BULB WATTAGE INCREASED FROM 30W TO 48W (1950's).

CAR 1440 AT SEASHORE ELECTRIC MUSEUM, KENNEBUNKPORT, MAINE.
ALL OTHER CARS SCRAPPED (1970's).

CAR NUMBERS: 1475 - 1549
TOTAL: 75 CARS
BUILT BY: PULLMAN STANDARD

DATE: 1937
AVERAGE COST PER CAR: N/A

# R7 (IND)

PULLMAN STANDARD

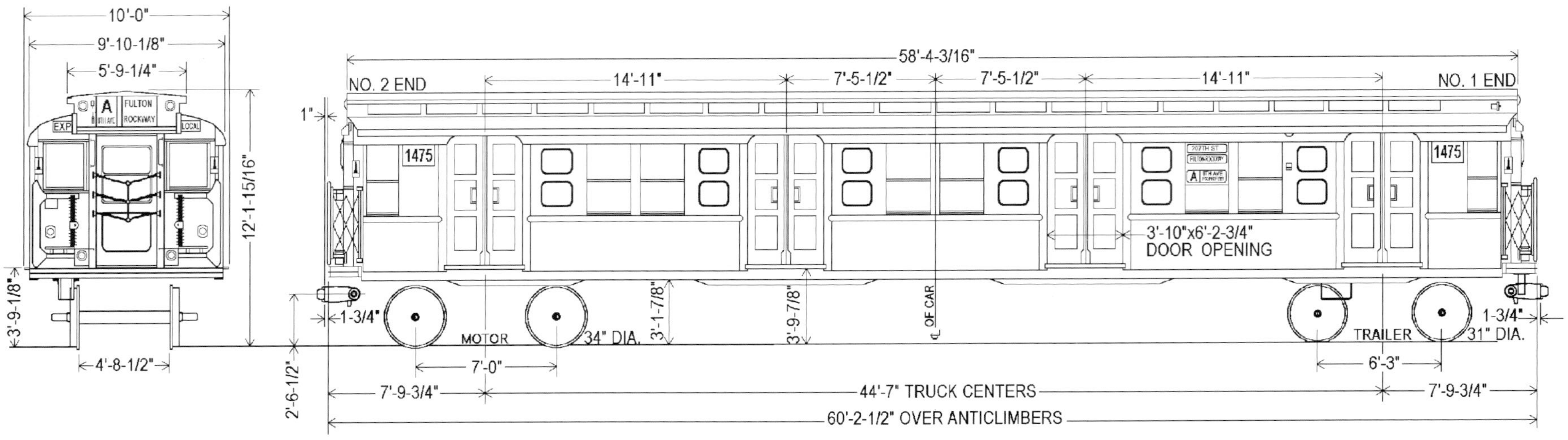

10'-0"
9'-10-1/8"
5'-9-1/4"
12'-1-15/16"
3'-9-1/8"
4'-8-1/2"
A FULTON ROCKWAY
EXP LOCAL

58'-4-3/16"
NO. 2 END
14'-11"
7'-5-1/2"
7'-5-1/2"
14'-11"
NO. 1 END
1"
1475
3'-10"x6'-2-3/4" DOOR OPENING
1-3/4"
MOTOR
34" DIA.
3'-1-7/8"
3'-9-7/8"
℄ OF CAR
TRAILER
1-3/4"
31" DIA.
2'-6-1/2"
7'-0"
6'-3"
7'-9-3/4"
44'-7" TRUCK CENTERS
7'-9-3/4"
60'-2-1/2" OVER ANTICLIMBERS

47,884 LBS

84,750 LBS TOTAL

36,866 LBS

Copyright © 1997 by NYC Transit

## CAR BODY - EQUIPMENT

| | | | |
|---|---|---|---|
| AIR COMPRESSOR | WABCO, D3F | HEATER | GOLD CAR HEATING , 414EA |
| BATTERY | EDISON, B4H, 24 CELL, WOOD BOX | LIGHTING | 30V, 1.0A INCANDESCENT [22] |
| BRAKE CYLINDER | WABCO, U-18x12 | MARKER LIGHT | ADAMS & WESTLAKE, SEMAPHORE |
| BRAKE EQUIPMENT TYPE | AMUE FRICTION | MASTER CONTROLLER | GE, 17KC15A1 |
| COUPLER | WABCO, H2A | PROPULSION CONTROL TYPE | GE, PC PC15C1 GROUP |
| DOOR HANGER | J.L. HOWARD | SASH | ADAMS & WESTLAKE |
| DOOR OPERATOR ENGINE | NATIONAL PNEUMATIC | SEAT ARRANGEMENT & CAPACITY | LONGITUDINAL & CROSS [56] |
| DRAFT GEAR | WAUGH, E8TV3 | SEAT CUSHION & BACK | HEYWOOD WAKEFIELD |
| ELECTRIC PORTION | WABCO, BL25A | SEAT MATERIAL | RATTAN |
| FAN | WH, 800-390 [5] | SIGNAL DEVICE | AIR WHISTLE |
| FIRE EXTINGUISHER | PYRENE | SWITCH PANEL | CONSOLIDATED CAR HEATING: #1 END, R344C; #2 END, R345B |
| HAND BRAKE | BLACKALL | TAIL LIGHT | TWIN SOCKET |

## TRUCK

| | | | |
|---|---|---|---|
| MANUFACTURER | PULLMAN STANDARD | BRAKE RIGGING | FOUNDATION SIMPLEX CLASP BRAKE |
| TYPE & QUANTITY | BUILT-UP, ARCH BAR MOTOR & TRAILER [75 EACH] | JOURNAL BEARING | 5X9 FRICTION |
| TRUCK NUMBERS | MOTOR & TRAILER: 1496 - 1570 | SHOCK CUSHION | MANHATTAN RUBBER |
| WEIGHT | MOTOR: 23,914 LBS; TRAILER: 12,270 LBS | TRACTION MOTOR | WH, 570D5 (190HP) NON-OIL SEALED [2/MOTOR TRUCK] |
| | | TRIP COCK | WABCO, 506418 |

CHANGED SWING HANGER ANGLE ON ALL TRUCKS.
ORIGINAL FIRE EXTINGUISHERS REPLACED (1957).
INSTALLED HEADLIGHTS (1962).
LIGHT BULB WATTAGE INCREASED FROM 30W TO 48W (1950's).
ALL CARS SCRAPPED (1970's).

CAR NUMBERS: 1550 - 1574, 1576 - 1599
TOTAL: 49 CARS
BUILT BY: PULLMAN STANDARD

DATE: 1938
AVERAGE COST PER CAR: $41,950

# R7A (IND)

GE EQUIPPED

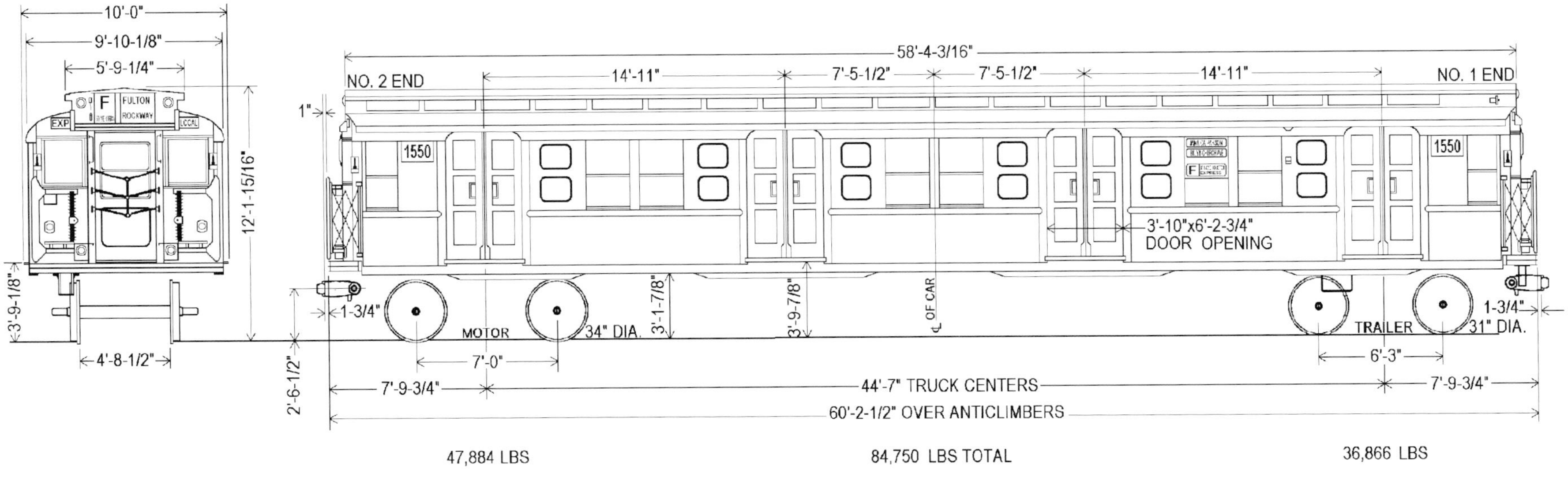

10'-0"
9'-10-1/8"
5'-9-1/4"
12'-1-15/16"
3'-9-1/8"
4'-8-1/2"
F FULTON ROCKWAY
EXP
LOCAL
58'-4-3/16"
NO. 2 END
14'-11"
7'-5-1/2"
7'-5-1/2"
14'-11"
NO. 1 END
1"
1550
1550
3'-10"x6'-2-3/4" DOOR OPENING
1-3/4"
MOTOR
34" DIA.
3'-1-7/8"
3'-9-7/8"
℄ OF CAR
TRAILER
1-3/4"
31" DIA.
2'-6-1/2"
7'-0"
6'-3"
7'-9-3/4"
44'-7" TRUCK CENTERS
7'-9-3/4"
60'-2-1/2" OVER ANTICLIMBERS

47,884 LBS

84,750 LBS TOTAL

36,866 LBS

Copyright © 1997 by NYC Transit

## CAR BODY - EQUIPMENT

| | | | |
|---|---|---|---|
| AIR COMPRESSOR | WABCO, D3F | HEATER | GOLD CAR HEATING, 414EA |
| BATTERY | EDISON, B4H, 24 CELL, WOOD BOX | LIGHTING | 30V, 1.0A INCANDESCENT [22] |
| BRAKE CYLINDER | WABCO, U-18x12 | MARKER LIGHT | ADAMS & WESTLAKE, SEMAPHORE |
| BRAKE EQUIPMENT TYPE | AMUE FRICTION | MASTER CONTROLLER | GE, 17KC15A1 |
| COUPLER | WABCO, H2A | PROPULSION CONTROL TYPE | GE, PC15C1 GROUP |
| DOOR HANGER | J.L. HOWARD | SASH | ADAMS & WESTLAKE |
| DOOR OPERATOR ENGINE | NATIONAL PNEUMATIC | SEAT ARRANGEMENT & CAPACITY | SEAT ARRANGEMENT & CAPACITY [56] |
| DRAFT GEAR | WAUGH, E8TV3 | SEAT CUSHION & BACK | J.G. BRILL |
| ELECTRIC PORTION | WABCO, BL25A | SEAT MATERIAL | RATTAN |
| FAN | WH, 800-390 [5] | SIGNAL DEVICE | AIR WHISTLE |
| FIRE EXTINGUISHER | PYRENE | SWITCH PANEL | CONSOLIDATED CAR HEATING: #1 END, R344C; #2 END, R345C |
| HAND BRAKE | BLACKALL | TAIL LIGHT | TWIN SOCKET |

## TRUCK

| | | | |
|---|---|---|---|
| MANUFACTURER | PULLMAN STANDARD | BRAKE RIGGING | FOUNDATION SIMPLEX CLASP BRAKE |
| TYPE & QUANTITY | BUILT-UP, ARCH BAR MOTOR & TRAILER [50 EACH] | JOURNAL BEARING | 5X9 FRICTION |
| TRUCK NUMBERS | MOTOR & TRAILER: 1571-1620 | SHOCK CUSHIONS | MANHATTAN RUBBER |
| WEIGHT | MOTOR: 23,914 LBS; TRAILER: 12,270 LBS | TRACTION MOTOR | WH, 570D5 (190HP), NON-OIL SEALED [2/MOTOR TRUCK] |
| | | TRIP COCK | WABCO, 506418 |

CHANGED SWING HANGER ANGLE ON ALL TRUCKS.
AIR BRAKE PIPE WAS COPPER TUBING ON CAR 1597.
CAR BODY ON CAR 1575 REBUILT BY ACF (12/1946 - 5/1947).
LIGHT BULB WATTAGE INCREASED FROM 30W TO 48W (1950's).

ORIGINAL FIRE EXTINGUISHERS REPLACED (1957).
INSTALLED HEADLIGHTS (1962).
ALL CARS SCRAPPED (1970's).

CAR NUMBER: 1575
TOTAL: 1 CAR
BUILT BY: PULLMAN STANDARD
DATE: 1938
AVERAGE COST PER CAR: $41,950

REBUILT BY: AMERICAN CAR FOUNDRY
DATE: 1947
REBUILT COST PER CAR: N/A

# R7A (IND)

R10 PROTOTYPE

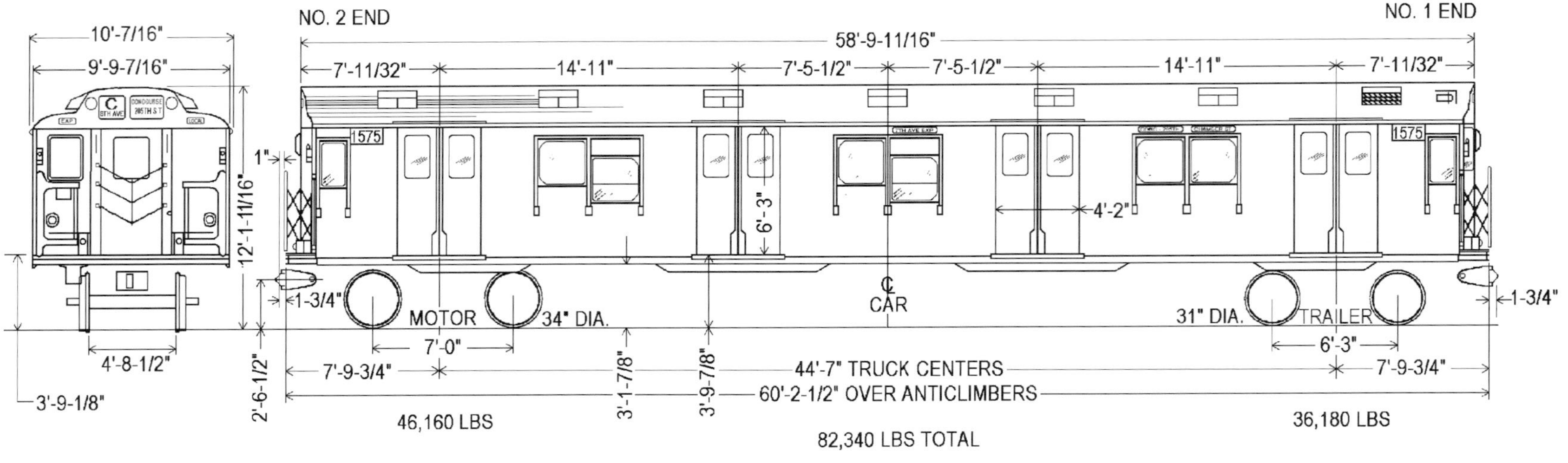

10'-7/16"
9'-9-7/16"
12'-1-11/16"
1"
4'-8-1/2"
3'-9-1/8"
2'-6-1/2"
NO. 2 END
NO. 1 END
58'-9-11/16"
7'-11/32" 14'-11" 7'-5-1/2" 7'-5-1/2" 14'-11" 7'-11/32"
1575
6'-3"
4'-2"
1-3/4"
MOTOR
34" DIA.
7'-0"
CAR
31" DIA.
TRAILER
6'-3"
1-3/4"
7'-9-3/4"
3'-1-7/8"
3'-9-7/8"
44'-7" TRUCK CENTERS
7'-9-3/4"
60'-2-1/2" OVER ANTICLIMBERS
46,160 LBS
82,340 LBS TOTAL
36,180 LBS

Copyright © 1997 by NYC Transit

## CAR BODY - EQUIPMENT

| | | | |
|---|---|---|---|
| AIR COMPRESSOR | WABCO, D3F | HEATER | RAILWAY UTILITY: CAR, H1171C; CAB, H1172A |
| BATTERY | EDISON, B4H, 24 CELL, WOOD BOX | LIGHTING | LUMINATOR 72" [24], 48" [2] |
| BRAKE CYLINDER | WABCO , U-18x12 | MARKER LIGHT | ADAMS & WESTLAKE |
| BRAKE EQUIPMENT TYPE | AMUE FRICTION | MASTER CONTROLLER | GE, 17KC15A1 |
| CONTROLLER | GE, PC-15C1 | PROPULSION CONTROL TYPE | GE, PC, PC15C1 GROUP |
| COUPLER | WABCO, H2A | SASH | HUNTER SASH |
| DOOR HANGER | ELLCON, BALL BEARING | SEAT ARRANGEMENT & CAPACITY | CROSS & LONGITUDINAL [56] |
| DOOR OPERATOR ENGINE | NATIONAL PNEUMATIC | SEAT CUSHION & BACK | H. WAKEFIELD |
| DRAFT GEAR | WAUGH, E8TV3 | SIGNAL DEVICE | AIR WHISTLE |
| ELECTRIC PORTION | WABCO, BL25A | SWITCH PANEL | SAFETY SWITCH PANEL, PS67A |
| FAN | GE, BRACKET [8] | TAIL LIGHT | TWIN SOCKET |
| FIRE EXTINGUISHER | PYRENE | THERMOSTAT | RAILWAY UTILITY, ARCOSTAT 690 |
| HAND BRAKE | NATIONAL BRAKE, 840XAL | VENTILATOR | RAILWAY UTILITY |

## TRUCK

| | | | |
|---|---|---|---|
| MANUFACTURER | PULLMAN STANDARD | BRAKE | FOUNDATION SIMPLEX CLASP BRAKE |
| TYPE & QUANTITY | BUILT-UP, ARCHBAR MOTOR & TRAILER [1 EACH] | JOURNAL BEARING | 5X9, BABBIT |
| TRUCK NUMBERS | MOTOR: 1613; TRAILER: 1696 | SHOCK CUSHION | MANHATTAN RUBBER |
| WEIGHT | MOTOR: 23,914 LBS; TRAILER: 12,270 LBS | TRACTION MOTOR | WH, 570D5 (190 HP), NON-OIL SEALED [2/MOTOR TRUCK] |
| | | TRIP COCK | WABCO, 506418 |

CHANGED SWING HANGER ANGLE ON ALL TRUCKS.
CAR WRECKED (1946).
REBUILT TO R10 PROTOTYPE.
EXCEPT FOR UNDERFRAME, THE TRUCKS AND EQUIPMENT AND THE SUPERSTRUCTURE IS NEW AND OF WELDED CONSTRUCTION. DELIVERED (6/1947). HANDBRAKE CHAIN WAS REPLACED FROM 1/4" CHAIN TO 3/8" CHAIN ( 12/4/1947).
EQUIPPED WITH NEW HONEYCOMB VENTILATORS (RAILWAY UTILITY).
THERMOSTAT REMOVED (9/10/1953).

FIRE EXTINGUISHER REPLACED (1957).
INSTALLED SHOCK ABSORBER (7/1/1947).
REPLACED 10" DIAMETER FANS TO 12" DIAMETER (7/21/1947).
PASSENGER SEAT CUSHION AND BACK EQUIPPED WITH NO SAG SPRINGS.
STORM DOORS FURNISHED WITH COUNTER BALANCES.
CAR AT NY TRANSIT MUSEUM.

CAR NUMBERS: 1600 - 1649
TOTAL: 50 CARS
BUILT BY: AMERICAN CAR FOUNDRY

DATE: 1938
AVERAGE COST PER CAR: $41,950

# R7A (IND)

WH EQUIPPED

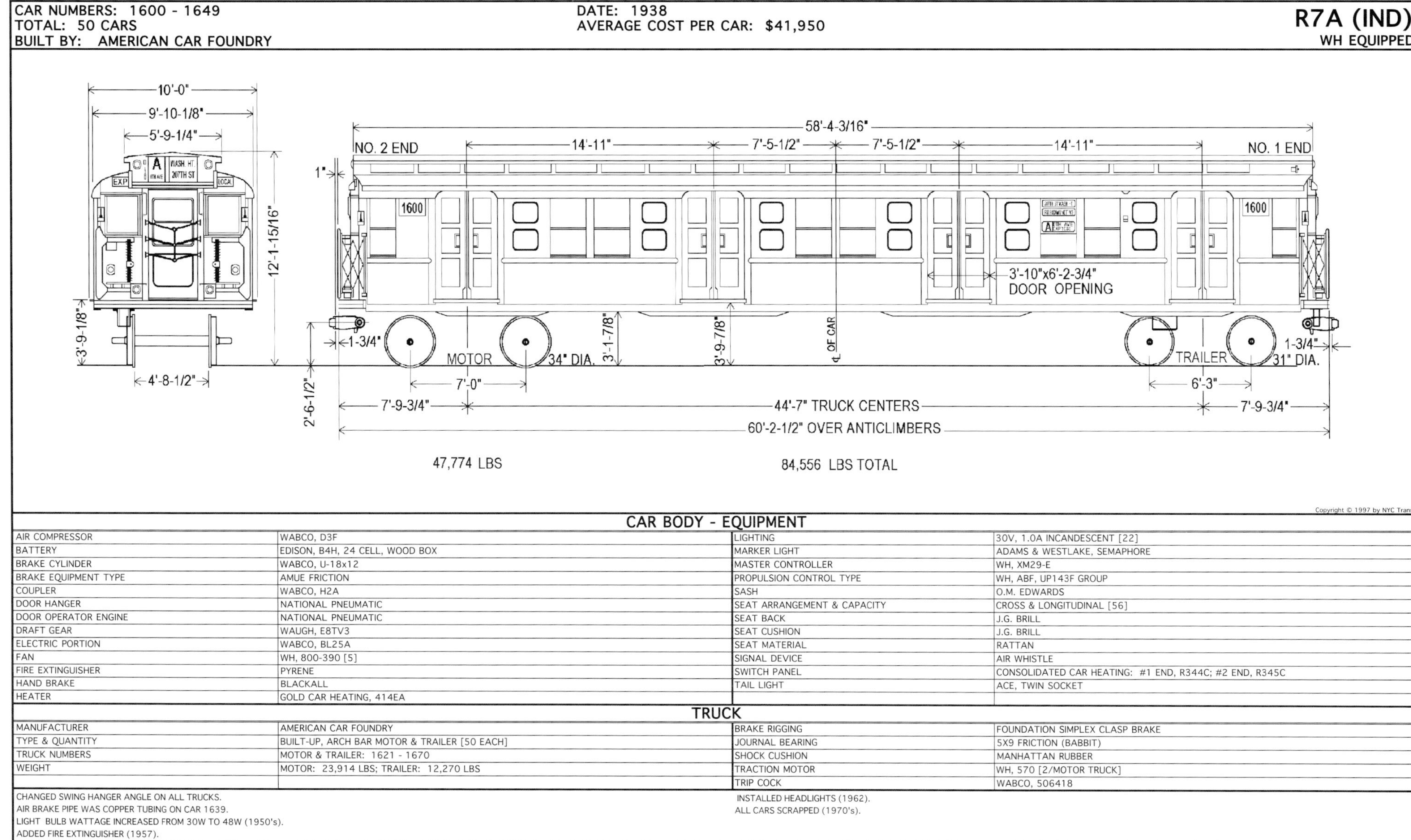


Copyright © 1997 by NYC Transit

## CAR BODY - EQUIPMENT

| | | | |
|---|---|---|---|
| AIR COMPRESSOR | WABCO, D3F | LIGHTING | 30V, 1.0A INCANDESCENT [22] |
| BATTERY | EDISON, B4H, 24 CELL, WOOD BOX | MARKER LIGHT | ADAMS & WESTLAKE, SEMAPHORE |
| BRAKE CYLINDER | WABCO, U-18x12 | MASTER CONTROLLER | WH, XM29-E |
| BRAKE EQUIPMENT TYPE | AMUE FRICTION | PROPULSION CONTROL TYPE | WH, ABF, UP143F GROUP |
| COUPLER | WABCO, H2A | SASH | O.M. EDWARDS |
| DOOR HANGER | NATIONAL PNEUMATIC | SEAT ARRANGEMENT & CAPACITY | CROSS & LONGITUDINAL [56] |
| DOOR OPERATOR ENGINE | NATIONAL PNEUMATIC | SEAT BACK | J.G. BRILL |
| DRAFT GEAR | WAUGH, E8TV3 | SEAT CUSHION | J.G. BRILL |
| ELECTRIC PORTION | WABCO, BL25A | SEAT MATERIAL | RATTAN |
| FAN | WH, 800-390 [5] | SIGNAL DEVICE | AIR WHISTLE |
| FIRE EXTINGUISHER | PYRENE | SWITCH PANEL | CONSOLIDATED CAR HEATING: #1 END, R344C; #2 END, R345C |
| HAND BRAKE | BLACKALL | TAIL LIGHT | ACE, TWIN SOCKET |
| HEATER | GOLD CAR HEATING, 414EA | | |

## TRUCK

| | | | |
|---|---|---|---|
| MANUFACTURER | AMERICAN CAR FOUNDRY | BRAKE RIGGING | FOUNDATION SIMPLEX CLASP BRAKE |
| TYPE & QUANTITY | BUILT-UP, ARCH BAR MOTOR & TRAILER [50 EACH] | JOURNAL BEARING | 5X9 FRICTION (BABBIT) |
| TRUCK NUMBERS | MOTOR & TRAILER: 1621 - 1670 | SHOCK CUSHION | MANHATTAN RUBBER |
| WEIGHT | MOTOR: 23,914 LBS; TRAILER: 12,270 LBS | TRACTION MOTOR | WH, 570 [2/MOTOR TRUCK] |
| | | TRIP COCK | WABCO, 506418 |

CHANGED SWING HANGER ANGLE ON ALL TRUCKS.
AIR BRAKE PIPE WAS COPPER TUBING ON CAR 1639.
LIGHT BULB WATTAGE INCREASED FROM 30W TO 48W (1950's).
ADDED FIRE EXTINGUISHER (1957).

INSTALLED HEADLIGHTS (1962).
ALL CARS SCRAPPED (1970's).

CAR NUMBERS: 1650 - 1701
TOTAL: 52 CARS
BUILT BY: AMERICAN CAR FOUNDRY

DATE: 1940
AVERAGE COST PER CAR: $41,200

# R9 (IND)

AMERICAN CAR FOUNDRY

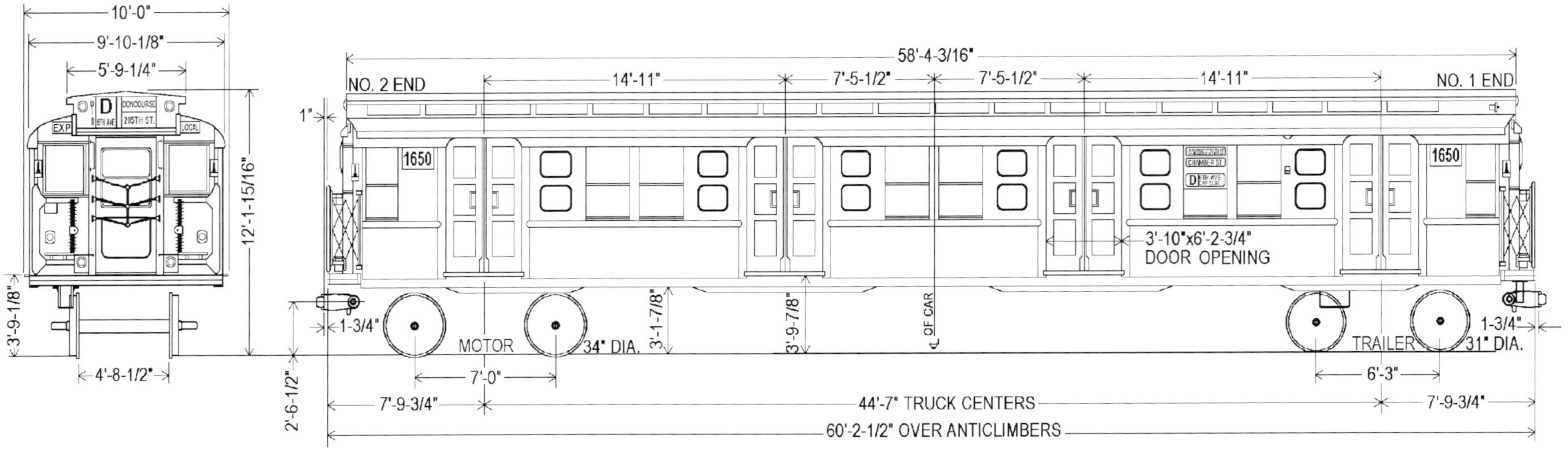

10'-0"
9'-10-1/8"
5'-9-1/4"
12'-1-15/16"
3'-9-1/8"
4'-8-1/2"
58'-4-3/16"
NO. 2 END
14'-11"
7'-5-1/2"
7'-5-1/2"
14'-11"
NO. 1 END
1"
3'-10"x6'-2-3/4"
DOOR OPENING
1-3/4"
MOTOR
34" DIA.
3'-1-7/8"
3'-9-7/8"
℄ OF CAR
TRAILER
1-3/4"
31" DIA.
2'-6-1/2"
7'-0"
6'-3"
7'-9-3/4"
44'-7" TRUCK CENTERS
7'-9-3/4"
60'-2-1/2" OVER ANTICLIMBERS

47,683 LBS
84,396 LBS TOTAL
36,713 LBS

Copyright © 1997 by NYC Transit

## CAR BODY - EQUIPMENT

| | | | |
|---|---|---|---|
| AIR COMPRESSOR | WABCO, D3F | LIGHTING | 30V, 1.0A INCANDESCENT [22] |
| BATTERY | EDISON, B4H, 24 CELL, WOOD BOX | MARKER LIGHT | ADAMS & WESTLAKE, SEMAPHORE |
| BRAKE CYLINDER | WABCO , U-18x12 | MASTER CONTROLLER | GE, 17KC15A1 |
| BRAKE EQUIPMENT TYPE | AMUE FRICTION | PROPULSION CONTROL TYPE | GE, PC, PC15C1 GROUP |
| COUPLER | WABCO, H2A | SASH | O.M. EDWARDS |
| DOOR HANGER | NATIONAL PNEUMATIC | SEAT ARRANGEMENT & CAPACITY | LONGITUDINAL & CROSS [56] |
| DOOR OPERATOR ENGINE | NATIONAL PNEUMATIC | SEAT BACK | J.G. BRILL |
| DRAFT GEAR | WAUGH, E8TV3 | SEAT CUSHION | J.G. BRILL |
| FAN | GE, DH [5] | SEAT MATERIAL | RATTAN |
| FIRE EXTINGUISHER | PYRENE | SIGNAL DEVICE | AIR WHISTLE |
| HAND BRAKE | BLACKALL | SWITCH PANEL | CONSOLIDATED CAR HEATING: #1 END, R334C; #2 END, R335B |
| HANDHOLD | ACF | TAIL LIGHT | TWIN SOCKET |
| HEATER | GOLD CAR HEATING, 414EA | | |

## TRUCK

| | | | |
|---|---|---|---|
| MANUFACTURER | AMERICAN CAR FOUNDRY | BRAKE RIGGING | FOUNDATION SIMPLEX CLASP BRAKE |
| TYPE & QUANTITY | BUILT-UP, ARCH BAR MOTOR & TRAILER [52 EACH] | JOURNAL BEARING | 5X9 FRICTION (BABBIT) |
| TRUCK NUMBERS | MOTOR & TRAILER: 1671 - 1722 | SHOCK CUSHION | MANHATTAN RUBBER |
| WEIGHT | MOTOR: 23,980 LBS; TRAILER: 12,010 LBS | TRACTION MOTOR | WH, 570D5; GE, 714D3 & D4 (190 HP) [2/MOTOR TRUCK] |
| | | TRIP COCK | WABCO, 506418 |

GE MOTORS ON TRUCKS 1671 - 1710 & 1721 - 1722.
WH MOTORS ON TRUCKS 1711 - 1720.
CHANGED SWING HANGER ANGLE ON ALL TRUCKS.
LIGHT BULB WATTAGE INCREASED FROM 30W TO 48W (1950's).

ORIGINAL FIRE EXTINGUISHERS REPLACED (1957).
INSTALLED HEADLIGHTS (1962).
CAR 1689 AT BRANFORD ELECTRIC MUSEUM.
ALL OTHER CARS SCRAPPED (1970's).

CAR NUMBERS: 1702 - 1802
TOTAL: 101 CARS
BUILT BY: PRESSED STEEL CAR

DATE: 1940
AVERAGE COST PER CAR: $41,200

# R9 (IND)

PRESSED STEEL CAR

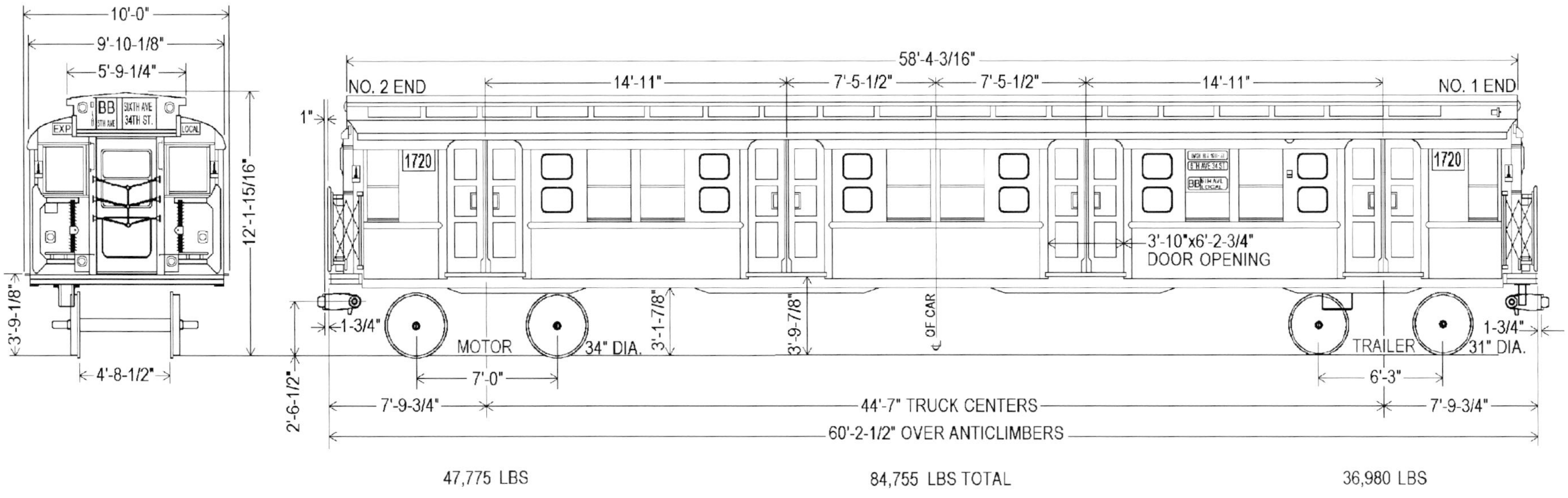

10'-0"
9'-10-1/8"
5'-9-1/4"
12'-1-15/16"
3'-9-1/8"
4'-8-1/2"

58'-4-3/16"
NO. 2 END
14'-11"
7'-5-1/2"
7'-5-1/2"
14'-11"
NO. 1 END
1"
1720
3'-10"x6'-2-3/4" DOOR OPENING
1-3/4"
MOTOR
34" DIA.
3'-1-7/8"
3'-9-7/8"
℄ OF CAR
TRAILER
1-3/4"
31" DIA.
2'-6-1/2"
7'-0"
6'-3"
7'-9-3/4"
44'-7" TRUCK CENTERS
7'-9-3/4"
60'-2-1/2" OVER ANTICLIMBERS

47,775 LBS
84,755 LBS TOTAL
36,980 LBS

Copyright © 1997 by NYC Transit

## CAR BODY - EQUIPMENT

| | | | |
|---|---|---|---|
| AIR COMPRESSOR | WABCO, D3F | LIGHTING | 30V, 1.0A INCANDESCENT [22] |
| BATTERY | EDISON, B4H, 24 CELL, WOOD BOX | MARKER LIGHT | ADAMS & WESTLAKE, SEMAPHORE |
| BRAKE CYLINDER | WABCO , U-18x12 | MASTER CONTROLLER | WH, XM29 |
| BRAKE EQUIPMENT TYPE | AMUE FRICTION | PROPULSION CONTROL TYPE | WH, ABF, UP143F GROUP |
| COUPLER | WABCO, H2A | SASH | O.M. EDWARDS |
| DOOR HANGER | NATIONAL PNEUMATIC | SEAT ARRANGEMENT & CAPACITY | LONGITUDINAL & CROSS [56] |
| DOOR OPERATOR ENGINE | NATIONAL PNEUMATIC | SEAT BACK | J.G. BRILL |
| DRAFT GEAR | WAUGH, E8TV3 | SEAT CUSHION | J.G. BRILL |
| FAN | WH, 800-390 [5] | SEAT MATERIAL | RATTAN |
| FIRE EXTINGUISHER | PYRENE | SIGNAL DEVICE | AIR WHISTLE |
| HAND BRAKE | BLACKALL | SWITCH PANEL | CONSOLIDATED CAR HEATING: #1 END, R334C; #2 END, R335B |
| HEATER | RAILWAY UTILITY 190 | TAIL LIGHT | TWIN SOCKET |

## TRUCK

| | | | |
|---|---|---|---|
| MANUFACTURER | PRESSED STEEL CAR | BRAKE RIGGING | FOUNDATION SIMPLEX CLASP BRAKE |
| TYPE & QUANTITY | ARCH BAR, MOTOR & TRAILER [101 EACH] | JOURNAL BEARING | 5X9 FRICTION (BABBIT) |
| TRUCK NUMBERS | MOTOR & TRAILER: 1723 - 1823 | SHOCK CUSHION | MANHATTAN RUBBER |
| WEIGHT | MOTOR: 23,980 LBS; TRAILER: 12,010 LBS | TRACTION MOTOR | WH, 570D5; GE, 714D4 & D5 (190 HP), NON-OIL SEALED [2/MOTOR TRUCK] |
| | | TRIP COCK | WABCO, 506418 |

GE MOTORS ON TRUCKS 1773-1822.
WH MOTORS ON TRUCKS 1723-1772.
CHANGED SWING HANGER ANGLE ON ALL TRUCKS.
LIGHT BULB WATTAGE INCREASED FROM 30W TO 48W (1950's).
ORIGINAL FIRE EXTINGUISHERS REPLACED (1957).

INSTALLED HEADLIGHTS (1962).
CAR 1802 SOLD TO RAILWAY PRESERVATION CORP.
CAR 1801 AT DEPARTMENT OF EDUCATION MALL, ALBANY, NEW YORK.
ALL OTHER CARS SCRAPPED (1970's).

# R10 (BMT/IND)

CAR NUMBERS: 2950 - 3349
TOTAL: 400 CARS
BUILT BY: AMERICAN CAR FOUNDRY

DATE: 1948 - 1949
AVERAGE COST PER CAR: $77,319 (BID)

NO. 2 END
NO. 1 END
10'-7/16"
9'-9-7/16"
12'-1-15/16"
4'-8-1/2"
3'-9-1/8"
2'-6-1/2"
1"
1-3/4"
58'-9-11/16"
7'-11/32" 14'-11" 7'-5-1/2" 7'-5-1/2" 14'-11" 7'-11/32"
4'-2"X6'-3" DOOR OPENING
34" DIA.
3'-3"
3'-10"
℄ CAR
6'-10"
7'-9-3/4"
44'-7" TRUCK CENTERS
60'-2-1/2" OVER ANTICLIMBERS
40,600 LBS
40,600 LBS
81,200 LBS TOTAL

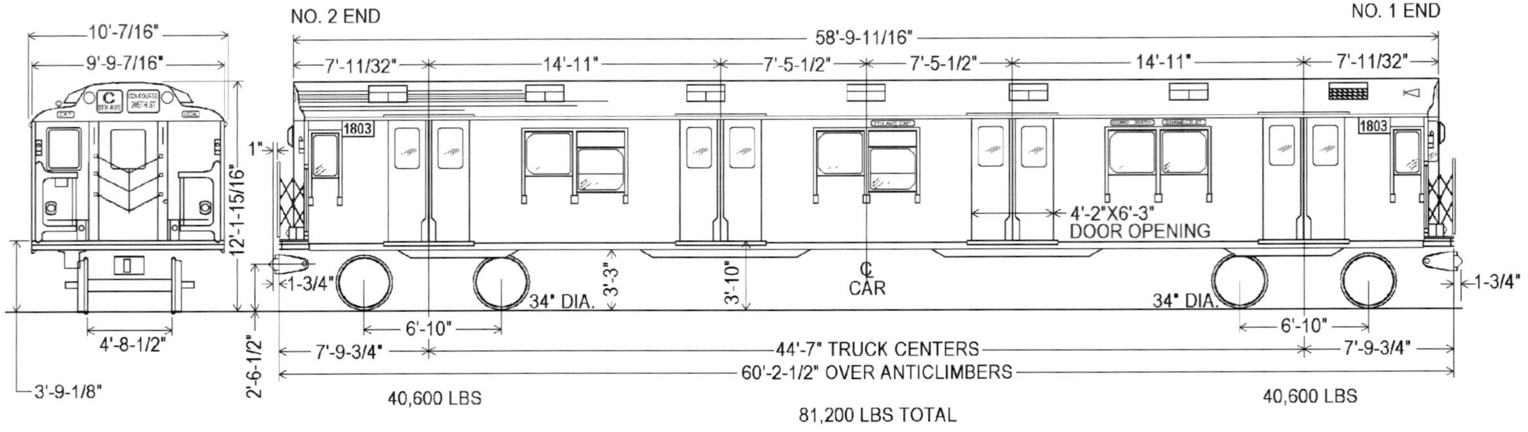

Copyright © 1997 by NYC Transit

## CAR BODY - EQUIPMENT

| | | | |
|---|---|---|---|
| AIR COMPRESSOR | WABCO, 3YC, 21 CFM | LIGHTING | LUMINATOR, 72" [24], 48" [2] |
| BATTERY | EDISON, B4H, 24 CELL, WOOD BOX | MARKER LIGHT SEMAPHORE | LOVELL DRESSEL |
| BRAKE CYLINDER | WABCO, 7x6, UAHT [2]/TRUCK | MASTER CONTROLLER | WH, XM29; GE, 17KC76A1 |
| BRAKE EQUIPMENT TYPE | WABCO, SMEE | MOTOR /GENERATOR | WH, YX300B; GE, GMG153A1 |
| BRAKE VALVE | ME42 | PROPULSION CONTROL TYPE | GE, PCM, 17MG116A GROUP |
| COUPLER | WABCO H2C | | WH, ABS, UP631A GROUP |
| DOOR HANGER | OTIS ELEVATOR, DIAMOND DOOR TRACK | SASH | HUNTER SASH |
| DOOR OPERATOR ENGINE | NATIONAL PNEUMATIC | SEAT BACK | ACF |
| DOOR PANEL | ACF | SEAT CUSHION | ACF |
| DRAFT GEAR | WAUGH WM-E4CL | SEAT MATERIAL | YELLOW VELON WITH BLUE STRIPES |
| ELECTRIC PORTION | WABCO BL25B | SEATING ARRANGEMENT & CAPACITY | LONGITUDINAL & CROSS[56] |
| FAN | GE, 12" TWIN BACKET FAN [8], ALL CARS | SIGN | HUNTER ILLUMINATED CAR SIGN |
| FIRE EXTINGUISHER | 2 LBS DRY POWDER | SIGNAL DEVICE | PNEUPHONIC HORN |
| HAND BRAKE | NATIONAL BRAKE, 840XAL | SWITCH PANEL | CONSOLIDATED CAR HEATING: #1 END, PS66A; #2 END, PS67A |
| HANDHOLD | ELLCON | TAIL LIGHT | ACF, TWIN SOCKET REVERSING CONTROLLER |
| HEATER | CONSOLIDATED CAR HEATING/GOLD CAR HEATING/RAILWAY UTILITY | THERMOSTAT | CONSOLIDATED CAR HEATING, MICROTHERM |

## TRUCK

| | | | | |
|---|---|---|---|---|
| MANUFACTURER | | G.S.C., W30156 (ORIGINAL), W30656 (REPLACED) | BRAKE RIGGING | SIMPLEX CLASP BRAKE |
| TRUCK TYPE & QUANTITY | | C.S.F. EQUALIZER BAR MOTOR [830]. (ORIGINAL FRAMES REPLACED) | GEAR UNIT | WH, WN44-45; GE, GA25A1 |
| TRUCK NUMBERS | | 3000-3825 (ORIGINAL) , 3000R-3829R (REPLACED) | JOURNAL BEARING | TIMKEN, 5x9 ROLLER |
| WEIGHT | #1 END | 18,900 LBS (ORIGINAL), 20,000 LBS (REPLACEMENT) | SHOCK ABSORBER | MONROE: LATERAL [1]; VERTICAL [2] |
| | #2 END | 15,600 LBS (ORIGINAL), 15,700 LBS (REPLACEMENT) | SIDE BEARING | STUCKI, ROLLER |
| | | | TRACTION MOTOR | WH, 1447A; GE, 1240A3 (100 HP) [2] |
| | | | TRIP COCK | WABCO, D1 |

WH CONTROL & MOTOR ON CARS 1803-1827, 3000-3049, 3100-3224.
GE CONTROL & MOTOR ON CARS 1828-1852, 3050-3099, 3225-3349.
CARS 2950-2999 WERE 1803-1852 (RENUMBERED 5/1970).
CARBODY IS OF WELDED CONSTRUCTION.
#2 ENDS OF CARS 3029 & 3240 REBUILT BY PULLMAN STANDARD (3/1959).
CAR 3240 EQUIPPED WITH R17 TYPE FLUORESCENT FIXTURES (3/1959).
CARS 3320-3349 TRANSFERRED TO BMT DIVISION (1954). RETURNED TO IND DIVISION (1/1959).
WH UNA CAM CONTROL ON CAR 3271 (5/17/1961).
ORIGINAL FIRE EXTINGUISHERS REPLACED (1957).
CAR 3138 HAD ADJUSTABLE LOUVERED VENTS AT LOWER THIRD OF ALL SIDE DOOR WINDOW OPENINGS.
CAR 3184 SOLD TO RAILWAY PRESERVATION CORP.
CAR 3189 AT PITKIN MAINT. SHOP.
INSTALLED HEADLIGHTS (1962).
ALL OTHER CARS WERE SCRAPPED (1980's).

CAR NUMBERS: 8010-8019
TOTAL: 10 CARS
BUILT BY: BUDD

DATE: 1949
AVERAGE COST PER CAR: $121,373 (BID)

# R11 (BMT/IND)

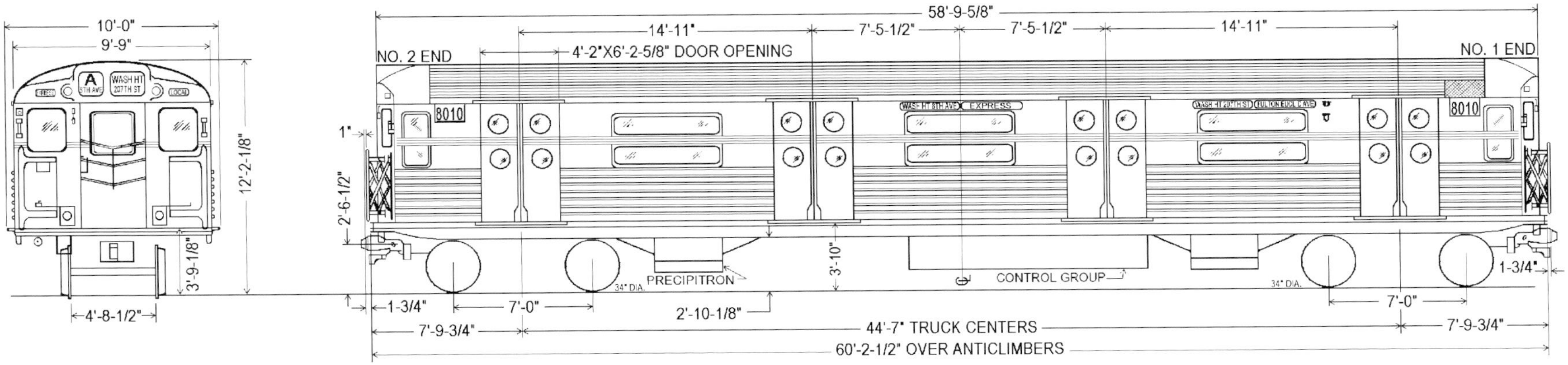

10'-0"
9'-9"
12'-2-1/8"
3'-9-1/8"
4'-8-1/2"

58'-9-5/8"
14'-11" 7'-5-1/2" 7'-5-1/2" 14'-11"
NO. 2 END
4'-2"X6'-2-5/8" DOOR OPENING
NO. 1 END
1"
2'-6-1/2"
3'-10"
PRECIPITRON
CONTROL GROUP
34" DIA.
1-3/4" 7'-0" 2'-10-1/8" 7'-0" 1-3/4"
7'-9-3/4" 44'-7" TRUCK CENTERS 7'-9-3/4"
60'-2-1/2" OVER ANTICLIMBERS

40,702 LBS
81,476 LBS TOTAL
40,774 LBS

Copyright © 1997 by NYC Transit

## CAR BODY - EQUIPMENT

| | | | |
|---|---|---|---|
| AIR COMPRESSOR | WABCO, 3YC, 21 CFM | MARKER LIGHT | LOVELL DRESSEL , SEMAPHORE |
| AIR DIFFUSER | MODULAR VANE | MASTER CONTROLLER | WH XM179 |
| BATTERY | EDISON, 24 CELL, B4H, WOOD BOX | MOTOR ALTERNATOR | WESTINGHOUSE ELECTRIC TYPE XF23A |
| BRAKE CYLINDER | WABCO TYPE C-3 | PROPULSION CONTROL TYPE | WH, ABS, UP631B GROUP |
| BRAKE EQUIPMENT TYPE | WABCO SMEE (DYNAMIC & FRICTION) | PUBLIC ADDRESS SYSTEM | RAYMOND ROSEN |
| BRAKE VALVE | WABCO, ME42 | SASH | ADAMS & WESTLAKE |
| COUPLER | WABCO, H2C, BL25B | SASH, END DOOR EQUATORIAL | HUNTER SASH |
| DOOR HANGER | OTIS ELEVATOR | SEAT BACK | AMERICAN CAR & FOUNDRY |
| DOOR OPERATOR MOTOR (ELECTRIC) | CONSOLIDATED CAR HEATING (8015-8019); NATIONAL PNEUMATIC (8010-8014) | SEAT CUSHION | AMERICAN CAR & FOUNDRY |
| DOOR PANEL | BUDD | SEAT MATERIAL | YELLOW VELON WITH TAN STRIPES |
| DRAFT GEAR | WAUGH, WM E3-4, E4-4 | SEATING ARRANGEMENT & CAPACITY | LONGITUDINAL & CROSS [56] |
| ELECTRIC PORTION | WABCO, BL25B | SIGN | HUNTER ILLUMINATED CAR SIGN |
| FIRE EXTINGUISHER | 2 LBS DRY POWDER | SIGNAL DEVICE | PNEUPHONIC HORN |
| HAND BAR | ELLCON | SWITCH PANEL | CONSOLIDATED CAR HEATING: #1 END, PS68A; #2 END, PS69A |
| HAND BRAKE | NATIONAL BRAKE 840XAL | TAIL LIGHT | TWIN SOCKET, REVERSER CONTROLLED |
| HEATER | CONSOLIDATED CAR HEATING | THERMOSTAT | "MICROTHERM" CONSOLIDATED CAR HEATING |
| LIGHTING | LUMINATOR, 72" [24]; 48" [2] | VENTILATION | WESTINGHOUSE EQUIPMENT DUAL BLOWER SYSTEMS |

## TRUCK

| | | | |
|---|---|---|---|
| MANUFACTURER | CLARK EQUIPMENT, TRUCK FRAME PATT. NO 55027 | GEAR UNIT | CLARK EQUIPMENT , HYPOID GEAR SPICER DRIVE |
| TRUCK TYPE & QUANTITY | C.S.F. EQUALIZER BAR MOTOR [22] | JOURNAL BEARING | SKF, 5x9 ROLLER, Z3890 |
| TRUCK NUMBERS | 8010-8031 | SHOCK ABSORBER | MONROE, LATERAL [1]; VERTICAL [2] |
| WEIGHT | #1 END: 18,110 LBS; #2 END: 18,110 LBS | TRACTION MOTOR | GE, 1240B, (100 HP) [2] |
| BRAKE RIGGING | DRUM BRAKE | TRIP COCK | WABCO D1 |

SHOT WELDED, STAINLESS STEEL CAR BODY.
OVERHAULED TO R34 CONTRACT (1964-65).
CARS HAD CENTRAL AIR VENTILATION WITH A PRECIPITRON UNIT TO PURIFY AIR.
CAR 8010 EQUIPMENT WAS ALTERED FOR COMPATIBILITY WITH COUPLED R16 CARS (1957).

INSTALLED CIRCULAR WINDOW ON END DOORS (1951).
CARS 8010-8016 TRANSFERRED TO BMT (6/14/1954).
CARS 8017-8019 TRANSFERRED TO BMT (6/21/1954).
CAR 8013 AT NY TRANSIT MUSEUM. ALL OTHER CARS WERE SOLD OR SCRAPPED (1980).

CAR NUMBERS: 5703-5802
TOTAL: 100 CARS
BUILT BY: AMERICAN CAR FOUNDRY

DATE: 1948
AVERAGE COST PER CAR: $71,487 (BID)

# R12 (IRT)

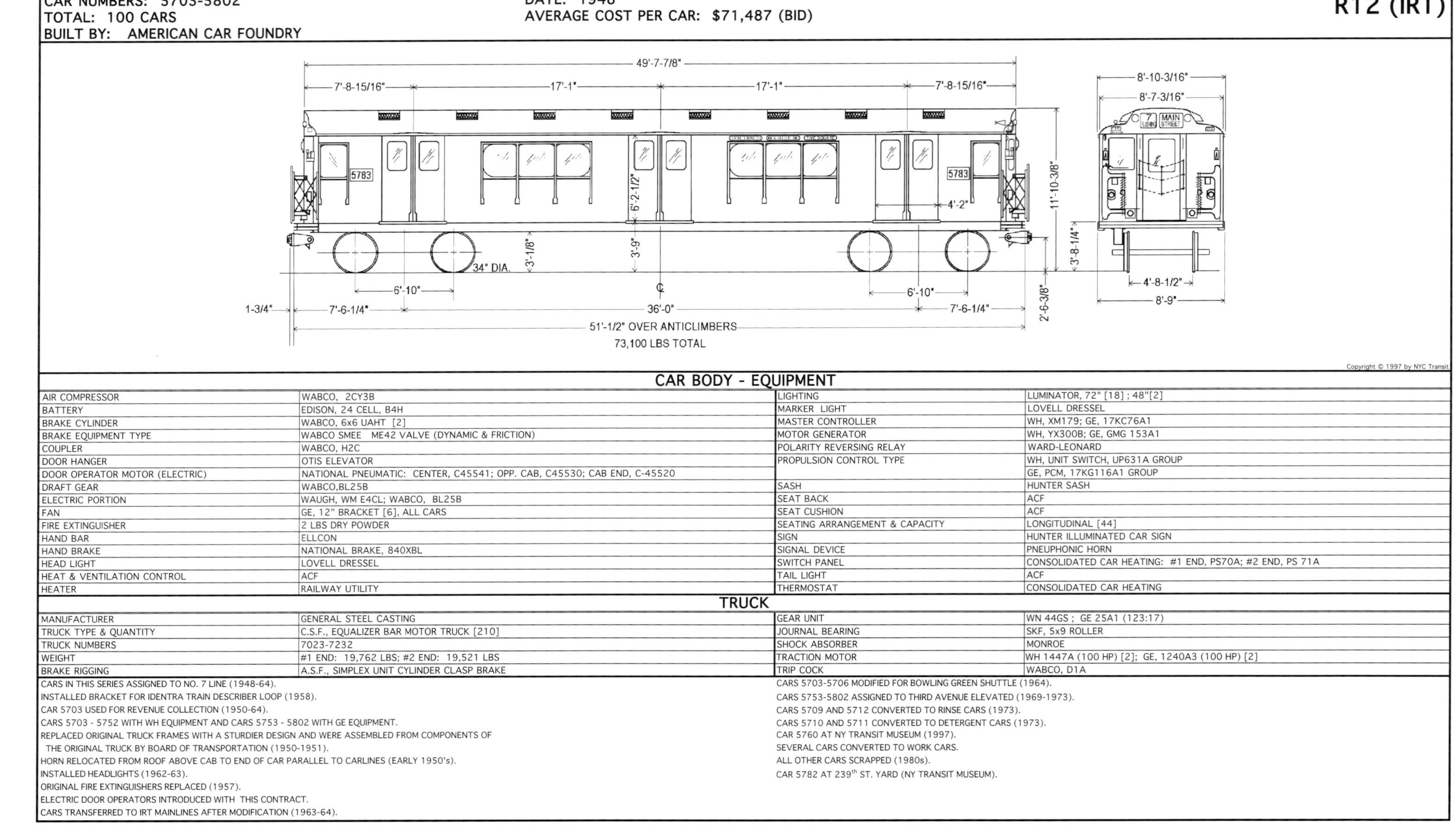


Copyright © 1997 by NYC Transit

## CAR BODY - EQUIPMENT

| | | | |
|---|---|---|---|
| AIR COMPRESSOR | WABCO, 2CY3B | LIGHTING | LUMINATOR, 72" [18] ; 48"[2] |
| BATTERY | EDISON, 24 CELL, B4H | MARKER LIGHT | LOVELL DRESSEL |
| BRAKE CYLINDER | WABCO, 6x6 UAHT [2] | MASTER CONTROLLER | WH, XM179; GE, 17KC76A1 |
| BRAKE EQUIPMENT TYPE | WABCO SMEE ME42 VALVE (DYNAMIC & FRICTION) | MOTOR GENERATOR | WH, YX300B; GE, GMG 153A1 |
| COUPLER | WABCO, H2C | POLARITY REVERSING RELAY | WARD-LEONARD |
| DOOR HANGER | OTIS ELEVATOR | PROPULSION CONTROL TYPE | WH, UNIT SWITCH, UP631A GROUP |
| DOOR OPERATOR MOTOR (ELECTRIC) | NATIONAL PNEUMATIC: CENTER, C45541; OPP. CAB, C45530; CAB END, C-45520 | | GE, PCM, 17KG116A1 GROUP |
| DRAFT GEAR | WABCO,BL25B | SASH | HUNTER SASH |
| ELECTRIC PORTION | WAUGH, WM E4CL; WABCO, BL25B | SEAT BACK | ACF |
| FAN | GE, 12" BRACKET [6], ALL CARS | SEAT CUSHION | ACF |
| FIRE EXTINGUISHER | 2 LBS DRY POWDER | SEATING ARRANGEMENT & CAPACITY | LONGITUDINAL [44] |
| HAND BAR | ELLCON | SIGN | HUNTER ILLUMINATED CAR SIGN |
| HAND BRAKE | NATIONAL BRAKE, 840XBL | SIGNAL DEVICE | PNEUPHONIC HORN |
| HEAD LIGHT | LOVELL DRESSEL | SWITCH PANEL | CONSOLIDATED CAR HEATING: #1 END, PS70A; #2 END, PS 71A |
| HEAT & VENTILATION CONTROL | ACF | TAIL LIGHT | ACF |
| HEATER | RAILWAY UTILITY | THERMOSTAT | CONSOLIDATED CAR HEATING |

## TRUCK

| | | | |
|---|---|---|---|
| MANUFACTURER | GENERAL STEEL CASTING | GEAR UNIT | WN 44GS ; GE 25A1 (123:17) |
| TRUCK TYPE & QUANTITY | C.S.F., EQUALIZER BAR MOTOR TRUCK [210] | JOURNAL BEARING | SKF, 5x9 ROLLER |
| TRUCK NUMBERS | 7023-7232 | SHOCK ABSORBER | MONROE |
| WEIGHT | #1 END: 19,762 LBS; #2 END: 19,521 LBS | TRACTION MOTOR | WH 1447A (100 HP) [2]; GE, 1240A3 (100 HP) [2] |
| BRAKE RIGGING | A.S.F., SIMPLEX UNIT CYLINDER CLASP BRAKE | TRIP COCK | WABCO, D1A |

CARS IN THIS SERIES ASSIGNED TO NO. 7 LINE (1948-64).
INSTALLED BRACKET FOR IDENTRA TRAIN DESCRIBER LOOP (1958).
CAR 5703 USED FOR REVENUE COLLECTION (1950-64).
CARS 5703 - 5752 WITH WH EQUIPMENT AND CARS 5753 - 5802 WITH GE EQUIPMENT.
REPLACED ORIGINAL TRUCK FRAMES WITH A STURDIER DESIGN AND WERE ASSEMBLED FROM COMPONENTS OF THE ORIGINAL TRUCK BY BOARD OF TRANSPORTATION (1950-1951).
HORN RELOCATED FROM ROOF ABOVE CAB TO END OF CAR PARALLEL TO CARLINES (EARLY 1950's).
INSTALLED HEADLIGHTS (1962-63).
ORIGINAL FIRE EXTINGUISHERS REPLACED (1957).
ELECTRIC DOOR OPERATORS INTRODUCED WITH THIS CONTRACT.
CARS TRANSFERRED TO IRT MAINLINES AFTER MODIFICATION (1963-64).

CARS 5703-5706 MODIFIED FOR BOWLING GREEN SHUTTLE (1964).
CARS 5753-5802 ASSIGNED TO THIRD AVENUE ELEVATED (1969-1973).
CARS 5709 AND 5712 CONVERTED TO RINSE CARS (1973).
CARS 5710 AND 5711 CONVERTED TO DETERGENT CARS (1973).
CAR 5760 AT NY TRANSIT MUSEUM (1997).
SEVERAL CARS CONVERTED TO WORK CARS.
ALL OTHER CARS SCRAPPED (1980s).
CAR 5782 AT 239th ST. YARD (NY TRANSIT MUSEUM).

CAR NUMBERS: 5803-5952
TOTAL: 150 CARS
BUILT BY: AMERICAN CAR FOUNDRY

DATE: 1949
AVERAGE COST PER CAR: $71,487 (BID)

# R14 (IRT)

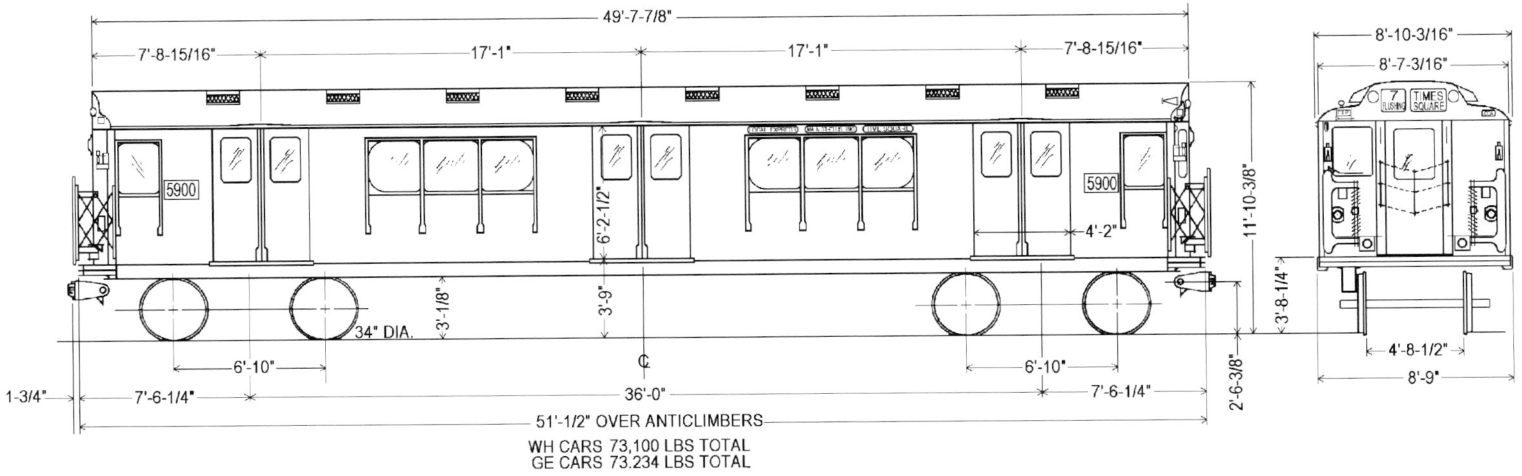

49'-7-7/8"
7'-8-15/16"
17'-1"
17'-1"
7'-8-15/16"
5900
6'-2-1/2"
4'-2"
11'-10-3/8"
3'-8-1/4"
3'-1/8"
3'-9"
34" DIA.
6'-10"
6'-10"
1-3/4"
7'-6-1/4"
36'-0"
7'-6-1/4"
2'-6-3/8"
51'-1/2" OVER ANTICLIMBERS
WH CARS 73,100 LBS TOTAL
GE CARS 73.234 LBS TOTAL
8'-10-3/16"
8'-7-3/16"
7 FLUSHING
TIMES SQUARE
4'-8-1/2"
8'-9"

Copyright © 1997 by NYC Transit

## CAR BODY - EQUIPMENT

| | | | |
|---|---|---|---|
| AIR COMPRESSOR | WABCO, 2CY3B | LIGHTING | LUMINATOR, 72" [18]; 48"[2] |
| BATTERY | EDISON, 24 CELL, B4H | MASTER CONTROLLER | WH XM179; GE 17KC76A1 |
| BRAKE CYLINDER | WABCO, 6x6 UAHT UNIT [2] | MOTOR/GENERATOR | WH YX300B; GE GMG 153A1 |
| BRAKE EQUIPMENT TYPE | WABCO, SMEE (DYNAMIC & FRICTION) | POLARITY REVERSING RELAY | WARD-LEONARD |
| COUPLER | WABCO H2C | PROPULSION CONTROL TYPE | GE, PCM, 17KG116A1 GROUP |
| DOOR HANGER | OTIS ELEVATOR | | WH, UNIT SWITCH, UP631 GROUP |
| DOOR OPERATOR MOTOR | NATIONAL PNEUMATIC: CENTER, C51600; OPP. CAB, C51620; CAB END, C51610 | SASH | HUNTER SASH |
| DRAFT GEAR | WAUGH WM E4CL | SEAT ARRANGEMENT & CAPACITY | LONGITUDINAL [44] |
| ELECTRIC PORTION | WABCO, BL25B | SEAT BACK | ACF |
| FAN | GE 12" BRACKET [6], ALL CARS | SEAT CUSHION & BACK | ACF |
| FIRE EXTINGUISHER | 2 LB DRY POWDER | SIGN | HUNTER ILLUMINATED CAR SIGN |
| HAND BAR | ELLCON | SIGNAL DEVICE | PNEUPHONIC HORN |
| HAND BRAKE | NATIONAL BRAKE 840XBL | SWITCH PANEL | CONSOLIDATED CAR HEATING: # 1 END, PS70A; # 2 END, PS71A |
| HEAD, MARKER & TAIL LIGHTS | LOVELL DRESSEL | THERMOSTAT | CONSOLIDATED |
| HEAT & VENTILATION CONTROL | ACF | VENTILATOR | RAILWAY UTILITY, HONEY COMB |
| HEATER | RAILWAY UTILITY | | |

## TRUCK

| | | | |
|---|---|---|---|
| MANUFACTURER | GENERAL STEEL CASTING | GEAR UNIT | WH, WN 44GS (123:17) |
| TRUCK TYPE & QUANTITY | C. S. F. EQUALIZER BAR MOTOR [310] | JOURNAL BEARING | SKF, 5x9 ROLLER |
| TRUCK NUMBERS | 7233-7542 | SHOCK ABSORBER | MONROE |
| WEIGHT | #1 END: 19,762 LBS; #2 END: 19,521 LBS | TRACTION MOTOR | WH, 1447A (100 HP) [2] |
| BRAKE RIGGING | A. S. F., SIMPLEX UNIT CYLINDER CLASP BRAKE | TRIP COCK | WABCO, D1A |

CARS IN THIS SERIES ASSIGNED TO FLUSHING LINE (1949-64).
CAR 5952 HAD ADJUSTABLE LOUVERED VENTS AT LOWER THIRD OF ALL SIDE DOOR WINDOW OPENINGS.
REPLACED ORIGINAL TRUCK FRAMES WITH A STURDIER DESIGN AND WERE ASSEMBLED FROM COMPONENTS OF THE ORIGINAL TRUCK BY BOARD OF TRANSPORTATION (1950-1951).
HORN RELOCATED FROM ROOF ABOVE CAB TO END OF CAR PARALLEL TO CARLINES (EARLY 1950's).
INSTALLED BRACKET FOR "IDENTRA" TRAIN DESCRIBER LOOP ( 1958).
CARS 5803 - 5877 WITH GE EQUIPMENT AND CARS 5878 - 5952 WITH WH EQUIPMENT.
INSTALLED HEADLIGHTS (1962-63).
ORIGINAL FIRE EXTINGUISHERS REPLACED (1957).
CARS TRANSFERRED TO IRT MAINLINES AFTER MODIFICATION (1963-64).
CAR 5871 AT 239th ST. YARD (NY TRANSIT MUSEUM CAR).
SOME CARS WERE CONVERTED TO WORK CARS.
ALL OTHER CARS WERE SCRAPPED (1980s)

CAR NUMBERS: 5953-5999, 6200-6252
TOTAL: 100 CARS
BUILT BY: AMERICAN CAR FOUNDRY

DATE: 1950
AVERAGE COST PER CAR: $77,587 (BID)

# R15 (IRT)

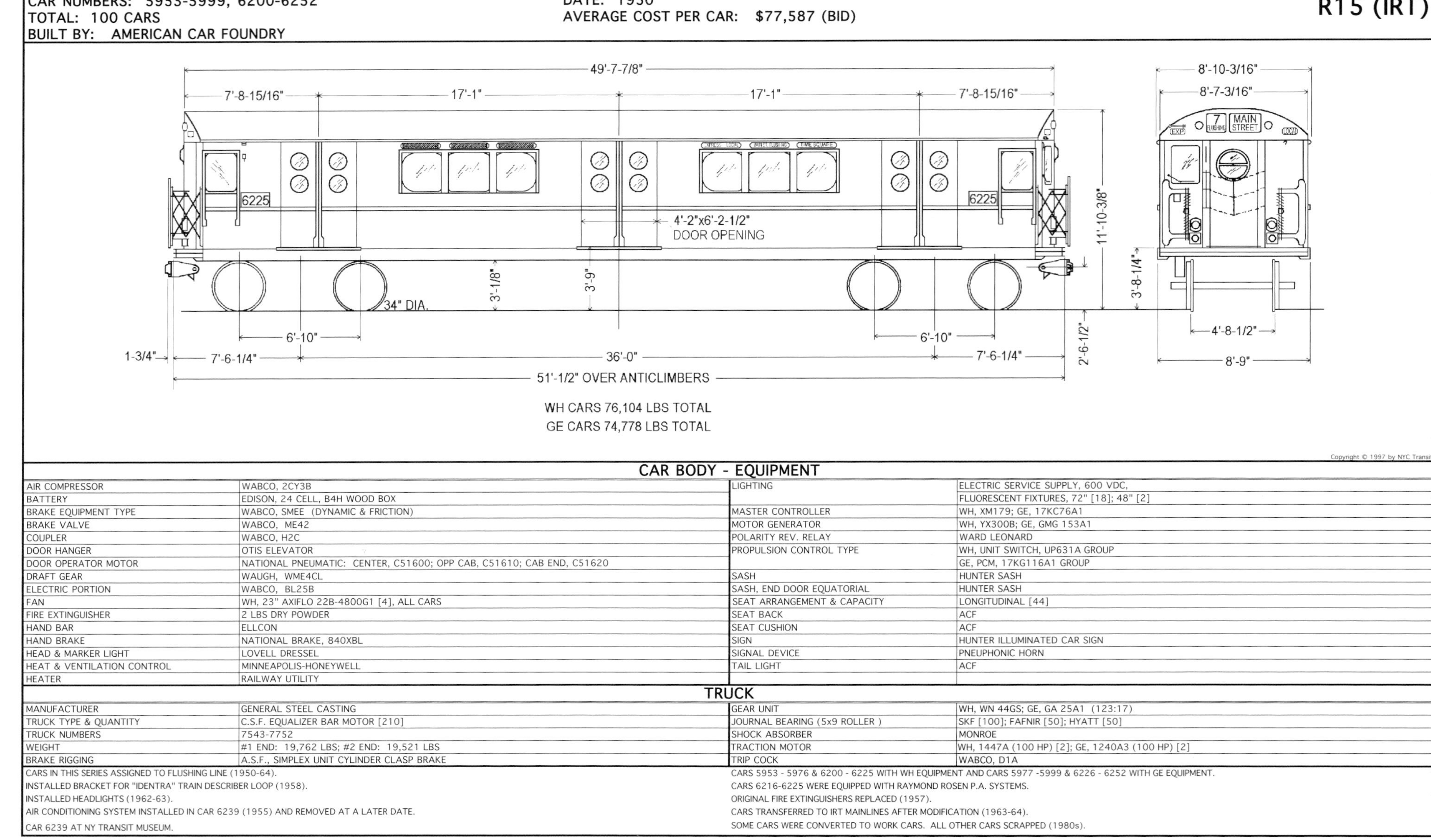


Copyright © 1997 by NYC Transit

## CAR BODY - EQUIPMENT

| | | | |
|---|---|---|---|
| AIR COMPRESSOR | WABCO, 2CY3B | LIGHTING | ELECTRIC SERVICE SUPPLY, 600 VDC, |
| BATTERY | EDISON, 24 CELL, B4H WOOD BOX | | FLUORESCENT FIXTURES, 72" [18]; 48" [2] |
| BRAKE EQUIPMENT TYPE | WABCO, SMEE (DYNAMIC & FRICTION) | MASTER CONTROLLER | WH, XM179; GE, 17KC76A1 |
| BRAKE VALVE | WABCO, ME42 | MOTOR GENERATOR | WH, YX300B; GE, GMG 153A1 |
| COUPLER | WABCO, H2C | POLARITY REV. RELAY | WARD LEONARD |
| DOOR HANGER | OTIS ELEVATOR | PROPULSION CONTROL TYPE | WH, UNIT SWITCH, UP631A GROUP |
| DOOR OPERATOR MOTOR | NATIONAL PNEUMATIC: CENTER, C51600; OPP CAB, C51610; CAB END, C51620 | | GE, PCM, 17KG116A1 GROUP |
| DRAFT GEAR | WAUGH, WME4CL | SASH | HUNTER SASH |
| ELECTRIC PORTION | WABCO, BL25B | SASH, END DOOR EQUATORIAL | HUNTER SASH |
| FAN | WH, 23" AXIFLO 22B-4800G1 [4], ALL CARS | SEAT ARRANGEMENT & CAPACITY | LONGITUDINAL [44] |
| FIRE EXTINGUISHER | 2 LBS DRY POWDER | SEAT BACK | ACF |
| HAND BAR | ELLCON | SEAT CUSHION | ACF |
| HAND BRAKE | NATIONAL BRAKE, 840XBL | SIGN | HUNTER ILLUMINATED CAR SIGN |
| HEAD & MARKER LIGHT | LOVELL DRESSEL | SIGNAL DEVICE | PNEUPHONIC HORN |
| HEAT & VENTILATION CONTROL | MINNEAPOLIS-HONEYWELL | TAIL LIGHT | ACF |
| HEATER | RAILWAY UTILITY | | |

## TRUCK

| | | | |
|---|---|---|---|
| MANUFACTURER | GENERAL STEEL CASTING | GEAR UNIT | WH, WN 44GS; GE, GA 25A1 (123:17) |
| TRUCK TYPE & QUANTITY | C.S.F. EQUALIZER BAR MOTOR [210] | JOURNAL BEARING (5x9 ROLLER ) | SKF [100]; FAFNIR [50]; HYATT [50] |
| TRUCK NUMBERS | 7543-7752 | SHOCK ABSORBER | MONROE |
| WEIGHT | #1 END: 19,762 LBS; #2 END: 19,521 LBS | TRACTION MOTOR | WH, 1447A (100 HP) [2]; GE, 1240A3 (100 HP) [2] |
| BRAKE RIGGING | A.S.F., SIMPLEX UNIT CYLINDER CLASP BRAKE | TRIP COCK | WABCO, D1A |

CARS IN THIS SERIES ASSIGNED TO FLUSHING LINE (1950-64).
INSTALLED BRACKET FOR "IDENTRA" TRAIN DESCRIBER LOOP (1958).
INSTALLED HEADLIGHTS (1962-63).
AIR CONDITIONING SYSTEM INSTALLED IN CAR 6239 (1955) AND REMOVED AT A LATER DATE.
CAR 6239 AT NY TRANSIT MUSEUM.

CARS 5953 - 5976 & 6200 - 6225 WITH WH EQUIPMENT AND CARS 5977 -5999 & 6226 - 6252 WITH GE EQUIPMENT.
CARS 6216-6225 WERE EQUIPPED WITH RAYMOND ROSEN P.A. SYSTEMS.
ORIGINAL FIRE EXTINGUISHERS REPLACED (1957).
CARS TRANSFERRED TO IRT MAINLINES AFTER MODIFICATION (1963-64).
SOME CARS WERE CONVERTED TO WORK CARS. ALL OTHER CARS SCRAPPED (1980s).

CAR NUMBERS: 6300-6499
TOTAL: 200 CARS
BUILT BY: AMERICAN CAR FOUNDRY

DATE: 1954 - 1955
AVERAGE COST PER CAR: $121,442 (BID)

# R16 (BMT/IND)

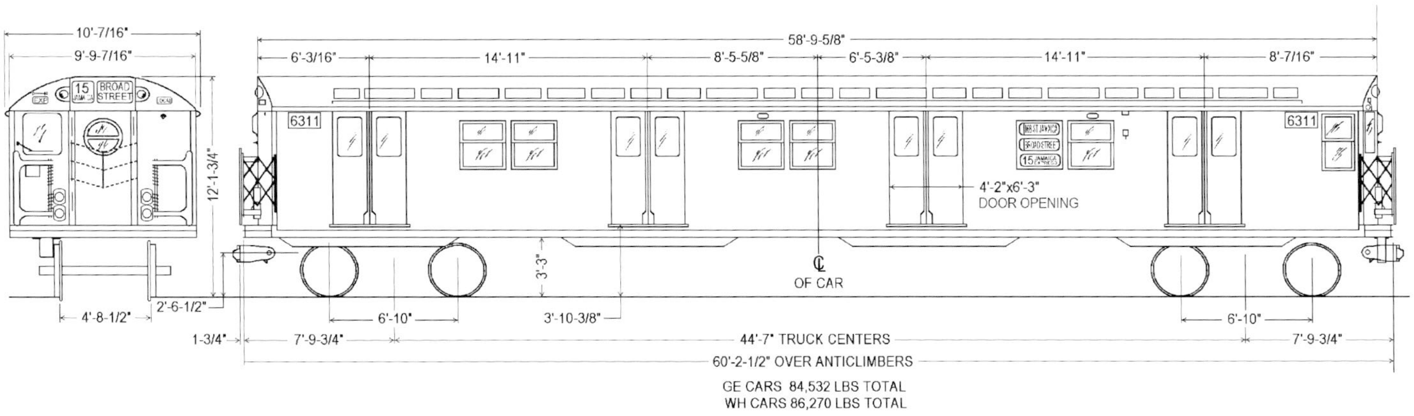

10'-7/16"
9'-9-7/16"
12'-1-3/4"
4'-8-1/2"
2'-6-1/2"
1-3/4"
58'-9-5/8"
6'-3/16" 14'-11" 8'-5-5/8" 6'-5-3/8" 14'-11" 8'-7/16"
4'-2"x6'-3" DOOR OPENING
3'-3"
℄ OF CAR
6'-10" 3'-10-3/8" 6'-10"
7'-9-3/4" 44'-7" TRUCK CENTERS 7'-9-3/4"
60'-2-1/2" OVER ANTICLIMBERS

GE CARS 84,532 LBS TOTAL
WH CARS 86,270 LBS TOTAL

Copyright © 1997 by NYC Transit

## CAR BODY - EQUIPMENT

| | | | |
|---|---|---|---|
| AIR COMPRESSOR | WABCO, 2CY3B | HEATER | GOLD CAR HEATING |
| BATTERY | EDISON, 24 CELL B4H WOOD BOX | LIGHTING | ELECTRIC SERVICE SUPPLY, 600 VDC FLUORESCENT FIXTURES 72" [18], 48" [4] |
| BRAKE CYLINDER | WABCO, 7X6 UAHT [2]/TRUCK | MASTER CONTROLLER | WH, XM179; GE, 17KC76A1 |
| BRAKE EQUIPMENT TYPE | WABCO, SMEE (DYNAMIC & FRICTION) | MOTOR GENERATOR | WH, YX300D; GE, GMG 153F1 |
| BRAKE VALVE | WABCO, ME42 | POLARITY REVERSING RELAY | WARD LEONARD |
| COUPLER | WABCO, H2C . | PROPULSION CONTROL TYPE | GE, MCM, 17KG1137D1 GROUP |
| DOOR HANGER | OTIS ELEVATOR | | WH, UNIT SWITCH, UPC631A GROUP |
| DOOR OPERATOR MOTOR | NATIONAL PNEUMATIC: CENTER, C-62710; OPP. CAB, C-62730; CAB END, C-62720 | PUBLIC ADDRESS SYSTEM | GRAYBAR , VACUUM TUBE |
| DRAFT GEAR | WAUGH, WME4CL | SASH | O.M. EDWARDS |
| ELECTRIC PORTION | WABCO, BL25B | SEAT ARRANGEMENT & CAPACITY | LONGITUDINAL & CROSS [54] |
| FAN | WH, 23" AXIFLO 22B-4800G1 [6], ALL CARS | SEAT BACK | ACF |
| FIRE EXTINGUISHER | 2 LBS DRY POWDER | SEAT CUSHION | ACF |
| HAND BRAKE | NATIONAL BRAKE, 840XBL | SIGN | HUNTER ILLUMINATED CAR SIGN |
| HANDHOLD | ELLCON | SIGNAL DEVICE | PNEUPHONIC HORN |
| HEAD, MARKER & TAIL LIGHTS | LOVELL DRESSEL | SWITCH PANEL | CONSOLIDATED CAR HEATING: #1 END, PS74A; #2 END, PS75A |
| HEAT & VENTILATION CONTROL | VAPOR | | |

## TRUCK

| | | | |
|---|---|---|---|
| MANUFACTURER | GENERAL STEEL CASTINGS | GEAR UNIT | WH, WN44BGS; GE, GA 25A1 (123.17) |
| TYPE & QUANTITY | C.S.F. EQUALIZER BAR MOTOR [400] | JOURNAL BEARING | SKF, 5X9 ROLLER |
| TRUCK NUMBERS | WH, 3830 - 4029; GE, 4030 - 4229 | SHOCK ABSORBER | MONROE |
| WEIGHT | #1 END: 19,762 LBS; #2 END: 19,521 LBS | TRACTION MOTOR | WH, 1447C (100 HP) [2]; GE, 1240A4 (100 HP) [2] |
| BRAKE RIGGING | A.S.F., SIMPLEX UNIT CYLINDER CLASP BRAKE | TRIP COCK | WABCO, D1 |

WH CARS: 6300 - 6399; GE CARS: 6400 - 6499.
CARS ASSIGNED TO BMT DIVISION.
STAGGERED DOOR CONCEPT INTRODUCED WITH THIS CONTRACT.
CARS 6300 - 6349 ASSIGNED TO IND DIVISION (6/1956 - 9/1958).
CARS 6400 - 6424 HAVE NO CENTER STANCHIONS.
CAR 6494 DESTROYED IN WRECK (9/26/1957).
CAR 6304 DESTROYED IN COLLISION 5/20/1970.
CAR 6452 AT NYC TRANSIT LEARNING CENTER (PS 248).
CARS 6318 AND 6463 PAINTED GOLD FOR THE FIFTIETH ANNIVERSARY OF THE FIFTH AVENUE ASSOCIATION (10/1957).

INSTALLED HEADLIGHTS (1961 - 1963).
ORIGINAL FIRE EXTINGUISHERS REPLACED (1957).
REPLACED DOOR OPERATOR ENGINE (10/CAR) WITH DOOR OPERATOR MOTOR (16/CAR, VAPOR, SLIM LINE) AND RELOCATED FROM UNDER SEAT TO (TAPERED) WALL PANELS ADJACENT TO DOORS (1973 - 1974).
CARS 6305 & 6339 SOLD TO RAILWAY PRESERVATION CORPORATION.
CAR 6387 AT NY TRANSIT MUSEUM.
CAR 6398 SOLD TO TROLLEY MUSEUM OF NEW YORK, KINGSTON, N.Y.
ALL OTHER CARS SCRAPPED (1978 - 1987).

CAR NUMBERS: 6500 - 6899
TOTAL: 400 CARS
BUILT BY: ST. LOUIS CAR

DATE: 1955 - 56
AVERAGE COST PER CAR: $121,442 (BID)

# R17 (IRT)

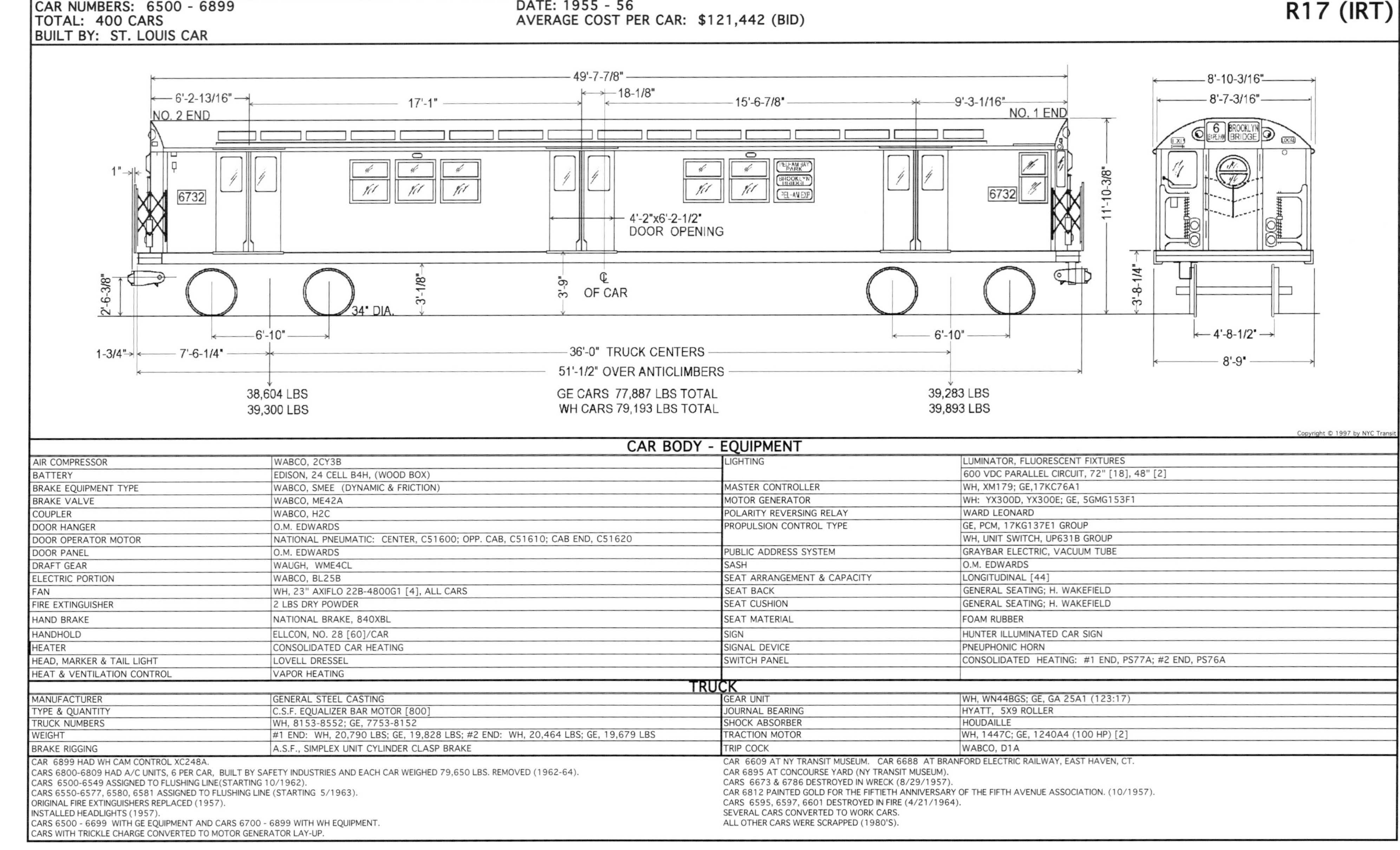


Copyright © 1997 by NYC Transit

## CAR BODY - EQUIPMENT

| | | | |
|---|---|---|---|
| AIR COMPRESSOR | WABCO, 2CY3B | LIGHTING | LUMINATOR, FLUORESCENT FIXTURES |
| BATTERY | EDISON, 24 CELL B4H, (WOOD BOX) | | 600 VDC PARALLEL CIRCUIT, 72" [18], 48" [2] |
| BRAKE EQUIPMENT TYPE | WABCO, SMEE (DYNAMIC & FRICTION) | MASTER CONTROLLER | WH, XM179; GE,17KC76A1 |
| BRAKE VALVE | WABCO, ME42A | MOTOR GENERATOR | WH: YX300D, YX300E; GE, 5GMG153F1 |
| COUPLER | WABCO, H2C | POLARITY REVERSING RELAY | WARD LEONARD |
| DOOR HANGER | O.M. EDWARDS | PROPULSION CONTROL TYPE | GE, PCM, 17KG137E1 GROUP |
| DOOR OPERATOR MOTOR | NATIONAL PNEUMATIC: CENTER, C51600; OPP. CAB, C51610; CAB END, C51620 | | WH, UNIT SWITCH, UP631B GROUP |
| DOOR PANEL | O.M. EDWARDS | PUBLIC ADDRESS SYSTEM | GRAYBAR ELECTRIC, VACUUM TUBE |
| DRAFT GEAR | WAUGH, WME4CL | SASH | O.M. EDWARDS |
| ELECTRIC PORTION | WABCO, BL25B | SEAT ARRANGEMENT & CAPACITY | LONGITUDINAL [44] |
| FAN | WH, 23" AXIFLO 22B-4800G1 [4], ALL CARS | SEAT BACK | GENERAL SEATING; H. WAKEFIELD |
| FIRE EXTINGUISHER | 2 LBS DRY POWDER | SEAT CUSHION | GENERAL SEATING; H. WAKEFIELD |
| HAND BRAKE | NATIONAL BRAKE, 840XBL | SEAT MATERIAL | FOAM RUBBER |
| HANDHOLD | ELLCON, NO. 28 [60]/CAR | SIGN | HUNTER ILLUMINATED CAR SIGN |
| HEATER | CONSOLIDATED CAR HEATING | SIGNAL DEVICE | PNEUPHONIC HORN |
| HEAD, MARKER & TAIL LIGHT | LOVELL DRESSEL | SWITCH PANEL | CONSOLIDATED HEATING: #1 END, PS77A; #2 END, PS76A |
| HEAT & VENTILATION CONTROL | VAPOR HEATING | | |

## TRUCK

| | | | |
|---|---|---|---|
| MANUFACTURER | GENERAL STEEL CASTING | GEAR UNIT | WH, WN44BGS; GE, GA 25A1 (123:17) |
| TYPE & QUANTITY | C.S.F. EQUALIZER BAR MOTOR [800] | JOURNAL BEARING | HYATT, 5X9 ROLLER |
| TRUCK NUMBERS | WH, 8153-8552; GE, 7753-8152 | SHOCK ABSORBER | HOUDAILLE |
| WEIGHT | #1 END: WH, 20,790 LBS; GE, 19,828 LBS; #2 END: WH, 20,464 LBS; GE, 19,679 LBS | TRACTION MOTOR | WH, 1447C; GE, 1240A4 (100 HP) [2] |
| BRAKE RIGGING | A.S.F., SIMPLEX UNIT CYLINDER CLASP BRAKE | TRIP COCK | WABCO, D1A |

CAR 6899 HAD WH CAM CONTROL XC248A.
CARS 6800-6809 HAD A/C UNITS, 6 PER CAR, BUILT BY SAFETY INDUSTRIES AND EACH CAR WEIGHED 79,650 LBS. REMOVED (1962-64).
CARS 6500-6549 ASSIGNED TO FLUSHING LINE(STARTING 10/1962).
CARS 6550-6577, 6580, 6581 ASSIGNED TO FLUSHING LINE (STARTING 5/1963).
ORIGINAL FIRE EXTINGUISHERS REPLACED (1957).
INSTALLED HEADLIGHTS (1957).
CARS 6500 - 6699 WITH GE EQUIPMENT AND CARS 6700 - 6899 WITH WH EQUIPMENT.
CARS WITH TRICKLE CHARGE CONVERTED TO MOTOR GENERATOR LAY-UP.

CAR 6609 AT NY TRANSIT MUSEUM. CAR 6688 AT BRANFORD ELECTRIC RAILWAY, EAST HAVEN, CT.
CAR 6895 AT CONCOURSE YARD (NY TRANSIT MUSEUM).
CARS 6673 & 6786 DESTROYED IN WRECK (8/29/1957).
CAR 6812 PAINTED GOLD FOR THE FIFTIETH ANNIVERSARY OF THE FIFTH AVENUE ASSOCIATION. (10/1957).
CARS 6595, 6597, 6601 DESTROYED IN FIRE (4/21/1964).
SEVERAL CARS CONVERTED TO WORK CARS.
ALL OTHER CARS WERE SCRAPPED (1980'S).

CAR NUMBERS: 7050 - 7299
TOTAL: 250 CARS
BUILT BY: ST. LOUIS CAR

DATE: 1956 - 57
AVERAGE COST PER CAR: $102,871 (BID)

# R21 (IRT)

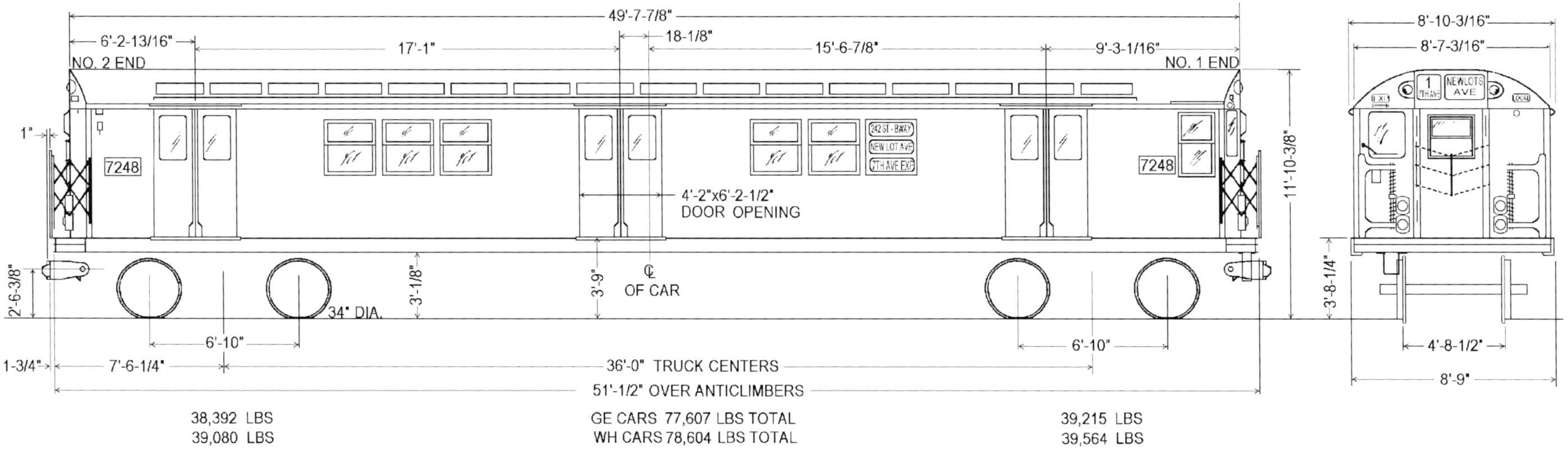

49'-7-7/8"
6'-2-13/16"
17'-1"
18-1/8"
15'-6-7/8"
9'-3-1/16"
NO. 2 END
NO. 1 END
1"
7248
242 ST - BWAY
NEW LOT AVE
7TH AVE EXP
4'-2"x6'-2-1/2" DOOR OPENING
11'-10-3/8"
2'-6-3/8"
3'-1/8"
34" DIA.
3'-9"
℄ OF CAR
6'-10"
6'-10"
1-3/4"
7'-6-1/4"
36'-0" TRUCK CENTERS
51'-1/2" OVER ANTICLIMBERS
8'-10-3/16"
8'-7-3/16"
1 7TH AVE
NEW LOTS AVE
3'-8-1/4"
4'-8-1/2"
8'-9"

38,392 LBS
39,080 LBS

GE CARS 77,607 LBS TOTAL
WH CARS 78,604 LBS TOTAL

39,215 LBS
39,564 LBS

Copyright © 1997 by NYC Transit

## CAR BODY - EQUIPMENT

| | | | |
|---|---|---|---|
| AIR COMPRESSOR | WABCO, 2CY3B | HEATER | CONSOLIDATED CAR HEATING |
| BATTERY | EDISON, B4H, 24 CELL WOOD BOX | HVAC CONTROL | VAPOR HEATING |
| BRAKE CYLINDER | WABCO, 6X6 UAHT [2]/TRUCK | LIGHTING | LUMINATOR, FLUORESCENT FIXTURES, 600 VDC PARALLEL CIRCUIT |
| BRAKE EQUIPMENT TYPE | WABCO, SMEE (DYNAMIC & FRICTION) | MASTER CONTROLLER | WH, XM179; GE, 17KC76A1 |
| BRAKE VALVE | WABCO, ME42A | MOTOR GENERATOR | WH, YX300D; GE, 5GMG153F1 |
| COUPLER | WABCO, H2C | PROPULSION CONTROL TYPE | GE, MCM, 17KG137F1 GROUP |
| DOOR HANGER | OTIS ELEVATOR | | WH, UNIT SWITCH, UPB31B GROUP |
| DOOR OPERATOR MOTOR | CONSOLIDATED CAR HEATING: CENTER, DE 93B; OPP. CAB, DE 95B; CAB END, DE 94B | PUBLIC ADDRESS SYSTEM | GRAYBAR ELECTRIC, VACUUM TUBE |
| DOOR PANEL | O.M. EDWARDS | SASH | O.M. EDWARDS |
| DRAFT GEAR | WAUGH, WME4CL | SEAT ARRANGEMENT & CAPACITY | LONGITUDINAL [44] |
| ELECTRIC PORTION | WABCO, BL25B | SEAT BACK | GENERAL SEATING; H. WAKEFIELD |
| FAN | WH, 23" AXIFLO, 52B580G4 [4], ALL CARS | SEAT CUSHION | GENERAL SEATING; H. WAKEFIELD |
| FIRE EXTINGUISHER | 2 LBS DRY POWDER | SIGN | ELECTRIC SERVICE |
| HAND BRAKE | NATIONAL BRAKE, 840XBL | SIGNAL DEVICE | PNEUPHONIC HORN |
| HEAD, MARKER & TAIL LIGHTS | LOVELL DRESSEL | SWITCH PANEL | CONSOLIDATED CAR HEATING: #1 END, PS77B; #2 END, PS76B |
| HANDHOLD | ELLCON, #32 [56] | | |

## TRUCK

| | | | |
|---|---|---|---|
| MANUFACTURER | GENERAL STEEL CASTING | GEAR UNIT | WH, WN-44BGS; GE, GA 25A1 (123:17) |
| TYPE | C.S.F. EQUALIZER BAR MOTOR | JOURNAL BEARING | TIMKEN, 5X9 ROLLER |
| TRUCK NUMBERS | WH, 8823-9072; GE, 8573-8822 | SHOCK ABSORBER | HOUDAILLE |
| WEIGHT | #1 END: 19,762 LBS; #2 END: 19,521 LBS | TRACTION MOTOR | WH, 1447C (100 HP) [2]; GE, 1240-A4 (100 HP) [2] |
| BRAKE RIGGING | A.S.F., SIMPLEX CYLINDER CLASP BRAKE | TRIP COCK | WABCO, D1A |

INSTALLED HEADLIGHTS (1958).
ORIGINAL FIRE EXTINGUISHERS REPLACED (1957).
CAB DOORS HINGED AT OPPOSITE SIDE OF END DOOR.
CARS 7050 - 7174 WITH GE EQUIPMENT AND CARS 7175 - 7299 WITH WH EQUIPMENT.

CAR 7267 AT CONCOURSE YARD (NY TRANSIT MUSEUM CAR).
SOME CARS WERE CONVERTED TO WORK CARS. REMAINING CARS WERE SCRAPPED.

CAR NUMBERS: 7300 - 7749
TOTAL: 450 CARS
BUILT BY: ST. LOUIS CAR

DATE: 1957 - 1958
AVERAGE COST PER CAR: $106,699 (BID)

# R22 (IRT)

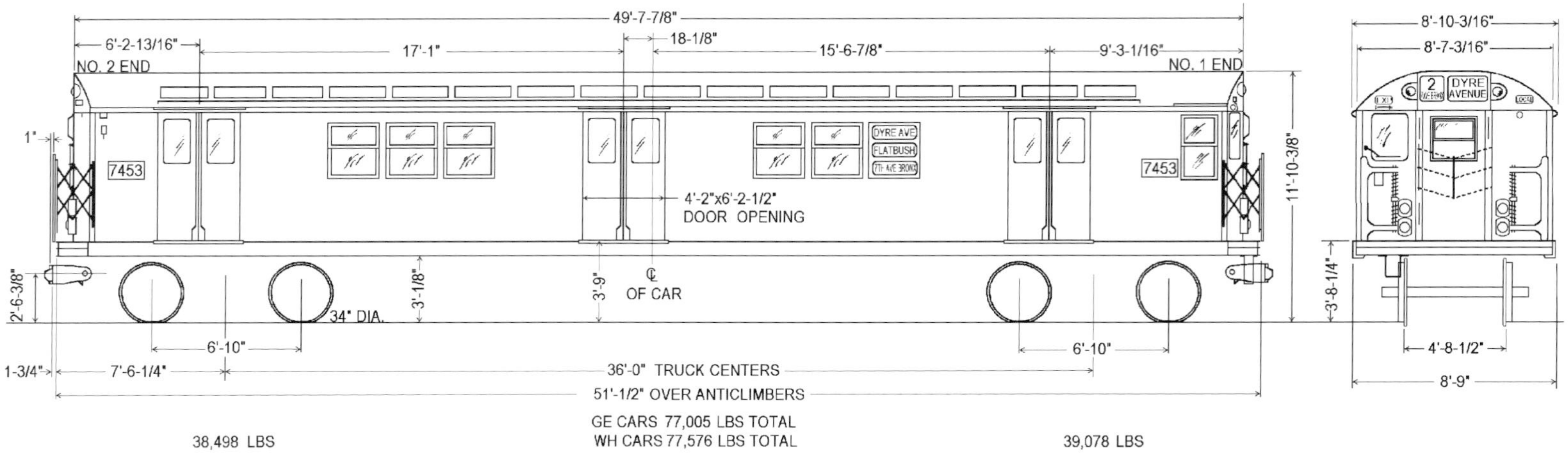

49'-7-7/8"
6'-2-13/16" 17'-1" 18-1/8" 15'-6-7/8" 9'-3-1/16"
NO. 2 END NO. 1 END
1"
7453
DYRE AVE
FLATBUSH
4'-2"x6'-2-1/2"
DOOR OPENING
11'-10-3/8"
2'-6-3/8"
34" DIA.
3'-1/8"
3'-9"
℄ OF CAR
3'-8-1/4"
6'-10" 6'-10"
1-3/4" 7'-6-1/4"
36'-0" TRUCK CENTERS
51'-1/2" OVER ANTICLIMBERS
GE CARS 77,005 LBS TOTAL
WH CARS 77,576 LBS TOTAL
38,498 LBS 39,078 LBS
8'-10-3/16"
8'-7-3/16"
2 DYRE AVENUE
4'-8-1/2"
8'-9"

Copyright © 1997 by NYC Transit

## CAR BODY - EQUIPMENT

| | | | |
|---|---|---|---|
| AIR COMPRESSOR | WABCO, 2CY3B | HEATER | CONSOLIDATED METAL |
| BATTERY | EDISON, 24 CELL ,B4H | HEATING & VENTILATING CONTROL | VAPOR HEATING |
| BRAKE CYLINDER | WABCO, 6X6 UAHT | LAY-UP CHARGE | MOTOR GENERATOR |
| BRAKE EQUIPMENT TYPE | WABCO, SMEE (DYNAMIC & FRICTION) | LIGHTING | LUMINATOR, 600 VDC FLUORESCENT FIXTURES |
| BRAKE VALVE | WABCO, ME42A | MASTER CONTROLLER | WH, XM329; GE, 17KC76A1 |
| COUPLER | WABCO, H2C | MOTOR GENERATOR | WH, YX303A; GE, 5GMG153H1, SEE NOTE 1 |
| DOOR HANGER | MIDLAND ROSS | PROPULSION CONTROL TYPE | GE, MCM, 17KG137F1 GROUP |
| DOOR OPERATOR MOTOR | CONSOLIDATED METAL: CENTER, DE93A; CAB END, DE95A; OPP. END, DE94A | | WH, UNIT SWITCH, UPE631A GROUP |
| DOOR PANEL | O.M. EDWARDS | PUBLIC ADDRESS SYSTEM | RAYMOND ROSEN; SOUND SYSTEMS |
| DRAFT GEAR | WAUGH, WM E4CL | SASH | O.M. EDWARDS |
| ELECTRIC PORTION | WABCO, BL25B | SEAT ARRANGEMENT & CAPACITY | LONGITUDINAL [44] |
| FAN | WH, 356C647G1, 23" AXIFLO [4], ALL CARS | SEAT BACK | GENERAL SEATING; H. WAKEFIELD |
| FIRE EXTINGUISHER | 2 LBS DRY POWDER | SEAT CUSHION | GENERAL SEATING; H. WAKEFIELD |
| HANDHOLD | CAR BUILDER [56]/CAR | SIGNAL DEVICE | PNEUPHONIC HORN |
| HAND BRAKE | NATIONAL BRAKE, 840XBL | SWITCH PANEL | CONSOLIDATED CAR HEATING: #1 END, PS77A; #2 END, PS76A |
| HEAD, MARKER & TAIL LIGHTS | LOVELL DRESSEL | | |

## TRUCK

| | | | |
|---|---|---|---|
| MANUFACTURER | GENERAL STEEL CASTING | JOURNAL BEARING | SKF, 5X9 ROLLER |
| TYPE & QUANTITY | C. S. F. EQUALIZER BAR MOTOR [900] | SHOCK ABSORBER | HOUDAILLE |
| WEIGHT | #1 END: WH, 20,300 LBS; #2 END: WH, 20,024 LBS | TRACTION MOTOR | WH, 1447C (100 HP) [2]; GE, 1240A4 |
| BRAKE RIGGING | A.S.F., SIMPLEX UNIT CYLINDER CLASP BRAKE | TRIP COCK | WABCO, D1A |
| GEAR UNIT | WH, WN44CGS (123:17) | | |

NOTE 1: CARS 7505 - 7524 EQUIPPED WITH WH CAM CONTROL/CONTROL GROUP XC248H AND M.G. SET YX303B.

CAR 7371 FOR PRESERVATION (2003).

FIRST CAR CONTRACT DELIVERED WITH HEADLIGHTS.
CARS 7515-7524 HAD A RIGIDIZED ALUMINUM INTERIOR WITH GREEN PLEXTONE PAINT SCHEME AND PINK MOLDED FIBERGLASS SEATS.
CARS 7300 - 7524 WITH WH EQUIPMENT AND CARS 7525 - 7749 WITH GE EQUIPMENT.
CARS 7325 - 7524 HAVE MINIATURE RELAYS IN DOOR CONTROL, PS77C & PS77D IN SWITCH PANELS.
AUTOMATED EQUIPMENT ADDED TO CARS 7513 (G.R.S) (TRUCKS 9779 & 9780), 7509 (G.R.S) (TRUCKS 9951 & 9934), AND 7516 (UNION SWITCH AND SIGNAL) (TRUCKS 9847 & 9948). AUTOMATED EQUIPMENT TESTED ON CARS 7659, 7675 & 7686.
SOME CARS CONVERTED TO WORK CARS. REMAINING CARS WERE SCRAPPED.

CAR NUMBERS: 7750-7859
TOTAL: 110 CARS
BUILT BY: AMERICAN CAR FOUNDRY

DATE: 1959-60
AVERAGE COST PER CAR: $107,157

# R26 (IRT)

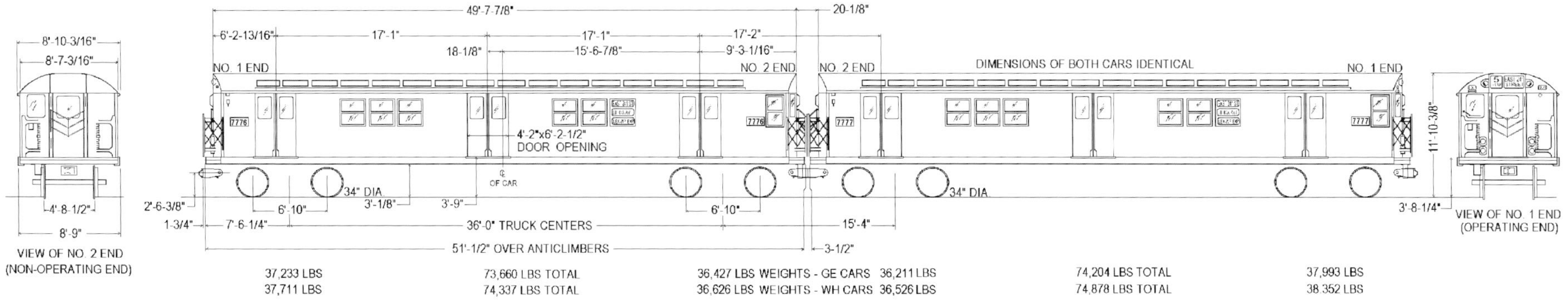

8'-10-3/16"
8'-7-3/16"
4'-8-1/2"
8'-9"
VIEW OF NO. 2 END
(NON-OPERATING END)

49'-7-7/8"
20-1/8"
6'-2-13/16"
17'-1"
18-1/8"
17'-1"
15'-6-7/8"
17'-2"
9'-3-1/16"
NO. 1 END
NO. 2 END
NO. 2 END
DIMENSIONS OF BOTH CARS IDENTICAL
NO. 1 END
4'-2"x6'-2-1/2"
DOOR OPENING
℄ OF CAR
34" DIA
34" DIA
2'-6-3/8"
1-3/4"
6'-10"
3'-1/8"
3'-9"
7'-6-1/4"
36'-0" TRUCK CENTERS
6'-10"
15'-4"
51'-1/2" OVER ANTICLIMBERS
3-1/2"
11'-10-3/8"
3'-8-1/4"
VIEW OF NO. 1 END
(OPERATING END)

37,233 LBS
37,711 LBS

73,660 LBS TOTAL
74,337 LBS TOTAL

36,427 LBS WEIGHTS - GE CARS 36,211 LBS
36,626 LBS WEIGHTS - WH CARS 36,526 LBS

74,204 LBS TOTAL
74,878 LBS TOTAL

37,993 LBS
38,352 LBS

Copyright © 1997 by NYC Transit

## CAR BODY - EQUIPMENT

| | | | |
|---|---|---|---|
| AUXILIARY CIRCUIT PROTECTION | CIRCUIT BREAKERS | HEAT & VENTILATION CONTROL | VAPOR HEATING |
| BRAKE EQUIPMENT TYPE | WABCO, SMEE (DYNAMIC & FRICTION) | HEATER | MIDLAND ROSS |
| BRAKE VALVE | WABCO ME42B | LIGHTING | LUMINATOR FLUORESCENT FIXTURES, 600 VDC PARALLEL CIRCUIT |
| DOOR HANGER & PANEL | O.M. EDWARDS | MASTER CONTROLLER | GE, 17KC76C1; WH, XM379 |
| DOOR OPERATOR MOTOR | MIDLAND ROSS: CENTER, DE 93C; CAB END, DE95E; OPP. CAB, DE94E | POLARITY REVERSING RELAY | WARD LEONARD |
| DRAFT GEAR | WAUGH, WM E4CL | PUBLIC ADDRESS SYSTEM | SOUND SYSTEMS, TRANSISTORIZED |
| ELECTRIC PORTION | WABCO, BL26B | SASH | O.M. EDWARDS |
| FAN | WH, 23" AXIFLO, [4] 356C647G01, ALL CARS | SEAT ARRANGEMENT & CAPACITY | LONGITUDINAL FIBERGLASS [44] |
| FIRE EXTINGUISHER | A. LAFRANCE, 2 LBS DRY POWDER | SEAT MANUFACTURER | AMERICAN SEATING |
| HAND BRAKE | NATIONAL BRAKE, 840XBL | SIGN | HUNTER ILLUMINATED CAR SIGN |
| HANDHOLD | ELLCON, #45. [56] | SIGNAL DEVICE | PNEUPHONIC HORN |
| HEAD, MARKER & TAIL LIGHTS | LOVELL DRESSEL | | |
| | **EVEN- NUMBERED CARS ONLY** | | **ODD-NUMBERED CARS ONLY** |
| BATTERY | EDISON, 24 CELL ,B4H WOOD BOX | AIR COMPRESSOR | WABCO, 2CY3B WITH A3 MOTOR |
| COUPLER | WABCO: #1 END, H2C; #2 END, H2CAR | COUPLER | WABCO: #1 END, H2C; #2 END, H2CAL |
| MOTOR GENERATOR | GE, 5GMG 153LI; WH, YX303B | PROPULSION CONTROL TYPE | GE, MCM, 17KG192B1 GROUP<br>WH, CAM, XC248K GROUP |
| PROPULSION CONTROL TYPE | GE, MCM, 17KG192A1 GROUP<br>WH, CAM, XC248J GROUP | SWITCH PANEL | MIDLAND ROSS: #1 END, 10126; #2 END, 10139 |
| SWITCH PANEL | MIDLAND ROSS: #1 END, 10140; #2 END, 10127 | | |

## TRUCK

| | | | |
|---|---|---|---|
| MANUFACTURER | G.S.C., 9973-10084 & 10125-10192; A.S.C., 10085-10124 | GEAR UNIT | WH WN44CGS (123:17); GE GA25A1 (123:17) |
| TRUCK TYPE & QUANTITY | C.S.F. EQUALIZER BAR MOTOR [220] | JOURNAL BEARING | SKF, 5X9, ROLLER |
| TRUCK NUMBERS | 9973-10080 | SPRING | PERELLI (RUBBER-COVERED) |
| WEIGHT | #1 END: 19,346 LBS; #2 END: 19,030 LBS | TRACTION MOTOR | WH 1447C(100 HP) [2]; GE 1240A4 (100 HP) [2] |
| BRAKE CYLINDER | WABCO, 6X6 UAHT UNIT CYLINDER [2] | TRIP COCK | WABCO, D1A |
| BRAKE RIGGING | A.S.F., SIMPLEX UNIT CYL CLASP BRAKE | | |

MARRIED PAIR CONSIST (TWO CAR UNIT) INTRODUCED WITH THIS CONTRACT.
CARS 7750 - 7803 WITH GE EQUIPMENT AND CARS 7804 - 7859 WITH WH EQUIPMENT.

CAR NUMBERS: 7750 - 7859
TOTAL: 110 CARS
BUILT BY: AMERICAN CAR FOUNDRY
DATE: 1959-60
AVERAGE COST PER CAR: $107,157

OVERHAULED BY: MK
DATE: 1985 - 1987
OVERHAUL COST PER CAR: $187,876

# R26 (IRT) GOH

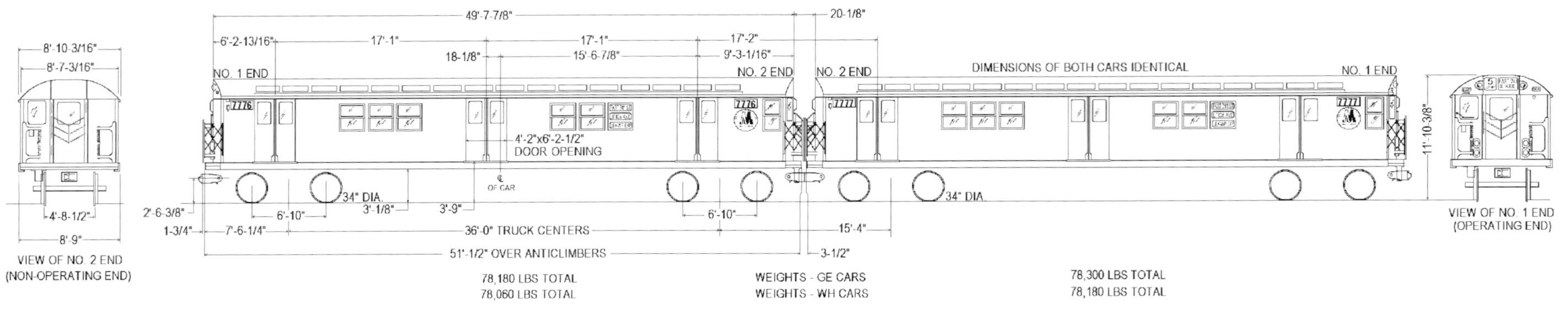

8'-10-3/16"
8'-7-3/16"
4'-8-1/2"
8'-9"
VIEW OF NO. 2 END (NON-OPERATING END)

49'-7-7/8"
20-1/8"
6'-2-13/16"
17'-1"
17'-1"
17'-2"
18-1/8"
15'-6-7/8"
9'-3-1/16"
NO. 1 END
NO. 2 END
NO. 2 END
DIMENSIONS OF BOTH CARS IDENTICAL
NO. 1 END
4'-2"x6'-2-1/2" DOOR OPENING
℄ OF CAR
34" DIA.
34" DIA.
2'-6-3/8"
1-3/4"
6'-10"
3'-1/8"
3'-9"
6'-10"
7'-6-1/4"
36'-0" TRUCK CENTERS
15'-4"
51'-1/2" OVER ANTICLIMBERS
3-1/2"
11'-10-3/8"
VIEW OF NO. 1 END (OPERATING END)

| | 78,180 LBS TOTAL | WEIGHTS - GE CARS | 78,300 LBS TOTAL |
|---|---|---|---|
| | 78,060 LBS TOTAL | WEIGHTS - WH CARS | 78,180 LBS TOTAL |

Copyright © 1997 by NYC Transit

## CAR BODY - EQUIPMENT

| | | | |
|---|---|---|---|
| AIR CONDITIONING | STONE SAFETY, 12 TONS (SEE NOTE 3); 6 TON UNIT [2] | HEATER | MIDLAND ROSS |
| AUXILIARY CIRCUIT PROTECTION | CIRCUIT BREAKERS | LIGHTING | LUMINATOR FLUORESCENT FIXTURES |
| BRAKE EQUIPMENT TYPE | NYAB (DYNAMIC & FRICTION) | | 600 VDC PARALLEL CIRCUIT |
| COUPLER | WABCO: #1 END, H2C; #2 END, LINKBAR | MASTER CONTROLLER | GE, 17KC76C1; WH, XM379 |
| DOOR HANGER | A.B. KING | PROPULSION CONTROL TYPE | GE, SCM, 17KG192AC2 GROUP |
| DOOR OPERATOR MOTOR | WESTCODE | PUBLIC ADDRESS SYSTEM | COMCO |
| DOOR PANEL | O.M. EDWARDS | SASH | O.M. EDWARDS |
| DRAFT GEAR | WAUGH, WM E4CL | SEAT ARRANGEMENT & CAPACITY | LONGITUDINAL [44] |
| ELECTRIC PORTION | #1 END: WABCO, BL26; #2 END: CANNON PLUG | SEAT MANUFACTURER | AMERICAN SEATING |
| HAND BRAKE | NATIONAL BRAKE, 840XBL | SEAT MATERIAL | FIBERGLASS |
| HANDHOLD | ELLCON, #45 [56]/CAR | SIGN | HUNTER ILLUMINATED CAR SIGN |
| HEAD, MARKER & TAIL LIGHTS | LOVELL DRESSEL | SIGNAL DEVICE | PNEUPHONIC HORN |
| HVAC CONTROL | STONE SAFETY | | |
| | **EVEN-NUMBERED CARS ONLY** | | **ODD-NUMBERED CARS ONLY** |
| BATTERY | EDISON, ED80,80 A.H. (METAL BOX) | AIR COMPRESSOR | WABCO, D4 |
| CONVERTER | TOSHIBA, COC001A0 (6 KW) | SWITCH PANEL | MIDLAND ROSS: #1 END, 10126; #2 END, 10139 |
| SWITCH PANEL | MIDLAND ROSS: #1 END, 10140; #2 END, 10127 | | |

## TRUCK

| | | | |
|---|---|---|---|
| MANUFACTURER | GENERAL STEEL CASTING & ADIRONDACK STEEL CASTING | BRAKE RIGGING | A.S.F., SIMPLEX UNIT CYLINDER CLASP BRAKE |
| TYPE & QUANTITY | C.S.F. EQUALIZER BAR MOTOR [220] | GEAR UNIT | WH, WN44CGS; GE, GA25A1 (123:17) |
| TRUCK NUMBERS | SEE NOTE 1 | JOURNAL BEARING | 5X9, ROLLER, SEE NOTE 2 |
| WEIGHT | #1 END: GE 19,580 LBS, WH 19,490 LBS ; #2 END: GE 19,270 LBS, WH 19,240 LBS | TRACTION MOTOR | WH, 1447J (115 HP) [2]; GE, 1257F1 (115 HP) [2] |
| BRAKE CYLINDER | WABCO, 6X6 UAHT [2]/TRUCK | TRIP COCK | WABCO, D1A |

NOTE 1: DUE TO TRUCK OVERHAUL PROGRAM, TRUCK NUMBERS VARIED.
NOTE 2: VARIED VENDORS. CAR 7804 - 7859 WERE ORIGINALLY WH EQUIPPED.
NOTE 3: ONE COMPRESSOR/CONDENSER AND TWO EVAPORATOR UNITS OVERHEAD.

SUBSEQUENT TO GOH, DROP-SASH WINDOWS (EXCEPT TRAIN OPERATOR'S & CONDUCTOR'S) REPLACED WITH LOUVERED WINDOWS ON MANY CARS BY NYCTA.
COUPLERS AT NO. 2 ENDS REPLACED WITH LINK BARS (1991 - 1992).

CARS 7770-7771 USED AS SCHOOL CARS AT CANARSIE YARD (2002).
CARS 7774, 7775 FOR PRESERVATION.
REMAINING CARS WERE SCRAPPED.

CAR NUMBERS: 8020-8249
TOTAL: 230 CARS
BUILT BY: ST. LOUIS CAR

DATE: 1960 - 61
AVERAGE COST PER CAR: $119,227

# R27 (BMT/IND)

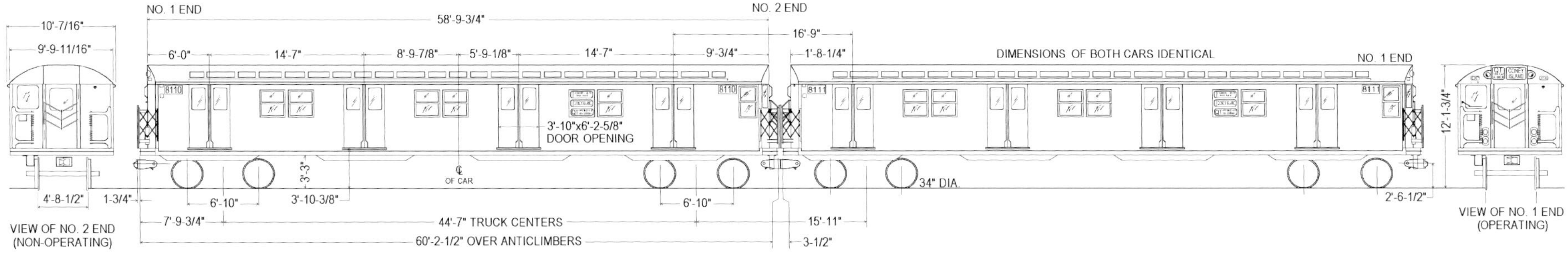

NO. 1 END
NO. 2 END
DIMENSIONS OF BOTH CARS IDENTICAL
NO. 1 END
10'-7/16"
9'-9-11/16"
58'-9-3/4"
6'-0" 14'-7" 8'-9-7/8" 5'-9-1/8" 14'-7" 9'-3/4"
16'-9"
1'-8-1/4"
3'-10"x6'-2-5/8" DOOR OPENING
3'-3"
℄ OF CAR
34" DIA.
12'-1-3/4"
2'-6-1/2"
4'-8-1/2" 1-3/4"
6'-10" 3'-10-3/8" 6'-10"
7'-9-3/4"
44'-7" TRUCK CENTERS
15'-11"
60'-2-1/2" OVER ANTICLIMBERS
3-1/2"
VIEW OF NO. 2 END (NON-OPERATING)
VIEW OF NO. 1 END (OPERATING)

40,361 LBS
39,918 LBS

79,837 LBS TOTAL
80,852 LBS TOTAL

39,476 LBS WEIGHTS - GE CARS 39,476 LBS
40,934 LBS WEIGHTS - WH CARS 40,934 LBS

80,546 LBS TOTAL
81,632 LBS TOTAL

40,665 LBS
40,368 LBS

Copyright © 1997 by NYC Transit

## CAR BODY - EQUIPMENT

| | | | |
|---|---|---|---|
| AUXILIARY CIRCUIT PROTECTION | CIRCUIT BREAKERS | HEATER | CONSOLIDATED |
| BRAKE EQUIPMENT TYPE | WABCO, SMEE (DYNAMIC & FRICTION) | LIGHTING | LUMINATOR FLUORESCENT FIXTURES |
| BRAKE VALVE | WABCO ME42B | | 600 VDC PARALLEL CIRCUIT |
| DOOR HANGER & PANEL | O.M. EDWARDS | MASTER CONTROLLER | GE, 17KC76C1; WH, XM429 |
| DOOR OPERATOR MOTOR | MIDLAND ROSS: CENTER, 500000; CAB END, 500002; OPP. CAB, 500001 | POLARITY REVERSING RELAY | WARD LEONARD |
| DRAFT GEAR | WAUGH, WM E4CL | PUBLIC ADDRESS SYSTEM | RAYMOND ROSEN, TRANSISTORIZED |
| FAN | WH, 23" AXIFLO 356C647G04 [6], ALL CARS | SASH | O.M. EDWARDS |
| FIRE EXTINGUISHER | A. LAFRANCE, 2.5 LBS DRY POWDER | SEAT ARRANGEMENT & CAPACITY | LONGITUDINAL [50] |
| HAND BRAKE | NATIONAL BRAKE, 840XBL | SEAT MANUFACTURER | HEYWOOD WAKEFIELD |
| HAND STRAP | ELLCON, #45 [68]/CAR | SEAT MATERIAL | FIBERGLASS |
| HEAD, MARKER & TAIL LIGHTS | LOVELL DRESSEL | SIGN | TELEWELD |
| HEAT & VENTILATION CONTROL | VAPOR HEATING | SIGNAL DEVICE | PNEUPHONIC HORN |
| | **EVEN- NUMBERED CARS ONLY** | | **ODD-NUMBERED CARS ONLY** |
| BATTERY | EDISON, (24 CELL) TYPE B4H | AIR COMPRESSOR | WABCO, 2CY3B WITH A3 MOTOR |
| COUPLER | WABCO: #1 END, H2C; #2 END, H2CAR | COUPLER | WABCO: #1 END, H2C; #2 END, H2CAL |
| MOTOR GENERATOR | GE, 5GMG 153LI; WH, YX304B | PROPULSION CONTROL TYPE | GE, MCM,17KG192B1 GROUP |
| PROPULSION CONTROL TYPE | GE, MCM, 17KG192A1 GROUP | | WH, CAM, XC248Q GROUP |
| | WH, CAM, XC248P GROUP | SWITCH PANEL | MIDLAND ROSS: #1 END, 10140; #2 END, 10127 |
| SWITCH PANEL | MIDLAND ROSS: #1 END, 10140; #2 END, 10127 | | |

## TRUCK

| | | | | |
|---|---|---|---|---|
| MANUFACTURER | | G.S.C., 10393-10832; A.S.C., 10833-10852 | BRAKE RIGGING | A.S.F., SIMPLEX UNIT CYLINDER CLASP BRAKE |
| TRUCK TYPE & QUANTITY | | C.S.F. EQUALIZER BAR MOTOR [460] | GEAR UNIT | WH, WN44CGS; GE, GA25A2 (123:17) |
| TRUCK NUMBERS | | WH: 10393-10624; GE: 10625-10852 | JOURNAL BEARING | HYATT, 5X9 ROLLER |
| WEIGHT | #1 END | WH: 20,148 LBS; GE: 19,451 LBS | SHOCK ABSORBER | HOUDAILLE |
| | #2 END | WH: 19,904 LBS; GE: 19,199 LBS | TRACTION MOTOR | WH, 1447C (100 HP) [2]; GE, 1240A4 (100 HP) [2] |
| BRAKE CYLINDER | | WABCO, 7X6 UAHT UNIT CYLINDER, [2]/TRUCK | TRIP COCK | WABCO, D1 |

CARS 8020 - 8135 WITH WH EQUIPMENT AND CARS 8136 - 8249 WITH GE EQUIPMENT.

CAR 8145 CONVERTED TO FIELD OFFICE (36 ST YARD) AND IS CURRENTLY A SCHOOL CAR AT PITKIN YARD (2003). ALL OTHER CARS SCRAPPED (1988 - 1989).

CAR NUMBERS: 7860-7959
TOTAL: 100 CARS
BUILT BY: AMERICAN CAR FOUNDRY

DATE: 1960-61
AVERAGE COST PER CAR: $114,495

# R28 (IRT)

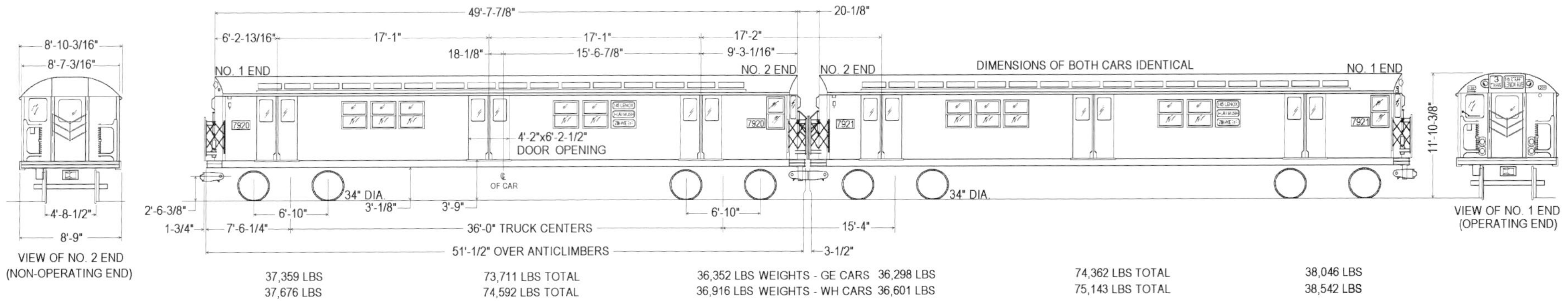

8'-10-3/16"
8'-7-3/16"
4'-8-1/2"
8'-9"
VIEW OF NO. 2 END
(NON-OPERATING END)

49'-7-7/8"
20-1/8"
6'-2-13/16"
17'-1"
18-1/8"
17'-1"
15'-6-7/8"
17'-2"
9'-3-1/16"
NO. 1 END
NO. 2 END
NO. 2 END
DIMENSIONS OF BOTH CARS IDENTICAL
NO. 1 END
4'-2"x6'-2-1/2"
DOOR OPENING
℄ OF CAR
34" DIA
34" DIA.
2'-6-3/8"
1-3/4"
7'-6-1/4"
6'-10"
3'-1/8"
3'-9"
36'-0" TRUCK CENTERS
6'-10"
15'-4"
51'-1/2" OVER ANTICLIMBERS
3-1/2"
11'-10-3/8"
VIEW OF NO. 1 END
(OPERATING END)

37,359 LBS
37,676 LBS

73,711 LBS TOTAL
74,592 LBS TOTAL

36,352 LBS WEIGHTS - GE CARS 36,298 LBS
36,916 LBS WEIGHTS - WH CARS 36,601 LBS

74,362 LBS TOTAL
75,143 LBS TOTAL

38,046 LBS
38,542 LBS

Copyright © 1997 by NYC Transit

## CAR BODY - EQUIPMENT

| | | | |
|---|---|---|---|
| AUXILIARY CIRCUIT PROTECTION | CIRCUIT BREAKERS | HEAT & VENTILATION CONTROL | VAPOR HEATING |
| BRAKE EQUIPMENT TYPE | WABCO, SMEE (DYNAMIC & FRICTION) | HEATER | MIDLAND ROSS |
| BRAKE VALVE | WABCO ME42B | LIGHTING | LUMINATOR, FLUORESCENT FIXTURES, 600 VDC PARALLEL CIRCUIT |
| DOOR HANGER | OTIS ELEVATOR | MASTER CONTROLLER | GE, 17KC76C1; WH, XM479 |
| DOOR OPERATOR MOTOR | MIDLAND ROSS: CENTER, 500000; CAB END, 500002; OPP. CAB, 500001 | POLARITY REVERSING RELAY | WARD LEONARD |
| DOOR PANEL | MORTON | PUBLIC ADDRESS SYSTEM | SOUND SYSTEMS, TRANSISTORIZED |
| DRAFT GEAR | WAUGH, WM E4CL | SASH | ADAMS & WESTLAKE |
| ELECTRIC PORTION | WABCO, BL26B | SEAT ARRANGEMENT & CAPACITY | LONGITUDINAL [44] |
| FAN | WH, 23" AXIFLO, 356C 647G03 [4], ALL CARS | SEAT MANUFACTURER | AMERICAN SEATING |
| FIRE EXTINGUISHER | A. LAFRANCE, 2.5 LBS DRY POWDER | SEAT MATERIAL | FIBERGLASS |
| HAND BRAKE | NATIONAL BRAKE, 840XBL | SIGN | HUNTER ILLUMINATED CAR SIGN |
| HANDHOLD | ELLCON, #45 [56] | SIGNAL DEVICE | PNEUPHONIC HORN |
| HEAD, MARKER & TAIL LIGHTS | LOVELL DRESSEL | | |
| | **EVEN-NUMBERED CARS ONLY** | | **ODD-NUMBERED CARS ONLY** |
| BATTERY(WOOD BOX) | EDISON, 24 CELL, B4H | COMPRESSOR UNIT | WABCO, 2CY3B WITH A3 MOTOR |
| COUPLERS | WABCO: #1 END, H2C; #2 END, H2CAR | COUPLERS | WABCO: #1 END, H2C; #2 END, H2CAL |
| MOTOR GENERATOR | GE, 5GMG153LI; WH, YX304B | PROPULSION CONTROL TYPE | GE, MCM, 17KG192B1 GROUP |
| PROPULSION CONTROL TYPE | GE, MCM, 17KG192A1 GROUP | | WH, CAM, XC248S GROUP |
| | WH, CAM, XC248R GROUP | SWITCH PANEL | MIDLAND ROSS: #1 END, 10140; #2 END, 10127 |
| SWITCH PANEL | MIDLAND ROSS: #1 END, 10140; #2 END, 10127 | | |

## TRUCK

| | | | | |
|---|---|---|---|---|
| MANUFACTURER | | G.S.C., 10193-10222, 10233-10392; L.F.M.10233-10232 | BRAKE RIGGING | A.S.F., SIMPLEX UNIT CYLINDER CLASP BRAKE |
| TRUCK TYPE & QUANTITY | | C.S.F. EQUALIZER BAR MOTOR [200] | GEAR UNIT | WH WN 44CGS; GE GA25A1 (123:17) |
| TRUCK NUMBERS | | 10193-10292, 10293-10392 | JOURNAL BEARING | SKF, 5X9 ROLLER |
| WEIGHT | #1 END | WH: 20,072 LBS; GE: 19,325 LBS | SHOCK ABSORBER | HOUDAILLE |
| | #2 END | WH: 19,802 LBS; GE: 19,185 LBS | TRACTION MOTOR | WH 1447C (100 HP) [2]; GE 1240A4 (100 HP) [2] |
| BRAKE CYLINDER | | WABCO, 6X6 UAHT [ 2]/TRUCK | TRIP COCK | WABCO, D1A |

CARS 7860 - 7909 WITH WH EQUIPMENT AND CARS 7910 - 7959 WITH GE EQUIPMENT.
CARS 7900-7909 HAD HONEYCOMB UNDERLAYMENT FLOOR.
REMOVED FLOOR HONEYCOMB UNDERLAYMENT AND REPLACED WITH CONVENTIONAL FLOOR BY ACF (1961-62).

CAR NUMBERS: 7860 - 7959
TOTAL: 100 CARS
BUILT BY: AMERICAN CAR FOUNDRY
DATE: 1960 - 61
AVERAGE COST PER CAR: $114,495

OVERHAULED BY: MK
DATE: 1985 - 87
OVERHAUL COST PER CAR: $172,000

# R28 (IRT) GOH

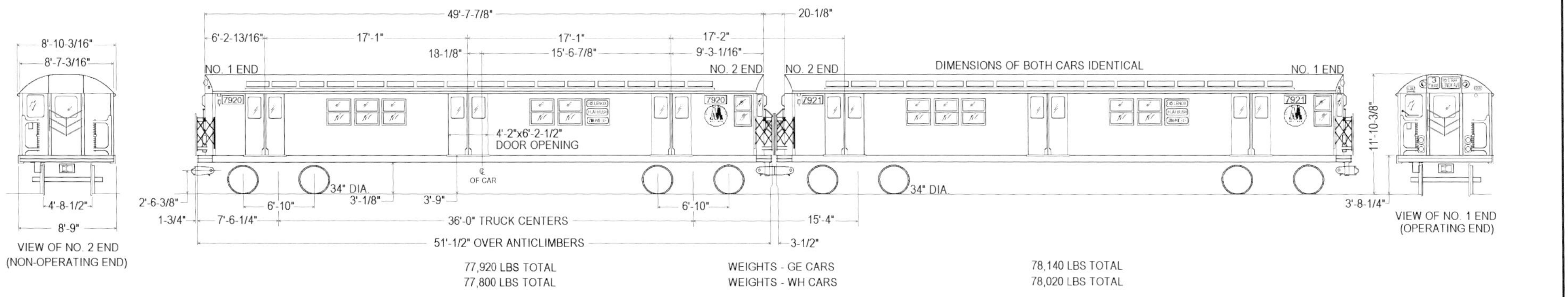

8'-10-3/16"
8'-7-3/16"
4'-8-1/2"
8'-9"
VIEW OF NO. 2 END
(NON-OPERATING END)

49'-7-7/8"
20-1/8"
6'-2-13/16"
17'-1"
17'-1"
17'-2"
18-1/8"
15'-6-7/8"
9'-3-1/16"
NO. 1 END
NO. 2 END
NO. 2 END
DIMENSIONS OF BOTH CARS IDENTICAL
NO. 1 END
4'-2"x6'-2-1/2"
DOOR OPENING
℄ OF CAR
34" DIA
34" DIA
2'-6-3/8"
1-3/4"
7'-6-1/4"
6'-10"
3'-1/8"
3'-9"
6'-10"
36'-0" TRUCK CENTERS
15'-4"
51'-1/2" OVER ANTICLIMBERS
3-1/2"
11'-10-3/8"
3'-8-1/4"
VIEW OF NO. 1 END
(OPERATING END)

| | 77,920 LBS TOTAL | 78,140 LBS TOTAL |
|---|---|---|
| WEIGHTS - GE CARS | 77,920 LBS TOTAL | 78,140 LBS TOTAL |
| WEIGHTS - WH CARS | 77,800 LBS TOTAL | 78,020 LBS TOTAL |

Copyright © 1997 by NYC Transit

## CAR BODY - EQUIPMENT

| | | | |
|---|---|---|---|
| AIR CONDITIONING | STONE SAFETY, 12 TONS (SEE NOTE 3) | HEATER | MIDLAND ROSS |
| AUXILIARY CIRCUIT PROTECTION | CIRCUIT BREAKERS | HVAC CONTROL | STONE SAFETY |
| BRAKE EQUIPMENT TYPE | NYAB (DYNAMIC & FRICTION) | LIGHTING | LUMINATOR FLUORESCENT FIXTURES |
| COUPLER | #1 END: WABCO, H2C; #2 END: WAUGH, LINK BAR | MASTER CONTROLLER | GE, K17KC76C1; WH, XM479 |
| DOOR HANGER | A.B. KING | PROPULSION CONTROL TYPE | GE, SCM, 17KG192AC2 GROUP |
| DOOR OPERATOR MOTOR | VAPOR | PUBLIC ADDRESS SYSTEM | COMCO |
| DOOR PANEL | MORTON | SASH | ADAMS WESTLAKE |
| DRAFT GEAR | WAUGH EQUIPMENT, WME4CL | SEAT ARRANGEMENT & CAPACITY | LONGITUDINAL [44] |
| ELECTRIC PORTION | #1 END: WABCO, BL26B; #2 END: CANNON PLUGS | SEAT MANUFACTURER | AMERICAN SEATING |
| HAND BRAKE | ELLCON NATIONAL, 840XBL | SIGN | HUNTER, ILLUMINATED CAR SIGN |
| HANDHOLD | ELLCON NATIONAL | SIGNAL DEVICE | PNEUPHONIC HORN |
| HEAD, MARKER & TAIL LIGHTS | LOVELL DRESSER | | |
| **EVEN- NUMBERED CARS ONLY** | | **ODD-NUMBERED CARS ONLY** | |
| BATTERY | EDISON, ED80 (80 AH) | AIR COMPRESSOR | WABCO, D4 |
| CONVERTER | TOSHIBA, COV001AO (6 KW) | SWITCH PANEL | MIDLAND ROSS: #1 END, 10126; #2 END, 10139 |
| SWITCH PANEL | MIDLAND ROSS: #1 END, 10140; #2 END, 10127 | | |

## TRUCK

| | | | | |
|---|---|---|---|---|
| MANUFACTURER | | GENERAL STEEL CASTING & LOCOMOTIVE FINISH MATERIAL | BRAKE RIGGING | A.S.F., SIMPLEX UNIT CYLINDER CLASP BRAKE |
| TYPE & QUANTITY | | C.S.F. EQUALIZER BAR MOTOR [200] | GEAR UNIT | GE, GA25A1; WH, WN44CGS (123:17) |
| TRUCK NUMBERS | | SEE NOTE 1 | JOURNAL BEARING | 5X9, AP ROLLER, SEE NOTE 2 |
| WEIGHT | #1 END | GE, 19,580 LBS; WH, 19,490 LBS | SHOCK ABSORBER | MONROE |
| | #2 END | GE, 19,270 LBS; WH, 19,240 LBS | TRACTION MOTOR | GE, 1257F1 (115 HP) [2]; WH, 1447J /1447JR (115 HP) [2] |
| BRAKE CYLINDER | | WABCO, 6X6 UAHT [2]/TRUCK | TRIP COCK | WABCO, D1A |

NOTE 1: DUE TO TRUCK OVERHAUL PROGRAM, TRUCK NUMBERS VARIED.
NOTE 2: VARIOUS VENDORS.
NOTE 3: ONE COMPRESSOR/CONDENSER UNDER CAR AND TWO EVAPORATOR UNITS OVERHEAD.

CARS 7860 - 7909 WERE ORIGINALLY WH EQUIPPED.
SUBSEQUENT TO GOHDROP SASH WINDOWS (EXCEPT TRAIN OPERATORS AND CONDUCTOR'S) REPLACED WITH LOUVERED WINDOWS ON MANY CARS BY NYCTA.
COUPLERS AT NO. 2 ENDS REPLACED WITH LINK BARS (1991 - 92).

CARS 7924, 7925 FOR PRESERVATION.
CARS 7926, 7927 TO ILLINOIS RAILWAY MUSEUM (07/11/2004).
REMAINING CARS WERE SCRAPPED.

CAR NUMBERS: 8570 - 8685 & 8688 - 8803
TOTAL: 232 CARS
BUILT BY: ST. LOUIS CAR

DATE: 1962
AVERAGE COST PER CAR: $110,842

# R29 (IRT)

STANDARD TRUCK
(CLASP BRAKE)

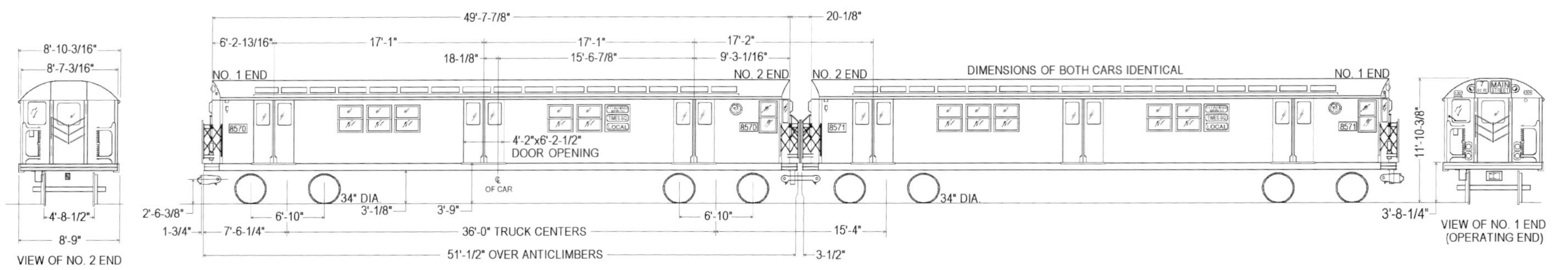

8'-10-3/16" 8'-7-3/16" 4'-8-1/2" 8'-9"
VIEW OF NO. 2 END (NON-OPERATING END)

49'-7-7/8" 20-1/8" 6'-2-13/16" 17'-1" 17'-1" 17'-2" 18-1/8" 15'-6-7/8" 9'-3-1/16"
NO. 1 END NO. 2 END NO. 2 END DIMENSIONS OF BOTH CARS IDENTICAL NO. 1 END
4'-2"x6'-2-1/2" DOOR OPENING
℄ OF CAR
34" DIA 34" DIA.
2'-6-3/8" 1-3/4" 6'-10" 3'-1/8" 3'-9" 6'-10" 7'-6-1/4" 36'-0" TRUCK CENTERS 15'-4" 51'-1/2" OVER ANTICLIMBERS 3-1/2"
11'-10-3/8" 3'-8-1/4"
VIEW OF NO. 1 END (OPERATING END)

36,980 LBS / 37,260 LBS — 72,000 LBS TOTAL / 72,700 LBS TOTAL — 35,120 LBS WEIGHTS - GE CARS 35,480 LBS / 35,440 LBS WEIGHTS - WH CARS 35,710 LBS — 72,840 LBS TOTAL / 73,580 LBS TOTAL — 37,360 LBS / 37,870 LBS

Copyright © 1997 by NYC Transit

## CAR BODY - EQUIPMENT

| | | | |
|---|---|---|---|
| AUXILIARY CIRCUIT PROTECTION | CIRCUIT BREAKERS | HEAT & VENTILATION CONTROL | VAPOR HEATING |
| BRAKE EQUIPMENT TYPE | WABCO, SMEE (FRICTION & DYNAMIC) | HEATER | MIDLAND ROSS |
| BRAKE VALVE | WABCO, ME42B | LIGHTING | LUMINATOR FLUORESCENT FIXTURES, |
| COUPLER | #1 END: WABCO, H2C; #2 END: WAUGH, LINKBAR 4F7579 | | 600 VDC PARALLEL CIRCUIT |
| DOOR OPERATOR MOTOR | MIDLAND ROSS: CENTER, 500003; CAB END, 500005; OPP. CAB, 500004 | MASTER CONTROLLER | GE, 17KC76C2; WH, XM579 |
| DOOR PANEL & HANGER | O.M. EDWARDS | POLARITY REVERSING RELAY | RELAY: VAPOR HEATING; TIMER: HORNE PRODUCTS |
| DRAFT GEAR | WAUGH, WM E4CL | PUBLIC ADDRESS SYSTEM | INDUSTRON |
| ELECTRIC PORTION | #1 END: WABCO, BL25B; #2 END: CANNON, PLUG | SASH | O.M. EDWARDS |
| FAN | WH, 23" AXIFLO, 356C 647G03 [4] ALL CARS | SEAT ARRANGEMENT & CAPACITY | LONGITUDINAL [44] |
| FIRE EXTINGUISHER | A. LAFRANCE 2.5 LBS DRY POWDER | SEAT MANUFACTURER | HEYWOOD WAKEFIELD |
| HAND BRAKE | NATIONAL BRAKE, 840XBL | SEAT MATERIAL | FIBERGLASS |
| HANDHOLD | ELLCON, #45 [56]/CAR | SIGN | HUNTER ILLUMINATED CAR SIGN |
| HEAD, MARKER & TAIL LIGHTS | LOVELL DRESSEL | SIGNAL DEVICE | PNEUPHONIC HORN |
| | EVEN- NUMBERED CARS ONLY | | ODD-NUMBERED CARS ONLY |
| BATTERY | EDISON, 24 CELL, B4H, WOOD BOX | AIR COMPRESSOR | WABCO, 2CY3B WITH A3 MOTOR |
| MOTOR GENERATOR | GE, 5GMG153LI; WH, YX304E | PROPULSION CONTROL TYPE | GE, SCM, 17KG192H3/H4 GROUP |
| PROPULSION CONTROL TYPE | GE, SCM, 17KG192F3/F4 GROUP | | WH, CAM, XCA248B GROUP |
| | WH, CAM, XCA248A GROUP | SWITCH PANEL | MIDLAND ROSS: #1 END,500007; #2 END, 500008 |
| SWITCH PANEL | MIDLAND ROSS: #1 END, 500009; #2 END, 500010 | | |

## TRUCK

| | | | |
|---|---|---|---|
| MANUFACTURER | GENERAL STEEL CASTING | JOURNAL BEARING (5 X 9 ROLLER) | TIMKEN - 11927-11958 |
| TYPE & QUANTITY | C.S.F. EQUALIZER BAR MOTOR [472] | | SKF SPECIAL 11885, 11886, 11891-11898, |
| TRUCK NUMBERS | 11493-11724, 11729-11960 | | & 11901-11916, 11919-11920, 11923-11926 |
| WEIGHT | #1 TRUCK: 19,762 LBS; #2 TRUCK: 19,521 LBS | | SKF, STANDARD ON REMAINDER |
| BRAKE CYLINDER | WABCO, 6X6 UAHT UNIT CYLINDERS [2]/TRUCK | SHOCK ABSORBER | HOUDAILLE |
| BRAKE RIGGING | A.S.F. SIMPLEX UNIT CYLINDER CLASP BRAKE | TRACTION MOTOR | GE, 1240A5 (100 HP) [2]; WH, 1447C (100 HP) [2] |
| GEAR UNIT | GE, GA25A2; WH, WN 44CGS (123:17) | TRIP COCK | WABCO, D1A |

CARS EQUIPPED WITH STATIC LIMIT RELAY. CAR 8684 EQUIPPED WITH CONTROL GROUP XC 248C AND CAR 8685 EQUIPPED WITH CONTROL GROUP XC248E.
CARS 8689-8803 (ODD NUMBERS) HAD COMPRESSOR CIRCUIT BREAKER. NO. 1 SWITCH PANEL IS M.R. 500007A
CARS 8570 - 8685 WITH WH EQUIPMENT AND CARS 8688 - 8803 WITH GE EQUIPMENT.
CARS 8690 - 8803 HAVE PLYMETAL SUB-FLOOR.
TRUCK 11606 HAD GE GEAR UNIT, TRUCK 11724 HAD GE MOTORS & GEAR UNIT.

CAR NUMBERS: 8686, 8687, 8804, 8805
TOTAL: 4 CARS
BUILT BY: ST. LOUIS CAR

DATE: 1962
AVERAGE COST PER CAR: $110,842

# R29 (IRT)

G70 TRUCK
(PACKAGE BRAKE)

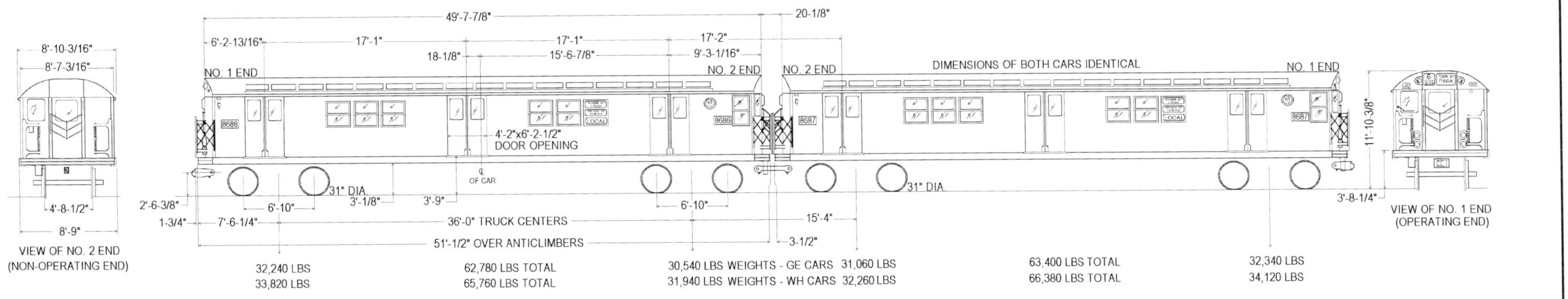

8'-10-3/16" 8'-7-3/16" 4'-8-1/2" 8'-9"
VIEW OF NO. 2 END
(NON-OPERATING END)

49'-7-7/8" 20-1/8" 6'-2-13/16" 17'-1" 18-1/8" 17'-1" 15'-6-7/8" 17'-2" 9'-3-1/16"
NO. 1 END NO. 2 END NO. 2 END DIMENSIONS OF BOTH CARS IDENTICAL NO. 1 END
4'-2"x6'-2-1/2" DOOR OPENING
℄ OF CAR
31" DIA. 3'-1/8" 3'-9" 31" DIA.
2'-6-3/8" 1-3/4" 6'-10" 7'-6-1/4" 36'-0" TRUCK CENTERS 6'-10" 15'-4" 51'-1/2" OVER ANTICLIMBERS 3-1/2"
11'-10-3/8" 3'-8-1/4"

32,240 LBS
33,820 LBS

62,780 LBS TOTAL
65,760 LBS TOTAL

30,540 LBS WEIGHTS - GE CARS 31,060 LBS
31,940 LBS WEIGHTS - WH CARS 32,260 LBS

63,400 LBS TOTAL
66,380 LBS TOTAL

32,340 LBS
34,120 LBS

VIEW OF NO. 1 END
(OPERATING END)

Copyright © 1997 by NYC Transit

## CAR BODY - EQUIPMENT

| | | | |
|---|---|---|---|
| AUXILIARY CIRCUIT PROTECTION | CIRCUIT BREAKERS | HEAT & VENTILATION CONTROL | VAPOR HEATING |
| BRAKE EQUIPMENT TYPE | WABCO, SMEE (DYNAMIC & FRICTION) | HEATER | MIDLAND ROSS |
| BRAKE VALVE | WABCO, ME42B | LIGHTING | LUMINATOR FLUORESCENT FIXTURES, 600 VDC PARALLEL CIRCUIT |
| COUPLING | #1 END: WABCO, H2C; #2 END: WAUGH, LINKBAR, 4F7579 | MASTER CONTROLLER | GE, 17KC76C2; WH, XM579 |
| DOOR HANGER & PANEL | O.M. EDWARDS | POLARITY REVERSING RELAY | VAPOR HEATING, RELAY; HORNE PRODUCTS, TIMER |
| DOOR OPERATOR MOTOR | MIDLAND ROSS: CENTER, 500003; CAB END, 500005; OPP. CAB, 500004 | PUBLIC ADDRESS SYSTEM | INDUSTRON |
| DRAFT GEAR | WAUGH, WM E4CL | SASH | O.M. EDWARDS |
| ELECTRIC PORTION | WABCO, BL26B | SEAT ARRANGEMENT & CAPACITY | LONGITUDINAL [44] |
| FAN | WH, 356C647G03, 23" AXIFLO, [4], ALL CARS | SEAT MANUFACTURER | HEYWOOD WAKEFIELD |
| FIRE EXTINGUISHER | A. LAFRANCE 2.5 LBS DRY POWDER | SEAT MATERIAL | FIBERGLASS |
| HAND BRAKE | NATIONAL BRAKE, 840XBL | SIGN | HUNTER ILLUMINATED CAR SIGN |
| HANDHOLD | ELLCON, #45 [ 56] | SIGNAL DEVICE | PNEUPHONIC HORN |
| HEAD, MARKER & TAIL LIGHTS | LOVELL DRESSEL | | |
| EVEN-NUMBERED CARS ONLY | | ODD-NUMBERED CARS ONLY | |
| BATTERY | EDISON, 24 CELL, B4H, WOOD BOX | COMPRESSOR UNIT | WABCO, 2CY3B WITH A3 MOTOR |
| MOTOR GENERATOR | WH, YX304E; GE, 5GMG 153LI | PROPULSION CONTROL TYPE | GE, SCM, 17KG192H3 GROUP |
| PROPULSION CONTROL TYPE | GE, SCM, 17KG192F3 GROUP | | WH, CAM, XCA248H GROUP |
| | WH, CAM, XCA248G GROUP | SWITCH PANEL | MIDLAND ROSS: #1 END, 500007; #2 END, 500008 |
| SWITCH PANEL | MIDLAND ROSS: #1 END, 500009; #2 END, 500010 | | |

## TRUCK

| | | | | |
|---|---|---|---|---|
| MANUFACTURER | | GENERAL STEEL CASTING | BRAKE RIGGING | SEE NOTE 1 |
| TYPE & QUANTITY | | CAST STEEL INSIDE FRAME MOTOR [8] | BRAKE SHOE | COMPOSITION/IRON |
| TRUCK NUMBERS | | 11725-28; 11961-64 | GEAR UNIT | WH, GU1001; GE, GA53A1 (123:17) |
| WEIGHT | #1 TRUCK | 11725-26, 14,240 LBS; 11727-28, 14,160 LBS; | JOURNAL BEARING | SKF, 5X9 ROLLER |
| | | 11961-62, 14,690 LBS; 11963-64, 14,610 LBS | SHOCK ABSORBER | HOUDAILLE |
| | #2 TRUCK | 11725-26, 14,060 LBS; 11727-28, 13,920 LBS; | TRACTION MOTOR | WH, 1455G (100 HP) [2]; GE, 1253A1 (100 HP) [2] |
| | | 11961-62, 14,510 LBS; 11963-64, 14,370 LBS | TRIP COCK | WABCO, D1A |

NOTE 1: CARS 8804 & 8805 HAVE PACKAGE BRAKE WITH CAST IRON BRAKE SHOES.
CARS 8686 & 8687 HAVE PACKAGE BRAKE WITH COMPOSITION BRAKE SHOES.

CARS 8686 & 8687 WITH WH EQUIPMENT AND CARS 8804 & 8805 WITH GE EQUIPMENT.

CAR NUMBERS: 8570 - 8805
TOTAL: 236 CARS
BUILT BY: ST. LOUIS CAR
DATE: 1962
AVERAGE COST PER CAR: $ 110,842

OVERHAULED BY: MK
DATE: 1985 - 87
OVERHAUL COST PER CAR: $306,000

# R29 (IRT) GOH

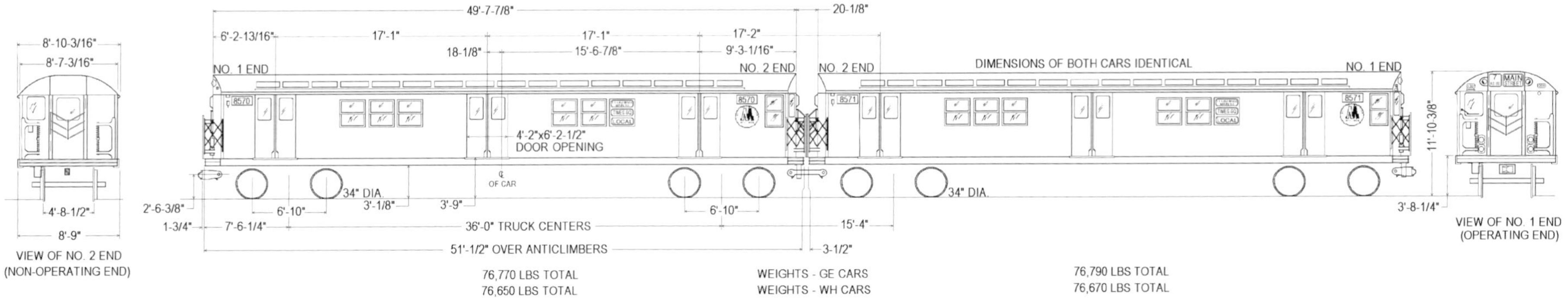

8'-10-3/16"
8'-7-3/16"
4'-8-1/2"
8'-9"
VIEW OF NO. 2 END (NON-OPERATING END)

49'-7-7/8"
20-1/8"
6'-2-13/16"
17'-1"
18-1/8"
17'-1"
15'-6-7/8"
17'-2"
9'-3-1/16"
NO. 1 END
NO. 2 END
NO. 2 END
DIMENSIONS OF BOTH CARS IDENTICAL
NO. 1 END
4'-2"x6'-2-1/2" DOOR OPENING
OF CAR
34" DIA.
34" DIA.
2'-6-3/8"
1-3/4"
7'-6-1/4"
6'-10"
3'-1/8"
3'-9"
6'-10"
36'-0" TRUCK CENTERS
15'-4"
51'-1/2" OVER ANTICLIMBERS
3-1/2"
11'-10-3/8"
3'-8-1/4"
VIEW OF NO. 1 END (OPERATING END)

| |  | |
|---|---|---|
| 76,770 LBS TOTAL | WEIGHTS - GE CARS | 76,790 LBS TOTAL |
| 76,650 LBS TOTAL | WEIGHTS - WH CARS | 76,670 LBS TOTAL |

Copyright © 1997 by NYC Transit

## CAR BODY - EQUIPMENT

| | | | |
|---|---|---|---|
| AIR CONDITIONING | STONE SAFETY, 12 TONS (SEE NOTE 3) | HVAC CONTROL | STONE SAFETY |
| AUXILIARY CIRCUIT PROTECTION | CIRCUIT BREAKERS | LIGHTING | LUMINATOR, FLUORESCENT FIXTURES |
| BRAKE EQUIPMENT TYPE | NYAB WITH WABCO BRAKE VALVE (DYNAMIC & FRICTION) | MASTER CONTROLLER | GE, 17KC76C2; WH, XM579 |
| COUPLER | #1 END: WABCO, H2C; #2 END: WAUGH, LINK BAR # 4F7579 | PROPULSION CONTROL TYPE | GE, SCM, 17KG192AEI GROUP |
| DOOR HANGER | A. B. KING | | WH, CAM, XCA248Y GROUP |
| DOOR OPERATOR MOTOR | VAPOR | PUBLIC ADDRESS SYSTEM | COMCO |
| DOOR PANEL | O.M. EDWARDS | SASH | O.M. EDWARDS |
| DRAFT GEAR | WAUGH, WME4CL | SEAT ARRANGEMENT & CAPACITY | LONGITUDINAL [44] |
| ELECTRIC PORTION | #1 END: WABCO, BL26B; #2 END: CANNON PLUG | SEAT MANUFACTURER | HEYWOOD WAKEFIELD |
| HAND BRAKE | ELLCON NATIONAL, 840XBL | SEAT MATERIAL | FIBERGLASS |
| HANDHOLD | ELLCON NATIONAL, # 45 [56] | SIGN | HUNTER ILLUMINATED CAR SIGN |
| HEAD, MARKER & TAIL LIGHTS | LOVELL DRESSEL | SIGNAL DEVICE | PNEUPHONIC HORN |
| HEATER | MIDLAND ROSS | | |
| | **EVEN-NUMBERED CARS ONLY** | | **ODD-NUMBERED CARS ONLY** |
| BATTERY | EDISON, ED80 (80 AH) | AIR COMPRESSOR | WABCO, D4 |
| CONVERTER | TOSHIBA, COV001A0, 6 KW | SWITCH PANEL | MIDLAND ROSS: #1 END, 500007; #2 END, 500008 |
| SWITCH PANEL | MIDLAND ROSS: #1 END, 500009; #2 END, 500010 | | |

## TRUCK

| | | | | |
|---|---|---|---|---|
| MANUFACTURER | | GENERAL STEEL CASTING | BRAKE RIGGING | ASF, SIMPLEX UNIT CYLINDER CLASP BRAKE |
| TYPE & QUANTITY | | C.S.F. EQUALIZER BAR MOTOR [472] | GEAR UNIT | GE, GA25A2; WH, WN44CGS (123:17) |
| TRUCK NUMBERS | | (SEE NOTE 1) | JOURNAL BEARING | 5X9, AP ROLLER (SEE NOTE 2) |
| WEIGHT | #1 TRUCK | GE, 19,580 LBS; WH, 19,490 LBS | SHOCK ABSORBER | MONROE |
| | #2 TRUCK | GE, 19,270 LBS; WH, 19,240 LBS | TRACTION MOTOR | GE, 1257E1 (115 HP) [2]; WH, 1447J/1447JR (115 HP) [2] |
| BRAKE CYLINDER | | WABCO, 6X6 UAHT [2] | TRIP COCK | WABCO HOUSING & NYAB COMPONENTS, D1A |

NOTE 1: DUE TO TRUCK OVERHAUL PROGRAM, TRUCK NUMBERS VARIED.
NOTE 2: VARIOUS VENDORS.
NOTE 3: ONE COMPRESSOR/CONDENSER AND TWO EVAPORATOR UNITS OVERHEAD. WH CARS 8570 - 8687, GE CARS 8688 - 8805.

CARS 8678, 8679 FOR PRESERVATION.
REMAINING CARS WERE SCRAPPED.

CARS 8690-8803 HAVE PLYMETAL SUB-FLOOR. CARS 8686, 8687, 8804 & 8805, FORMERLY G70 TRUCKS, CONVERTED TO STANDARD TRUCKS AND CLASP BRAKES (1970).
TRUCK 11606 HAD GE GEAR UNITS, TRUCK 11724 HAD GE MOTORS & GEAR UNITS.
CAR 8684 EQUIPPED WITH STATIC LIMIT RELAY CONTROL GROUP XC 248C & CAR 8685 EQUIPPED WITH CONTROL GROUP XC248E.
CARS 8689-8803 (ODD NUMBERS) HAD COMPRESSOR CIRCUIT BREAKER. SWITCH PANEL IS M.R. 500007A.
SUBSEQUENT TO GOH, DROP SASH WINDOWS (EXCEPT TRAIN OPERATOR'S & CONDUCTOR'S) REPLACED WITH LOUVERED WINDOWS ON MANY CARS BY NYCTA.

CAR NUMBERS: 8250-8351 & 8412-8569
TOTAL: 260 CARS
BUILT BY: ST. LOUIS CAR

DATE: 1961 - 62
AVERAGE COST PER CAR: $121,663

# R30 (BMT/IND)

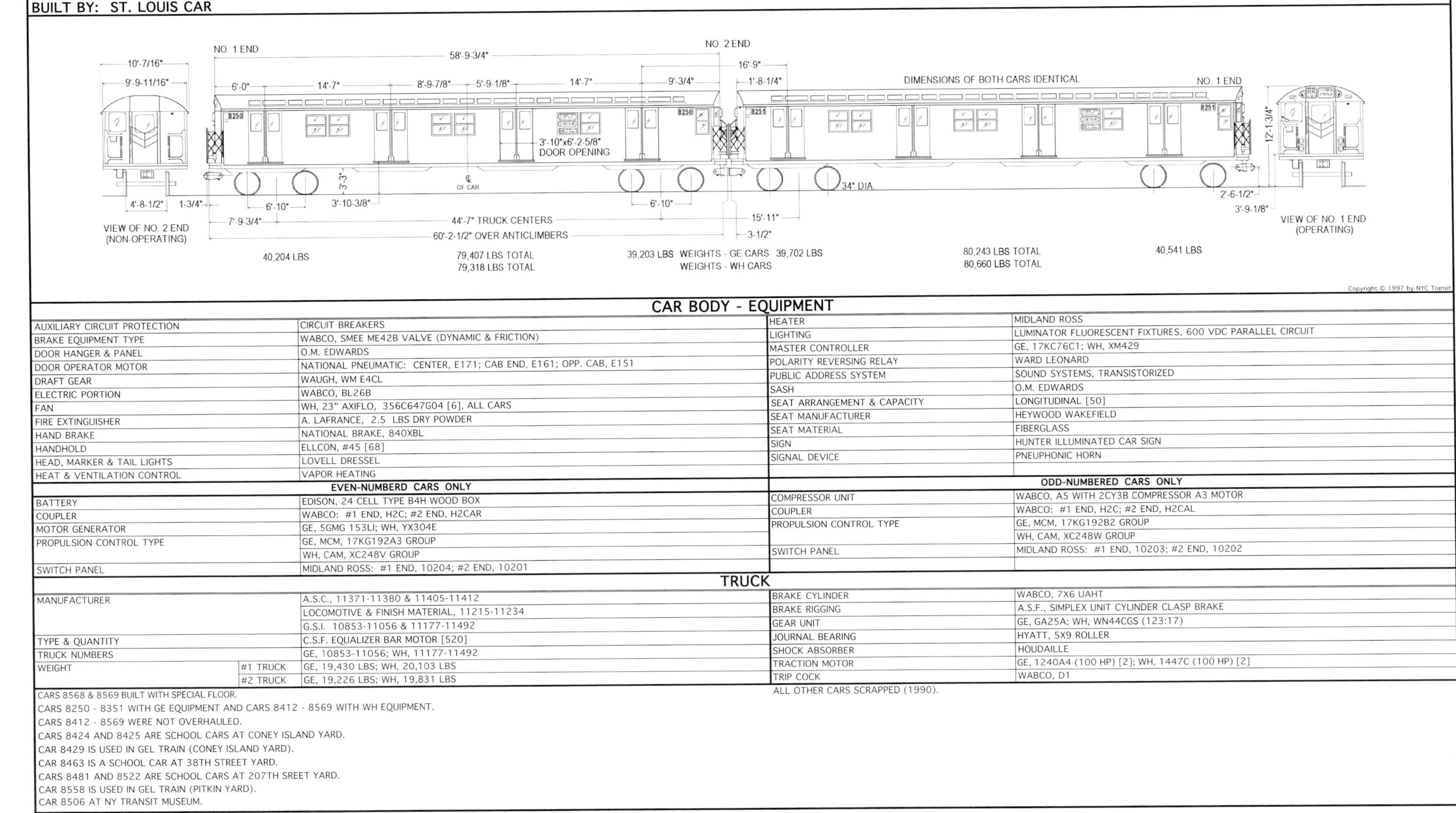


Copyright © 1997 by NYC Transit

## CAR BODY - EQUIPMENT

| | | | |
|---|---|---|---|
| AUXILIARY CIRCUIT PROTECTION | CIRCUIT BREAKERS | HEATER | MIDLAND ROSS |
| BRAKE EQUIPMENT TYPE | WABCO, SMEE ME42B VALVE (DYNAMIC & FRICTION) | LIGHTING | LUMINATOR FLUORESCENT FIXTURES, 600 VDC PARALLEL CIRCUIT |
| DOOR HANGER & PANEL | O.M. EDWARDS | MASTER CONTROLLER | GE, 17KC76C1; WH, XM429 |
| DOOR OPERATOR MOTOR | NATIONAL PNEUMATIC: CENTER, E171; CAB END, E161; OPP. CAB, E151 | POLARITY REVERSING RELAY | WARD LEONARD |
| DRAFT GEAR | WAUGH, WM E4CL | PUBLIC ADDRESS SYSTEM | SOUND SYSTEMS, TRANSISTORIZED |
| ELECTRIC PORTION | WABCO, BL26B | SASH | O.M. EDWARDS |
| FAN | WH, 23" AXIFLO, 356C647G04 [6], ALL CARS | SEAT ARRANGEMENT & CAPACITY | LONGITUDINAL [50] |
| FIRE EXTINGUISHER | A. LAFRANCE, 2.5 LBS DRY POWDER | SEAT MANUFACTURER | HEYWOOD WAKEFIELD |
| HAND BRAKE | NATIONAL BRAKE, 840XBL | SEAT MATERIAL | FIBERGLASS |
| HANDHOLD | ELLCON, #45 [68] | SIGN | HUNTER ILLUMINATED CAR SIGN |
| HEAD, MARKER & TAIL LIGHTS | LOVELL DRESSEL | SIGNAL DEVICE | PNEUPHONIC HORN |
| HEAT & VENTILATION CONTROL | VAPOR HEATING | | |
| | **EVEN-NUMBERED CARS ONLY** | | **ODD-NUMBERED CARS ONLY** |
| BATTERY | EDISON, 24 CELL TYPE B4H WOOD BOX | COMPRESSOR UNIT | WABCO, A5 WITH 2CY3B COMPRESSOR A3 MOTOR |
| COUPLER | WABCO: #1 END, H2C; #2 END, H2CAR | COUPLER | WABCO: #1 END, H2C; #2 END, H2CAL |
| MOTOR GENERATOR | GE, 5GMG 153LI; WH, YX304E | PROPULSION CONTROL TYPE | GE, MCM, 17KG192B2 GROUP |
| PROPULSION CONTROL TYPE | GE, MCM, 17KG192A3 GROUP | | WH, CAM, XC248W GROUP |
| | WH, CAM, XC248V GROUP | SWITCH PANEL | MIDLAND ROSS: #1 END, 10203; #2 END, 10202 |
| SWITCH PANEL | MIDLAND ROSS: #1 END, 10204; #2 END, 10201 | | |

## TRUCK

| | | | | |
|---|---|---|---|---|
| MANUFACTURER | | A.S.C., 11371-11380 & 11405-11412 | BRAKE CYLINDER | WABCO, 7X6 UAHT |
| | | LOCOMOTIVE & FINISH MATERIAL, 11215-11234 | BRAKE RIGGING | A.S.F., SIMPLEX UNIT CYLINDER CLASP BRAKE |
| | | G.S.I. 10853-11056 & 11177-11492 | GEAR UNIT | GE, GA25A; WH, WN44CGS (123:17) |
| TYPE & QUANTITY | | C.S.F. EQUALIZER BAR MOTOR [520] | JOURNAL BEARING | HYATT, 5X9 ROLLER |
| TRUCK NUMBERS | | GE, 10853-11056; WH, 11177-11492 | SHOCK ABSORBER | HOUDAILLE |
| WEIGHT | #1 TRUCK | GE, 19,430 LBS; WH, 20,103 LBS | TRACTION MOTOR | GE, 1240A4 (100 HP) [2]; WH, 1447C (100 HP) [2] |
| | #2 TRUCK | GE, 19,226 LBS; WH, 19,831 LBS | TRIP COCK | WABCO, D1 |

CARS 8568 & 8569 BUILT WITH SPECIAL FLOOR.
CARS 8250 - 8351 WITH GE EQUIPMENT AND CARS 8412 - 8569 WITH WH EQUIPMENT.
CARS 8412 - 8569 WERE NOT OVERHAULED.
CARS 8424 AND 8425 ARE SCHOOL CARS AT CONEY ISLAND YARD.
CAR 8429 IS USED IN GEL TRAIN (CONEY ISLAND YARD).
CAR 8463 IS A SCHOOL CAR AT 38TH STREET YARD.
CARS 8481 AND 8522 ARE SCHOOL CARS AT 207TH SREET YARD.
CAR 8558 IS USED IN GEL TRAIN (PITKIN YARD).
CAR 8506 AT NY TRANSIT MUSEUM.

ALL OTHER CARS SCRAPPED (1990).

CAR NUMBERS: 8352-8411
TOTAL: 60 CARS
BUILT BY: ST. LOUIS CAR

DATE: 1961
AVERAGE COST PER CAR: $121,563

# R30A (BMT/IND)

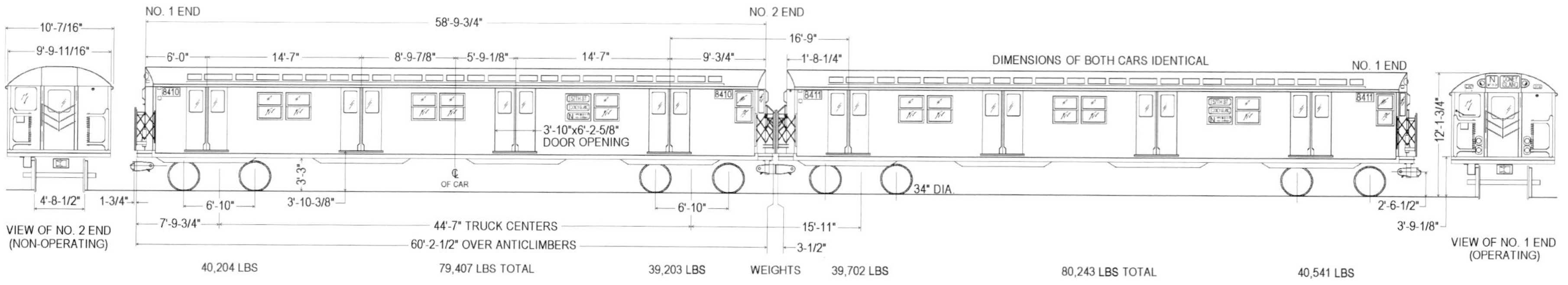

10'-7/16"
9'-9-11/16"
4'-8-1/2"
1-3/4"
VIEW OF NO. 2 END (NON-OPERATING)
NO. 1 END
58'-9-3/4"
NO. 2 END
16'-9"
6'-0"
14'-7"
8'-9-7/8"
5'-9-1/8"
14'-7"
9'-3/4"
1'-8-1/4"
DIMENSIONS OF BOTH CARS IDENTICAL
NO. 1 END
12'-1-3/4"
3'-10"x6'-2-5/8" DOOR OPENING
3'-3"
℄ OF CAR
34" DIA.
6'-10"
3'-10-3/8"
6'-10"
2'-6-1/2"
7'-9-3/4"
44'-7" TRUCK CENTERS
60'-2-1/2" OVER ANTICLIMBERS
15'-11"
3-1/2"
3'-9-1/8"
VIEW OF NO. 1 END (OPERATING)
40,204 LBS
79,407 LBS TOTAL
39,203 LBS
WEIGHTS
39,702 LBS
80,243 LBS TOTAL
40,541 LBS

Copyright © 1997 by NYC Transit

## CAR BODY - EQUIPMENT

| | | | | | |
|---|---|---|---|---|---|
| AUXILIARY CIRCUIT PROTECTION | | CIRCUIT BREAKERS | HEAT & VENTILATION CONTROL | | VAPOR HEATING |
| BRAKE EQUIPMENT TYPE | | WABCO, SMEE (DYNAMIC & FRICTION) | HEATER | | MIDLAND ROSS |
| BRAKE VALVE | | WABCO, ME42B | LIGHTING | | LUMINATOR FLUORESCENT FIXTURES, 600 VDC PARALLEL CIRCUIT |
| DOOR HANGER & PANEL | | O.M. EDWARDS | MASTER CONTROLLER | | GE, 17KC76C1; WH, XM429 |
| DOOR OPERATOR MOTOR | | NATIONAL PNEUMATIC: CENTER, E171; CAB END, E161; OPP. CAB, E151 | POLARITY REVERSING RELAY | | WARD LEONARD |
| DRAFT GEAR | | WAUGH, WM E4CL | PUBLIC ADDRESS SYSTEM | | SOUND SYSTEM, TRANSISTORIZED |
| ELECTRIC PORTION | | WABCO, BL26B | SASH | | O.M. EDWARDS |
| FAN | | WH, 23" AXIFLO [6], 356C647G04 | SEAT ARRANGEMENT & CAPACITY | | LONGITUDINAL [50] |
| FIRE EXTINGUISHER | | A. LAFRANCE, 2.5 LBS DRY POWDER | SEAT MANUFACTURER | | HEYWOOD WAKEFIELD |
| HAND BRAKE | | NATIONAL BRAKE, 840XBL | SEAT MATERIAL | | FIBERGLASS |
| HANDHOLD | | ELLCON, #45 [68] | SIGN | | HUNTER ILLUMINATING CAR SIGN |
| HEAD, MARKER & TAIL LIGHTS | | LOVELL DRESSEL | SIGNAL DEVICE | | PNEUPHONIC HORN |
| EVEN- NUMBERED CARS ONLY | | | ODD-NUMBERED CARS ONLY | | |
| BATTERY | | EDISON, 24 CELL TYPE B4H | AIR COMPRESSOR | | WABCO, 2CY3B WITH A3 MOTOR |
| COUPLERS | | WABCO: #1 END, H2C; #2 END, H2CAR | COUPLER | | WABCO: #1 END, H2C; #2 END, H2CAL |
| MOTOR GENERATOR | | GE, 5GMG 153LI; WH, YX304E | PROPULSION CONTROL TYPE | | GE, MCM, 17KG192B2 GROUP |
| PROPULSION CONTROL TYPE | | GE, MCM, 17KG192A3 GROUP | SWITCH PANEL | #1 END | MIDLAND ROSS/GE, 10203; WH, 10203A |
| SWITCH PANEL | #1 END | MIDLAND ROSS/GE, 10204; WH, 1024A | | #2 END | MIDLAND ROSS, 10202 |
| | #2 END | MIDLAND ROSS, 10201 | | | |

## TRUCK

| | | | | | |
|---|---|---|---|---|---|
| MANUFACTURER | | A.S.C., 11371-11380 & 11405-11412 | BRAKE CYLINDER | | WABCO, 7X6 UAHT |
| | | LOCOMOTIVE & FINISH MATERIAL, 11215-11234 | BRAKE RIGGING | | A.S.F., SIMPLEX UNIT CYLINDER CLASP BRAKE |
| | | G.S.I., 10853-11056 & 11177-11492 | GEAR UNIT | | GE, GA25A; WH, WN44CGS (123:17) |
| TYPE & QUANTITY | | C.S.F. EQUALIZER BAR MOTOR [520] | JOURNAL BEARING | | HYATT, 5X9 ROLLER |
| TRUCK NUMBERS | | GE, 10853-11056; WH, 11177-11492 | SHOCK ABSORBER | | HOUDAILLE |
| WEIGHT | #1 TRUCK | GE, 19,430 LBS; WH, 20,103 LBS | TRACTION MOTOR | | GE, 1240A4 (100 HP) [2]; WH, 1447C (100 HP) [2] |
| | #2 TRUCK | GE, 19,226 LBS; WH, 19,831 LBS | TRIP COCK | | WABCO, D1 |

CAR NUMBERS: 8250 - 8411
TOTAL: 162 CARS
BUILT BY: ST. LOUIS CAR
DATE: 1961 - 62
AVERAGE COST PER CAR: $121,644, SEE NOTE 2

OVERHAULED BY: NYC TRANSIT AUTHORITY
DATE: 1985 - 1986
OVERHAUL COST PER CAR: $317,000

# R30/30A (BMT/IND) GOH

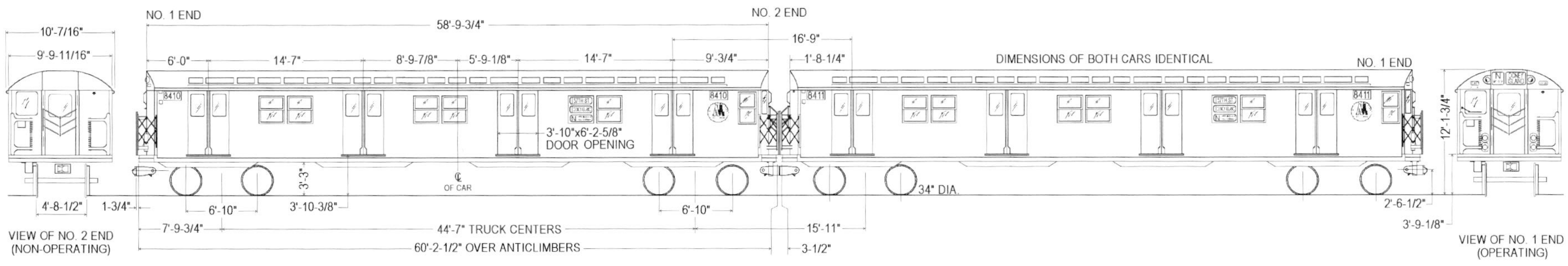

NO. 1 END — NO. 2 END — DIMENSIONS OF BOTH CARS IDENTICAL — NO. 1 END

10'-7/16" · 9'-9-11/16" · 4'-8-1/2" · 1-3/4" · 58'-9-3/4" · 6'-0" · 14'-7" · 8'-9-7/8" · 5'-9-1/8" · 14'-7" · 9'-3/4" · 16'-9" · 1'-8-1/4" · 3'-3" · 6'-10" · 3'-10-3/8" · ℄ OF CAR · 6'-10" · 34" DIA. · 3'-10"x6'-2-5/8" DOOR OPENING · 7'-9-3/4" · 44'-7" TRUCK CENTERS · 60'-2-1/2" OVER ANTICLIMBERS · 15'-11" · 3-1/2" · 2'-6-1/2" · 3'-9-1/8" · 12'-1-3/4"

VIEW OF NO. 2 END (NON-OPERATING)

VIEW OF NO. 1 END (OPERATING)

Copyright © 1997 by NYC Transit

## CAR BODY - EQUIPMENT

| | | | |
|---|---|---|---|
| AUXILIARY CIRCUIT PROTECTION | CIRCUIT BREAKERS | HEAT & VENTILATION CONTROL | VAPOR HEATING |
| BRAKE EQUIPMENT TYPE | WABCO, SMEE (DYNAMIC & FRICTION) | HEATER | MIDLAND ROSS |
| BRAKE VALVE | WABCO, ME42B | LIGHTING | LUMINATOR, FLUORESCENT FIXTURES, 600 VDC PARALLEL CIRCUIT |
| DOOR HANGER & PANEL | O.M. EDWARDS | MASTER CONTROLLER | GE, 17KC76C1 |
| DOOR OPERATOR MOTOR | WESTCODE | POLARITY REVERSING RELAY | WARD LEONARD |
| DRAFT GEAR | WAUGH, WM E4CL | PUBLIC ADDRESS SYSTEM | COMCO |
| ELECTRIC PORTION | WABCO, BL26B | SASH | O.M. EDWARDS |
| FAN | WH, 23" AXIFLO 356C647G04 [6] | SEAT ARRANGEMENT & CAPACITY | LONGITUDINAL [50] |
| HAND BRAKE | NATIONAL BRAKE, 840XBL | SEAT MANUFACTURER | HEYWOOD WAKEFIELD |
| HANDHOLD | ELLCON, #45 [68]/CAR | SIGN | HUNTER ILLUMINATING CAR SIGN |
| HEAD, MARKER & TAIL LIGHTS | LOVELL DRESSEL | SIGNAL DEVICE | PNEUPHONIC HORN |
| | **EVEN- NUMBERED CARS ONLY** | | **ODD-NUMBERED CARS ONLY** |
| BATTERY | EDISON, 24 CELL (80 AH) | COMPRESSOR UNIT | WABCO, D4 |
| CONVERTER | GE, 5GMG 153LI; WH, YX304E | COUPLER | WABCO: #1 END, H2C; #2 END, H2CAL |
| COUPLER | WABCO: #1 END, H2C; #2 END, H2CAR | PROPULSION CONTROL TYPE | GE, SCM, 17KG192AC3 GROUP |
| PROPULSION CONTROL TYPE | GE, SCM, 17KG192AC3 GROUP | SWITCH PANEL #1 END | MIDLAND ROSS, 10203 |
| SWITCH PANEL #1 END | MIDLAND ROSS/GE, 10204 | #2 END | MIDLAND ROSS, 10202 |
| #2 END | MIDLAND ROSS, 10201 | | |

## TRUCK

| | | | |
|---|---|---|---|
| MANUFACTURER | ADIRONDACK STEEL CASTING | BRAKE CYLINDER | WABCO, 7X6 UAHT |
| | LOCOMOTIVE & FINISH MATERIAL | BRAKE RIGGING | A.S.F., SIMPLEX UNIT CYLINDER CLASP BRAKE |
| | GENERAL STEEL CASTING | GEAR UNIT | GE, GA25A; WH, WN44CGS (123:17) |
| TYPE & QUANTITY | C.S.F. EQUALIZER BAR MOTOR [520] | JOURNAL BEARING | HYATT, 5X9 ROLLER (SEE NOTE 1) |
| TRUCK NUMBERS | N/A | SHOCK ABSORBER | HOUDAILLE |
| TRUCK WEIGHT | N/A | TRACTION MOTOR | GE, 1257F1 (115 HP) [2]; WH, 1447JR (115 HP) [2] |
| | | TRIP COCK | WABCO, D1 |

NOTE 1: VARIOUS VENDORS.
NOTE 2: COMBINED AVERAGE COST OF R30 & R30A.
CARS 8265 AND 8386 USED FOR RTO TRAINING (CONCOURSE YARD).
CARS 8289 & 8290 ARE USED FOR POLICE TRAINING (CONEY ISLAND YARD).
CAR 8337 AT TRANSIT TECH HIGH SCHOOL.
CARS 8392 & 8401 USED FOR FIRE TRAINING (FIRE SCHOOL) AT CONEY ISLAND YARD.
SOME CARS WERE CONVERTED TO WORK CARS. ALL OTHER CARS SCRAPPED (1991 - 1992).

CAR NUMBERS: 3350 - 3649
TOTAL: 300 CARS
BUILT BY: BUDD

DATE: 1964 - 65
AVERAGE COST PER CAR: $114,857

# R32A (BMT/IND)

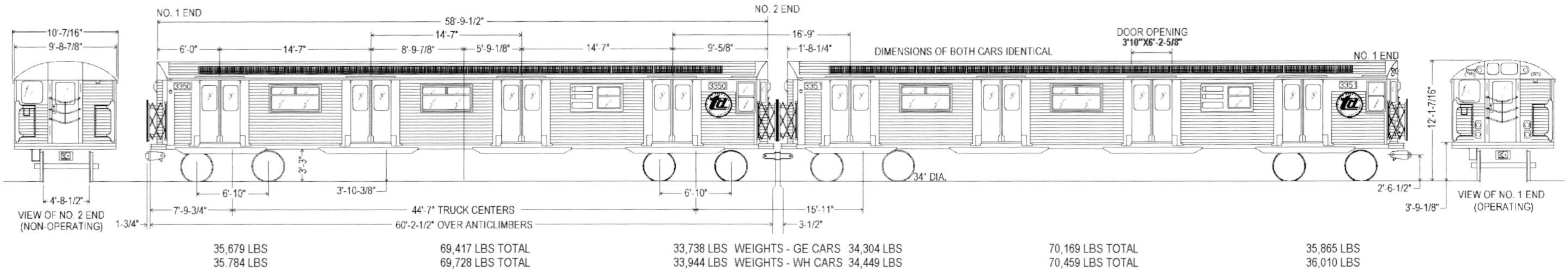

NO. 1 END — NO. 2 END — NO. 1 END
58'-9-1/2" · 14'-7" · 6'-0" · 14'-7" · 8'-9-7/8" · 5'-9-1/8" · 14'-7" · 9'-5/8" · 16'-9" · 1'-8-1/4"
DIMENSIONS OF BOTH CARS IDENTICAL
DOOR OPENING 3'10"X6'-2-5/8"
10'-7/16" · 9'-8-7/8" · 12'-1-7/16"
3'-3" · 3'-10-3/8" · 6'-10" · 6'-10" · 34" DIA. · 2'-6-1/2" · 3'-9-1/8"
4'-8-1/2" · 7'-9-3/4" · 44'-7" TRUCK CENTERS · 15'-11"
1-3/4" · 60'-2-1/2" OVER ANTICLIMBERS · 3-1/2"
VIEW OF NO. 2 END (NON-OPERATING)
VIEW OF NO. 1 END (OPERATING)

| 35,679 LBS | 69,417 LBS TOTAL | 33,738 LBS WEIGHTS - GE CARS 34,304 LBS | 70,169 LBS TOTAL | 35,865 LBS |
|---|---|---|---|---|
| 35,784 LBS | 69,728 LBS TOTAL | 33,944 LBS WEIGHTS - WH CARS 34,449 LBS | 70,459 LBS TOTAL | 36,010 LBS |

Copyright © 1997 by NYC Transit

## CAR BODY - EQUIPMENT

| | | | | |
|---|---|---|---|---|
| AUXILIARY CIRCUIT PROTECTION | | CIRCUIT BREAKERS | HEAD, MARKER & TAIL LIGHTS | LOVELL DRESSEL |
| BRAKE EQUIPMENT TYPE | | WABCO, SMEE (DYNAMIC & FRICTION) | HEAT & VENTILATION CONTROL | VAPOR HEATING |
| BRAKE VALVE | | WABCO, ME42B | HEATER | MIDLAND ROSS, CARS 3350-3549; VAPOR, CARS 3550-3649 |
| COUPLER | | WABCO: #1 END, H2C; #2 END, LINK BAR | LIGHTING | LUMINATOR, FLUORESCENT FIXTURES 600 VDC PARALLEL CIRCUIT |
| DOOR HANGER & PANEL | | O.M. EDWARDS | MASTER CONTROLLER | GE, 17KC76C2; WH, XM729 |
| DOOR OPERATOR MOTOR | CENTER | MIDLAND ROSS, 500003A, CARS 3350-3479 [130] | POLARITY REVERSING TIMER | INDUSTRON. SOLID STATE |
| | | VAPOR, 56370014, CARS 3480-3481 [2] | PUBLIC ADDRESS SYSTEM | INDUSTRON-TRANSISTOR |
| | OPP. CAB | MIDLAND ROSS, 500005B CARS 3482-3551 [70] | SASH | O.M. EDWARDS |
| | | VAPOR, 56370015, CARS 3552-3649 [98] | SEAT ARRANGEMENT & CAPACITY | LONGITUDINAL [50] |
| | CAB END | MIDLAND ROSS, 500005C; VAPOR, 56370016 | SEAT MANUFACTURER | GENERAL SEATING |
| DRAFT GEAR | | WAUGH, WM E4CL | SEAT MATERIAL | FIBERGLASS |
| ELECTRIC PORTION | | #1 END: WABCO, BL26B; #2 END: CANNON PLUG | SIGN | TRANSIGN |
| FAN | | WH, 23" AXIFLO 4 SPEED [6]/, ALL CARS | SIGNAL DEVICE | PNEUPHONIC HORN |
| HANDHOLD | | ELLCON NATIONAL, # 45 | STANCHION | ELLCON-NATIONAL |
| FIRE EXTINGUISHER | | W. KIDDE,1.5-2.75 LB DRY POWDER | WINDOW | O.M. EDWARDS |
| HAND BRAKE | | ELLCON-NATIONAL, 840XBL | | |
| | | **EVEN-NUMBERED CARS ONLY** | | **ODD-NUMBERED CARS ONLY** |
| BATTERY | | EDISON, 24 CELL B4H (FIBERGLASS BOX) | AIR COMPRESSOR | WABCO, 2CY3B |
| MOTOR GENERATOR | | GE, 5GMG 153LI; WH, YX304E1 | AIR COMPRESSOR MOTOR | WABCO, A3 |
| PROPULSION CONTROL TYPE | | GE, SCM, 17KG192F6 GROUP | AIR COMPRESSOR UNIT | WABCO, A6 |
| | | WH, CAM, XCA248P GROUP | PROPULSION CONTROL TYPE | GE, SCM, 17KG192H6 GROUP |
| SWITCH PANEL | | MIDLAND ROSS: #1 END, 500009C; #2 END, 500010B | | WH, CAM, 1XCA-2480 GROUP |
| | | | SWITCH PANEL | MIDLAND ROSS: #1 END, 500007D; #2 END, 500008B |

## TRUCK

| | | | | |
|---|---|---|---|---|
| MANUFACTURERS | | GENERAL STEEL CASTING | AUTOMATIC SLACK ADJUSTER | WABCO, C15D1 |
| | | LOCOMOTIVE & FINISH MATERIAL | BRAKE CYLINDER | WABCO, 7X6 UNIT CYLINDERS, UAHT [2] |
| | | ADIRONDACK STEEL CASTING | BRAKE RIGGING | A.S.F, SIMPLEX UNIT CYLINDER CLASP BRAKE |
| TYPE | | C.S.F. EQUALIZER BAR MOTOR | GEAR UNIT | WH, WN44CGS; GE, 7GA25A2, (123:17) |
| TRUCK NUMBERS | | WH, 14193-14492; GE, 13893-14192 | JOURNAL BEARING | SKF, 5X9 ROLLER |
| WEIGHT | #1 TRUCK | WH, 19,345 LBS; GE, 19,145 LBS | SHOCK ABSORBER | HOUDAILLE |
| | #2 TRUCK | N/A | TRACTION MOTOR | WH, 1447C (100 HP) [2]; GE, 1240A5 (100 HP) [2] |
| | | | TRIP COCK | WABCO, D1 |

CARS 3500 - 3649 WITH GE EQUIPMENT AND CARS 3350 -3499 WITH WH EQUIPMENT.
CARS 3350-3499 HAVE HORNS MOUNTED ON FRONT OF CAR NEAR ROOF. BALANCE OF CARS HAVE HORNS MOUNTED BELOW ANTICLIMBERS ON FRONT (LEFT SIDE).
CARS 3498 - 3501 HAVE WATT-HOUR METERS.

CAR NUMBERS: 3650-3945
TOTAL: 296 CARS
BUILT BY: BUDD

DATE: 1965
AVERAGE COST PER CAR: $114,951

# R32 (BMT/IND)

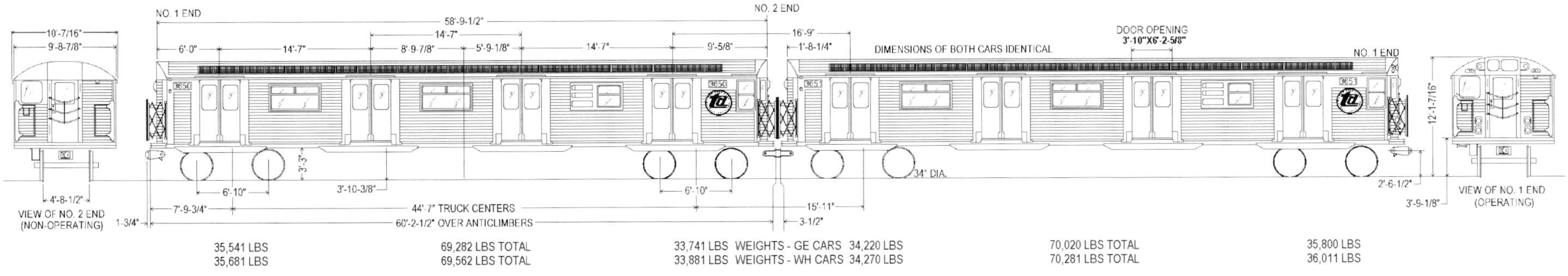

NO. 1 END — NO. 2 END — DIMENSIONS OF BOTH CARS IDENTICAL — DOOR OPENING 3'-10"X6'-2-5/8" — NO. 1 END

VIEW OF NO. 2 END (NON-OPERATING) — 10'-7/16" — 9'-8-7/8" — 4'-8-1/2" — 1-3/4"

58'-9-1/2" — 14'-7" — 6'-0" — 14'-7" — 8'-9-7/8" — 5'-9-1/8" — 14'-7" — 9'-5/8" — 16'-9" — 1'-8-1/4" — 3'-3" — 3'-10-3/8" — 6'-10" — 6'-10" — 7'-9-3/4" — 44'-7" TRUCK CENTERS — 60'-2-1/2" OVER ANTICLIMBERS — 15'-11" — 3-1/2" — 34" DIA. — 2'-6-1/2" — 3'-9-1/8" — 12'-1-7/16"

VIEW OF NO. 1 END (OPERATING)

35,541 LBS / 35,681 LBS — 69,282 LBS TOTAL / 69,562 LBS TOTAL — 33,741 LBS WEIGHTS - GE CARS 34,220 LBS / 33,881 LBS WEIGHTS - WH CARS 34,270 LBS — 70,020 LBS TOTAL / 70,281 LBS TOTAL — 35,800 LBS / 36,011 LBS

Copyright © 1997 by NYC Transit

## CAR BODY - EQUIPMENT

| | | | |
|---|---|---|---|
| AUXILIARY CIRCUIT PROTECTION | CIRCUIT BREAKERS | HEAT & VENTILATION CONTROL | VAPOR HEATING |
| BRAKE EQUIPMENT TYPE | WABCO, SMEE (DYNAMIC & FRICTION) | LIGHTING | LUMINATOR, FLUORESCENT FIXTURES, |
| BRAKE VALVE | WABCO, ME42B | | 600 VDC PARALLEL CIRCUIT |
| COUPLER | WABCO: #1 END, H2C; #2 END, LINK BAR | MASTER CONTROLLER | GE, 17KC76C2; WH, XM729 |
| DOOR HANGER & PANEL | O.M. EDWARDS | POLARITY REVERSING TIMER | INDUSTRON, SOLID STATE |
| DOOR OPERATOR MOTOR | VAPOR: CENTER, 56370014; OPP. CAB, 56370015; CAB END, 56370016 | PUBLIC ADDRESS SYSTEM | INDUSTRON, TRANSISTOR |
| DRAFT GEAR | WAUGE, WM E4CL | SASH (CAB) | O.M. EDWARDS |
| ELECTRIC PORTION | #1 END: WABCO, BL26B; #2 END: PYLE NATIONAL, PLUG | SEAT ARRANGEMENT & CAPACITY | LONGITUDINAL [50] |
| FAN | WH, 23" AXIFLO 4 SPEED [6], ALL CARS | SEAT MANUFACTURER | GENERAL SEATING |
| FIRE EXTINGUISHER | W. KIDDE,1.5-2.75 LB DRY POWDER | SEAT MATERIAL | FIBERGLASS |
| HAND BRAKE | ELLCON-NATIONAL, 840XBL | SIGN | TRANSIGN |
| HANDHOLD | ELLCON NATIONAL, #45 | SIGNAL DEVICE | PNEUPHONIC HORN |
| HEAD, MARKER & TAIL LIGHTS | LOVELL DRESSEL | STANCHION | ELLCON-NATIONAL |
| HEATER | VAPOR | WINDOW | O.M. EDWARDS |
| **EVEN-NUMBERED CARS ONLY** | | **ODD-NUMBERED CARS ONLY** | |
| BATTERY | EDISON, 24 CELL B4H FIBERGLASS BOX | AIR COMPRESSOR | WABCO, 2CY3B |
| MOTOR GENERATOR | GE, 5GMG 153LI; WH, YX304E1 | AIR COMPRESSOR MOTOR | WABCO, A3 |
| PROPULSION CONTROL TYPE | GE, SCM, 17KG192F6 GROUP | PROPULSION CONTROL TYPE | GE, SCM, 17KG192H6 GROUP |
| | WH, CAM, XCA248P GROUP | | WH, CAM, 1XCA-2480 GROUP |
| SWITCH PANEL | MIDLAND ROSS: #1 END, 50007-D; #2 END, 500008B | SWITCH PANEL | MIDLAND ROSS: #1 END, 500007D; #2 END, 500008B |

## TRUCK

| | | | | |
|---|---|---|---|---|
| MANUFACTURER | | GENERAL STEEL CASTING | BRAKE RIGGING | A.S.F. SIMPLEX UNIT CYLINDER CLASP BRAKE |
| TYPE | | C.S.F. EQUALIZER BAR MOTOR | GEAR UNIT | WH, WN44CGS; GE, 7GA25A2 (123:17) |
| TRUCK NUMBERS | | WH, 14793-15084; GE, 14493-14792 | JOURNAL BEARING | SKF, 5X9 ROLLER |
| WEIGHT | #1 TRUCK | WH, 19,345 LBS; GE, 19,145 LBS | SHOCK ABSORBER | HOUDAILLE |
| | #2 TRUCK | N/A | TRACTION MOTOR | WH, 1447C (100 HP) [2]; GE, 1240A5 (100 HP) [2] |
| AUTO. SLACK ADJUSTER | | WABCO, C15D1 | TRIP COCK | WABCO, D1A |
| BRAKE CYLINDER | | WABCO, 7X6 UAHT [2] | | |

CARS 3800-3945 EQUIPPED WITH BACKLIGHTED AD CARDS WITH [2] 28" LIGHT FIXTURES TO ILLUMINATE END AD CARDS.
CARS 3650-3799 HAVE WH, 887D09G01 (ANNULAR CONVECTION VANES).
CARS 3800-3945 HAVE WH, 888D290G0 (PERIPHERAL DISCHARGE AIR DIFFUSERS)
CARS 3650 - 3799 WITH GE EQUIPMENT AND CARS 3800 - 3945 WITH WH EQUIPMENT.
CAR 3945 WAS EQUIPPED WITH INVERTER FOR LIGHTING TEST. TIMER NOT REQUIRED.
ALL CARS HAVE HORNS MOUNTED BELOW ANTICLIMBERS ON FRONT (LEFT SIDE).

CAR NUMBERS: 3946 - 3949
TOTAL: 4 CARS
BUILT BY: BUDD

DATE: 1965
AVERAGE COST PER CAR: $114,951

# R32 (BMT/IND)
PIONEER TRUCKS

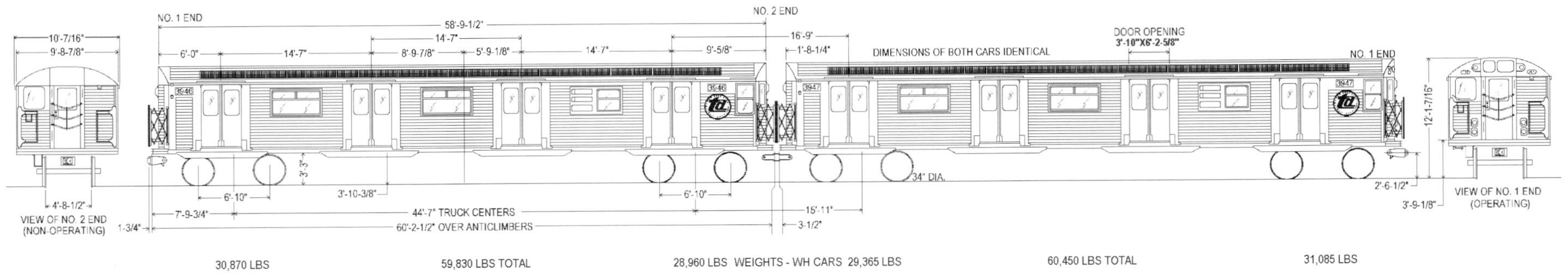

VIEW OF NO. 2 END (NON-OPERATING)
NO. 1 END
NO. 2 END
DIMENSIONS OF BOTH CARS IDENTICAL
DOOR OPENING 3'-10"X6'-2-5/8"
NO. 1 END
VIEW OF NO. 1 END (OPERATING)
10'-7/16" 9'-8-7/8" 4'-8-1/2" 1-3/4" 6'-0" 14'-7" 58'-9-1/2" 14'-7" 8'-9-7/8" 5'-9-1/8" 14'-7" 9'-5/8" 16'-9" 1'-8-1/4" 3'-3" 3'-10-3/8" 6'-10" 7'-9-3/4" 44'-7" TRUCK CENTERS 60'-2-1/2" OVER ANTICLIMBERS 6'-10" 15'-11" 3-1/2" 34" DIA. 12'-1-7/16" 2'-6-1/2" 3'-9-1/8"

30,870 LBS | 59,830 LBS TOTAL | 28,960 LBS WEIGHTS - WH CARS 29,365 LBS | 60,450 LBS TOTAL | 31,085 LBS

Copyright © 1997 by NYC Transit

## CAR BODY - EQUIPMENT

| | | | |
|---|---|---|---|
| AUXILIARY CIRCUIT PROTECTION | CIRCUIT BREAKERS | HEAT & VENTILATION CONTROL | VAPOR HEATING |
| BRAKE EQUIPMENT TYPE | WABCO, SMEE (DYNAMIC & FRICTION) | LIGHTING | LUMINATOR, FLUORESCENT FIXTURES, 600 VDC PARALLEL CIRCUIT |
| BRAKE VALVE | WABCO, ME42B | MASTER CONTROLLER | WH, XM729 |
| COUPLER | WABCO: #1 END, H2C; #2 END, LINK BAR | POLARITY REVERSING TIMER | INDUSTRON SOLID STATE |
| DOOR HANGER & PANEL | O.M. EDWARDS | PUBLIC ADDRESS SYSTEM | INDUSTRON, TRANSISTOR |
| DOOR OPERATOR MOTOR | VAPOR: CENTER, 56370014; OPP. CAB, 56370015; CAB END, 56370016 | RADIO | GE |
| DRAFT GEAR | WAUGH, WM E4CL | SASH (CAB) | O.M. EDWARDS |
| ELECTRIC PORTION | #1 END: WABCO, BL26B; #2 END: PYLE NATIONAL, CANNON PLUG | SEAT CAPACITY & TYPE | LONGITUDINAL [50] |
| FAN | WH, 23" AXIFLO 4 SPEED [6] | SEAT MANUFACTURER | GENERAL SEATING |
| FIRE EXTINGUISHER | W. KIDDE,1.5-2.75 LB DRY POWDER | SEAT MATERIAL | FIBERGLASS |
| HAND BRAKE | ELLCON-NATIONAL, 840XBL | SIGN | TRANSIGN |
| HANDHOLD | ELLCON NATIONAL, #45 | SIGNAL DEVICE | PNEUPHONIC HORN |
| HEAD, MARKER & TAIL LIGHTS | LOVELL DRESSEL | STANCHION | ELLCON-NATIONAL |
| HEATER | VAPOR | WINDOW | O.M. EDWARDS |
| **EVEN-NUMBERED CARS ONLY** | | **ODD-NUMBERED CARS ONLY** | |
| PROPULSION CONTROL TYPE | WH, CAM, XCA248P GROUP | PROPULSION CONTROL TYPE | WH, CAM, 1XCA2480 GROUP |
| MOTOR GENERATOR | GE, 5GMG153LI; WH, YX304E1 | AIR COMPRESSOR MOTOR | WABCO, A3 |
| BATTERY | EDISON, 24 CELL B4H (FIBERGLASS BOX) | AIR COMPRESSOR | WABCO, 2CY3B |
| SWITCH PANEL | MIDLAND ROSS: #1 END, 50007-D; #2 END, 500008B | SWITCH PANEL | MIDLAND ROSS: #1 END, 500007D; #2 END, 500008B |

## TRUCK

| | | | |
|---|---|---|---|
| MANUFACTURER | BUDD | GEAR UNIT | WGU100-1 (123:17) |
| TYPE | PIONEER III | JOURNAL BEARING | SKF, INBOARD BEARINGS |
| TRUCK NUMBERS | WH, 15084 -15092 | SHOCK ABSORBER | HOUDAILLE |
| WEIGHT | N/A | TRACTION MOTOR | WH, 1455H (100 HP) [2] |
| BRAKE CYLINDER | BUDD, D8611801 | TRIP COCK | WABCO, D1 |
| BRAKE RIGGING | DISC BRAKE | | |

PIONEER TRUCKS SCRAPPED AND CARS EQUIPPED WITH STANDARD TRUCKS (1976).
CAR 3946 WAS EQUIPPED WITH INVERTER FOR LIGHTING TEST WITHOUT TIMER.

CARS 3946-3949 EQUIPPED WITH BACKLIGHTED AD CARDS WITH [2] 28" LIGHT FIXTURES TO ILLUMINATE END AD CARDS.

CAR NUMBERS: 3594-5, 3880-1, 3892-3, 3934-7
TOTAL: 10 CARS
BUILT BY: BUDD
DATE: 1965
AVERAGE COST PER CAR: $114,951

OVERHAULED BY: BUFFALO TRANSIT SERVICES (GE)
DATE: 1988
OVERHAUL COST PER CAR: $576,000

# R32 (BMT/IND) GOH

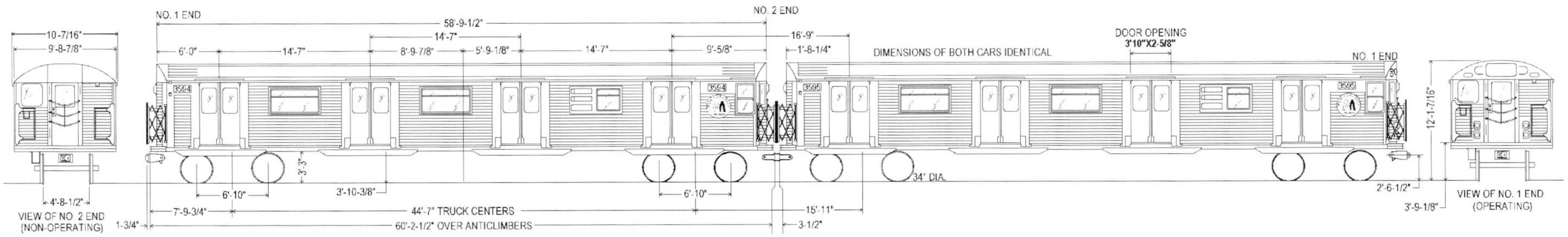

NO. 1 END
NO. 2 END
NO. 1 END
58'-9-1/2"
14'-7"
6'-0"
14'-7"
8'-9-7/8"
5'-9-1/8"
14'-7"
9'-5/8"
16'-9"
1'-8-1/4"
DIMENSIONS OF BOTH CARS IDENTICAL
DOOR OPENING 3'10"X2-5/8"
10-7/16"
9'-8-7/8"
12'-1-7/16"
3'-3"
6'-10"
3'-10-3/8"
6'-10"
34" DIA.
2'-6-1/2"
4'-8-1/2"
7'-9-3/4"
44'-7" TRUCK CENTERS
15'-11"
3'-9-1/8"
1-3/4"
60'-2-1/2" OVER ANTICLIMBERS
3-1/2"
VIEW OF NO. 2 END (NON-OPERATING)
VIEW OF NO. 1 END (OPERATING)

Copyright © 1997 by NYC Transit

## CAR BODY - EQUIPMENT

| | | | |
|---|---|---|---|
| AIR CONDITIONING | SIGMA | LIGHTING | LUMINATOR, FLUORESCENT FIXTURES |
| AUXILIARY CIRCUIT PROTECTION | CIRCUIT BREAKERS | MASTER CONTROLLER | 17KC76AE3 |
| BRAKE EQUIPMENT TYPE | WABCO, RT2 (DYNAMIC & FRICTION) | PROPULSION CONTROL TYPE | GE, SCM, 17KG192A3 GROUP |
| COUPLER | WABCO: #1 END, H2C; #2 END, LINK BAR | PUBLIC ADDRESS SYSTEM | COMCO |
| DOOR HANGER | A.B. KING | SASH (CAB) | BUFFALO TRANSIT SERVICES |
| DOOR OPERATOR MOTOR | WESTCODE | SEAT ARRANGEMENT & CAPACITY | LONGITUDINAL [50] |
| DOOR PANEL | EBONEX | SEAT MANUFACTURER | GENERAL SEATING |
| DRAFT GEAR | WAUGH, WN E4CL | SEAT MATERIAL | FIBERGLASS |
| ELECTRIC PORTION | #1 END: WABCO, BL33F; #2 END: PYLE NATIONAL, CANNON PLUG | SIGN | TRANSIGN |
| HAND BAR | BUFFALO TRANSIT SERVICES | SIGNAL DEVICE | PNEUPHONIC HORN |
| HAND BRAKE | ELLCON NATIONAL, 840XBL | STANCHION | ELLCON NATIONAL |
| HEAD, MARKER & TAIL LIGHTS | LOVELL DRESSEL | WINDOW | BUFFALO TRANSIT SERVICES |
| | **EVEN-NUMBERED CARS ONLY** | | **ODD-NUMBERED CARS ONLY** |
| BATTERY | MCGRAW EDISON, FIBERGLASS BOX | AIR COMPRESSOR | WABCO, D4S |
| CONVERTER | TOSHIBA | SWITCH PANEL | MIDLAND ROSS; VAPOR |
| SWITCH PANEL | VAPOR | | |

## TRUCK

| | | | | |
|---|---|---|---|---|
| MANUFACTURER | | GENERAL STEEL CASTING/ADIRONDACK STEEL CASTING | BRAKE RIGGING | A.S.F., SIMPLEX UNIT CYLINDER CLASP BRAKE |
| TYPE & QUANTITY | | C.S.F EQUALIZER BAR MOTOR [592] | GEAR UNIT | WH, NCG44; GE, GA64C1 (7.235:11) |
| TRUCK NUMBERS | | N/A | JOURNAL BEARING | 5X9 AP ROLLER (SEE NOTE 1) |
| WEIGHT | #1 TRUCK | N/A | SHOCK ABSORBER | MONROE |
| | #2 TRUCK | N/A | TRACTION MOTOR | WH, 1447J & JR (100 HP) [2]; GE, 1257F (100 HP) [2] |
| AUTOMATIC SLACK ADJUSTER | | WABCO, C15D1 | TRIP COCK | WABCO, D1 |
| BRAKE CYLINDER | | WABCO, 7X6 UAHT[2] | | |

NOTE 1: VARIOUS VENDORS. PILOT PROJECT.

CARS 3594 & 3595 WERE ORIGINALLY CONTRACT R32A.
INSTALLED DOPPLER SPEEDOMETER BY EDO (6/1996 - 12/1996).
AMETEK ALTERNATING CURRENT BLOWER MOTORS/INVERTERS INSTALLED (2002).

CAR NUMBERS: 3350-3593, 3596-3879, 3882-3891, 3894-3933, 3938-3949
TOTAL: 584 CARS
BUILT BY: BUDD
DATE: 1964-65
AVERAGE COST PER CAR: $114,951

OVERHAULED BY: MK
DATE: 1988 - 90
OVERHAUL COST PER CAR: $476,000

# R32 (BMT/IND)

PHASE I & II

GOH

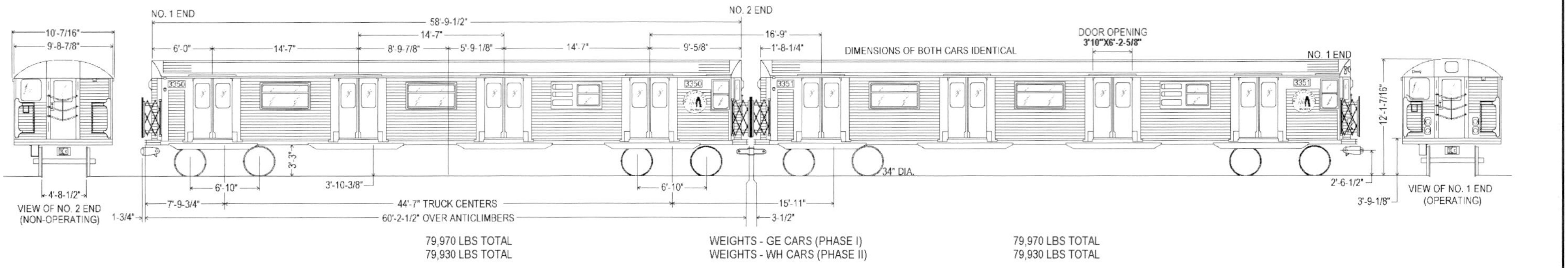

Copyright © 1997 by NYC Transit

## CAR BODY - EQUIPMENT

| | | | |
|---|---|---|---|
| AIR CONDITIONING | STONE SAFETY/THERMO KING | LIGHTING | LUMINATOR, FLUORESCENT FIXTURES, INVERTER BALLAST |
| AUXILIARY CIRCUIT PROTECTION | CIRCUIT BREAKERS | MASTER CONTROLLER | GE, 17KC76AE3 |
| BRAKE EQUIPMENT TYPE | WABCO, RT2 (DYNAMIC & FRICTION) | PROPULSION CONTROL TYPE | GE, SCM, 17KG192E3 GROUP |
| COUPLER | WABCO: #1 END, H2C; #2 END, LINK BAR | PUBLIC ADDRESS SYSTEM | COMCO |
| DOOR HANGER | A. B. KING | ROUTE SIGN | LUMINATOR |
| DOOR OPERATOR MOTOR | VAPOR (PHASE 1); WESTCODE (PHASE II) | SASH (CAB) | EBONEX |
| DOOR PANEL | EBONEX | SEAT ARRANGEMENT & CAPACITY | LONGITUDINAL [50] |
| DRAFT GEAR | WAUGH, WM E4CL | SEAT MANUFACTURER | GENERAL SEATING |
| ELECTRIC PORTION | #1 END: WABCO, BL33F; #2 END: PYLE NATIONAL, CANNON PLUG | SEAT MATERIAL | FIBERGLASS |
| HAND BAR | PHILADELPHIA PIPE | SIGN | TRANSIGN |
| HAND BRAKE | ELLCON-NATIONAL, 840XBL | SIGNAL DEVICE | PNEUPHONIC HORN |
| HEAD, MARKER & TAIL LIGHTS | LOVELL DRESSEL | STANCHION | ELLCON-NATIONAL |
| HEATER | MIDLAND ROSS | WINDOW | EBONEX |
| HVAC CONTROL | VAPOR | | |
| EVEN-NUMBERED CARS ONLY | | ODD-NUMBERED CARS ONLY | |
| BATTERY | MCGRAW EDISON | AIR COMPRESSOR | WABCO, D4S |
| CONVERTER | WH | SWITCH PANEL | MIDLAND ROSS; VAPOR |
| SWITCH PANEL | VAPOR | | |

## TRUCK

| | | | | |
|---|---|---|---|---|
| MANUFACTURER | | GENERAL STEEL CASTING<br>ADIRONDACK STEEL CASTING | BRAKE CYLINDER | WABCO, 7X6 UAHT [2] |
| | | | BRAKE RIGGING | A.S.F., SIMPLEX UNIT CYL CLASP BRAKE |
| TYPE & QUANTITY | | C.S.F. EQUALIZER BAR MOTOR (592) | GEAR UNIT | WH, NGG44; GE, GA64C (17.235:11) |
| TRUCK NUMBERS | | N/A | JOURNAL BEARING | 5X9 AP ROLLER (SEE NOTE 1) |
| WEIGHT | #1 END | GE, 19,050 LBS; WH, 18,990 LBS | SHOCK ABSORBER | MONROE |
| | #2 END | GE, 19,470 LBS; WH, 19,410 LBS | TRACTION MOTOR | WH, 1447JR (115 HP) [2]; GE, 1257E1 (115 HP) [2] |
| AUTOMATIC SLACK ADJUSTER | | WABCO, C15D1 | TRIP COCK | WABCO, D1 |

NOTE 1: VARIOUS VENDORS. SIX CARS SCRAPPED.

CARS 3350 - 3593 AND 3596 - 3649 WERE ORIGINALLY CONTRACT R32A.
CARS 3616, 3629, 3651, 3669(I) & 3766 WERE SCRAPPED PRIOR TO GOH PROGRAM.
CAR 3620 IS OUT OF SERVICE AT C.I. YARD (2003).
INSTALLED DOPPLER SPEEDOMETER BY EDO (6/1996 - 12/1996).
AMETEK ALTERNATING CURRENT BLOWER MOTORS/INVERTERS INSTALLED ON ALL CARS (2003).

CAR 3668 CONVERTED TO COMPRESSOR CAR & RE-NUMBERED 3669(II) ca. 5/82.
CAR 3659 CONVERTED FROM COMPRESSOR TO STATIC CONVERTER CAR AND RENUMBERED 3348 (LATE 1994).
INTERIOR BACK-LIT AD RACKS REMOVED IN CARS 3800-3879, 3882-3891, 3894-3933, 3938-3949 DURING GOH.

CAR NUMBERS: 8806 - 9305
TOTAL: 500 CARS
BUILT BY: ST. LOUIS CAR

DATE: 1962 - 63
AVERAGE COST PER CAR: $108,500

# R33 (IRT)

DROP SASH

8'-10-3/16" 8'-7-3/16" 4'-8-1/2" 8'-9"
VIEW OF NO. 2 END (NON-OPERATING END)

49'-7-7/8" 20-1/8" 6'-2-13/16" 17'-1" 18-1/8" 17'-1" 15'-6-7/8" 17'-2" 9'-3-1/16"
NO. 1 END NO. 2 END NO. 2 END DIMENSIONS OF BOTH CARS IDENTICAL NO. 1 END
4'-2"x6'-2-1/2" DOOR OPENING
34" DIA 3'-1/8" 3'-9" 2'-6-3/8" 1-3/4" 6'-10" 7'-6-1/4" 36'-0" TRUCK CENTERS 6'-10" 15'-4" 51'-1/2" OVER ANTICLIMBERS 3-1/2" 34" DIA
11'-10-3/8"
VIEW OF NO. 1 END (OPERATING END)

37,150 LBS / 37,510 LBS — 72,800 LBS / 73,270 LBS — 35,650 LBS WEIGHTS - GE CARS 35,640 LBS / 35,760 LBS WEIGHTS - WH CARS 35,720 LBS — 72,980 LBS TOTAL / 73,420 LBS TOTAL — 37,340 LBS / 37,700 LBS

Copyright © 1997 by NYC Transit

## CAR BODY - EQUIPMENT

| | | | | | |
|---|---|---|---|---|---|
| AUXILIARY CIRCUIT PROTECTION | | CIRCUIT BREAKERS | HANDHOLD | | ELLCON NATIONAL, #45 [56] |
| BRAKE EQUIPMENT TYPE | | WABCO, SMEE (DYNAMIC & FRICTION) | HEAD, MARKER & TAIL LIGHTS | | LOVELL DRESSEL |
| BRAKE VALVE | | WABCO, ME42B | HEAT & VENTILATION CONTROL | | VAPOR, CARS 8806-9295; WH, CARS 9296-9305 |
| COUPLER | #1 END | WABCO, H2C | HEATER | | GE |
| | #2 END | WAUGH, LINKBAR, 2F7631/WABCO, 565150 | LIGHTING | | LUMINATOR, FLUORESCENT FIXTURES |
| DOOR HANGER & PANEL | | O.M. EDWARDS | MASTER CONTROLLER | | GE, 17KC76C1; WH, XM 579 |
| DOOR OPERATOR MOTOR | CENTER | MIDLAND ROSS, 500003; VAPOR HEAT, 56270010, SEE NOTE 1 | PUBLIC ADDRESS SYSTEM | | INDUSTRON |
| | CAB END | MIDLAND ROSS, 500005; VAPOR HEAT, 56270011, SEE NOTE 1 | POLARITY REVERSING RELAY | | HORNE |
| | OPP END | MIDLAND ROSS, 500004; VAPOR HEAT, 56270012, SEE NOTE 1 | PROPULSION CONTROL TYPE | | GE, SCM, CARS 8806-9075; WH, CAM, CARS 9076-9305 |
| DRAFT GEAR | | WAUGH, WM E4CL | SASH | | O.M. EDWARDS |
| ELECTRIC PORTION | | #1 END: WABCO, BL26C; #2 END: PYLE NATIONAL, CANNON PLUG | SEAT ARRANGEMENT & CAPACITY | | LONGITUDINAL [44] |
| FANS (23" AXIFLOW) | | TRANE, CARS 9266-9274, 9281-9292, 9302-9305 | SEAT MANUFACTURER | | HEYWOOD WAKEFIELD/AMERICAN SEATING |
| | | WH, #356C647G05 (ON REMAINDER OF CARS) | SEAT MATERIAL | | FIBERGLASS |
| FIRE EXTINGUISHER | | A. LAFRANCE, 2 1/2 LBS DRY POWDER | SIGN | | HUNTER ILLUMINATED CAR SIGN |
| HAND BRAKE | | ELLCON NATIONAL, 840 XBL | SIGNAL DEVICE | | PNEUPHONIC HORN |
| | | **EVEN-NUMBERED CARS ONLY** | | | **ODD-NUMBERED CARS ONLY** |
| BATTERY | | SAFT NIFE, 52 A.H. | AIR COMPRESSOR | | WABCO, 2CY3B |
| MOTOR GENERATOR | | GE, 5GMG153L1; WH, YX304E | AIR COMPRESSOR MOTOR | | WABCO, A3 |
| PROPULSION CONTROL TYPE | | GE, MCM, 17KG192F4 GROUP | PROPULSION CONTROL TYPE | | GE, MCM, 17KG192H4 GROUP |
| | | WH, CAM, XCA248E GROUP | | | WH, CAM, XCA248F GROUP |
| SWITCH PANEL | #1 END | MIDLAND ROSS, 500009; VAPOR HEAT, 56270014, SEE NOTE 1 | SWITCH PANEL | #1 END | MIDLAND ROSS, 500007A; VAPOR, 56270013 |
| | #2 END | MIDLAND ROSS, 5000010A; VAPOR HEAT, 56270016, SEE NOTE 1 | | #2 END | MIDLAND ROSS, 500008A; VAPOR, 56270015 |

## TRUCK

| | | | | |
|---|---|---|---|---|
| MANUFACTURER | | GENERAL STEEL CASTING | BRAKE RIGGING | A.S.F., SIMPLEX UNIT CYLINDER CLASP BRAKE |
| TYPE & QUANTITY | | C.S.F. EQUALIZER BAR MOTOR [1000] | GEAR UNIT | WH, WN44CGS, (123:17); GE, GA25A2 (123:17) |
| TRUCK NUMBERS | | 11965-12504 & 12505-12966 | JOURNAL BEARING | 5X9 ROLLER: HYATT, 1600; SKF, 1600; TIMKEN, 800 |
| WEIGHT | #1 END | WH, 16,240 LBS; GE, 16,340 LBS | SHOCK ABSORBER | MONROE |
| | #2 END | WH, 16,000 LBS; GE, 16,080 LBS | TRACTION MOTOR | WH, 1447C (100 HP) [2]; GE, 1240A5 (100 HP) [2] |
| BRAKE CYLINDER | | WABCO, 6X6 UAHT [2] | TRIP COCK | WABCO, D1A |

NOTE 1: MIDLAND ROSS, CARS 8806 - 8899, 9076 - 9207, 9212 - 9255 (330 CARS).
VAPOR HEATING, CARS 9000 - 9075, 9208 -9211, 9256 - 9305 (170 CARS).

| ORIGINALLY ON CARS | MF'G. | TRUCK NUMBERS | TYPE OF SHOES |
|---|---|---|---|
| 8896-8899 | WABCO | 12309-12316 | COMPOSITION |
| 8900-8903 | A.S.F. | 12317-12324 | COMPOSITION |
| 9136-37, 9160, 9161(REMOVED 1964) | A.S.F. | 12669-12676 | CAST IRON |
| 9162, 9165(REMOVED 1964) | WABCO | 12677-12684 | CAST IRON |

WATT HOUR METERS ON CARS 8896, 8897, 9164, 9165.
PLYMETAL SUBFLOOR INTRODUCED WITH THIS CONTRACT.
CARS 8806 - 9075 WITH GE EQUIPMENT AND CARS 9076 - 9305 WITH WH EQUIPMENT.
REPLACED PACKAGE BRAKES WITH CLASP BRAKES ON TRUCKS FOR CARS 8896 - 8899, 9162 - 9165, 9136, 9187, 9160 AND 9161.

CAR NUMBERS: 8806-9305
TOTAL: 494 CARS
BUILT BY: ST. LOUIS CAR
DATE:1962 - 63
AVERAGE COST PER CAR: $108,500

OVERHAUL BY: NYC TRANSIT AUTHORITY
DATE: 1986 - 91
OVERHAUL COST PER CAR: $381,000

# R33 (IRT)
DROP SASH
GOH

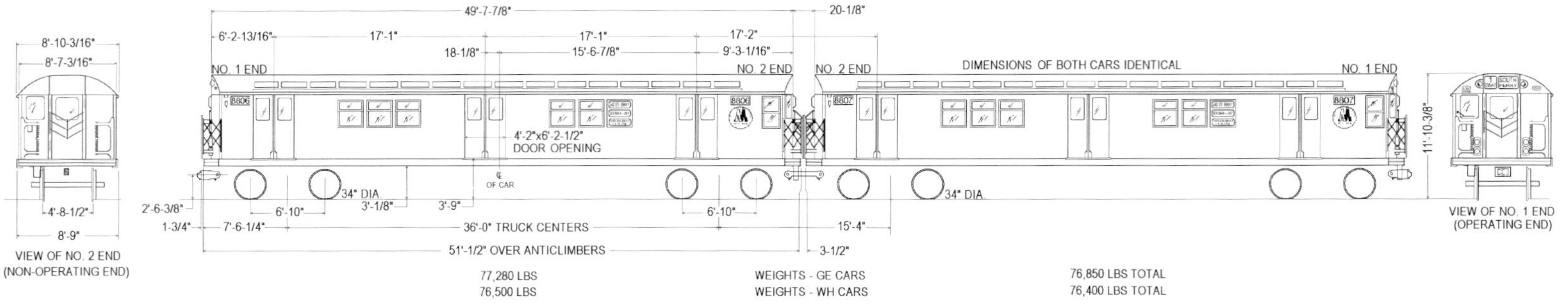

8'-10-3/16" 8'-7-3/16" 4'-8-1/2" 8'-9"
VIEW OF NO. 2 END (NON-OPERATING END)

49'-7-7/8" 20-1/8" 6'-2-13/16" 17'-1" 17'-1" 17'-2" 18-1/8" 15'-6-7/8" 9'-3-1/16"
NO. 1 END NO. 2 END NO. 2 END DIMENSIONS OF BOTH CARS IDENTICAL NO. 1 END
4'-2"x6'-2-1/2" DOOR OPENING
2'-6-3/8" 1-3/4" 7'-6-1/4" 6'-10" 34" DIA 3'-1/8" 3'-9" ℄ OF CAR 6'-10" 15'-4" 34" DIA
36'-0" TRUCK CENTERS
51'-1/2" OVER ANTICLIMBERS 3-1/2"

| | | |
|---|---|---|
| 77,280 LBS | WEIGHTS - GE CARS | 76,850 LBS TOTAL |
| 76,500 LBS | WEIGHTS - WH CARS | 76,400 LBS TOTAL |

11'-10-3/8"
VIEW OF NO. 1 END (OPERATING END)

Copyright © 1997 by NYC Transit

## CAR BODY - EQUIPMENT

| | | | | | |
|---|---|---|---|---|---|
| AIR CONDITIONING | | STONE SAFETY/VAPOR | HEATER | | GE |
| AUXILIARY CIRCUIT PROTECTION | | CIRCUIT BREAKERS | HVAC CONTROL | | STONE SAFETY |
| BRAKE EQUIPMENT TYPE | | WABCO, SMEE (DYNAMIC & FRICTION) | LIGHTING | | LUMINATOR FLUORESCENT FIXTURES |
| BRAKE VALVE | | WABCO, ME42B | MASTER CONTROLLER | | GE, 17KC76AD1; WH, XM 579 |
| COUPLER | #1 END | WABCO, H2C W/BL26C ELEC. PORTION | PANTOGRAPH GATE | | DARO |
| | #2 END | WAUGH, 2F7631; WABCO, 565150 | PROPULSION CONTROL TYPE | | GE, SCM, SEE NOTE 1 |
| DOOR HANGER | | O.M. EDWARDS | PUBLIC ADDRESS SYSTEM | | COMCO |
| DOOR OPERATOR MOTOR | | VAPOR HEAT: CENTER, 56166429; CAB END, 56166430; OPP. CAB, 56166431 | SASH | | O.M. EDWARDS & J.T. NELSON |
| DOOR PANEL | | O.M. EDWARDS & L.M.T. | SEAT MATERIAL | | FIBERGLASS |
| DRAFT GEAR | | WAUGH, WM E4CL | SEAT ARRANGEMENT & CAPACITY | | LONGITUDINAL [44] |
| HAND BRAKE | | ELLCON NATIONAL, 840 XBL | SEAT MANUFACTURER | | HEYWOOD WAKEFIELD & AMERICAN SEATING |
| HANDHOLD | | ELLCON NATIONAL, #45 [56] | SIGN | | HUNTER ILLUMINATED CAR SIGN |
| HEAD, MARKER & TAIL LIGHTS | | LOVELL DRESSEL | SIGNAL DEVICE | | PNEUPHONIC HORN |
| | | EVEN-NUMBERED CARS ONLY | | | ODD-NUMBERED CARS ONLY |
| BATTERY | | SAFT NIFE, 52 A.H. | BRAKE EQUIPMENT | | WABCO, E2 OPERATING SYSTEM (CARS 9076-9305) |
| BRAKE EQUIPMENT SYSTEM | | WABCO, E2E OPERATING SYSTEM (CARS 9076-9305) | | | NYAB, NEWTRAN BRAKE SYSTEM (CARS 8806-9075) |
| | | NYAB, NEWTRAN BRAKE SYSTEM (CARS 8606-9075) | COMPRESSOR UNIT | | WABCO, D4S |
| CONVERTER | | TOSHIBA, COV001 (CARS 9076-9305) | SWITCH PANEL | #1 END | MIDLAND ROSS, 500007A; VAPOR HEAT, 56270013 |
| | | STONE SAFETY (CARS 8806-9075) | | #2 END | MIDLAND ROSS, 500008A; VAPOR HEAT, 56270015 |
| SWITCH PANEL | #1 END | MIDLAND ROSS, 500009; VAPOR HEAT, 56270014 | | | |
| | #2 END | MIDLAND ROSS, 5000010A; VAPOR HEAT, 56270016 | | | |

## TRUCK

| | | | | | |
|---|---|---|---|---|---|
| MANUFACTURER | | GENERAL STEEL CASTING | GEAR UNIT | | WH, WN44CGS ; GE, GA25A2 (123:17) |
| TYPE & QUANTITY | | C.S.F. EQUALIZER BAR MOTOR [1000] | JOURNAL BEARING | | 5X9 CLASS D, AP ROLLER (SEE NOTE 2) |
| WEIGHT | #1 TRUCK | GE, 19,580 LBS; WH, 19,300 LBS | TRACTION MOTOR | | WH, 1447JR (100 HP) [2]; GE, 1257E1 (100 HP) [2] |
| | #2 TRUCK | GE, 19,270 LBS; WH, 19,190 LBS | TRIP COCK | | WABCO, D1A |
| BRAKE RIGGING | | A.S.F., SIMPLEX UNIT CYLINDER CLASP BRAKE | SHOCK ABSORBER | | MONROE |
| BRAKE CYLINDER | | WABCO, 6X6 UAHT [2] | | | |

NOTE 1: GE, 17KG 192 AE3 FOR CARS 8806-9075 AND 17 KG 192 AE2 FOR CARS 9076-9305.
NOTE 2: VARIOUS VENDORS.

PACKAGE BRAKES WERE ON TRUCKS FOR CARS 8896 - 8899, 8900 - 8903, 9162 - 9165, 9136, 9137, 9160, 9161 (REMOVED 1964).
ALL CARS HAD DROP-SASH WINDOWS REPLACED (EXCEPT TRAIN OPERATOR'S AND CONDUCTOR'S) WITH LOUVERED WINDOWS SUBSEQUENT TO GOH BY NYCT.
CAR 8885 IS USED IN GEL TRAIN (E. 180th STREET YARD).
CARS 8912, 8913 DELIVERED TO TIFFANY ST. IRON SHOP (1/12/2003).
CARS 9156, 9157 ARE USED FOR POLICE TRAINING AT FLOYD BENNETT FIELD (2000).
EDO SPEEDOMETERS INSTALLED ON ALL CARS (1997).

CARS 8968, 8969, 9114, 9131, 9213 & 9224 WERE SCRAPPED.
CARS 8950, 8951 FOR PRESERVATION
SOME CARS CONVERTED TO WORK CARS. REMAINING CARS WERE SCRAPPED.

CAR NUMBERS. 9306-9345
TOTAL : 40 CARS
BUILT BY: ST. LOUIS CAR

DATE: 1963
AVERAGE COST PER CAR: $108,500

# R33S (IRT)

PICTURE WINDOW

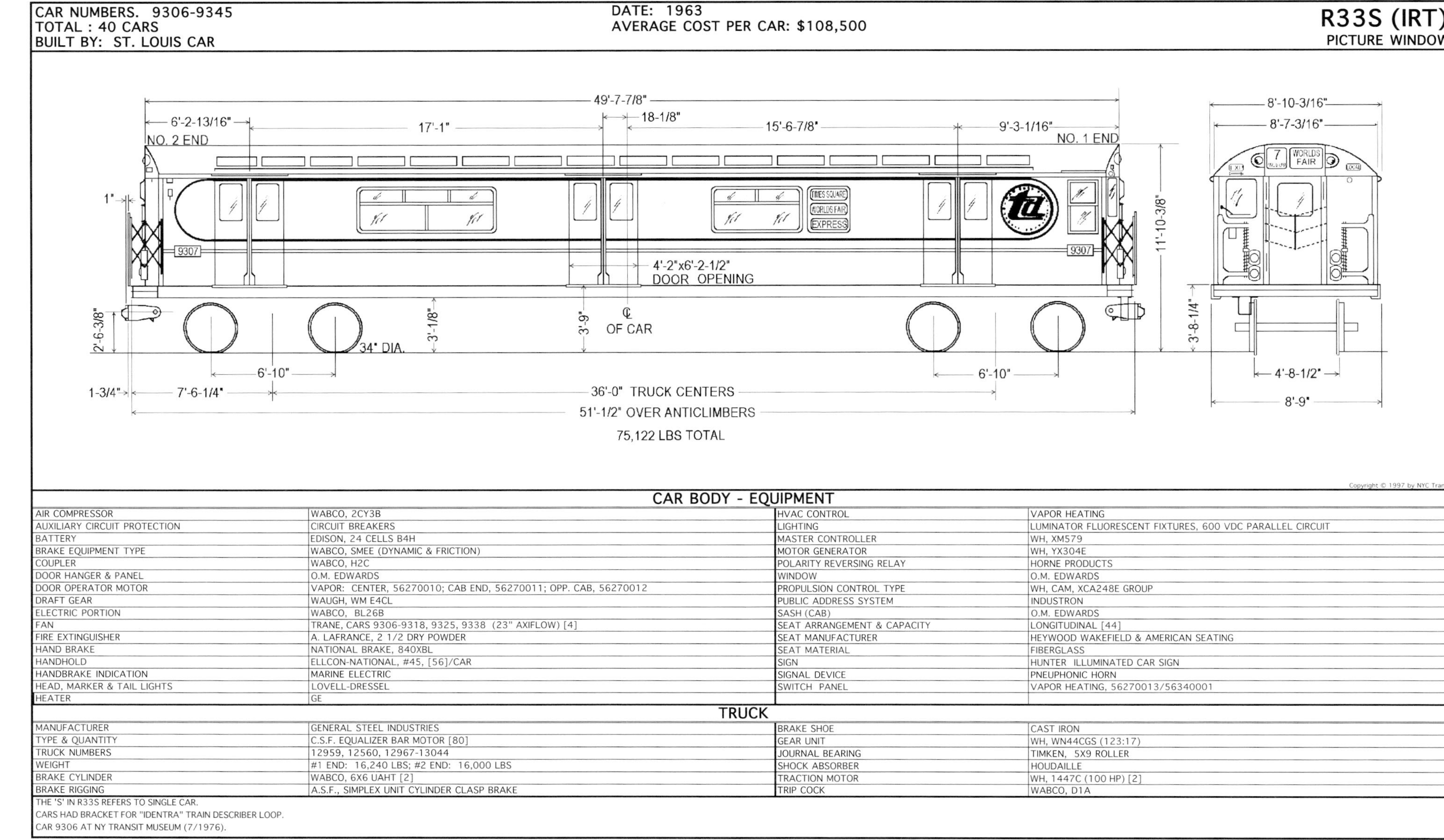


Copyright © 1997 by NYC Transit

## CAR BODY - EQUIPMENT

| | | | |
|---|---|---|---|
| AIR COMPRESSOR | WABCO, 2CY3B | HVAC CONTROL | VAPOR HEATING |
| AUXILIARY CIRCUIT PROTECTION | CIRCUIT BREAKERS | LIGHTING | LUMINATOR FLUORESCENT FIXTURES, 600 VDC PARALLEL CIRCUIT |
| BATTERY | EDISON, 24 CELLS B4H | MASTER CONTROLLER | WH, XM579 |
| BRAKE EQUIPMENT TYPE | WABCO, SMEE (DYNAMIC & FRICTION) | MOTOR GENERATOR | WH, YX304E |
| COUPLER | WABCO, H2C | POLARITY REVERSING RELAY | HORNE PRODUCTS |
| DOOR HANGER & PANEL | O.M. EDWARDS | WINDOW | O.M. EDWARDS |
| DOOR OPERATOR MOTOR | VAPOR: CENTER, 56270010; CAB END, 56270011; OPP. CAB, 56270012 | PROPULSION CONTROL TYPE | WH, CAM, XCA248E GROUP |
| DRAFT GEAR | WAUGH, WM E4CL | PUBLIC ADDRESS SYSTEM | INDUSTRON |
| ELECTRIC PORTION | WABCO, BL26B | SASH (CAB) | O.M. EDWARDS |
| FAN | TRANE, CARS 9306-9318, 9325, 9338 (23" AXIFLOW) [4] | SEAT ARRANGEMENT & CAPACITY | LONGITUDINAL [44] |
| FIRE EXTINGUISHER | A. LAFRANCE, 2 1/2 DRY POWDER | SEAT MANUFACTURER | HEYWOOD WAKEFIELD & AMERICAN SEATING |
| HAND BRAKE | NATIONAL BRAKE, 840XBL | SEAT MATERIAL | FIBERGLASS |
| HANDHOLD | ELLCON-NATIONAL, #45, [56]/CAR | SIGN | HUNTER ILLUMINATED CAR SIGN |
| HANDBRAKE INDICATION | MARINE ELECTRIC | SIGNAL DEVICE | PNEUPHONIC HORN |
| HEAD, MARKER & TAIL LIGHTS | LOVELL-DRESSEL | SWITCH PANEL | VAPOR HEATING, 56270013/56340001 |
| HEATER | GE | | |

## TRUCK

| | | | |
|---|---|---|---|
| MANUFACTURER | GENERAL STEEL INDUSTRIES | BRAKE SHOE | CAST IRON |
| TYPE & QUANTITY | C.S.F. EQUALIZER BAR MOTOR [80] | GEAR UNIT | WH, WN44CGS (123:17) |
| TRUCK NUMBERS | 12959, 12560, 12967-13044 | JOURNAL BEARING | TIMKEN, 5X9 ROLLER |
| WEIGHT | #1 END: 16,240 LBS; #2 END: 16,000 LBS | SHOCK ABSORBER | HOUDAILLE |
| BRAKE CYLINDER | WABCO, 6X6 UAHT [2] | TRACTION MOTOR | WH, 1447C (100 HP) [2] |
| BRAKE RIGGING | A.S.F., SIMPLEX UNIT CYLINDER CLASP BRAKE | TRIP COCK | WABCO, D1A |

THE 'S' IN R33S REFERS TO SINGLE CAR.
CARS HAD BRACKET FOR "IDENTRA" TRAIN DESCRIBER LOOP.
CAR 9306 AT NY TRANSIT MUSEUM (7/1976).

CAR NUMBERS: 9306-9345
TOTAL: 40 CARS
BUILT BY: ST. LOUIS CAR
DATE: 1963
AVERAGE COST PER CAR: $108,500

OVERHAULED BY: NYC TRANSIT AUTHORITY
DATE: 1985
OVERHAUL COST PER CAR: $195,600

# R33S (IRT)

PICTURE WINDOW

GOH

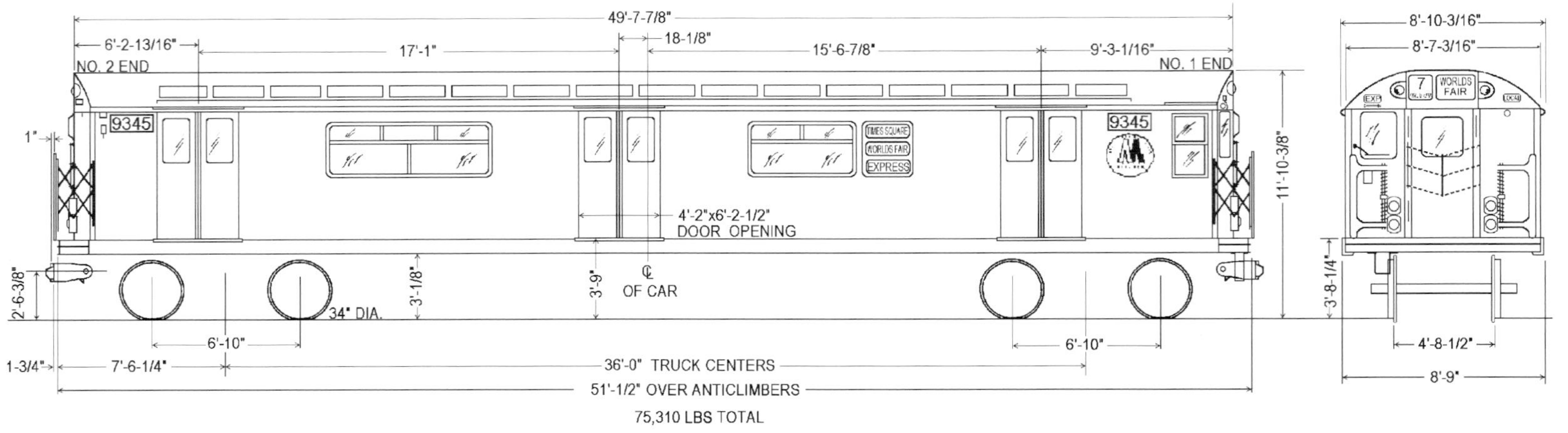

49'-7-7/8"
6'-2-13/16"
17'-1"
18-1/8"
15'-6-7/8"
9'-3-1/16"
NO. 2 END
NO. 1 END
9345
1"
TIMES SQUARE
WORLDS FAIR
EXPRESS
9345
11'-10-3/8"
4'-2"x6'-2-1/2"
DOOR OPENING
2'-6-3/8"
3'-1/8"
3'-9"
℄
OF CAR
34" DIA.
6'-10"
6'-10"
1-3/4"
7'-6-1/4"
36'-0" TRUCK CENTERS
51'-1/2" OVER ANTICLIMBERS
75,310 LBS TOTAL
8'-10-3/16"
8'-7-3/16"
7
WORLDS FAIR
3'-8-1/4"
4'-8-1/2"
8'-9"

Copyright © 1997 by NYC Transit

## CAR BODY - EQUIPMENT

| | | | |
|---|---|---|---|
| AIR COMPRESSOR | WABCO, D4 | HEATER | GE |
| AUXILIARY CIRCUIT PROTECTION | CIRCUIT BREAKERS | HVAC CONTROL | VAPOR HEATING |
| BATTERY | EDISON, 24 CELLS B4H | LIGHTING | LUMINATOR FLUORESCENT FIXTURES, 600 VDC PARALLEL CIRCUIT |
| BRAKE EQUIPMENT TYPE | WABCO, SMEE (DYNAMIC & FRICTION) | MASTER CONTROLLER | WH, XM579 |
| CONVERTER | McGRAW EDISON/TOSHIBA | POLARITY REVERSING RELAY | HORNE PRODUCTS |
| COUPLER | WABCO, H2C | PROPULSION CONTROL TYPE | WH, CAM, XCA248E GROUP |
| DOOR HANGER & PANEL | O.M. EDWARDS | PUBLIC ADDRESS SYSTEM | INDUSTRON |
| DOOR OPERATOR MOTOR | VAPOR; WESTCODE | SASH (CAB) | O.M. EDWARDS |
| DRAFT GEAR | WAUGH, WM E4CL | SEAT ARRANGEMENT & CAPACITY | LONGITUDINAL [44] |
| ELECTRIC PORTION | WABCO BL26B | SEAT MANUFACTURER | HEYWOOD WAKEFIELD/AMERICAN SEATING |
| FAN | 23" AXIFLO, [4] | SEAT MATERIAL | FIBERGLASS |
| HAND BRAKE | NATIONAL BRAKE, 840XBL | SIGN | HUNTER ILLUMINATED CAR SIGN |
| HANDHOLD | ELLCON-NATIONAL, #45, [56] | SWITCH PANEL | VAPOR HEATING, 56270013; 56340001 |
| HANDBRAKE INDICATION | MARINE ELECTRIC | WINDOW | O.M. EDWARDS |
| HEAD, MARKER & TAIL LIGHTS | LOVELL-DRESSEL | SIGNAL DEVICE | PNEUPHONIC HORN |

## TRUCK

| | | | |
|---|---|---|---|
| MANUFACTURER | GENERAL STEEL INDUSTRIES | BRAKE SHOE | COMPOSITION TYPE |
| TYPE & QUANTITY | C.S.F. EQUALIZER BAR MOTOR [80] | GEAR UNIT | WH, WN44CGS (123:17) |
| TRUCK NUMBERS | 12959, 12560, 12967-13044 | JOURNAL BEARING | 5X9 AP ROLLER, SEE NOTE 1 |
| WEIGHT | #1 END: 19,580 LBS; #2 END: 19,270 LBS | SHOCK ABSORBER | MONROE (PISTON TYPE) |
| BRAKE CYLINDER | WABCO, 6X6 UAHT [ 2] | TRACTION MOTOR | WH, 1447C (100 HP) [2] |
| BRAKE RIGGING | A.S.F. SIMPLEX UNIT CYLINDER CLASP BRAKE | TRIP COCK | WABCO, D1A |

NOTE 1: VARIOUS VENDORS.

THE 'S' IN R33S REFERS TO SINGLE CAR.
CARS HAD BRACKET FOR "IDENTRA" TRAIN DESCRIBER LOOP.
CAR 9306 WAS NOT OVERHAULED AND IS AT NY TRANSIT MUSEUM (7/1976).
OVERHAULED AT C.I. SHOP.

CAR 9321 WAS SCRAPPED (2001).
CAR 9327 FOR PRESERVATION.
ALL OTHER CARS IN WORK SERVICE (2003).

CAR NUMBERS: 8010-8019
TOTAL: 10 CARS
BUILT BY: BUDD
DATE: 1949
AVERAGE COST PER CAR: $121,373 (BID)

OVERHAULED BY: NYC TRANSIT AUTHORITY
DATE: 1964 - 65, SEE NOTE 1
OVERHAUL COST PER CAR: N/A

# R34 (BMT/IND)

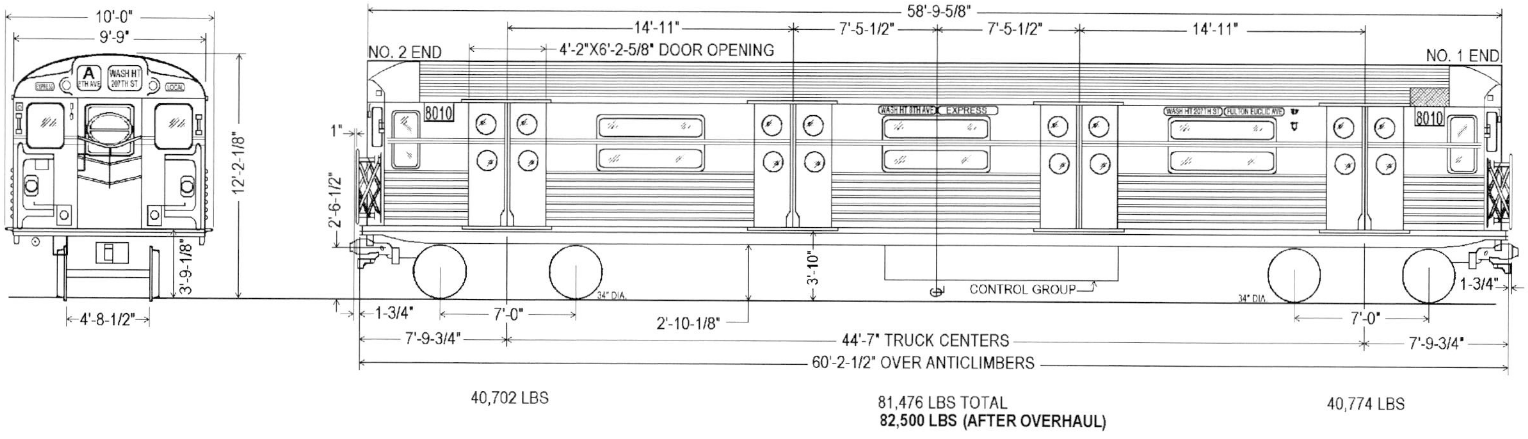

10'-0"
9'-9"
12'-2-1/8"
3'-9-1/8"
4'-8-1/2"

58'-9-5/8"
14'-11"
7'-5-1/2"
7'-5-1/2"
14'-11"
NO. 2 END
4'-2"X6'-2-5/8" DOOR OPENING
NO. 1 END
1"
2'-6-1/2"
3'-10"
CONTROL GROUP
34" DIA.
1-3/4"
7'-0"
2'-10-1/8"
7'-0"
7'-9-3/4"
44'-7" TRUCK CENTERS
7'-9-3/4"
60'-2-1/2" OVER ANTICLIMBERS

40,702 LBS

81,476 LBS TOTAL
**82,500 LBS (AFTER OVERHAUL)**

40,774 LBS

Copyright © 1997 by NYC Transit

## CAR BODY - EQUIPMENT

| | | | |
|---|---|---|---|
| AIR COMPRESSOR | WABCO, 3YC (21 CFM) | LIGHTING | LUMINATOR, 72" LONG [24] , 48" LONG [2] |
| AIR DIFFUSER | MODULAR VANES | MARKER LIGHT | LOVELL DRESSEL (SEMAPHORE) |
| BATTERY | EDISON, 24 CELL B4H WOOD BOX | MASTER CONTROLLER | WH XM179 |
| BRAKE CYLINDER | WABCO, C5 | MOTOR GENERATOR | YX304E |
| BRAKE EQUIPMENT TYPE | WABCO, SMEE (DYNAMIC & FRICTION) | PROPULSION CONTROL TYPE | WH, ABS, UP631B GROUP |
| BRAKE VALVE | WABCO, ME42 | PUBLIC ADDRESS SYSTEM | RAYMOND ROSEN |
| COUPLER | WABCO, H2C, BL25B | SASH (CAB) | ADAMS & WESTLAKE |
| DOOR HANGER | OTIS ELEVATOR | SASH, END DOOR EQUATORIAL | HUNTER SASH |
| DOOR OPERATOR MOTOR | NATIONAL PNEUMATIC, CARS 8010-8014<br>CONSOLIDATED CAR HEATING, CARS 8015-8016 | SEAT ARRANGEMENT & CAPACITY | LONGITUDINAL & CROSS [56] |
| | | SEAT CUSHION & BACK | BUDD CAR |
| DRAFT GEAR | WAUGH, WM E3-4, E4-4 | SIGN | HUNTER ILLUMINATED CAR SIGN |
| ELECTRIC PORTION | WABCO BL25B | SIGNAL DEVICE | PNEUPHONIC HORN |
| FIRE EXTINGUISHER | 2 LBS DRY POWDER | SWITCH PANEL | CONSOLIDATED CAR HEATING: #1 END, PS68A; #2 END, PS69A |
| HAND BAR | ELLCON NATIONAL | TAIL LIGHT | TWIN SOCKET |
| HAND BRAKE | NATIONAL BRAKE 840XAL | THERMOSTAT | VAPOR |
| HEAT & VENTILATION CONTROL | VAPOR | VENTILATION | AXIFLO |
| HEATER | CONSOLIDATED CAR HEATING | WINDOW | ADAMS & WESTLAKE |

## TRUCK

| | | | |
|---|---|---|---|
| MANUFACTURER | CLARK EQUIPMENT (G.S.C.)<br>TRUCK FRAME PATT. NO 55027 | GEAR UNIT | CLARK EQUIPMENT , HYPOID GEAR SPICER DRIVE |
| | | JOURNAL BEARING | SKF, 5x9 ROLLER |
| TYPE & QUANTITY | C.S.F. EQUALIZER BAR MOTOR [22] | SHOCK ABSORBER | MONROE, LATERAL [1] & VERTICAL [2] |
| TRUCK NUMBERS | 8010-8031 | TRACTION MOTOR | GE, 1240B (100 HP) [2] |
| WEIGHT | #1 END: 18,110 LBS; #2 END; 18,110 LBS | TRIP COCK | WABCO, D1 |
| BRAKE RIGGING | A.S.F., SIMPLEX UNIT CYLINDER CLASP BRAKE | | |

NOTE 1: OVERHAULED AT C.I. SHOP FROM R11 CONTRACT.

HEADLIGHTS INSTALLED UPON OVERHAUL.
CARS 8010-8016 TRANSFERRED TO BMT (6/14/1954).
CARS 8017-8019 TRANSFERRED TO BMT (6/21/1954).

REMOVED PRECIPITRON UNIT AND REPLACED DRUM BRAKE WITH DISC BRAKE DURING OVERHAUL.
SHOT WELDED, STAINLESS STEEL CAR BODY.
CAR 8013 AT NY TRANSIT MUSEUM.
ALL OTHER CARS WERE SOLD OR SCRAPPED (1980).

CAR NUMBERS: 9346 - 9523, 9558 - 9769
TOTAL: 390 CARS
BUILT BY: ST. LOUIS CAR

DATE: 1963 - 64
AVERAGE COST PER CAR: $110,563

# R36 (IRT)
PICTURE WINDOW

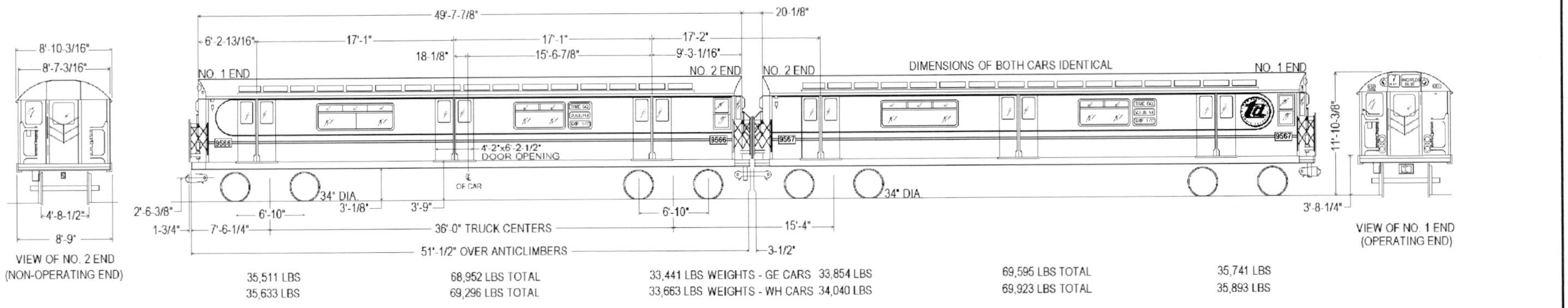

8'-10-3/16"
8'-7-3/16"
4'-8-1/2"
8'-9"
VIEW OF NO. 2 END
(NON-OPERATING END)

49'-7-7/8"
20-1/8"
6'-2-13/16"
17'-1"
17'-1"
17'-2"
18-1/8"
15'-6-7/8"
9'-3-1/16"
NO. 1 END
NO. 2 END
NO. 2 END
DIMENSIONS OF BOTH CARS IDENTICAL
NO. 1 END
4'-2"x6'-2-1/2" DOOR OPENING
℄ OF CAR
34" DIA.
34" DIA.
2'-6-3/8"
1-3/4"
6'-10"
3'-1/8"
3'-9"
6'-10"
7'-6-1/4"
36'-0" TRUCK CENTERS
15'-4"
51'-1/2" OVER ANTICLIMBERS
3-1/2"
11'-10-3/8"
3'-8-1/4"
VIEW OF NO. 1 END
(OPERATING END)

35,511 LBS
35,633 LBS
68,952 LBS TOTAL
69,296 LBS TOTAL
33,441 LBS WEIGHTS - GE CARS 33,854 LBS
33,663 LBS WEIGHTS - WH CARS 34,040 LBS
69,595 LBS TOTAL
69,923 LBS TOTAL
35,741 LBS
35,893 LBS

Copyright © 1997 by NYC Transit

## CAR BODY - EQUIPMENT

| | | | | |
|---|---|---|---|---|
| AUXILIARY CIRCUIT PROTECTION | | CIRCUIT BREAKERS | HEAT & VENT. CONTROL | VAPOR |
| BRAKE EQUIPMENT TYPE | | WABCO, SMEE (DYNAMIC & FRICTION) | HEATER | BRYANT |
| BRAKE VALVE | | WABCO ME42B | LIGHTING | LUMINATOR, FLUORESCENT FIXTURES, 600 VDC PARALLEL CIRCUIT |
| COUPLER | | WABCO: #1 END, H2C; #2 END, LINK BAR 565150 | MASTER CONTROLLER | GE, 17KC76C2; WH, XM 579 |
| DOOR HANGER & PANEL | | O.M. EDWARDS | POLARITY REVERSING RELAY | HORNE PRODUCTS |
| DOOR OPERATOR MOTOR | CENTER | MIDLAND ROSS, 500003; VAPOR HEAT, 56370002, SEE NOTE 1 | PUBLIC ADDRESS SYSTEM | INDUSTRON |
| | CAB END | MIDLAND ROSS, 500005; VAPOR HEAT, 56370004, SEE NOTE 1 | SASH (CAB) | O.M. EDWARDS |
| | OPP. END | MIDLAND ROSS, 500004; VAPOR HEAT, 56370003, SEE NOTE 1 | SEAT ARRANGEMENT & CAPACITY | LONGITUDINAL [44] |
| DRAFT GEAR | | WAUGH, WM E4CL | SEAT MANUFACTURER | GENERAL SEATING (T/O) |
| ELECTRIC PORTION | | WABCO BL26B | | HEYWOOD WAKEFIELD (PASSENGER) |
| FAN | | WH, #356C647G05, 23 " AXIFLO, [4], ALL CARS | SEAT MATERIAL | FIBERGLASS |
| FIRE EXTINGUISHER | | A. LAFRANCE, 2.5 LBS DRY POWDER | SIGN | HUNTER ILLUMINATED CAR SIGN |
| HEAD, MARKER & TAIL LIGHTS | | LOVELL DRESSEL | SIGNAL DEVICE | PNEUPHONIC HORN |
| HAND BRAKE | | NATIONAL BRAKE, 840XBL | WINDOW | O.M. EDWARDS |
| HANDHOLD | | ELLCON NATIONAL, #45 [56] | | |

| EVEN-NUMBERED CARS ONLY | | | ODD-NUMBERED CARS ONLY | | |
|---|---|---|---|---|---|
| BATTERY | | EDISON, 24 CELL B4H WOOD BOX | AIR COMPRESSOR | | WABCO, 2CY3B |
| MOTOR/GENERATOR | | GE, 5GMG 153LI WH, YX304E | AIR COMPRESSOR MOTOR | | WABCO, A3 |
| PROPULSION CONTROL TYPE | | GE, SCM, 17KG192F5 GROUP | AIR COMPRESSOR SYSTEM | | WABCO, A6A |
| | | WH, CAM, XCA248I GROUP | PROPULSION CONTROL TYPE | | GE, SCM, 17KG192H5 GROUP |
| SWITCH PANEL | #1 END | MIDLAND ROSS, 500009B; VAPOR, 56270014, SEE NOTE 1 | | | WH, CAM XCA248J GROUP |
| | #2 END | MIDLAND ROSS, 5000010A; VAPOR, 56340006, SEE NOTE 1 | SWITCH PANEL | #1 END | MIDLAND ROSS, 500007B; VAPOR, 56270013, SEE NOTE 1 |
| | | | | #2 END | MIDLAND ROSS, 500008A; VAPOR, 56340005, SEE NOTE 1 |

## TRUCK

| | | | |
|---|---|---|---|
| MANUFACTURER | GENERAL STEEL INDUSTRIES | GEAR UNIT | WH WN44CGS; GE GA25A2 (123:17) |
| TYPE & QUANTITY | C.S.F. EQUALIZER BAR MOTOR [7800] | JOURNAL BEARING | SKF, 5X9 ROLLER |
| TRUCK NUMBERS | WH: 13045-13396,13399-13402; GE: 13469 - 13892 | SHOCK ABSORBER | HOUDAILLE |
| WEIGHT | #1 END: 16,240 LBS; #2 END: 16,000 LBS | TRACTION MOTOR | WH 1447C (100 HP) [2] ; GE 1240A5 (100 HP) [2] |
| BRAKE CYLINDER | WABCO, 6X6 UAHT [2] | TRIP COCK | WABCO, D1A |
| BRAKE RIGGING | A.S.F. SIMPLEX UNIT CYLINDER CLASP BRAKE | | |

NOTE 1: MIDLAND ROSS:9396 - 9501, 9503, 9505, 9514 - 9523, 9558 -9686, 9688, 9738 - 9739. (250 CARS)
VAPOR: 9346 - 9395, 9502, 9504, 9506 - 9513, 9687, 9689 - 9737, 9740 - 9769. (140 CARS)

CARS 9346 - 9523 WITH WH EQUIPMENT AND CARS 9558 - 9769 WITH GE EQUIPMENT.
CARS EQUIPPED WITH BRACKET FOR "IDENTRA" TRAIN DESCRIBER LOOP.

CAR NUMBERS: 9346 - 9523, 9558 - 9769
TOTAL: 390 CARS
BUILT BY: ST. LOUIS CAR
DATE: 1963 - 64
AVERAGE COST PER CAR: $110,563

OVERHAUL BY: MK, NAB, & NYC TRANSIT AUTHORITY
DATE: 1982 - 85
OVERHAUL COST PER CAR: SEE NOTE 1

# R36 (IRT)

PICTURE WINDOW

GOH

8'-10-3/16"
8'-7-3/16"
4'-8-1/2"
8'-9"
VIEW OF NO. 2 END (NON-OPERATING END)
49'-7-7/8"
20-1/8"
6'-2-13/16"
17'-1"
17'-1"
17'-2"
18 1/8"
15'-6-7/8"
9'-3-1/16"
NO. 1 END
NO. 2 END
NO. 2 END
DIMENSIONS OF BOTH CARS IDENTICAL
NO. 1 END
4'-2"x6'-2-1/2" DOOR OPENING
℄ OF CAR
34" DIA
34" DIA
2'-6-3/8"
1-3/4"
6'-10"
3'-1/8"
3'-9"
6'-10"
7'-6-1/4"
36'-0" TRUCK CENTERS
15'-4"
51'-1/2" OVER ANTICLIMBERS
3-1/2"
11'-10-3/8"
3'-8-1/4"
VIEW OF NO. 1 END (OPERATING END)
73,530 LBS TOTAL
72,910 LBS TOTAL
GE CARS
WH CARS
73,100 LBS TOTAL
73,080 LBS TOTAL

Copyright © 1997 by NYC Transit

## CAR BODY - EQUIPMENT

| | | | | | |
|---|---|---|---|---|---|
| AIR CONDITIONING | | STONE SAFETY | HEATER | | BRYANT |
| AUXILIARY CIRCUIT PROTECTION | | CIRCUIT BREAKERS | HVAC CONTROL | | STONE SAFETY |
| BRAKE EQUIPMENT TYPE | | WABCO, SMEE (DYNAMIC & FRICTION) | LIGHTING | | LUMINATOR, FLUORESCENT FIXTURES, 600 VDC PARALLEL CIRCUIT |
| BRAKE VALVE | | WABCO, ME42B | MASTER CONTROLLER | | GE, 17KC76C2; WH, XM 579 |
| COUPLER | | WABCO: #1 END, H2C; #2 END, LINK BAR 565150 | PUBLIC ADDRESS SYSTEM | | COMCO |
| DOOR HANGER & PANEL | | O.M. EDWARDS | SASH (CAB) | | O.M. EDWARDS |
| DOOR OPERATOR MOTOR | | SEE NOTE 4 | SEAT MATERIAL | | FIBERGLASS |
| DRAFT GEAR | | WAUGH, WM E4CL | SEAT ARRANGEMENT & CAPACITY | | LONGITUDINAL [44] |
| ELECTRIC PORTION | | WABCO, BL26B | SEAT MANUFACTURER | | GENERAL SEATING (MOTORMAN) HEYWOOD WAKEFIELD (PASSENGER) |
| HAND BRAKE | | NATIONAL BRAKE, 840XBL | SIGN | | HUNTER ILLUMINATED CAR SIGN |
| HANDHOLD | | ELLCON NATIONAL, #45 [56] | SIGNAL DEVICE | | PNEUPHONIC HORN |
| HEAD, MARKER & TAIL LIGHTS | | LOVELL DRESSEL | WINDOW | | O.M. EDWARDS |
| EVEN-NUMBERED CARS ONLY | | | ODD-NUMBERED CARS ONLY | | |
| BATTERY | | HOPPECKE/EDISON | AIR COMPRESSOR | | WABCO, D4 |
| CONVERTER | | TOSHIBA/STONE SAFETY | PROPULSION CONTROL TYPE | | GE, SCM, 17KG192H5 GROUP |
| PROPULSION CONTROL TYPE | | GE, SCM, 17KG192F5 GROUP | | | WH, CAM, XCA-248J GROUP |
| | | WH, CAM, XCA248I GROUP | SWITCH PANEL | #1 END | MIDLAND ROSS, 500007B; VAPOR, 56270013, SEE NOTES 4 |
| SWITCH PANEL | #1 END | MIDLAND ROSS, 500009B; VAPOR, 56270014, SEE NOTES 4 | | #2 END | MIDLAND ROSS, 500008A; VAPOR, 56340005, SEE NOTES 4 |
| | #2 END | MIDLAND ROSS, 5000010A; VAPOR, 56340006, SEE NOTES 4 | | | |

## TRUCK

| | | | | | |
|---|---|---|---|---|---|
| MANUFACTURER | | GENERAL STEEL INDUSTRIES | BRAKE RIGGING | | A.S.F. SIMPLEX UNIT CYLINDER CLASP BRAKE |
| TYPE & QUANTITY | | C.S.F. EQUALIZER BAR MOTOR [780] | GEAR UNIT | | WH, WN44CGS; GE, GA23A2 (123:17) |
| TRUCK NUMBERS | | N/A | JOURNAL BEARING | | 5X9 AP ROLLER, SEE NOTE 3 |
| WEIGHT | #1 END | GE, 16,270 LBS; WH, 19,490 LBS | SHOCK ABSORBER | | MONROE |
| | #2 END | GE, 19,090 LBS; WH, 19,190 LBS | TRACTION MOTOR | | WH, 1447C (100 HP) [2]; GE, 1240A5 (100 HP) [2] |
| BRAKE CYLINDER | | WABCO, 6X6 UAHT [2] | TRIP COCK | | WABCO, D1A |

NOTE 1: $163,000 FOR NYAB, AND MK; $164,000 FOR NYC TRANSIT AUTHORITY.
NOTE 2: THREE DIFFERENT BATTERIES DEPENDING ON PROPULSION CONTROL GROUP MANUFACTURER.
NOTE 3: VARIOUS VENDORS.
NOTE 4: DOOR CONTROL SYSTEM, OPERATORS & SWITCH PANELS BY: MIDLAND ROSS: CARS 9396 - 9501, 9503, 9505, 9514 - 9523, 9558 -9686, 9688, 9738 - 9739 (250 CARS).
VAPOR: CARS 9346 - 9395, 9502, 9504, 9506 - 9513, 9687, 9689 - 9737, 9740 - 9769 (140 CARS).

CARS 9400, 9401 FOR PRESERVATION.

THESE CARS EQUIPPED WITH BRACKET FOR "IDENTRA" TRAIN DESCRIBER LOOP.
REPLACED PROPULSION CONTROL BOX GROUP WITH GE 17KG192A (1995).
CARS 9346 - 9523 WITH WH EQUIPMENT AND CARS 9558 - 9769 WITH GE EQUIPMENT.
MOST CARS IN THIS SERIES WERE SCRAPPED (2003).

CAR NUMBERS: 9524 - 9557
TOTAL: 34 CARS
BUILT BY: ST. LOUIS CAR

DATE: 1964
AVERAGE COST PER CAR: $110,563

# R36 (IRT)

DROP SASH

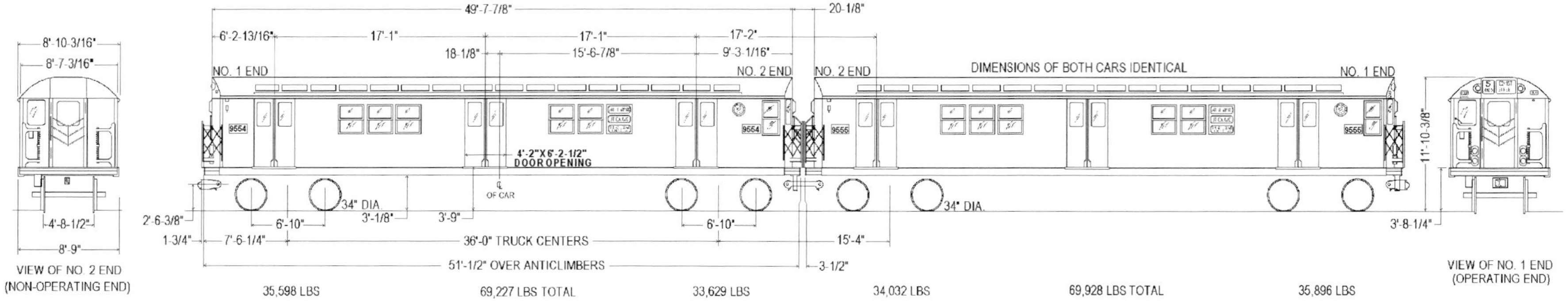

8'-10-3/16"
8'-7-3/16"
4'-8-1/2"
8'-9"
VIEW OF NO. 2 END (NON-OPERATING END)

49'-7-7/8"
20-1/8"
6'-2-13/16"
17'-1"
17'-1"
17'-2"
18-1/8"
15'-6-7/8"
9'-3-1/16"
NO. 1 END
NO. 2 END
NO. 2 END
DIMENSIONS OF BOTH CARS IDENTICAL
NO. 1 END
4'-2"X6'-2-1/2" DOOR OPENING
℄ OF CAR
34" DIA
3'-1/8"
3'-9"
34" DIA.
2'-6-3/8"
6'-10"
6'-10"
1-3/4"
7'-6-1/4"
36'-0" TRUCK CENTERS
15'-4"
51'-1/2" OVER ANTICLIMBERS
3-1/2"
11'-10-3/8"
3'-8-1/4"
35,598 LBS
69,227 LBS TOTAL
33,629 LBS
34,032 LBS
69,928 LBS TOTAL
35,896 LBS
VIEW OF NO. 1 END (OPERATING END)

Copyright © 1997 by NYC Transit

## CAR BODY - EQUIPMENT

| | | | |
|---|---|---|---|
| AUXILIARY CIRCUIT PROTECTION | CIRCUIT BREAKERS | HEAT & VENTILATION CONTROL | VAPOR |
| BRAKE EQUIPMENT TYPE | WABCO SMEE (DYNAMIC & FRICTION) | HEATER | BRYANT |
| BRAKE VALVE | WABCO, ME42B; | LIGHTING | LUMINATOR, FLUORESCENT FIXTURES, 600 VDC PARALLEL CIRCUIT |
| COUPLER | WABCO: #1 END, H2C; #2 END, LINK BAR 565150 | MASTER CONTROLLER | WH, XM 579 |
| DOOR HANGER & PANEL | O.M. EDWARDS | POLARITY REVERSING RELAY | HORNE PRODUCTS |
| DOOR OPERATOR MOTOR | MIDLAND ROSS: CENTER, 500003; CAB END, 500005; OPP. END, 500004 | PUBLIC ADDRESS SYSTEM | INDUSTRON |
| DRAFT GEAR | WAUGH, WM E4CL | SASH | O.M. EDWARDS |
| ELECTRIC PORTION | WABCO, BL26B | SEAT ARRANGEMENT & CAPACITY | LONGITUDINAL [44] |
| FAN | WH, #356C647G05, 23" AXIFLO [4] | SEAT MANUFACTURER | GENERAL SEATING (MOTORMAN), HEYWOOD WAKEFIELD (PASSENGER) |
| FIRE EXTINGUISHER | A. LA FRANCE, 2.5 LBS DRY POWDER | SEAT MATERIAL | FIBERGLASS |
| HAND BRAKE | NATIONAL BRAKE, 840XBL | SIGN | HUNTER ILLUMINATED CAR SIGN |
| HANDHOLD | ELLCON NATIONAL, #45 [56] | SIGNAL DEVICE | PNEUPHONIC HORN |
| HEAD, MARKER & TAIL LIGHTS | LOVELL DRESSEL | | |
| | EVEN-NUMBERED CARS ONLY | | ODD-NUMBERED CARS ONLY |
| BATTERY | EDISON, 24 CELL B4H WOOD BOX | AIR COMPRESSOR | WABCO, 2CY3B |
| MOTOR GENERATOR | WH, YX304E | AIR COMPRESSOR MOTOR | A3 |
| PROPULSION CONTROL TYPE | WH, CAM, XCA248I GROUP | AIR COMPRESSOR SYSTEM | WABCO, A6A |
| SWITCH PANEL | MIDLAND ROSS: #1 END, 500009B; #2 END, 5000010A | PROPULSION CONTROL TYPE | WH, CAM, XCA248J GROUP |
| | | SWITCH PANEL | MIDLAND ROSS: #1 END, 500007C; #2 END, 500008A |

## TRUCK

| | | | |
|---|---|---|---|
| MANUFACTURER | GENERAL STEEL INDUSTRIES | GEAR UNIT | WH, WN44CGS (123:17) |
| TYPE & QUANTITY | C.S.F. EQUALIZER BAR MOTOR [68] | JOURNAL BEARING | SKF, 5X9 ROLLER |
| TRUCK NUMBERS | 13397, 13398,13403 - 13468 | SHOCK ABSORBER | HOUDAILLE |
| WEIGHT | #1 END: 16,240 LBS; #2 END: 16,000 LBS | TRACTION MOTOR | WH, 1447C (100 HP) [2] |
| BRAKE CYLINDER | WABCO, 6X6 UAHT [2] | TRIP COCK | WABCO, D1A |
| BRAKE RIGGING | A.S.F. SIMPLEX UNIT CYLINDER CLASP BRAKE | | |

CAR NUMBERS: 9524 - 9557
TOTAL: 34 CARS
BUILT BY: ST. LOUIS CAR
DATE: 1964
AVERAGE COST PER CAR: $110,563

OVERHAULED BY: MK, NAB & NYC TRANSIT AUTHORITY
DATE: 1982 - 85
OVERHAUL COST PER CAR: SEE NOTE 1

# R36 (IRT)
DROP SASH
GOH

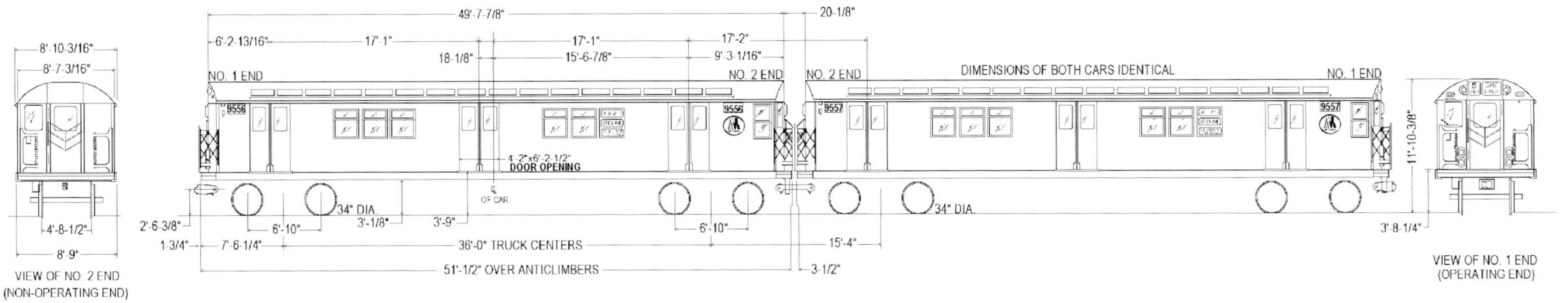

8'-10-3/16"
8'-7-3/16"
4'-8-1/2"
8'-9"
VIEW OF NO. 2 END
(NON-OPERATING END)

49'-7-7/8"
20-1/8"
6'-2-13/16"
17'-1"
18-1/8"
17'-1"
15'-6-7/8"
17'-2"
9'-3-1/16"
NO. 1 END
NO. 2 END
NO. 2 END
DIMENSIONS OF BOTH CARS IDENTICAL
NO. 1 END
9556
9556
9557
9557
4-2"x6'-2-1/2" DOOR OPENING
℄ OF CAR
34" DIA
34" DIA.
2'-6-3/8"
1-3/4"
6'-10"
3'-1/8"
3'-9"
6'-10"
7'-6-1/4"
36'-0" TRUCK CENTERS
15'-4"
51'-1/2" OVER ANTICLIMBERS
3-1/2"
11'-10-3/8"
3'-8-1/4"
VIEW OF NO. 1 END
(OPERATING END)

Copyright © 1997 by NYC Transit

## CAR BODY - EQUIPMENT

| | | | |
|---|---|---|---|
| AIR CONDITIONING | STONE SAFETY, 12 TON | HEATER | BRYANT; GE |
| AUXILIARY CIRCUIT PROTECTION | CIRCUIT BREAKERS | HVAC CONTROL | STONE SAFETY |
| BRAKE EQUIPMENT TYPE | WABCO, SMEE (DYNAMIC & FRICTION) | LIGHTING | LUMINATOR, FLUORESCENT FIXTURES, 600 VDC PARALLEL CIRCUIT |
| BRAKE VALVE | WABCO, ME42B; | MASTER CONTROLLER | WH, XM 579 |
| COUPLER | WABCO: #1 END, H2C; #2 END, LINK BAR 565150 | POLARITY REVERSING RELAY | HORNE PRODUCTS |
| DOOR HANGER & PANEL | O.M. EDWARDS | PUBLIC ADDRESS SYSTEM | COMCO |
| DOOR OPERATOR MOTOR | VAPOR | SASH | O.M. EDWARDS |
| DRAFT GEAR | WAUGH, WM E4CL | SEAT ARRANGEMENT & CAPACITY | LONGITUDINAL [44] |
| ELECTRIC PORTION | WABCO, BL26B | SEAT MANUFACTURER | GENERAL SEATING (TRAIN OPERATOR), HEYWOOD WAKEFIELD (PASSENGER) |
| HAND BRAKE | NATIONAL BRAKE, 840XBL | SEAT MATERIAL | FIBERGLASS |
| HANDHOLD | ELLCON NATIONAL, #45 [56] | SIGN | HUNTER ILLUMINATED CAR SIGN |
| HEAD, MARKER & TAIL LIGHTS | LOVELL DRESSEL | SIGNAL DEVICE | PNEUPHONIC HORN |
| EVEN-NUMBERED CARS ONLY | | ODD-NUMBERED CARS ONLY | |
| BATTERY | HOPPECKE; EDISON | AIR COMPRESSOR | WABCO, D4S |
| CONVERTER | MCGRAW; EDISON | PROPULSION CONTROL TYPE | WH, CAM, XCA248J GROUP |
| PROPULSION CONTROL TYPE | WH, CAM, XCA248I GROUP | SWITCH PANEL | MIDLAND ROSS: #1 END, 500007B; #2 END, 500008A |
| SWITCH PANEL | MIDLAND ROSS: #1 END, 500009B; #2 END, 5000010A | | |

## TRUCK

| | | | |
|---|---|---|---|
| MANUFACTURER | GENERAL STEEL INDUSTRIES | GEAR UNIT | WH, WN44CGS, 123:17 |
| TYPE & QUANTITY | C.S.F. EQUALIZER BAR MOTOR [68] | JOURNAL BEARING | 5X9 AP ROLLER, SEE NOTE 3 |
| TRUCK NUMBERS | N/A | SHOCK ABSORBER | MONROE |
| WEIGHT | #1 END: 16,240 LBS; #2 END: 16,000 LBS | TRACTION MOTOR | WH, 1447C (100 HP) [2] |
| BRAKE CYLINDER | WABCO, 6X6 UAHT [2] | TRIP COCK | WABCO, D1A |
| BRAKE RIGGING | A.S.F. SIMPLEX UNIT CYLINDER CLASP BRAKE | | |

NOTE 1: $163,000 FOR NAB, AND MK; $164,000 FOR NYC TRANSIT AUTHORITY.
NOTE 2: THREE DIFFERENT BATTERIES DEPENDING ON PROPULSION CONTROL GROUP MANUFACTURER.
NOTE 3: VARIOUS VENDORS.

CARS 9542 AND 9543 ARE IN NY TRANSIT MUSEUM COLLECTION (2003).

SUBSEQUENT TO GOH, DROP-SASH WINDOWS (EXCEPT TRAIN OPERATOR'S AND CONDUCTOR'S) REPLACED WITH LOUVERED WINDOWS BY NYCT.
INSTALLED CONDUCTOR'S EMERGENCY VALVE ALARM BY EDO.
EDO SPEEDOMETERS INSTALLED ON 178 CARS (1997).
MOST CARS IN THIS SERIES HAVE BEEN SCRAPPED (2003).

CAR NUMBERS: 3950 - 4139
TOTAL: 190 CARS
BUILT BY: ST. LOUIS CAR

DATE: 1966 - 67
AVERAGE COST PER CAR: $111,733

# R38 (BMT/IND)

NON-AIR CONDITIONED

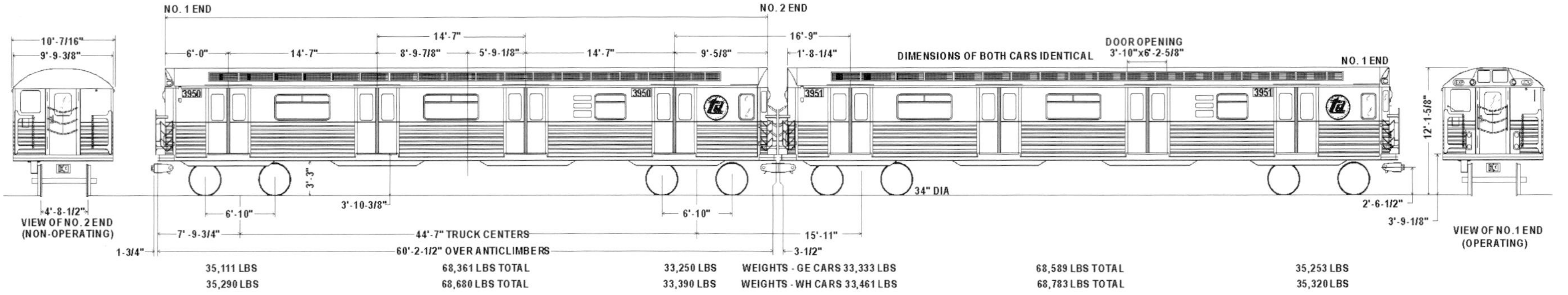

NO. 1 END
NO. 2 END
NO. 1 END
DIMENSIONS OF BOTH CARS IDENTICAL
DOOR OPENING 3'-10"x6'-2-5/8"
10'-7/16"
9'-9-3/8"
4'-8-1/2"
VIEW OF NO. 2 END (NON-OPERATING)
6'-0" 14'-7" 14'-7" 8'-9-7/8" 5'-9-1/8" 14'-7" 9'-5/8" 16'-9" 1'-8-1/4"
3'-3" 3'-10-3/8" 6'-10" 6'-10" 34" DIA
1-3/4" 7'-9-3/4" 44'-7" TRUCK CENTERS 60'-2-1/2" OVER ANTICLIMBERS 15'-11" 3-1/2"
12'-1-5/8" 2'-6-1/2" 3'-9-1/8"
VIEW OF NO. 1 END (OPERATING)

35,111 LBS / 35,290 LBS — 68,361 LBS TOTAL / 68,680 LBS TOTAL — 33,250 LBS / 33,390 LBS
WEIGHTS - GE CARS 33,333 LBS
WEIGHTS - WH CARS 33,461 LBS
68,589 LBS TOTAL / 68,783 LBS TOTAL — 35,253 LBS / 35,320 LBS

Copyright © 1997 by NYC Transit

## CAR BODY - EQUIPMENT

| | | | | | |
|---|---|---|---|---|---|
| AUXILIARY CIRCUIT PROTECTION | | CIRCUIT BREAKERS | HEAT & VENTILATION CONTROL | | VAPOR |
| BRAKE EQUIPMENT TYPE | | WABCO, SMEE (DYNAMIC & FRICTION) | HEATER | | ST. LOUIS CAR |
| BRAKE VALVE | | WABCO, ME43 | LIGHTING | | LUMINATOR, FLUORESCENT FIXTURES, 600 VDC PARALLEL CIRCUIT, |
| COUPLER | | WABCO: #1 END, H2C; #2 END, LINK BAR | | | GLASS LENS & BACKLIGHTED 'AD' CARD |
| DOOR HANGER & PANEL | | O.M.EDWARDS | MASTER CONTROLLER | | GE, GEMKC76AD1; WH, XM 779 |
| DOOR OPERATOR MOTOR | CENTER | 4100-4139: MIDLAND ROSS, 500003B; CARS 3950-4099: VAPOR HEAT, 56570127E | POLARITY REVERSING RELAY | | INDUSTRON, SOLID STATE |
| | CAB END | CARS 4100-4139: MIDLAND ROSS, 500005D; CARS 3950-4099: VAPOR HEAT, 56570129D | PUBLIC ADDRESS SYSTEM | | SOUND SYSTEM, TRANSISTOR |
| | OPP END | CARS 4100-4139: MIDLAND ROSS, 500004C; CARS 3950-4099: VAPOR HEAT, 56570128D | SASH (CAB) | | ADAMS & WESTLAKE (CARS 4050-4139); O.M.EDWARDS (CARS 3950-4049) |
| DOOR PANEL | | O.M.EDWARDS (CARS 3950-4049); MORTON (CARS 4050-4139) | SEAT ARRANGEMENT & CAPACITY | | LONGITUDINAL [50] |
| DRAFT GEAR | | WAUGH, WM E4CL | SEAT MANUFACTURER | | GENERAL SEATING |
| ELECTRIC PORTION | #1 END | WABCO, ELECTRIC PORTION, BL26B | SEAT MATERIAL | | FIBERGLASS |
| | #2 END | PYLE NATIONAL COUPLER | SIGN | | TRANSIGN |
| FAN | | WH, AXIFLOW 4 SPEED, PERIPHERAL DISCHARGE DIFFUSERS [6], ALL CARS | SIGNAL DEVICE | | PNEUPHONIC HORN |
| FIRE EXTINGUISHER | | W. KIDDE, 1.5 - 2.75 LBS DRY POWDER | STANCHION | | ELLCON NATIONAL |
| HAND BRAKE | | NATIONAL BRAKE, 840XBL | WINDOW | | O.M.EDWARDS (CARS 3950-4049); ADAMS & WESTLAKE (CARS 4050-4139) |
| HANDHOLD | | ELLCON NATIONAL, #45 [56] | WINDSHIELD WIPER | #1 END | SPRAGUE DEVICES, PNEUMATIC |
| HEAD, MARKER & TAIL LIGHTS | | LOVELL DRESSEL | | | |
| **EVEN-NUMBERED CARS ONLY** | | | **ODD-NUMBERED CARS ONLY** | | |
| BATTERY | | EDISON, 24 CELL B4H (FIBERGLASS BOX) | COMPRESSOR UNIT | | WABCO, D3 |
| MOTOR GENERATOR | | GE, GMG 153LI; WH, YX304E1 | PROPULSION CONTROL TYPE | | GE, SCM, 17KG192H7 GROUP |
| PROPULSION CONTROL TYPE | | GE, SCM, 17KG192F7GROUP | | | WH, CAM, XCA248S GROUP |
| | | WH, CAM, XCA248T GROUP | SWITCH PANEL | | MIDLAND ROSS, 500007E; VAPOR HEAT, 56550132F |
| SWITCH PANEL | | MIDLAND ROSS, 500009D; VAPOR HEAT., 56550131F | | | |

## TRUCK

| | | | | | |
|---|---|---|---|---|---|
| MANUFACTURER | | GENERAL STEEL CASTING & ADIRONDACK STEEL CASTING | BRAKE RIGGING | | A.S.F. SIMPLEX UNIT CYLINDER CLASP BRAKE |
| TYPE & QUANTITY | | C.S.F. EQUALIZER BAR MOTOR [380] | GEAR UNIT | | WH, WN44CGS; GE, GA25A2 (7.235:1) |
| TRUCK NUMBERS | | WH, 15293 - 15472; GE, 15093 - 15292 | JOURNAL BEARING | | SKF, 5X9 ROLLER |
| WEIGHT | #1 END | WH, 19,512 LBS; GE, 19,228 LBS | SHOCK ABSORBER | | HOUDAILLE |
| | #2 END | WH, 19,088 LBS; GE, 18,784 LBS | TRACTION MOTOR | | WH, 1447C (100 HP) [2]; GE, 1240A5 (100 HP) [2] |
| AUTO SLACK ADJUSTER | | WABCO, C15D1 | TRIP COCK | | WABCO, D1 |
| BRAKE CYLINDER | | WABCO, 7X6 UAHT [2] | | | |

CARS 3950 - 4049 WITH GE EQUIPMENT AND CARS 4050 - 4139 WITH WH EQUIPMENT.
CARS ARE NON-AIR CONDITIONED.
CARS 3990, 3991, 4000, 4001 WERE SCRAPPED.

CAR NUMBERS: 4140 - 4149
TOTAL: 10 CARS
BUILT BY: ST. LOUIS CAR

DATE: 1967
AVERAGE COST PER CAR: $111,733

# R38 (BMT/IND)

AIR CONDITIONED

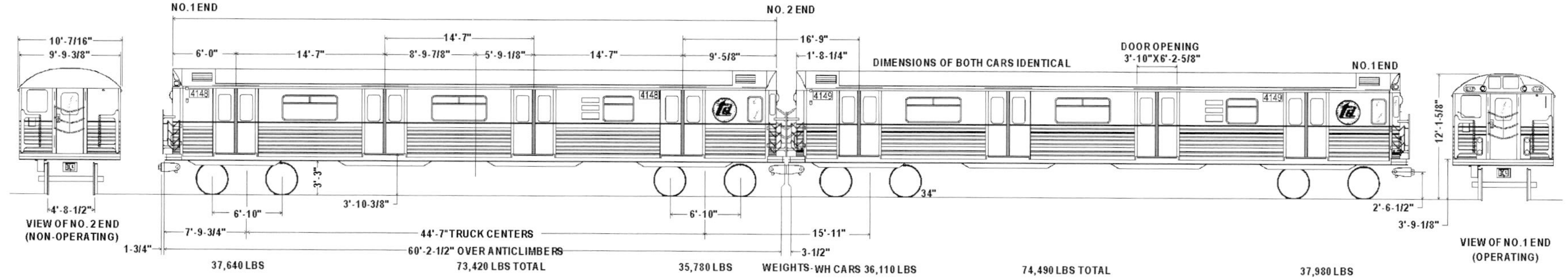

NO. 1 END
NO. 2 END
10'-7/16"
9'-9-3/8"
4'-8-1/2"
VIEW OF NO. 2 END (NON-OPERATING)
1-3/4"
6'-0"
14'-7"
14'-7"
8'-9-7/8"
5'-9-1/8"
14'-7"
9'-5/8"
3'-3"
3'-10-3/8"
6'-10"
7'-9-3/4"
44'-7" TRUCK CENTERS
60'-2-1/2" OVER ANTICLIMBERS
6'-10"
16'-9"
1'-8-1/4"
DIMENSIONS OF BOTH CARS IDENTICAL
DOOR OPENING 3'-10"X6'-2-5/8"
NO. 1 END
34"
15'-11"
3-1/2"
12'-1-5/8"
2'-6-1/2"
3'-9-1/8"
VIEW OF NO. 1 END (OPERATING)
37,640 LBS
73,420 LBS TOTAL
35,780 LBS
WEIGHTS-WH CARS 36,110 LBS
74,490 LBS TOTAL
37,980 LBS

Copyright © 1997 by NYC Transit

## CAR BODY - EQUIPMENT

| | | | |
|---|---|---|---|
| AIR CONDITIONING | TRANE, 10 TON UNITS [2] CARS 4140 - 4144 | HVAC CONTROL | VAPOR |
| | SAFETY, 9 TON UNITS [2] CARS 4145 - 4149 | LIGHTING | LUMINATOR, FLUORESCENT FIXTURES, 600 VDC PARALLEL CIRCUIT |
| AUXILIARY CIRCUIT PROTECTION | CIRCUIT BREAKERS | MASTER CONTROLLER | WH, XM 779 |
| BRAKE EQUIPMENT TYPE | WABCO, SMEE (DYNAMIC & FRICTION) | POLARITY REVERSING RELAY | INDUSTRON, SOLID STATE |
| BRAKE VALVE | WABCO, ME43 | PROPULSION CONTROL TYPE | WH, CAM, XCA248T GROUP |
| COUPLER | WABCO, H2C; WABCO, LINK BAR | PUBLIC ADDRESS SYSTEM | TRANSISTOR, SOUND SYSTEM |
| DOOR HANGER & PANEL | MORTON | SASH (CAB) | ADAMS & WESTLAKE |
| DOOR OPERATOR MOTOR | MIDLAND ROSS: CENTER, 500003B; OPP. END, 500005D; CAB END, 500004C | SEAT ARRANGEMENT & CAPACITY | LONGITUDINAL [50] |
| DRAFT GEAR | WAUGH, WM E4CL | SEAT MANUFACTURER | GENERAL SEATING |
| ELECTRIC PORTION | #1 END: WABCO, BL26B; #2 END: PYLE NATIONAL , CANNON PLUG | SEAT MATERIAL | FIBERGLASS |
| FIRE EXTINGUISHER | W. KIDDE, 1.5 - 2.75 LBS DRY POWDER | SIGN | TRANSIGN |
| HAND BRAKE | NATIONAL BRAKE, 840XBL | SIGNAL DEVICE | PNEUPHONIC HORN |
| HANDHOLD | ELLCON NATIONAL | STANCHION | ELLCON NATIONAL |
| HEAD, MARKER & TAIL LIGHTS | LOVELL DRESSEL | WINDOW | ADAMS & WESTLAKE |
| HEATER | ST. LOUIS CAR | WINDSHIELD WIPER | SPRAGUE DEVICES ,PNEUMATIC |
| EVEN-NUMBERED CARS ONLY | | ODD-NUMBERED CARS ONLY | |
| BATTERY | EDISON, 24 CELL B4H (FIBERGLASS BOX) | AIR COMPRESSOR | WABCO, D3 |
| MOTOR/GENERATOR | WH, YX304E1 | SWITCH PANEL | MIDLAND ROSS, 500007E |
| SWITCH PANEL | MIDLAND ROSS, 500009D | | |

## TRUCK

| | | | |
|---|---|---|---|
| MANUFACTURER | GENERAL STEEL CASTING | BRAKE RIGGING | A.S.F. SIMPLEX UNIT CYLINDER CLASP BRAKE |
| TYPE & QUANTITY | C.S.F. EQUALIZER BAR MOTOR [380] | GEAR UNIT | WH, WN44CGS (7.235:1) |
| TRUCK NUMBERS | 15473 - 15492 | JOURNAL BEARING | SKF, 5X9 ROLLER |
| WEIGHT | #1 END: 19,512 LBS; #2 END: 19,088 LBS | SHOCK ABSORBER | HOUDAILLE |
| AUTO SLACK ADJUSTER | WABCO, C15D1 | TRACTION MOTOR | WH, 1477C (100 HP) [2] |
| BRAKE CYLINDER | WABCO, 7X6 UAHT [2] | TRIP COCK | WABCO, D1 |

CAR NUMBERS: 3950 - 4149
TOTAL: 196 CARS
BUILT BY: ST. LOUIS CAR
DATE: 1966 - 67
AVERAGE COST PER CAR: $111,733

OVERHAUL BY: BUFFALO TRANSIT SERVICES (GE)
DATE: 1987 - 88
OVERHAUL COST PER CAR: $400,000

# R38 (BMT/IND) GOH

NO. 1 END
NO. 2 END
NO. 1 END
10'-7/16"
9'-9-3/8"
6'-0"
14'-7"
14'-7"
8'-9-7/8"
5'-9-1/8"
14'-7"
8'-9-7/8"
16'-9"
1'-8-1/4"
DIMENSIONS OF BOTH CARS IDENTICAL
DOOR OPENING 3'-10"x6'-2-5/8"
3950
3951
12'-1-5/8"
3'-3"
3'-10-3/8"
34" DIA
2'-6-1/2"
3'-9-1/8"
4'-8-1/2"
VIEW OF NO.2 END (NON-OPERATING)
VIEW OF NO.1 END (OPERATING)
6'-10"
6'10"
7'-9-3/4"
44'-7" TRUCK CENTERS
15'-11"
1-3/4"
60'-2-1/2" OVER ANTICLIMBERS
3-1/2"
77,410 LBS TOTAL
77,430 LBS TOTAL

Copyright © 1997 by NYC Transit

## CAR BODY - EQUIPMENT

| | | | |
|---|---|---|---|
| AIR CONDITIONING | STONE SAFETY | HVAC CONTROL | STONE SAFETY |
| AUXILIARY CIRCUIT PROTECTION | CIRCUIT BREAKERS | LIGHTING | LUMINATOR, FLUORESCENT FIXTURES |
| BRAKE EQUIPMENT TYPE | WABCO, E2 (DYNAMIC & FRICTION) | MARKER & TAIL LIGHTS | BUFFALO TRANSIT SERVICES |
| BRAKE VALVE | WABCO, ME43 | MASTER CONTROLLER | GE, 17KC76AE2 |
| COUPLER | WABCO: #1 END, H2C; #2 END, LINK BAR | PUBLIC ADDRESS SYSTEM | MIDWEST |
| DOOR HANGER | A.B. KING | SASH (CAB) | J.T. NELSON |
| DOOR OPERATOR MOTOR | WESTCODE | SEAT ARRANGEMENT & CAPACITY | LONGITUDINAL [50] |
| DOOR PANEL | EBONEX | SEAT MANUFACTURER | GENERAL SEATING |
| DRAFT GEAR | WAUGH, WM E4CL | SEAT MATERIAL | FIBERGLASS |
| ELECTRIC PORTION | #1 END: WABCO,BL33F; #2 END: PYLE NATIONAL COUPLER | SIGN | TRANSIGN |
| HAND BAR | PHILADELPHIA PIPE | SIGNAL DEVICE | PNEUPHONIC HORN |
| HAND BRAKE | ELLCON NATIONAL, 840XBL | STANCHION | ELLCON NATIONAL |
| HEAD LIGHT | LOVELL DRESSEL | WINDOW | J.T. NELSON |
| HEATER | ST. LOUIS CAR | WINDSHIELD WIPER | UNITED TECHNOLOGIES (ELECTRIC) |
| EVEN-NUMBERED CARS ONLY | | ODD-NUMBERED CARS ONLY | |
| BATTERY | EDISON, 24 CELL FIBERGLASS BOX | AIR COMPRESSOR | WABCO, D4S |
| CONVERTER | TOSHIBA | PROPULSION CONTROL TYPE | GE, SCM, 17KG192H7 GROUP |
| PROPULSION CONTROL TYPE | GE, SCM, 17KG192AE2 GROUP | SWITCH PANEL | MIDLAND ROSS, 500007E; VAPOR HEAT, 56550132F |
| SWITCH PANEL | MIDLAND ROSS, 500009D; VAPOR HEAT, 56550131F | | |

## TRUCK

| | | | |
|---|---|---|---|
| MANUFACTURER | GENERAL STEEL CASTING & ADIRONDACK STEEL CASTING | BRAKE RIGGING | A.S.F. SIMPLEX UNIT CYLINDER CLASP BRAKE |
| TYPE & QUANTITY | C.S.F. EQUALIZER BAR MOTOR [380] | GEAR UNIT | GE, 76ALAC1 (7.235:1) |
| TRUCK NUMBERS | N/A | JOURNAL BEARING | 5X9, AP ROLLER, SEE NOTE 1 |
| WEIGHT | #1 END: 19,470 LBS; #2 END: 19,050 LBS | SHOCK ABSORBER | MONROE |
| AUTO SLACK ADJUSTER | WABCO, C15D1 | TRACTION MOTOR | GE, 1257E1 (115 HP) [2] |
| BRAKE CYLINDER | WABCO, 7X6 UAHT [2] | TRIP COCK | WABCO, D1 |

NOTE 1: VARIOUS VENDORS. CAR NUMBERS INCLUDES OVERHAULING OF ORIGINAL 10 AIR CONDITIONED CARS.

CARS 3950 - 4139 WERE NON-AIR CONDITIONED PRIOR TO OVERHAUL.
INSTALLED DOPPLER SPEEDOMETER BY EDO (1996).

CAR NUMBERS: 4150-4349 (SEE NOTE 1)
TOTAL: 200 CARS
BUILT BY: ST. LOUIS CAR

DATE: 1968-69
AVERAGE COST PER CAR: $111,793

# R40 (BMT/IND)

SLOPE END - CLASP BRAKE
NON-AIR CONDITIONED

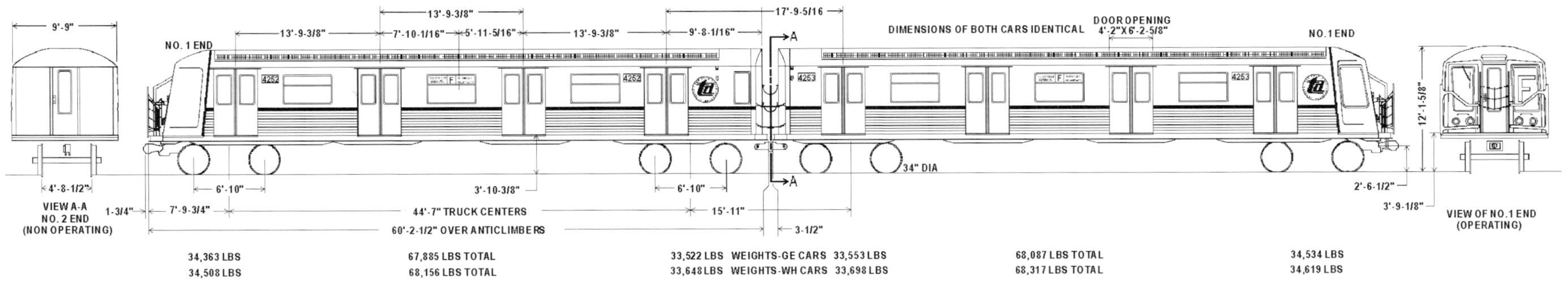

9'-9"
NO. 1 END
13'-9-3/8"
7'-10-1/16"
5'-11-5/16"
13'-9-3/8"
9'-8-1/16"
17'-9-5/16
DIMENSIONS OF BOTH CARS IDENTICAL
DOOR OPENING 4'-2"X6'-2-5/8"
NO. 1 END
12'-1-5/8"
4'-8-1/2"
6'-10"
3'-10-3/8"
6'-10"
34" DIA
2'-6-1/2"
3'-9-1/8"
VIEW A-A NO. 2 END (NON OPERATING)
1-3/4"
7'-9-3/4"
44'-7" TRUCK CENTERS
15'-11"
60'-2-1/2" OVER ANTICLIMBERS
3-1/2"
VIEW OF NO. 1 END (OPERATING)

34,363 LBS / 34,508 LBS
67,885 LBS TOTAL / 68,156 LBS TOTAL
33,522 LBS WEIGHTS-GE CARS 33,553 LBS
33,648 LBS WEIGHTS-WH CARS 33,698 LBS
68,087 LBS TOTAL / 68,317 LBS TOTAL
34,534 LBS / 34,619 LBS

Copyright © 1997 by NYC Transit

## CAR BODY - EQUIPMENT

| | | | |
|---|---|---|---|
| AUXILIARY CIRCUIT PROTECTION | CIRCUIT BREAKERS | HEATER | ST. LOUIS CAR |
| BRAKE EQUIPMENT TYPE | WABCO, SMEE (DYNAMIC & FRICTION) | LIGHTING | LUMINATOR, ENCLOSED GLASS, LENS & BACKLIGHTED "AD" CARDS |
| BRAKE VALVE | WABCO, ME43 | MASTER CONTROLLER | GE, 17KC76E1 WH, XM-829 |
| COUPLER | WABCO: # 1 END, H2C; # 2 END, LINKBAR | POLARITY REVERSING | VAPOR, SOLID STATE |
| DOOR HANGER & PANEL | O.M. EDWARDS | PUBLIC ADDRESS SYSTEM | SOUND SYSTEMS |
| DOOR OPERATOR MOTOR | VAPOR: CENTER, 56670462; OPP. CAB, 56670463; CAB END, 56670464 | SASH (CAB) | O.M. EDWARDS |
| DRAFT GEAR | WAUGH, WM E4CL | SEAT ARRANGEMENT & CAPACITY | LONGITUDINAL [44] |
| ELECTRIC PORTION | #1 END: WABCO, BL26B; #2 END: PYLE NATIONAL, PLUG | SEAT MANUFACTURER | GENERAL SEATING |
| FAN | WH, 23" AXIFLO 4-SPEED ALL CARS | SEAT MATERIAL | FIBERGLASS |
| FIRE EXTINGUISHER | W. KIDDE, 1 1/2-2 3/4 LB DRY POWDER | SIGN | TRANSLITE |
| HAND BRAKE | NATIONAL BRAKE, 840-XB-L | SIGNAL DEVICE | PNEUPHONIC HORN |
| HANDHOLD | ELLCON NATIONAL [56] | STANCHION | ELLCON NATIONAL [6] |
| HEAD, MARKER & TAIL LIGHT | LOVELL DRESSEL | WINDOW | O.M. EDWARDS |
| HEAT & VENTILATION CONTROL | VAPOR | WINDSHIELD WIPER | SPRAGUE DEVICES, PNEUMATIC |
| | **EVEN-NUMBERED CARS** | | **ODD-NUMBERED CARS** |
| BATTERY | ELECTRIC STORAGE BATTERY FIBERGLASS BOX | AIR COMPRESSOR | WABCO, D-3 |
| MOTOR/GENERATOR | GE, 5GMG 153LI; WH, YX304E1 | PROPULSION CONTROL TYPE | GE, SCM, 17KG192H8 GROUP |
| PROPULSION CONTROL TYPE | GE, SCM, 17KG192H8 GROUP | | WH, CAM, XCA248S GROUP |
| | WH, CAM, XCA248T GROUP | SWITCH PANEL | VAPOR, 56660444, NO. 2 CAB |
| SWITCH PANEL | VAPOR, 56660443, NO.2 CAB | | |

## TRUCK

| | | | |
|---|---|---|---|
| MANUFACTURER | GENERAL STEEL INDUSTRIES & ADIRONDACK STEEL CASTING | BRAKE RIGGING | A.S.F., SIMPLEX UNIT CYLINDER CLASP BRAKE |
| TYPE & QUANTITY | C. S. F. EQUALIZER BAR MOTOR [400] | GEAR UNIT | GE, WN44; WH, 7GA25 (7.235:1) |
| TRUCK NUMBERS | WH, 15493 - 15692; GE, 15893 - 16092 | JOURNAL BEARING | SKF, 5X9 ROLLER |
| WEIGHT | #1 END: 19,008 LBS; #2 END: 18,622 LBS | SHOCK ABSORBER | HOUDAILLE |
| AUTO. SLACK ADJUSTER | WABCO, C15D1 | TRACTION MOTOR | GE, 1447C (100 HP) [2]; WH, 1240A2 (100 HP) [2] |
| BRAKE CYLINDER | WABCO, 7X6 UAHT [2] | TRIP COCK | WABCO, D1 |

NOTE 1: CARS 4250 - 4349 WERE ORIGINALLY DELIVERED AS CARS 4350 - 4449. RENUMBERED (1970).

CARS 4250 - 4349 WITH GE EQUIPMENT AND CARS 4150 - 4249 WITH WH EQUIPMENT.
CARS 4200, 4201 SCRAPPED PRIOR TO GOH PROGRAM.

CAR NUMBERS: 4150 - 4349 (SEE NOTE 2)
TOTAL: 198 CARS
BUILT BY: ST. LOUIS CAR
DATE: 1968 - 69
AVERAGE COST PER CAR : $111,793

OVERHAULED BY: SUMITOMO
DATE: 1987 - 89
OVERHAUL COST PER CAR: $399,000

# R40 (BMT/IND)

SLOPE END - CLASP BRAKE
AIR CONDITIONED
GOH

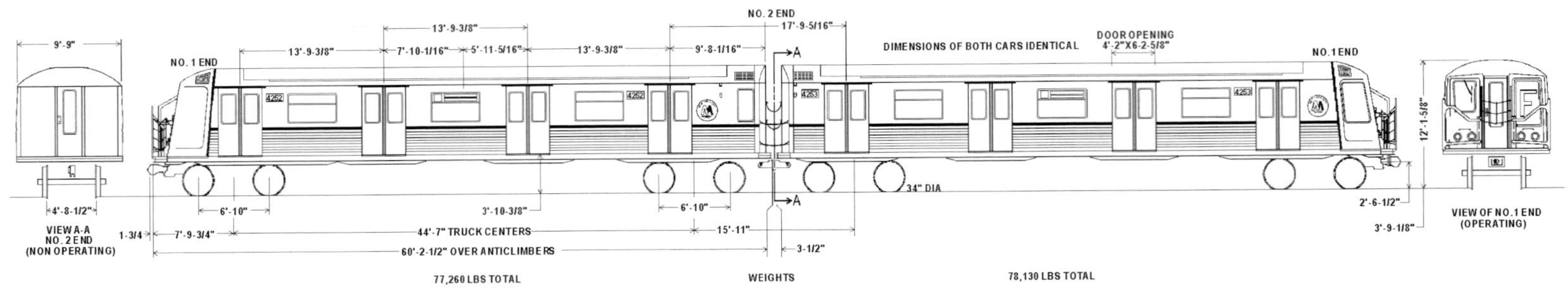

9'-9"
VIEW A-A
NO. 2 END
(NON OPERATING)
4'-8-1/2"
NO. 1 END
13'-9-3/8"
7'-10-1/16"
5'-11-5/16"
13'-9-3/8"
13'-9-3/8"
9'-8-1/16"
NO. 2 END
17'-9-5/16"
DIMENSIONS OF BOTH CARS IDENTICAL
DOOR OPENING
4'-2"X6-2-5/8"
NO. 1 END
12'-1-5/8"
VIEW OF NO. 1 END
(OPERATING)
1-3/4
7'-9-3/4"
6'-10"
3'-10-3/8"
44'-7" TRUCK CENTERS
60'-2-1/2" OVER ANTICLIMBERS
6'-10"
15'-11"
3-1/2"
34" DIA
2'-6-1/2"
3'-9-1/8"
77,260 LBS TOTAL
WEIGHTS
78,130 LBS TOTAL

Copyright © 1997 by NYC Transit

## CAR BODY - EQUIPMENT

| | | |
|---|---|---|
| AIR CONDITIONING | | THERMO KING, 9 TON UNITS |
| AUXILIARY CIRCUIT PROTECTION | | CIRCUIT BREAKERS |
| BRAKE EQUIPMENT TYPE | | WABCO, SMEE (DYNAMIC & FRICTION) |
| BRAKE VALVE | | WABCO, ME43 |
| COUPLER | | WABCO: #1 END, H2C; #2 END, LINKBAR |
| DOOR HANGER | | A.B. KING |
| DOOR OPERATOR MOTOR | | VAPOR; WESTCODE |
| DOOR PANEL | | EBONEX |
| DRAFT GEAR | | WAUGH, WME4CL |
| ELECTRIC PORTION | | # 1 END: WABCO, BL37; # 2 END: VEAM, CONNECTOR |
| HAND BRAKE | | NATIONAL BRAKE, 840XBL |
| HANDHOLD | | ELLCON NATIONAL [47] |
| HEAD & MARKER LIGHTS | | LOVELL DRESSEL |
| HEATER | | CHROMOLOX; EMERSON; FABER |
| HVAC CONTROL | | VAPOR |
| LIGHTING | | LUMINATOR, ENCLOSED PLASTIC LENS & BACK LIGHTED "AD" CARDS W/LIGHTING INVERTER BALLAST |
| MASTER CONTROLLER | | GE, 17KC76E1 |
| PROPULSION CONTROL TYPE | | GE, SCM, 17KG192AE2 GROUP |
| PUBLIC ADDRESS SYSTEM | | COMCO |
| SASH (CAB) | | EBONEX |
| SEAT ARRANGEMENT & CAPACITY | | LONGITUDINAL [44] |
| SEAT MANUFACTURER | | NIAGARA SEATING/ARTCRAFT |
| SEAT MATERIAL | | FIBERGLASS |
| SIGN | | TRANSLITE/TELEWELD |
| SIGNAL DEVICE | | PNEUPHONIC HORN |
| STANCHION | | ELLCON NATIONAL [6] |
| TAIL LIGHT | | ST. LOUIS CAR |
| WINDOW | | EBONEX |
| WINDSHIELD WIPER | NO. 1 END | SYRACUSE DIESEL, ELECTRIC |

**EVEN-NUMBERED CARS ONLY**

| | |
|---|---|
| BATTERY | HOPPECKE; NIFE (FIBERGLASS BOX) |
| CONVERTER | STONE SAFETY, 7 KW |
| SWITCH PANEL | VAPOR, NO. 2 CAB |

**ODD-NUMBERED CARS ONLY**

| | |
|---|---|
| AIR COMPRESSOR | WABCO, D4S |
| SWITCH PANEL | VAPOR, NO.2 CAB |

## TRUCK

| | | |
|---|---|---|
| MANUFACTURER | | GENERAL STEEL INDUSTRIES & ADIRONDACK STEEL CASTING |
| TYPE & QUANTITY | | C.S.F. EQUALIZER BAR MOTOR [200] |
| TRUCK SERIAL NUMBERS | | N/A |
| WEIGHT | #1 END | 19,470 LBS |
| | #2 END | 19,050 LBS |
| AUTO. SLACK ADJUSTER | | WABCO, C15D1 |
| BRAKE CYLINDER | | WABCO, 7X6 UAHT [2] |
| BRAKE RIGGING | | A.S.F. SIMPLEX UNIT CYLINDER CLASP BRAKE |
| GEAR UNIT | | GE, 7GA25 (7.235:1) |
| JOURNAL BEARING | | 5X9, AP ROLLER, SEE NOTE 1 |
| SHOCK ABSORBER | | MONROE |
| TRACTION MOTOR | | GE, 1257E1 (115 HP) [2] |
| TRIP COCK | | WABCO, D1 |

NOTE 1: VARIOUS VENDORS.
NOTE 2: CARS 4250 -4349 WERE ORIGINALLY DELIVERED AS CARS 4350 - 4449. RENUMBERED (1970).

CARS WERE NON-AIR CONDITIONED PRIOR TO OVERHAUL.
CARS 4200, 4201 SCRAPPED PRIOR TO GOH PROGRAM.
CAR 4260 TO RANDALL'S ISLAND 2004.
INSTALLED DOPPLER SPEEDOMETER BY EDO (6/1996 - 12/1996).

CAR NUMBERS: 4350 - 4449 (SEE NOTE 1)
TOTAL: 100 CARS
BUILT BY: ST. LOUIS CAR

DATE: 1968 - 69
AVERAGE COST PER CAR: $137,382

# R40 (BMT/IND)

SLOPE END - CLASP BRAKE
AIR CONDITIONED

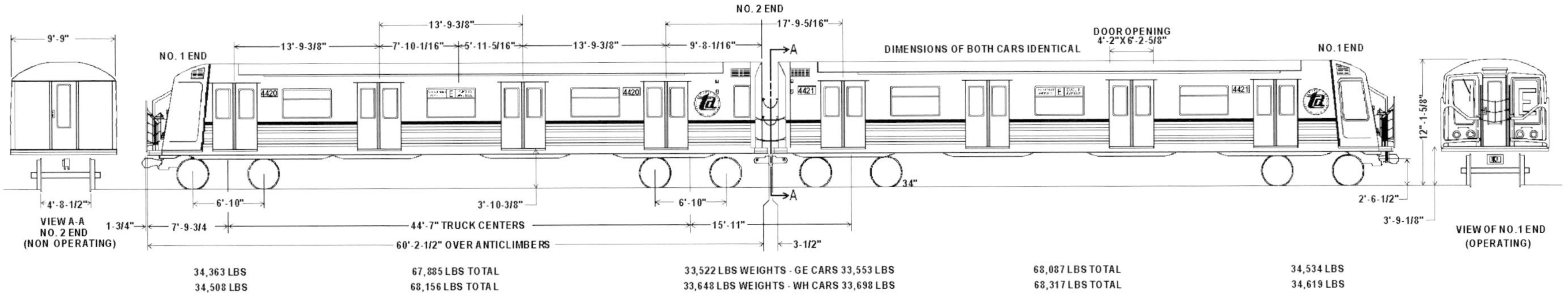

9'-9"
4'-8-1/2"
VIEW A-A
NO. 2 END
(NON OPERATING)
NO. 1 END
13'-9-3/8"
13'-9-3/8"
7'-10-1/16"
5'-11-5/16"
13'-9-3/8"
9'-8-1/16"
NO. 2 END
17'-9-5/16"
A
DIMENSIONS OF BOTH CARS IDENTICAL
DOOR OPENING
4'-2"X6'-2-5/8"
NO. 1 END
12"-1-5/8"
1-3/4"
7'-9-3/4
6'-10"
3'-10-3/8"
44'-7" TRUCK CENTERS
60'-2-1/2" OVER ANTICLIMBERS
6'-10"
15'-11"
3-1/2"
34"
2'-6-1/2"
3'-9-1/8"
VIEW OF NO. 1 END
(OPERATING)

34,363 LBS
34,508 LBS
67,885 LBS TOTAL
68,156 LBS TOTAL
33,522 LBS WEIGHTS - GE CARS 33,553 LBS
33,648 LBS WEIGHTS - WH CARS 33,698 LBS
68,087 LBS TOTAL
68,317 LBS TOTAL
34,534 LBS
34,619 LBS

Copyright © 1997 by NYC Transit

## CAR BODY - EQUIPMENT

| | | | |
|---|---|---|---|
| AIR CONDITIONING | SAFETY, 9 TON UNITS [2], VAPOR CYCLE | HVAC CONTROL | VAPOR |
| AUXILIARY CIRCUIT PROTECTION | CIRCUIT BREAKERS | LIGHTING | LUMINATOR, ENCLOSED GLASS LENS & BACKLIGHTED "AD" CARDS |
| BRAKE EQUIPMENT TYPE | WABCO, SMEE (DYNAMIC & FRICTION) | MASTER CONTROLLER | GE, 17KC76E1 |
| BRAKE VALVE | WABCO, ME43 | POLARITY REVERSING RELAY | VAPOR, SOLID STATE |
| COUPLER | WABCO: #1 END, H2C; #2 END, LINKBAR | PUBLIC ADDRESS SYSTEM | SOUND SYSTEMS |
| DOOR HANGER & PANEL | O.M. EDWARDS | SASH (CAB) | TRANSPORTATION SASH |
| DOOR OPERATOR MOTOR | VAPOR HEAT: CENTER, 56670462; OPP. CAB, 56670463; CAB END, 56670464 | SEAT ARRANGEMENT & CAPACITY | LONGITUDINAL [44] |
| DRAFT GEAR | WAUGH, WM E4CL | SEAT MANUFACTURER | GENERAL SEATING |
| ELECTRIC PORTION | WABCO: #1 END, BL26B; #2 END, PYLE NATIONAL, PLUG | SEAT MATERIAL | FIBERGLASS |
| FIRE EXTINGUISHER | W. KIDDE, 1 1/2 - 2 3/4 LBS DRY POWDER | SIGNAL DEVICE | PNEUPHONIC HORN |
| HAND BRAKE | NATIONAL BRAKE, 840-XB-L | SIGNS | TRANSLITE |
| HANDHOLD | ELLCON NATIONAL [56] | STANCHION | ELLCON - NATIONAL [6] |
| HEAD, MARKER & TAIL LIGHTS | LOVELL DRESSEL | WINDOW | TRANSPORTATION SASH |
| HEATER | ST. LOUIS CAR | WINDSHIELD WIPER | SPRAGUE DEVICES, PNEUMATIC |
| EVEN-NUMBERED CARS | | ODD-NUMBERED CARS | |
| BATTERY | ELECTRIC STORAGE BATTERY FIBERGLASS BOX | AIR COMPRESSOR | WABCO, D3 |
| MOTOR GENERATOR | GE, GMG153LI; WH, YXA248T | PROPULSION CONTROL TYPE | GE, SCM, 17KG192H8 GROUP |
| PROPULSION CONTROL TYPE | GE, SCM, 17KG192F8 GROUP | SWITCH PANEL | VAPOR, 5680029A, #2 CAB |
| SWITCH PANEL | VAPOR, 5680028A, #2 CAB | | |

## TRUCK

| | | | |
|---|---|---|---|
| MANUFACTURER | GENERAL STEEL INDUSTRIES & ADIRONDACK STEEL CASTING | BRAKE RIGGING | A.S.F. SIMPLEX UNIT CYLINDER CLASP BRAKE |
| TYPE & QUANTITY | C.S.F. EQUALIZER BAR MOTOR [200] | GEAR UNIT | GE, 7GA25 (7.235:1) |
| TRUCK SERIAL NUMBERS | GE, 16092 - 16291 | JOURNAL BEARING | SKF, 5X9 ROLLER |
| WEIGHT | #1 END: 19,008 LBS; #2 END: 18,622 LBS | SHOCK ABSORBER | HOUDAILLE |
| AUTO. SLACK ADJUSTER | WABCO, C15D1 | TRACTION MOTOR | GE, 1240A5 (100 HP) [2] |
| BRAKE CYLINDER | WABCO, 7X6 UAHT [2] | TRIP COCK | WABCO, D1 |

NOTE 1: CARS 4350 - 4449 WERE ORIGINALLY DELIVERED AS CARS 4450 - 4549. RENUMBERED (1970).

CARS 4420 AND 4421 SCRAPPED PRIOR TO GOH PROGRAM.

CAR NUMBERS: 4350 - 4449 (SEE NOTE 2)
TOTAL: 98 CARS
BUILT BY: ST. LOUIS CAR
DATE: 1968
AVERAGE COST PER CAR: $137,382

OVERHAULED BY: SUMITOMO
DATE: 1988 - 89
OVERHAUL COST PER CAR: $399,000

# R40 (BMT/IND)

SLOPE END - CLASP BRAKE
AIR CONDITIONED
**GOH**

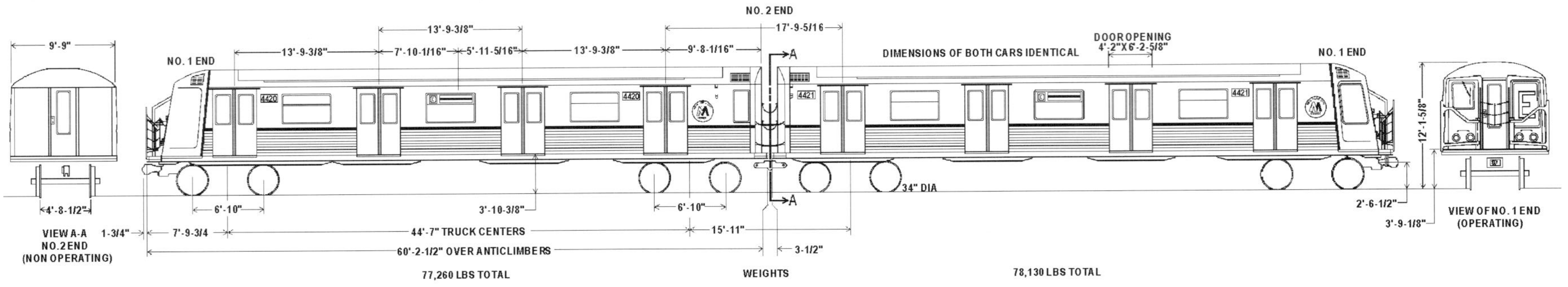

9'-9"
VIEW A-A NO.2 END (NON OPERATING)
4'-8-1/2"
1-3/4"
NO. 1 END
13'-9-3/8"
13'-9-3/8"
7'-10-1/16"
5'-11-5/16"
13'-9-3/8"
9'-8-1/16"
NO. 2 END
17'-9-5/16
DIMENSIONS OF BOTH CARS IDENTICAL
DOOR OPENING 4'-2"X6'-2-5/8"
NO. 1 END
12'-1-5/8"
6'-10"
3'-10-3/8"
6'-10"
34" DIA
2'-6-1/2"
3'-9-1/8"
VIEW OF NO. 1 END (OPERATING)
7'-9-3/4
44'-7" TRUCK CENTERS
15'-11"
60'-2-1/2" OVER ANTICLIMBERS
3-1/2"
77,260 LBS TOTAL
WEIGHTS
78,130 LBS TOTAL

Copyright © 1997 by NYC Transit

## CAR BODY - EQUIPMENT

|  | |  |
|---|---|---|
| AIR CONDITIONING | | THERMO KING, 9 TON UNITS [2] |
| AUXILIARY CIRCUIT PROTECTION | | CIRCUIT BREAKERS |
| BRAKE EQUIPMENT TYPE | | WABCO, SMEE (DYNAMIC & FRICTION) |
| BRAKE VALVE | | WABCO, ME43 |
| COUPLER | | WABCO: #1 END, H2C; #2 END, LINKBAR |
| DOOR HANGER | | A.B. KING |
| DOOR OPERATOR MOTOR | | WESTCODE |
| DOOR PANEL | | EBONEX |
| DRAFT GEAR | | WAUGH, WME4CL |
| ELECTRIC PORTION | | #1 END: WABCO, BL37; #2 END: VEAM, CONNECTOR |
| HAND BRAKE | | NATIONAL BRAKE, 840XBL |
| HANDHOLD | | ELLCON NATIONAL [47] |
| HEAD & MARKER LIGHTS | | LOVELL DRESSEL |
| HEATER | | CHROMOLOX/EMERSON/FABER |
| HVAC CONTROL | | VAPOR |
| **EVEN-NUMBERED CARS ONLY** | | |
| BATTERY | | HOPPECKE; NIFE (FIBERGLASS BOX) |
| CONVERTER | | STONE SAFETY, 7 KW |
| SWITCH PANEL | #2 CAB | VAPOR |

|  | |  |
|---|---|---|
| LIGHTING | | LUMINATOR, ENCLOSED PLASTIC LENS & BACK |
| | | LIGHTED "AD" CARDS W/LIGHTING INVERTER BALLAST |
| MASTER CONTROLLER | | GE, 17KC76E1 |
| PROPULSION CONTROL TYPE | | GE, SCM, 17KG192AE2 |
| PUBLIC ADDRESS SYSTEM | | COMCO |
| SASH (CAB) | | EBONEX |
| SEAT ARRANGEMENT & CAPACITY | | LONGITUDINAL [44] |
| SEAT MANUFACTURER | | NIAGARA SEATING; ARTCRAFT |
| SEAT MATERIAL | | FIBERGLASS |
| SIGN | | TRANSLITE/TELEWELD |
| SIGNAL DEVICE | | PNEUPHONIC HORN |
| STANCHION | | ELLCON NATIONAL [6] |
| TAIL LIGHT | | ST. LOUIS CAR |
| WINDOW | | EBONEX |
| WINDSHIELD WIPER | #1 END | SYRACUSE DIESEL, ELECTRIC |
| **ODD-NUMBERED CARS ONLY** | | |
| AIR COMPRESSOR | | WABCO, D4S |
| SWITCH PANEL | #2 CAB | VAPOR |

## TRUCK

|  | |  |
|---|---|---|
| MANUFACTURER | | GENERAL STEEL INDUSTRIES & ADIRONDACK STEEL CASTING |
| TYPE & QUANTITY | | C.S.F. EQUALIZER BAR MOTOR [200] |
| TRUCK SERIAL NUMBERS | | N/A |
| WEIGHT | #1 END | 19,470 LBS |
| | #2 END | 19,050 LBS |
| AUTO. SLACK ADJUSTER | | WABCO, C15D1 |
| GEAR UNIT | | GE, 7GA25 (7.235:1) |

|  |  |
|---|---|
| BRAKE CYLINDER | WABCO, 7X6 UAHT [2] |
| BRAKE RIGGING | A.S.F. SIMPLEX UNIT CYLINDER CLASP BRAKE |
| JOURNAL BEARING | 5X9, AP ROLLER, SEE NOTE 1 |
| SHOCK ABSORBER | MONROE |
| TRACTION MOTOR | GE, 1257E1 (115 HP) [2] |
| TRIP COCK | WABCO, D1 |

NOTE 1: VARIOUS VENDORS.
NOTE 2: CARS 4350 - 4449 WERE ORIGINALLY DELIVERED AS CARS 4450 - 4549. RENUMBERED (1971).

CARS 4427 AND 4428 SCRAPPED (1996).
INSTALLED DOPPLER SPEEDOMETER BY EDO (6/1996 - 12/1996).

CAR NUMBERS: 4450 - 4517 (SEE NOTE 1)
TOTAL: 68 CARS
BUILT BY: ST. LOUIS CAR

DATE: 1968 - 69
AVERAGE COST PER CAR: $137,382

# R40 (BMT/IND)

STRAIGHT END
AIR CONDITIONED

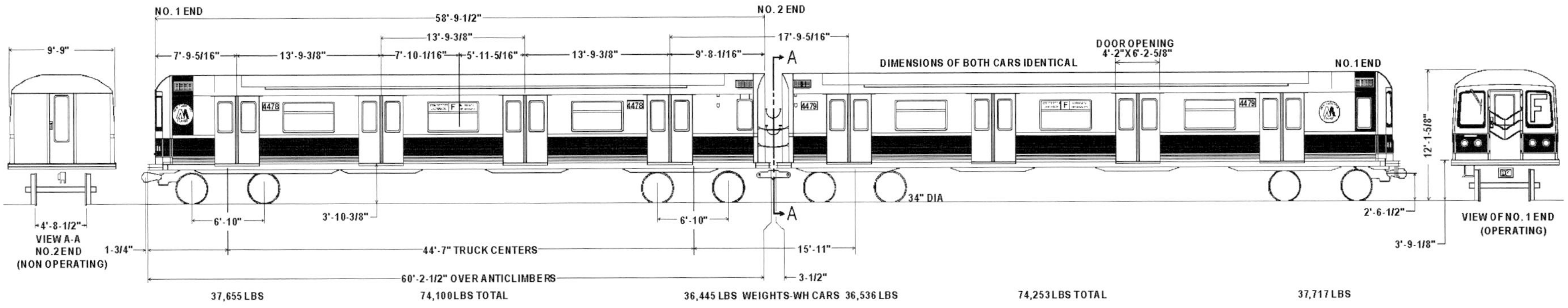

NO. 1 END — NO. 2 END — 58'-9-1/2" — 13'-9-3/8" — 7'-9-5/16" — 13'-9-3/8" — 7'-10-1/16" — 5'-11-5/16" — 13'-9-3/8" — 9'-8-1/16" — 17'-9-5/16" — A

DIMENSIONS OF BOTH CARS IDENTICAL — DOOR OPENING 4'-2"X6'-2-5/8" — NO. 1 END

9'-9" — 4'-8-1/2" — VIEW A-A NO. 2 END (NON OPERATING) — 1-3/4" — 6'-10" — 3'-10-3/8" — 6'-10" — 34" DIA — 44'-7" TRUCK CENTERS — 15'-11" — 60'-2-1/2" OVER ANTICLIMBERS — 3-1/2" — 12'-1-5/8" — 2'-6-1/2" — 3'-9-1/8" — VIEW OF NO. 1 END (OPERATING)

37,655 LBS — 74,100 LBS TOTAL — 36,445 LBS WEIGHTS-WH CARS 36,536 LBS — 74,253 LBS TOTAL — 37,717 LBS

Copyright © 1997 by NYC Transit

## CAR BODY - EQUIPMENT

| | | | |
|---|---|---|---|
| AIR CONDITIONING | SAFETY, 9 TON UNITS [2], CARS 4450-4479; TRANE,10 TON UNITS [2], CARS 4480-4517 | HEATER | ST. LOUIS CAR |
| | | HVAC CONTROL | VAPOR |
| AUXILIARY CIRCUIT PROTECTION | CIRCUIT BREAKERS | LIGHTING | LUMINATOR, ENCLOSED GLASS LENS & BACKLIGHTED "AD" CARDS |
| BRAKE EQUIPMENT TYPE | WABCO, SMEE (DYNAMIC & FRICTION) | MASTER CONTROLLER | WH, XM-829 |
| BRAKE VALVE | WABCO, ME43 | POLARITY REVERSING | VAPOR, SOLID STATE |
| COUPLER | WABCO: #1 END, H2C; #2 END, LINKBAR | PUBLIC ADDRESS SYSTEM | SOUND SYSTEMS |
| DOOR HANGER | MORTON | SASH (CAB) | TRANSPORTATION SASH, SEE NOTE 2 |
| DOOR OPERATOR MOTOR | MID. ROSS: CENTER, 500003C; OPP. CAB, 500004D; CAB END, 500005E | SEAT ARRANGEMENT & CAPACITY | LONGITUDINAL [44] |
| DOOR PANEL | MORTON; O.M. EDWARDS | SEAT MANUFACTURER | GENERAL SEATING |
| DRAFT GEAR | WAUGH, WM E4CL | SEAT MATERIAL | FIBERGLASS |
| ELECTRIC PORTION | #1 END: WABCO, BL26B; #2 END, PYLE NATIONAL, PLUG | SIGN | TRANSLITE |
| FIRE EXTINGUISHER | W. KIDDE, 1 1/2-2 3/4 LB DRY POWDER | SIGNAL DEVICE | PNEUPHONIC HORN |
| HAND BRAKE | NATIONAL BRAKE, 840XBL | STANCHION | ELLCON NATIONAL [6] |
| HANDHOLD | ELLCON NATIONAL [47] | WINDOW | TRANSPORTATION SASH |
| HEAD, MARKER & TAIL LIGHTS | LOVELL DRESSEL | WINDSHIELD WIPER #1 END | SPRAGUE DEVICE, PNEUMATIC |
| **EVEN-NUMBERED CARS** | | **ODD-NUMBERED CARS** | |
| BATTERY | ELECTRIC STORAGE BATTERY (FIBERGLASS BOX) | AIR COMPRESSOR | WABCO, D-3 |
| MOTOR GENERATOR | WH, YX304E1 | PROPULSION CONTROL TYPE | WH, CAM, XCA248S GROUP |
| PROPULSION CONTROL TYPE | WH, CAM, XCA248T GROUP | SWITCH PANEL #2 CAB | MID. ROSS, 500007F |
| SWITCH PANEL #2 CAB | VAPOR, 56660443 | | |

## TRUCK

| | | | |
|---|---|---|---|
| MANUFACTURER | GENERAL STEEL INDUSTRIES; ADIRONDACK STEEL CASTING | BRAKE RIGGING | A.S.F., SIMPLEX UNIT CYLINDER CLASP BRAKE |
| TYPE & QUANTITY | C.S.F. EQUALIZER BAR MOTOR TRUCK [136] | GEAR UNIT | WH, WN44 (7.235:1) |
| TRUCK SERIAL NUMBERS | WH, 15693 - 15828 | JOURNAL BEARING | SKF, 5X9 ROLLER |
| WEIGHT | N/A | SHOCK ABSORBER | HOUDAILLE |
| AUTO. SLACK ADJUSTER | WABCO, C15D1 | TRACTION MOTOR | WH, 1447C (100 HP) [2] |
| BRAKE CYLINDER | WABCO, 7X6 UAHT [2] | TRIP COCK | WABCO, D1 |

NOTE 1: CARS 4450 -4517 WERE ORIGINALLY DELIVERED AS CARS 4250 - 4317. RENUMBERED (1970).
NOTE 2: CARS 4450 - 4525 (76) HAVE SPECIAL GLASS & SASH AT CAR END.

CAR 4461 SCRAPPED (1996).

CAR NUMBERS: 4518 - 4549 (SEE NOTE 1)
TOTAL: 32 CARS
BUILT BY: ST. LOUIS CAR

DATE: 1968 - 69
AVERAGE COST PER CAR: $137,382

# R40 (BMT/IND)

STRAIGHT END - PACKAGE BRAKE
AIR CONDITIONED

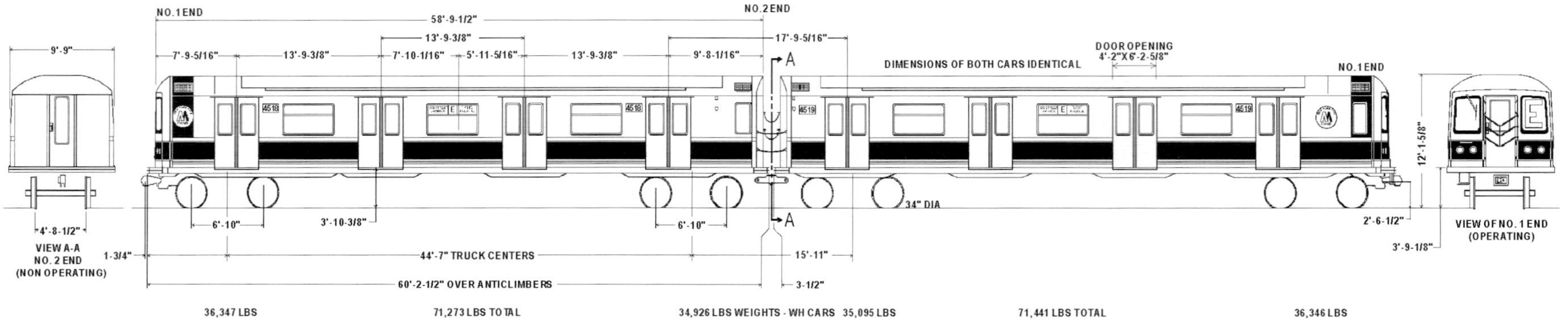

NO.1 END — NO.2 END — NO.1 END
58'-9-1/2"
13'-9-3/8"
9'-9"
7'-9-5/16" 13'-9-3/8" 7'-10-1/16" 5'-11-5/16" 13'-9-3/8" 9'-8-1/16"
17'-9-5/16"
DIMENSIONS OF BOTH CARS IDENTICAL
DOOR OPENING 4'-2"X6'-2-5/8"
12'-1-5/8"
34" DIA
4'-8-1/2"
6'-10" 3'-10-3/8" 6'-10"
2'-6-1/2"
3'-9-1/8"
VIEW A-A NO. 2 END (NON OPERATING)
VIEW OF NO. 1 END (OPERATING)
1-3/4"
44'-7" TRUCK CENTERS
15'-11"
60'-2-1/2" OVER ANTICLIMBERS
3-1/2"
36,347 LBS 71,273 LBS TOTAL 34,926 LBS WEIGHTS - WH CARS 35,095 LBS 71,441 LBS TOTAL 36,346 LBS

Copyright © 1997 by NYC Transit

## CAR BODY - EQUIPMENT

| | | | | | |
|---|---|---|---|---|---|
| AIR CONDITIONING | | TRANE,10 TON UNITS [2]; VAPOR CYCLE | HEATER | | ST. LOUIS CAR |
| AUX. CIRCUIT PROTECTION | | CIRCUIT BREAKERS | LIGHTING | | LUMINATOR, ENCLOSED GLASS LENS & BACKLIGHTED "AD" CARDS |
| BRAKE EQUIPMENT TYPE | | WABCO, SMEE (DYNAMIC & FRICTION) | MASTER CONTROLLER | | WH XM829 |
| BRAKE VALVE | | WABCO, ME-43 | POLARITY REVERSING RELAY | | VAPOR, SOLID STATE |
| COUPLER | | WABCO: #1 END, H2C; #2 END, LINK BAR | PUBLIC ADDRESS SYSTEM | | SOUND SYSTEMS |
| DOOR OPERATOR MOTOR | | MID. ROSS: CENTER, 500003C; OPP. CAB, 500004D; CAB END, 500005E | SASH (CAB) | | TRANSPORTATION SASH |
| DOOR HANGER & PANEL | | MORTON | SEAT ARRANGEMENT & CAPACITY | | LONGITUDINAL [44] |
| DRAFT GEAR | | #1 END: WAUGH, WM-E4-CL | SEAT MANUFACTURER | | GENERAL SEATING |
| ELECTRIC PORTION | | #1 END: WABCO, BL26B; #2 END: PYLE NATIONAL, PLUG | SEAT MATERIAL | | FIBERGLASS |
| FIRE EXTINGUISHER | | W. KIDDE, 1.5 - 2.75 LB DRY POWDER | SIGN | | TRANSLITE |
| HANDHOLD | | ELLCON NATIONAL [47] | SIGNAL DEVICE | | PNEUPHONIC HORN |
| HANDBRAKE | | ELLCON NATIONAL 840XBL | STANCHION | | ELLCON NATIONAL [6] |
| HEAD, MARKER & TAIL LIGHTS | | LOVELL DRESSEL | WINDOW | | TRANSPORTATION SASH |
| HVAC CONTROL | | VAPOR | WINDSHIELD WIPER | #1 END | SPRAGUE DEVICES, PNEUMATIC |
| | | EVEN-NUMBERED CARS ONLY | | | ODD-NUMBERED CARS ONLY |
| BATTERY | | ELECTRIC STORAGE BATTERY FIBERGLASS BOX | AIR COMPRESSOR | | WABCO, D3 |
| MOTOR GENERATOR | | WH, YX304E1 | PROPULSION CONTROL TYPE | | WH, CAM, XCA248S GROUP |
| PROPULSION CONTROL TYPE | | WH, CAM, XCA248T GROUP | SWITCH PANEL | # 2 CAB | MID. ROSS, 500007F |
| SWITCH PANEL | #2 CAB | MID. ROSS, 500009E | | | |

## TRUCK

| | | | |
|---|---|---|---|
| MANUFACTURER | GENERAL STEEL INDUSTRIES & ADIRONDACK STEEL CASTING | BRAKE CYLINDER | WABCO, 7X6 UAHT [2] |
| TYPE & QUANTITY | C.S.F. EQUALIZER MOTOR [64] | GEAR UNIT | WIN44 (7.235:1) |
| TRUCK SERIAL NUMBERS | WABCO: 15829-15850; 15867-15876 | JOURNAL BEARING | SKF, 5X9 ROLLER |
| | ASF: 15851-15866; 15877-15892 | SHOCK ABSORBER | HOUDAILLE |
| WEIGHT | #1 END: 18,084 LBS; #2 END: 17,472 LBS | TRACTION MOTOR | 1447C (100 HP) [4] |
| AUTO. SLACK ADJUSTER | WABCO, C15D1 | TRIP COCK | WABCO, D1 |

NOTE 1: CARS 4518 -4549 WERE ORIGINALLY DELIVERED AS CARS 4318 - 4349. RENUMBERED (1970).

CAR NUMBERS: 4450 - 4549 (SEE NOTE 2)
TOTAL: 100 CARS
BUILT BY: ST. LOUIS CAR
DATE: 1968 - 69
AVERAGE COST PER CAR: $137,382

OVERHAULED BY: SUMITOMO
DATE: 1987-89
OVERHAUL COST PER CAR: $399,000

# R40 (BMT/IND)

STRAIGHT END
AIR CONDITIONED
GOH

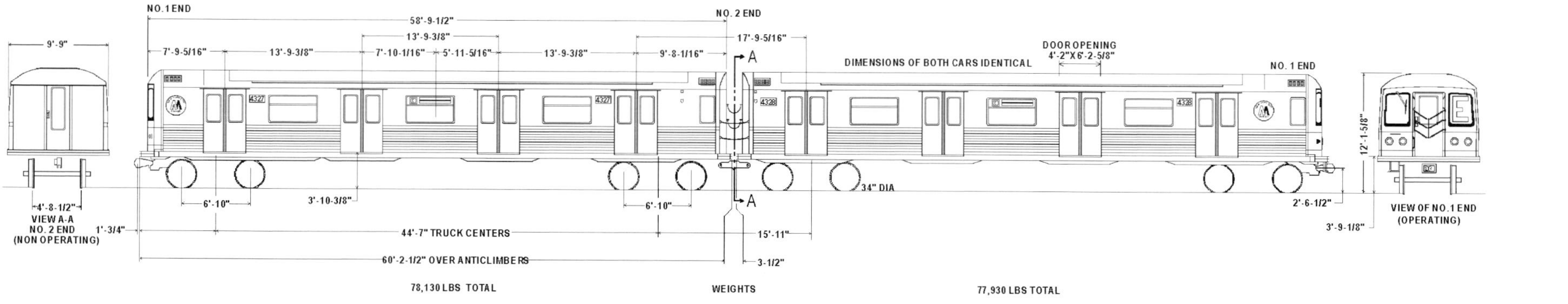

9'-9"
4'-8-1/2"
VIEW A-A
NO. 2 END
(NON OPERATING)
1'-3/4"
NO. 1 END
58'-9-1/2"
13'-9-3/8"
7'-9-5/16"
13'-9-3/8"
7'-10-1/16"
5'-11-5/16"
13'-9-3/8"
9'-8-1/16"
NO. 2 END
17'-9-5/16"
A
DIMENSIONS OF BOTH CARS IDENTICAL
DOOR OPENING
4'-2"X6'-2-5/8"
NO. 1 END
12'-1-5/8"
6'-10"
3'-10-3/8"
6'-10"
34" DIA
2'-6-1/2"
3'-9-1/8"
VIEW OF NO. 1 END
(OPERATING)
44'-7" TRUCK CENTERS
15'-11"
60'-2-1/2" OVER ANTICLIMBERS
3-1/2"
78,130 LBS TOTAL
WEIGHTS
77,930 LBS TOTAL

Copyright © 1997 by NYC Transit

## CAR BODY - EQUIPMENT

| | | | |
|---|---|---|---|
| AIR CONDITIONING | THERMO KING, 9 TON UNITS [2] | LIGHTING | ENCLOSED PLASTIC LENS & BACKLIGHTED |
| AUXILIARY CIRCUIT PROTECTION | CIRCUIT BREAKERS | | "AD" CARDS W/LIGHTING INVERTER BALLAST |
| BRAKE EQUIPMENT TYPE | WABCO, SMEE (DYNAMIC & FRICTION) | MASTER CONTROLLER | GE, 17KC76E1 |
| BRAKE VALVE | WABCO, ME43 | PROPULSION CONTROL TYPE | GE, SCM, 17KG192AE2 GROUP |
| COUPLER | WABCO: #1 END, H2C; #2 END, LINKBAR | PUBLIC ADDRESS SYSTEM | COMCO |
| DOOR HANGER | A.B. KING | SASH (CAB) | EBONEX |
| DOOR OPERATOR MOTOR | WESTCODE | SEAT ARRANGEMENT & CAPACITY | LONGITUDINAL [44] |
| DOOR PANEL | EBONEX | SEAT MANUFACTURER | NIAGARA SEATING/ARTCRAFT |
| DRAFT GEAR | WAUGH, WME4CL | SEAT MATERIAL | FIBERGLASS |
| ELECTRIC PORTION | #1 END: WABCO, BL37; # 2 END: VEAM, CONNECTOR | SIGN | TRANSLITE/TELEWELD |
| HAND BRAKE | NATIONAL BRAKE, 840XBL | SIGNAL DEVICE | PNEUPHONIC HORN |
| HANDHOLD | ELLCON NATIONAL [47] | STANCHION | ELLCON NATIONAL [6] |
| HEAD LIGHT | LOVELL DRESSEL | TAIL LIGHT | ST. LOUIS CAR |
| HEATER | CHROMOLOX/EMERSON/FABER LUMINATOR | WINDOW | EBONEX |
| HVAC CONTROL | VAPOR | WINDSHIELD WIPER #1 END | SYRACUSE DIESEL, ELECTRIC |
| **EVEN-NUMBERED CARS ONLY** | | **ODD-NUMBERED CARS ONLY** | |
| BATTERY | HOPPECKE/NIFE (FIBER GLASS BOX) | AIR COMPRESSOR | WABCO, D4S |
| CONVERTER | STONE SAFETY, 7 KW | SWITCH PANEL #2 CAB | VAPOR |
| SWITCH PANEL #2 CAB | VAPOR | | |

## TRUCK

| | | | |
|---|---|---|---|
| MANUFACTURER | GENERAL STEEL INDUSTRIES & ADIRONDACK STEEL CASTING | BRAKE RIGGING | A.S.F. SIMPLEX UNIT CYLINDER CLASP BRAKES |
| TYPE & QUANTITY | C.S.F. EQUALIZER BAR MOTOR [200] | GEAR UNIT | GE, 7GA25 (7.235:1) |
| TRUCK NUMBERS | N/A | JOURNAL BEARING | 5X9, AP ROLLER, SEE NOTE 1 |
| WEIGHT | #1 END: ; #2 END | SHOCK ABSORBER | MONROE |
| AUTO. SLACK ADJUSTER | WABCO, C15D1 | TRACTION MOTOR | GE, 1257E1 (115 HP) [2] |
| BRAKE CYLINDER | WABCO, 7"X16" UAHT [2] | TRIP COCK | WABCO, D1 |

NOTE 1: VARIOUS VENDORS.
NOTE 2: CARS 4450 - 4549 WERE ORIGINALLY DELIVERED AS CARS 4250 - 4349. RENUMBERED (1970).

CARS 4518 - 4549 HAD PACKAGE BRAKES PRIOR TO OVERHAUL.
INSTALLED DOPPLER SPEEDOMETER BY EDO (1996).
CAR 4461 SCRAPPED (1996).

CAR NUMBERS: 4550 - 4949
TOTAL: 400 CARS
BUILT BY: ST. LOUIS CAR

DATE: 1969 - 70
AVERAGE COST PER CAR: $132,670

# R42 (BMT/IND)

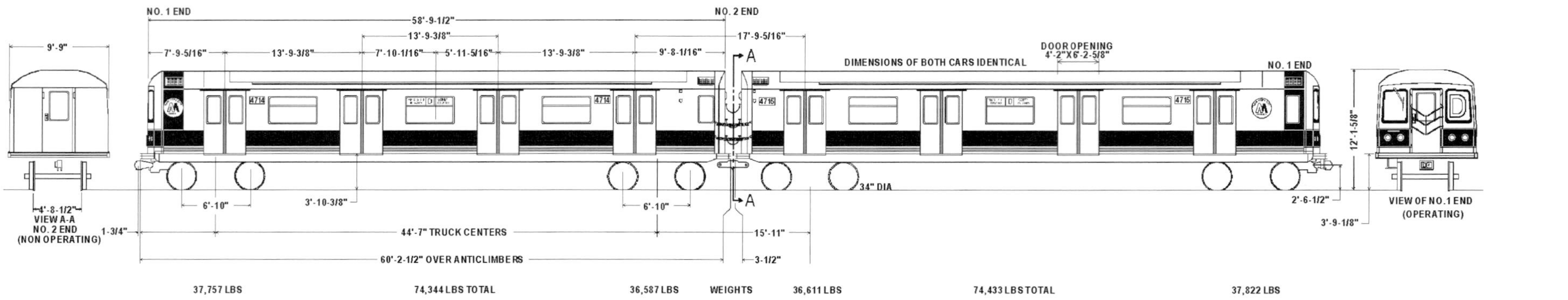

9'-9"
4'-8-1/2"
VIEW A-A
NO. 2 END
(NON OPERATING)
1-3/4"
NO. 1 END
58'-9-1/2"
13'-9-3/8"
7'-9-5/16"
13'-9-3/8"
7'-10-1/16"
5'-11-5/16"
13'-9-3/8"
9'-8-1/16"
NO. 2 END
17'-9-5/16"
A
DIMENSIONS OF BOTH CARS IDENTICAL
DOOR OPENING
4'-2"X6'-2-5/8"
NO. 1 END
12'-1-5/8"
6'-10"
3'-10-3/8"
6'-10"
34" DIA
2'-6-1/2"
3'-9-1/8"
VIEW OF NO. 1 END
(OPERATING)
44'-7" TRUCK CENTERS
15'-11"
60'-2-1/2" OVER ANTICLIMBERS
3-1/2"

37,757 LBS
74,344 LBS TOTAL
36,587 LBS
WEIGHTS
36,611 LBS
74,433 LBS TOTAL
37,822 LBS

Copyright © 1997 by NYC Transit

## CAR BODY - EQUIPMENT

| | | | |
|---|---|---|---|
| AIR CONDITIONING | SAFETY/TRANE, 10-TON UNITS [2] | HEATER | ST. LOUIS CAR/BRYANT |
| AUXILIARY CIRCUIT PROTECTION | CIRCUIT BREAKERS | HVAC CONTROL | VAPOR |
| BRAKE EQUIPMENT TYPE | WABCO, SMEE (DYNAMIC & FRICTION) | LIGHTING | LUMINATOR, ENCLOSED GLASS LENS & BACKLIGHTED "AD" CARDS |
| BRAKE VALVE | WABCO, ME43 | MASTER CONTROLLER | WH. , XM829 |
| COMMUNICATION | MOTOROLA, SP3211703 | PROPULSION CONTROL TYPE | WH, CAM, XCD2486 GROUP |
| COUPLER | WABCO; WAUGH: #1 END, H2C; #2 END, LINK BAR | PUBLIC ADDRESS SYSTEM | SOUND SYSTEMS, TRANSISTOR |
| DOOR HANGER & PANEL | O.M. EDWARDS | SEAT ARRANGEMENT & CAPACITY | LONGITUDINAL [44] |
| DOOR OPERATOR MOTOR | VAPOR: R.H., 56970501; L.H., 56970502 | SEAT MANUFACTURER | GENERAL SEATING |
| DRAFT GEAR | WAUGH , WE4CL | SEAT MATERIAL | FIBERGLASS |
| ELECTRIC PORTION | WALTON | SIGN | TRANSLITE |
| END SIGN | TRANSLITE | SIGNAL DEVICE | PNEUPHONIC HORN |
| FIRE EXTINGUISHER | W. KIDDE, 2-3/4 LB PRESS. DRY POWDER | STANCHION | ELLCON NATIONAL [8] |
| HAND BRAKE | ELLCON NATIONAL, 840XBL | WINDOW | O.M. EDWARDS |
| HEAD & TAIL LIGHT | ST. LOUIS CAR | WINDSHIELD WIPER | SPRAGUE , PNEUMATIC |
| HANDHOLD | ELLCON NATIONAL | | |
| | **EVEN-NUMBERED CARS ONLY** | | **ODD-NUMBERED CARS ONLY** |
| AIR COMPRESSOR | WABCO, D3 | SWITCH PANEL | VAPOR,56761118 (CARS 4750 - 4949) |
| BATTERY | EDISON, ED120; EXIDE, GP3125 (NICKEL CADMIUM 26 CELL) | | MID. ROSS, 500007G (CARS 4550 - 4749) |
| CONVERTER | WH, TE158, 5 KW | | |
| STATIC CONVERTER | WH, TE209 | | |
| SWITCH PANEL | VAPOR, 56761117 (CARS 4750 - 4949) | | |
| | MID. ROSS, 500009F (CARS 4550 - 4749) | | |

## TRUCK

| | | | |
|---|---|---|---|
| MANUFACTURER | GENERAL STEEL INDUSTRIES | BRAKE RIGGING | AMERICAN STEEL FOUNDRY, |
| TYPE & QUANTITY | C.S.F. EQUALIZER BAR MOTOR [820] | | WABCO, UNIT CYLINDER BRAKE |
| TRUCK SERIAL NUMBERS | N/A | GEAR UNIT | WH, WN44CGS (7.235:1) |
| WEIGHT | #1 END: 18,700 LBS; #2 END: 18,100 LBS | JOURNAL BEARING | SKF, 5x9 ROLLER |
| AUTO. SLACK ADJUSTER | WABCO, C15D1 | SHOCK ABSORBER | HOUDAILLE |
| BRAKE CYLINDER | WABCO, 7X6 UAHT [2] | TRACTION MOTOR | WH, 1447C (115 HP) [2] |
| | | TRIP COCK | WABCO, D1 |

PANTOGRAPH GATES WERE APPLIED TO #1 ENDS OF ALL CARS (1977 - EARLY 1981).

CAR NUMBERS: 4550 - 4839
TOTAL: 282 CARS
BUILT BY: ST. LOUIS CAR
DATE: 1969 - 70
AVERAGE COST PER CAR: $132,670

OVERHAULED BY: MK
DATE: 1988 - 89
OVERHAUL COST PER CAR: $428,000

# R42 (BMT/IND)

MK OVERHAULED

GOH

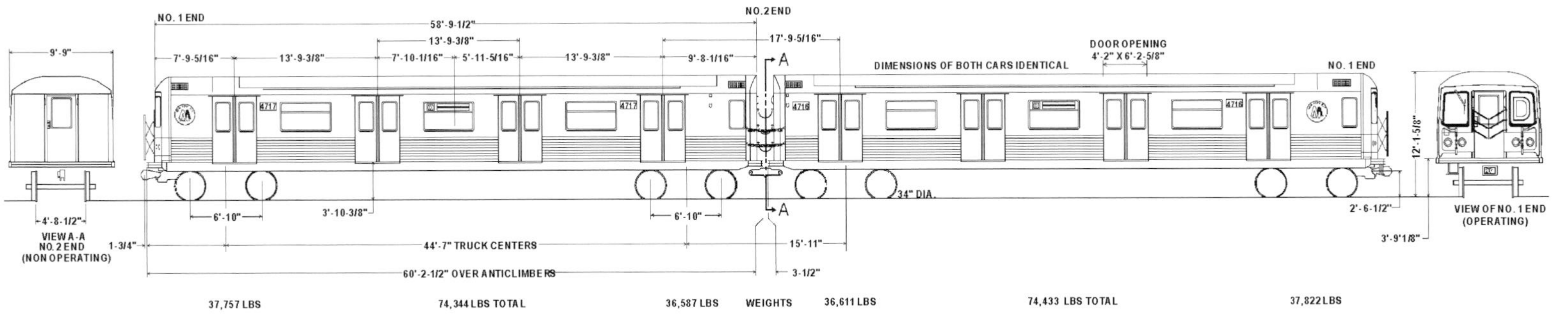

NO. 1 END
58'-9-1/2"
13'-9-3/8"
7'-9-5/16"
13'-9-3/8"
7'-10-1/16"
5'-11-5/16"
13'-9-3/8"
9'-8-1/16"
NO. 2 END
17'-9-5/16"
A
DOOR OPENING 4'-2" X 6'-2-5/8"
DIMENSIONS OF BOTH CARS IDENTICAL
NO. 1 END
9'-9"
4717
4716
12'-1-5/8"
34" DIA.
4'-8-1/2"
VIEW A-A NO. 2 END (NON OPERATING)
1-3/4"
6'-10"
3'-10-3/8"
6'-10"
2'-6-1/2"
VIEW OF NO. 1 END (OPERATING)
44'-7" TRUCK CENTERS
15'-11"
3'-9'1/8"
60'-2-1/2" OVER ANTICLIMBERS
3-1/2"
37,757 LBS
74,344 LBS TOTAL
36,587 LBS
WEIGHTS
36,611 LBS
74,433 LBS TOTAL
37,822 LBS

Copyright © 1997 by NYC Transit

## CAR BODY - EQUIPMENT

| | | | |
|---|---|---|---|
| AIR CONDITIONING | 9 TON UNITS (2). SEE NOTE 2 | LIGHTING | LUMINATOR, FLUORESCENT FIXTURES |
| AUXILIARY CIRCUIT PROTECTION | CIRCUIT BREAKERS | LIGHTING INVERTER | WH, TE5D |
| BRAKE EQUIPMENT TYPE | NYAB (DYNAMIC & FRICTION) | MASTER CONTROLLER | WH, XM829 |
| BRAKE VALVE | NYAB NEWTRAN | PUBLIC ADDRESS SYSTEM | SOUND SYSTEM, TRANSISTOR |
| COUPLER | WABCO: #1 END, H2C; #2 END, LINK BAR | SASH (CAB) | ELLCON NATIONAL |
| DOOR HANGER | A.B. KING | SEAT ARRANGEMENT & CAPACITY | LONGITUDINAL [44] |
| DOOR OPERATOR MOTOR | VAPOR | SEAT MANUFACTURER | GENERAL SEATING |
| DOOR PANEL | EBONEX | SIGN | TRANSLITE |
| DRAFT GEAR | WAUGH, WE4CL | SIGNAL DEVICE | PNEUPHONIC HORN |
| HAND BRAKE | ELLCON NATIONAL, 840XBL | STANCHION | ELLCON NATIONAL [6] |
| HANDHOLD | EBONEX | TRAIN LINE CIRCUIT | END #1: WABCO, BL37; END #2: PYLE NATIONAL |
| HEAD & TAIL LIGHTS | MK | WINDOW | ELLCON NATIONAL |
| HEATER | BRYANT ELECTRIC | WINDSHIELD WIPER | AMERICAN BOSCH, ELECTRIC |
| HVAC CONTROL | VAPOR | | |
| **EVEN-NUMBERED CARS ONLY** | | **ODD-NUMBERED CARS ONLY** | |
| BATTERY | SAFT NIFE | AIR COMPRESSOR | WABCO, D4S |
| PROPULSION CONTROL TYPE | GE, SCM, 17KG192A1E1 GROUP | PROPULSION CONTROL TYPE | GE, SCM, 17KG192AE2 GROUP |
| STATIC CONVERTER | WH, TE159 (7KW) | SWITCH PANEL #2 CAB | VAPOR |
| SWITCH PANEL #2 CAB | VAPOR | | |

## TRUCK

| | | | |
|---|---|---|---|
| MANUFACTURER | GENERAL STEEL INDUSTRIES & ADIRONDACK STEEL CASTING | AUTO SLACK ADJUSTER | WABCO, C15D1 |
| TYPE & QUANTITY | C.S.F. EQUALIZER BAR MOTOR [820] | BRAKE CYLINDER | WABCO, 7X 6 UNIT CYLINDERS UAHT [2] |
| TRUCK SERIAL NO. | 16293-17112 (20 SPARES) | BRAKE RIGGING | A.S.F. SIMPLEX UNIT CYLINDER CLASP BRAKE |
| WEIGHT | #1 END: ; #2 END: | JOURNAL BEARING | 5X9, CLASS D, AP ROLLER, SEE NOTE 1 |
| TRACTION MOTOR | WH, 1447J (115 HP) [2] | SHOCK ABSORBER | MONROE |
| GEAR UNIT | WH, WNT44-1 (7.235:1) | TRIP COCK | WABCO, D1 |

NOTE 1: VARIOUS VENDORS.
NOTE 2: STONE SAFETY CARS 4550 - 4749, THERMO KING CARS 4750 - 4840.

INTERIOR BACK-LIT AD RACKS REMOVED DURING GOH.
INSTALLED DOPPLER SPEEDOMETER BY EDO (1996).
AMETEK ALTERNATING CURRENT BLOWER MOTORS/INVERTERS INSTALLED ON ALL CARS (2000).
4680-81, 4685, 4714-15, 4726, 4766-67 WERE NEVER OVERHAULED (SCRAPPED, 1988). CAR 4664 SCRAPPED (9/1996).

CAR NUMBERS: 4840 - 4949
TOTAL: 110 CARS
BUILT BY: ST. LOUIS CAR
DATE: 1969 - 70
AVERAGE COST PER CAR: $132,670

OVERHAULED BY: NYC TRANSIT AUTHORITY
DATE: 1988 - 89
OVERHAUL COST PER CAR: $688,000

# R42 (BMT/IND)

NYCTA OVERHAULED

## GOH

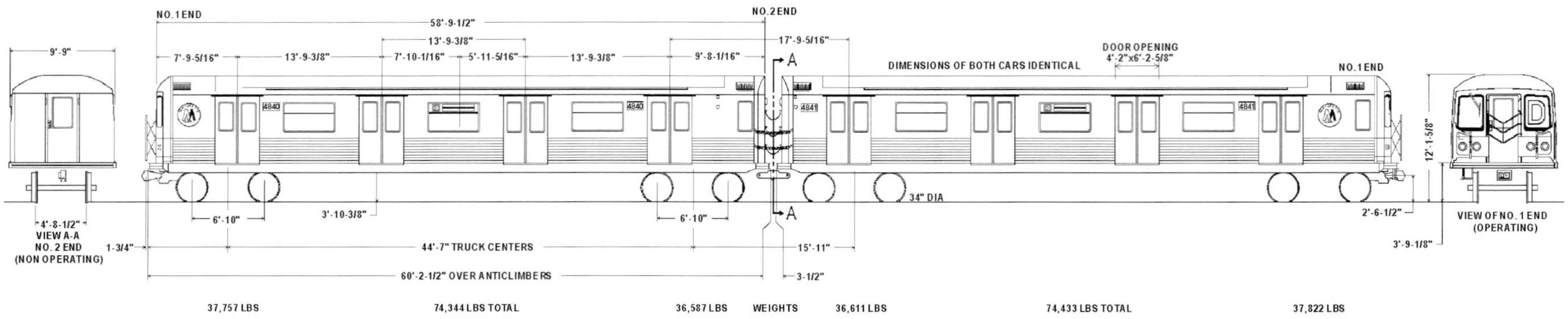

NO. 1 END — NO. 2 END — NO. 1 END

58'-9-1/2" · 13'-9-3/8" · 7'-9-5/16" · 13'-9-3/8" · 7'-10-1/16" · 5'-11-5/16" · 13'-9-3/8" · 9'-8-1/16" · 17'-9-5/16"

DIMENSIONS OF BOTH CARS IDENTICAL

DOOR OPENING 4'-2"x6'-2-5/8"

9'-9" · 4'-8-1/2" · VIEW A-A NO. 2 END (NON OPERATING) · 1-3/4" · 6'-10" · 3'-10-3/8" · 44'-7" TRUCK CENTERS · 60'-2-1/2" OVER ANTICLIMBERS · 6'-10" · 15'-11" · 3-1/2" · 34" DIA · 2'-6-1/2" · 12'-1-5/8" · 3'-9-1/8" · VIEW OF NO. 1 END (OPERATING)

WEIGHTS: 37,757 LBS · 74,344 LBS TOTAL · 36,587 LBS | 36,611 LBS · 74,433 LBS TOTAL · 37,822 LBS

Copyright © 1997 by NYC Transit

## CAR BODY - EQUIPMENT

| | | | |
|---|---|---|---|
| AIR CONDITIONING | STONE SAFETY, 9 TON UNITS [2] | LIGHTING | LUMINATOR, FLUORESCENT FIXTURES |
| AUXILIARY CIRCUIT PROTECTION | CIRCUIT BREAKERS | LIGHTING INVERTER | WH, TE5D |
| BRAKE EQUIPMENT TYPE | NYAB, NEWTRAN (DYNAMIC & FRICTION) | MASTER CONTROLLER | GE, 17KC76AE1 |
| CONTROL GROUP | GE, 17KG192AE2 | PROPULSION CONTROL TYPE | GE, SCM, 17KG192A1E1 GROUP |
| COUPLER | WABCO: #1 END, H2C; #2 END, LINK BAR | PUBLIC ADDRESS SYSTEM | COMCO |
| DOOR HANGER & PANEL | EBONEX | SASH (CAB) | ELLCON NATIONAL |
| DOOR OPERATOR MOTOR | WESTCODE | SEAT ARRANGEMENT & CAPACITY | LONGITUDINAL [44] |
| DRAFT GEAR | WAUGH, WE4CL | SEAT MANUFACTURER | GENERAL SEATING |
| ELECTRIC PORTION | #1 END: WABCO, BL37; #2 END: PYLE NATIONAL, PLUG | SIGN | TRANSLITE |
| HAND BRAKE | ELLCON NATIONAL, 840XBL | SIGNAL DEVICE | PNEUPHONIC HORN |
| HANDHOLD | EBONEX | STANCHION | ELLCON NATIONAL [6] |
| HEAD & TAIL LIGHTS | ST. LOUIS CAR | WINDOW | J.T. NELSON |
| HEATER | BRYANT ELECTRIC | WINDSHIELD WIPER | AMERICAN BOSCH, ELECTRIC |
| HVAC CONTROL | STONE SAFETY | | |
| **EVEN-NUMBERED CARS ONLY** | | **ODD-NUMBERED CARS ONLY** | |
| BATTERY | SAFT NIFE | AIR COMPRESSOR | WABCO, D4S |
| STATIC CONVERTER | STONE SAFETY, 7 KW | SWITCH PANEL #2 CAB | VAPOR |
| SWITCH PANEL #2 CAB | VAPOR | | |

## TRUCK

| | | | |
|---|---|---|---|
| MANUFACTURER | GENERAL STEEL INDUSTRIES & ADIRONDACK STEEL CASTING | BRAKE RIGGING | A.S.F. SIMPLEX UNIT CYLINDER CLASP BRAKE |
| TYPE & QUANTITY | C.S.F. EQUALIZER BAR MOTOR [820] | GEAR UNIT | WH, WNT44-1 (7.235:1) |
| TRUCK SERIAL NUMBERS | 16293 - 17112 (20 SPARES) | JOURNAL BEARING | 5X9 CLASS D, AP ROLLER, SEE NOTE 1 |
| WEIGHT | #1 END: ; #2 END: | SHOCK ABSORBER | MONROE |
| AUTO SLACK ADJUSTER | WABCO, C15D1 | TRACTION MOTOR | WH, 1447J (115 HP) [2] |
| BRAKE CYLINDER | WABCO, 7X 6 UAHT [2] | TRIP COCK | WABCO, D1 |

NOTE 1: VARIOUS VENDORS.
NOTE 2: STONE SAFETY CARS 4550 - 4749, THERMO KING CARS 4750 - 4840.

INTERIOR BACK-LIT AD RACKS REMOVED DURING GOH.
INSTALLED DOPPLER SPEEDOMETER BY EDO (1996).
AMETEK ALTERNATING CURRENT BLOWER MOTORS/INVERTERS INSTALLED ON ALL CARS (2000).

CAR NUMBERS: 100 - 399
TOTAL : 300 CARS
BUILT BY: ST. LOUIS CAR

DATE: 1971 - 73
AVERAGE COST PER CAR: $211,850

# R44 (BMT/IND)

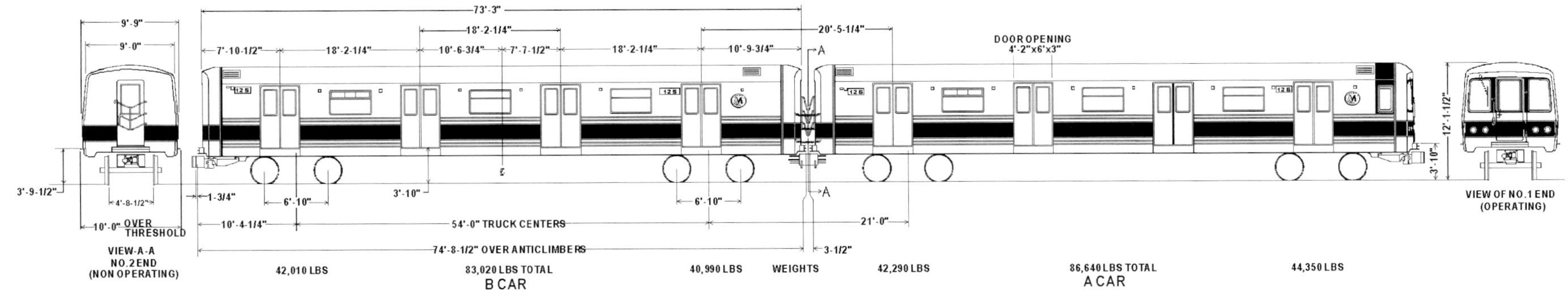

9'-9"
9'-0"
3'-9-1/2"
4'-8-1/2"
10'-0" OVER THRESHOLD
VIEW-A-A
NO.2 END
(NON OPERATING)

73'-3"
18'-2-1/4"
20'-5-1/4"
7'-10-1/2" 18'-2-1/4" 10'-6-3/4" 7'-7-1/2" 18'-2-1/4" 10'-9-3/4"
1-3/4" 6'-10" 3'-10" 6'-10"
10'-4-1/4" 54'-0" TRUCK CENTERS 21'-0"
74'-8-1/2" OVER ANTICLIMBERS 3-1/2"
DOOR OPENING
4'-2"x6'x3"
12'-1-1/2"
3'-10"
VIEW OF NO.1 END
(OPERATING)

WEIGHTS: 42,010 LBS; 40,990 LBS — 83,020 LBS TOTAL, B CAR; 42,290 LBS; 44,350 LBS — 86,640 LBS TOTAL, A CAR

Copyright © 1997 by NYC Transit

## CAR BODY - EQUIPMENT

| | | |
|---|---|---|
| AIR CONDITIONING | | SAFETY, 10-TON UNITS [2] |
| AUXILIARY CIRCUIT PROTECTION | | CIRCUIT BREAKERS |
| BRAKE EQUIPMENT TYPE | | WABCO, RT5C (DYNAMIC & FRICTION) |
| DRAFT GEAR | #1 END | "A" CAR: OHIO BRASS, FORM 70 WITH CENTERING DEVICE |
| | | "B" CAR: OHIO BRASS, FORM 70 W/O CENTERING DEVICE |
| | #2 END | "A" & "B" CAR: OHIO BRASS, FORM 70 W/O CENTERING DEVICE |
| DOOR HANGER & PANEL | | O.M. EDWARDS |
| DOOR OPERATOR MOTOR | | VAPOR: R.H., 56970501; L.H., 56970502 |
| ELECTRIC PORTION | | WALTON |
| END SIGN | | TRANSLITE |
| FIRE EXTINGUISHER | | 2-3/4 LBS DRY POWDER |
| HAND BAR | | ELLCON NATIONAL |
| HAND BRAKE | | ELLCON NATIONAL |
| HEAD & TAIL LIGHTS | | LUMINATOR |
| HEATER | | GE |
| **EVEN-NUMBERED CARS ONLY (A CARS)** | | |
| AIR COMPRESSOR | | WABCO, D3 |
| BATTERY | | EDISON, ED 120; EXIDE, GPS126 NICKEL CADMIUM, 26 CELL |
| STATIC CONVERTER | | WH, TE209 |
| SWITCH PANEL | #1 END | VAPOR |

| | | |
|---|---|---|
| HVAC CONTROL | | SAFETY ELECTRIC/VAPOR |
| LIGHTING | | LUMINATOR, ENCLOSED GLASS LENS & BACKLIGHTED "AD" CARDS |
| LIGHTING INVERTER | | WH, TE5E |
| MASTER CONTROLLER | | WH, XMA23TE5E |
| PROPULSION CONTROL TYPE | | WH, CAM, XCD2486 GROUP |
| PUBLIC ADDRESS SYSTEM | | SOUND SYSTEMS, TRANSISTOR |
| SASH (CAB) | | O.M. EDWARDS |
| SEAT ARRANGEMENT & CAPACITY | | CROSS & LONGITUDINAL, "A" CAR [72]; "B" CAR [76] |
| SEAT MANUFACTURER | | AMERICAN SEATING |
| SEAT MATERIAL | | FIBERGLASS |
| SIGN | | TRANSIGN |
| SIGNAL DEVICE | | PNEUPHONIC HORN |
| STANCHION | | ELLCON NATIONAL [8] |
| WINDOW | | O.M. EDWARDS |
| WINDSHIELD WIPER | | AMERICAN BOSCH, ELECTRIC |
| **ODD-NUMBER CARS ONLY (B CARS)** | | |
| SWITCH PANEL | #1 END | VAPOR |

## TRUCK

| | |
|---|---|
| MANUFACTURER | GENERAL STEEL INDUSTRIES |
| TYPE | C.S.F. EQUALIZER BAR MOTOR |
| WEIGHT | #1 END: 18,700 LBS; #2 END: 18,100 LBS |
| BRAKE RIGGING | A.S.F. WABCO, UNIT CYLINDER BRAKE |
| GEAR UNIT | WH, WN44EGS (117:23) |
| JOURNAL BEARING | SKF, 5x9 ROLLER |
| SHOCK ABSORBER | MONROE |
| TRACTION MOTOR | WH, 1447F (115 HP) [2] |
| TRIP COCK | WABCO, D1 |

CAR CONSIST: A-B-B-A. NO. 1 END 'A' CAR HAS TRAIN OPERATOR/CONDUCTOR CAB.
CARS 115 & 116 WERE CLEARANCE-TEST CARS (SHELLS ONLY).
CARS 328-335 WERE DELIVERED WITH CARPET FLOORS (REMOVED IN 1975).
INSTALLATION OF INTER-CAR SAFETY BARRIERS (BOLOGNA SPRINGS) COMPLETED BY 12/31/1984.
CAR 227 TRANSFERRED TO FIRE TRAINING ACADEMY ON RANDALLS ISLAND.

CARS 109, 120, 132, 176, 215, 248, 288, 315, 385 WERE SCRAPPED.
CARS 388 - 399 TRANSFERRED TO SIRTOA (1986).
ORIGINAL SPEEDOMETERS REMOVED. REINSTALLED BY EDO (1996).

CAR NUMBERS: 5202 - 5479, SEE NOTE 5
TOTAL: 278 CARS
BUILT BY: ST. LOUIS CAR
DATE: 1971 - 73
AVERAGE COST PER CAR: $211,850

OVERHAULED BY: MK AND NYC TRANSIT AUTHORITY
DATE: 1991 - 92
OVERHAUL COST PER CAR: SEE NOTE 1

# R44 (BMT/IND) GOH

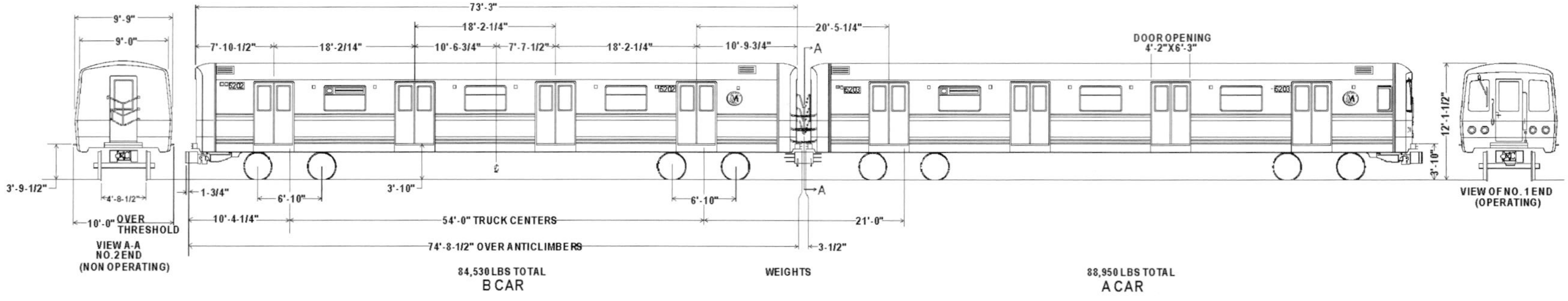

9'-9"
9'-0"
3'-9-1/2"
4'-8-1/2"
10'-0" OVER THRESHOLD
VIEW A-A NO.2 END (NON OPERATING)
73'-3"
18'-2-1/4"
20'-5-1/4"
7'-10-1/2" 18'-2/14" 10'-6-3/4" 7'-7-1/2" 18'-2-1/4" 10'-9-3/4"
1-3/4" 6'-10" 3'-10" 6'-10"
10'-4-1/4" 54'-0" TRUCK CENTERS 21'-0"
74'-8-1/2" OVER ANTICLIMBERS 3-1/2"
DOOR OPENING 4'-2"X6'-3"
12'-1-1/2"
3'-10"
VIEW OF NO. 1 END (OPERATING)
84,530 LBS TOTAL
B CAR
WEIGHTS
88,950 LBS TOTAL
A CAR

Copyright © 1997 by NYC Transit

## CAR BODY - EQUIPMENT

| | | | | | | |
|---|---|---|---|---|---|---|
| AIR CONDITIONING | | STONE SAFETY, 10-TON UNITS [2] | HVAC CONTROL | | STONE SAFETY |
| AUXILIARY CIRCUIT PROTECTION | | CIRCUIT BREAKERS | MASTER CONTROLLER | | WESTCODE (SEE NOTE 2) |
| BRAKE EQUIPMENT TYPE | | WESTCODE (DYNAMIC & FRICTION) | PROPULSION CONTROL TYPE | | WH, ECAM, XCA448F GROUP |
| DOOR HANGER & PANEL | | MORTON | PUBLIC ADDRESS SYSTEM | | COMCO |
| DOOR OPERATOR MOTOR (SEE NOTE 2) | | VAPOR: R.H., 56970501-60; L.H., 56970502-60 | SASH (CAB) | | J.T. NELSON |
| DRAFT GEAR | #1 END | "A" CAR: DRESSER, WITH CENTER DEVICE | SEAT ARRANGEMENT & CAPACITY | | CROSS & LONGITUDINAL, "A" CAR [72]; "B" CAR [76] |
| | | "B" CAR: LINKBAR | SEAT MANUFACTURER | | AMERICAN SEATING |
| | #2 END | "A" & "B" CAR: LINK BAR | SEAT MATERIAL | | FIBERGLASS |
| HAND BRAKE | | ELLCON NATIONAL | SIGN | | LUMINATOR/TELEWELD |
| HEAD & TAIL LIGHTS | | LUMINATOR | STANCHION | | ELLCON NATIONAL [8] |
| HEATER | | GE | WINDOW | | J.T. NELSON |
| LIGHTING | | LUMINATOR, ENCLOSED GLASS LENS & BACK LIGHTED "AD" CARDS | WINDSHIELD WIPER | | AMERICAN BOSCH, ELECTRIC |
| | | **EVEN-NUMBERED CARS ONLY (A CARS)** | | | **ODD-NUMBERED CARS ONLY (B CARS)** |
| AIR COMPRESSOR | | KNORR | SWITCH PANEL | #1 END | VAPOR |
| BATTERY | | ALCAD | | | |
| STATIC CONVERTER | | WH, TE409C | | | |
| SWITCH PANEL | #1 END | VAPOR | | | |

## TRUCK

| | | | |
|---|---|---|---|
| MANUFACTURER | GENERAL STEEL CASTING | GEAR UNIT | WH, WNT44-IMS (117.23) |
| TYPE | C.S.F. EQUALIZER BAR MOTOR | JOURNAL BEARING | 5x9, AP ROLLER, SEE NOTE 3 |
| TRUCKS NUMBERS | (SEE NOTE 4) | SHOCK ABSORBER | MONROE |
| WEIGHT | #1 END: 18,190 LBS; #2 END: 17,880 LBS | TRACTION MOTOR | WH, 1447F (115 HP) [2] |
| BRAKE RIGGING | WABCO TREAD BRAKE UNIT | TRIP COCK | WABCO, D1 |

NOTE 1: $548,000 BY MK, $612,000 BY NYCTA. MK (CARS 5202-5341); NYCTA - 207th ST. AND CONEY ISLAND (CARS 5342-5479).
NOTE 2: REMANUFACTURED BY WESTCODE.
NOTE 3: VARIOUS VENDORS.
NOTE 4: DUE TO TRUCK OVERHAUL PROGRAM, TRUCK NUMBERS VARIED.
NOTE 5: ORIGINAL CAR NUMBERS WERE 100-399 (300 CARS). TEN CARS SCRAPPED PRIOR TO OVERHAUL.

SIDE CURTAIN SIGNS REPLACED WITH DIGITAL ELECTRONIC SIGNS DURING GOH.
MATERIAL FOR THE NYCTA OVERHAULED CARS WAS SUPPLIED BY MK.
CAR CONSIST: A-B-B-A. NO. 1 END A CAR HAS MOTOR OPERATOR/CONDUCTOR CAB. NYAB LINK BAR WITH PLUGS BETWEEN CONSIST, MECHANICAL COUPLER W/ELECTRIC PORTION AT OPEN ENDS.
ORIGINAL SPEEDOMETERS REMOVED. REINSTALLED GEAR BOX SPEEDOMETER BY BROCKSOPP (1992).
INSTALLED DOOR OBSTRUCTION SENSING DEVICE BY WESTCODE (1995).
CARS 5282, 5283, 5284, 5285 AND 5402 SCRAPPED (1997). CAR 5319 SCRAPPED.

CAR NUMBERS: 400 - 435 CONSECUTIVE, 436 - 466 EVEN
TOTAL: 52 CARS
BUILT BY: ST. LOUIS CAR

DATE: 1971 - 73
AVERAGE COST PER CAR: $211,850

# R44 (SIRTOA)

WEIGHTS: 42,010 LBS — 83,020 LBS TOTAL — 40,990 LBS (B CAR); 42,290 LBS — 86,640 LBS TOTAL — 44,350 LBS (A CAR)

Copyright © 1997 by NYC Transit

## CAR BODY - EQUIPMENT

| | | |
|---|---|---|
| AIR CONDITIONING | | SAFETY ELECTRIC, 10-TON UNITS [2] |
| AUXILIARY CIRCUIT PROTECTION | | CIRCUIT BREAKERS |
| BRAKE EQUIPMENT TYPE | | WABCO (DYNAMIC & FRICTION) |
| DOOR HANGER & PANEL | | O.M. EDWARDS |
| DOOR OPERATOR MOTOR | | VAPOR: L.H., 56970502; R.H., 56970501 |
| DRAFT GEAR | #1 END | "A" CAR: OHIO BRASS, FORM 70 WITH CENTER DEVICE |
| | | "B" CAR: OHIO BRASS, FORM 70 W/O CENTER DEVICE |
| | #2 END | "A" & "B" CAR: OHIO BRASS, FORM 70 W/O CENTER DEVICE |
| ELECTRIC PORTION | | WALTON |
| FIRE EXTINGUISHER | | 2-3/4 LBS DRY POWDER |
| HAND BRAKE | | ELLCON NATIONAL |
| HEAD & TAIL LIGHTS | | LUMINATOR |
| HEATER | | GE |
| HVAC CONTROL | | STONE SAFETY/VAPOR |
| **EVEN-NUMBERED CARS ONLY (A CARS)** | | |
| AIR COMPRESSOR | | WABCO, D3 |
| BATTERY | | EDISON, ED 120; EXIDE, GP3126 NICKEL CADMIUM 26 CELL |
| STATIC CONVERTER | | WH, TE209 |
| SWITCH PANEL | #1 END | VAPOR |

| | | |
|---|---|---|
| LIGHTING | | LUMINATOR, ENCLOSED GLASS LENS & BACK LIGHTED "AD" CARDS |
| LIGHTING INVERTER | | WH, TE5E |
| MASTER CONTROLLER | | WH, XMA25 |
| PROPULSION CONTROL TYPE | | WH, CAM, XCD2486 GROUP |
| PUBLIC ADDRESS SYSTEM | | SOUND SYSTEMS, TRANSISTOR |
| SASH (CAB) | | O.M. EDWARDS |
| SEAT MANUFACTURER | | AMERICAN SEATING |
| SEAT MATERIAL | | FIBERGLASS |
| SEATING ARRANGEMENT & CAPACITY | | CROSS & LONGITUDINAL, "A" CAR [72]; "B" CAR [76] |
| SIGN | | TRANSIGN/TRANSLITE |
| SIGNAL DEVICE | | PNEUPHONIC HORN |
| STANCHION | | ELLCON NATIONAL [8] |
| WINDOW | | O.M. EDWARDS |
| WINDSHIELD WIPER | | AMERICAN BOSCH, ELECTRIC |
| **ODD-NUMBERED CARS ONLY (B CARS)** | | |
| SWITCH PANEL | #1 END | VAPOR |

## TRUCK

| | |
|---|---|
| MANUFACTURER | GENERAL STEEL INDUSTRIES |
| TYPE | C.S.F. EQUALIZER BAR MOTOR |
| WEIGHT | #1 END: 19,000 LBS; #2 END: 18,400 LBS |
| BRAKE RIGGING | TREAD BRAKE UNIT |
| TRACTION MOTOR | WH, 1447F (115 HP) [2] |
| GEAR UNIT | WH, WN44EGS (117:23) |
| JOURNAL BEARING | SKF, 5x9 ROLLER |
| SHOCK ABSORBER | MONROE |

CAR CONSIST: A-B-B-A.
CARS 388 - 399 TRANSFERRED FROM NYCTA (1986).
ORIGINAL SPEEDOMETERS REMOVED. REINSTALLED GEAR BOX SPEEDOMETER BY BROCKSOPP (1992).

CAR NUMBERS: 388 - 435 CONSECUTIVE, 436 - 466 EVEN
TOTAL: 64 CARS
BUILT BY: ST. LOUIS CAR
DATE: 1971 - 73
AVERAGE COST PER CAR: $211,850

OVERHAULED BY: NYC TRANSIT AUTHORITY
DATE: 1988
OVERHAUL COST PER CAR: $612,000

# R44 (SIR)
## GOH

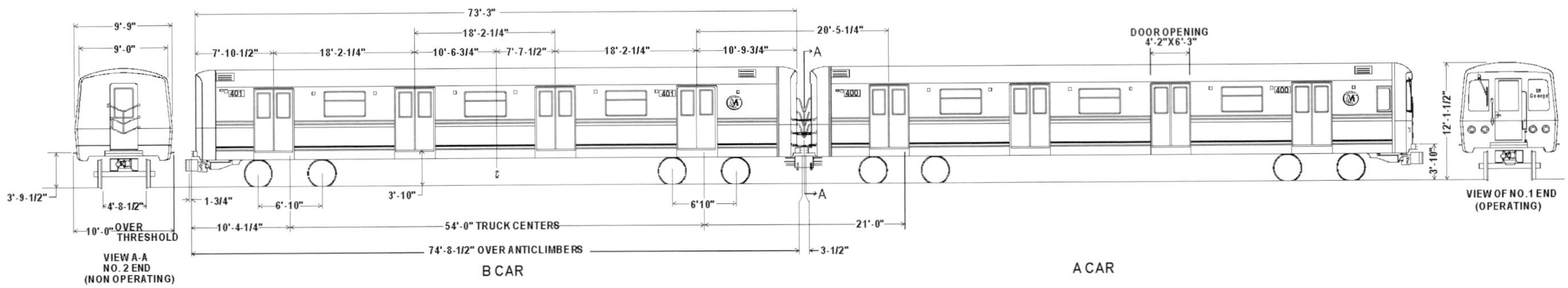

9'-9"
9'-0"
3'-9-1/2"
4'-8-1/2"
10'-0" OVER THRESHOLD
VIEW A-A
NO. 2 END
(NON OPERATING)

73'-3"
18'-2-1/4"
7'-10-1/2"
18'-2-1/4"
10'-6-3/4"
7'-7-1/2"
18'-2-1/4"
10'-9-3/4"
20'-5-1/4"
1-3/4"
6'-10"
3'-10"
6'10"
10'-4-1/4"
54'-0" TRUCK CENTERS
21'-0"
74'-8-1/2" OVER ANTICLIMBERS
3-1/2"
B CAR
A CAR
DOOR OPENING 4'-2"X6'-3"
12'-1-1/2"
3'-10"
VIEW OF NO.1 END
(OPERATING)

Copyright © 1997 by NYC Transit

## CAR BODY - EQUIPMENT

| | | | | |
|---|---|---|---|---|
| AIR CONDITIONING | | STONE SAFETY, 10-TON UNITS [2] | LIGHTING | LUMINATOR |
| AUXILIARY CIRCUIT PROTECTION | | CIRCUIT BREAKERS | MASTER CONTROLLER | WABCO CINESTON |
| BRAKE EQUIPMENT TYPE | | WABCO RT5C | PROPULSION CONTROL TYPE | GE, CAM ,17KG192A1 GROUP |
| DOOR HANGER & PANEL | | O.M. EDWARDS | PUBLIC ADDRESS SYSTEM | COMCO |
| DOOR OPERATOR MOTOR | | VAPOR: R.H., 56970501; L.H., 56970502 | SASH (CAB) | J. T. NELSON |
| DRAFT GEAR | #1 END | "A" CAR: OHIO BRASS, FORM 70 | SEAT ARRANGEMENT & CAPACITY | CROSS & LONGITUDINAL, "A" CAR [72]; "B" CAR [76] |
| | | "B" CAR: OHIO BRASS, FORM 70 | SEAT MANUFACTURER | AMERICAN SEATING |
| | #2 END | "A" & "B" CAR: OHIO BRASS, FORM 70 | SEAT MATERIAL | FIBERGLASS |
| ELECTRIC PORTION | | OHIO BRASS | SIGN | N/A |
| HAND BRAKE | | ELLCON NATIONAL | STANCHION | ELLCON NATIONAL [8] |
| HEAD & TAIL LIGHTS | | LUMINATOR | WINDOW | J. T. NELSON/TRANSLITE |
| HEATER | | GE | WINDSHIELD WIPER | AMERICAN BOSCH, ELECTRIC |
| HVAC CONTROL | | STONE SAFETY | SIGNAL DEVICE | PNEUPHONIC HORN |

| EVEN-NUMBERED CARS ONLY (A CARS) | | | ODD-NUMBERED CARS ONLY (B CARS) | | |
|---|---|---|---|---|---|
| AIR COMPRESSOR | | WABCO, D4S | SWITCH PANEL | #1 END | VAPOR |
| BATTERY | | SAFT-NIFE | | | |
| STATIC CONVERTER | | GE, 17KG417B1 (12 KW) | | | |
| SWITCH PANEL | #1 END | VAPOR | | | |

## TRUCK

| | | | |
|---|---|---|---|
| MANUFACTURER | GENERAL STEEL INDUSTRIES | GEAR UNIT | GE, 76A64D1 (117:23) |
| TYPE | C.S.F. EQUALIZER BAR MOTOR | JOURNAL BEARING | TIMKEN, AP ROLLER |
| WEIGHT | #1 END: 19,000 LBS; #2 END: 18,400 LBS | SHOCK ABSORBER | MONROE |
| BRAKE RIGGING | WABCO, TREAD BRAKE UNIT | TRACTION MOTOR | GE, 1257E1 (115 HP) [2] |

SIR - STATEN ISLAND RAILWAY.
CONSIST: A-B-B-A.
MOTOR OPERATOR'S WINDSHIELD GLASS HAS EMBEDDED WIRES FOR HEATING SURFACE.
MATERIAL FOR THE NYCTA OVERHAULED CARS WAS SUPPLIED BY MK.
NYC TRANSIT AUTHORITY COMPLETED OVERHAUL OF THE ORIGINALLY STARTED CONTRACT BY AMERICAN COASTAL IND.
ORIGINAL SPEEDOMETERS REMOVED. REINSTALLED GEAR BOX SPEEDOMETER BY BROCKSOPP (1992).

CAR NUMBERS: 500 - 1278, SEE NOTE 1
TOTAL: 754 CARS
BUILT BY: PULLMAN STANDARD

DATE: 1975 - 78
AVERAGE COST PER CAR: $275,381(BID)

# R46 (BMT/IND)

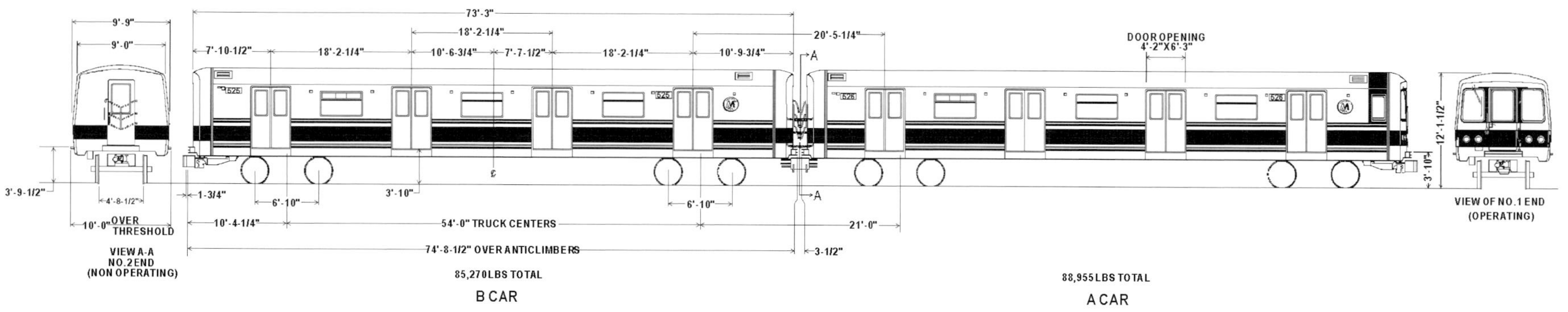

9'-9" 9'-0" 3'-9-1/2" 4'-8-1/2" 10'-0" OVER THRESHOLD

VIEW A-A
NO. 2 END
(NON OPERATING)

73'-3" 18'-2-1/4" 7'-10-1/2" 18'-2-1/4" 10'-6-3/4" 7'-7-1/2" 18'-2-1/4" 10'-9-3/4" 20'-5-1/4" 1-3/4" 6'-10" 3'-10" 6'-10" 10'-4-1/4" 54'-0" TRUCK CENTERS 21'-0" 74'-8-1/2" OVER ANTICLIMBERS 3-1/2"

DOOR OPENING 4'-2"X6'-3"

12'-1-1/2" 3'-10"

85,270 LBS TOTAL
B CAR

88,955 LBS TOTAL
A CAR

VIEW OF NO. 1 END
(OPERATING)

Copyright © 1997 by NYC Transit

## CAR BODY - EQUIPMENT

| | | | | | |
|---|---|---|---|---|---|
| AIR CONDITIONING | | TRANE,10 TON UNIT [2] | HVAC CONTROL | | TRANE/VAPOR |
| AUXILIARY CIRCUIT PROTECTION | | CIRCUIT BREAKERS | LIGHTING | | LUMINATOR, ENCLOSED GLASS LENS & BACK LIGHTED "AD" CARDS |
| BRAKE EQUIPMENT TYPE | | GE; WESTCODE (DYNAMIC & FRICTION) | LIGHTING INVERTER | | GE, 352020BV105 |
| COMMUNICATIONS | | MOTOROLA, SP3211703 | MASTER CONTROLLER | | GE, 17KC111A1 |
| DOOR HANGER & PANEL | | MORTON | PROPULSION CONTROL TYPE | | GE, SCM, 17KG327A1 GROUP |
| DOOR OPERATOR MOTOR | | VAPOR: R.H., 57141136; L.H., 57141137 | PUBLIC ADDRESS SYSTEM | | SOUND SYSTEMS, TRANSISTOR |
| DRAFT GEAR | #1END | "A" CAR: DRESSER, WITH CENTERING DEVICE | SASH (CAB) | | O.M. EDWARDS |
| | | "B" CAR: DRESSER, W/O CENTERING DEVICE | SEAT ARRANGEMENT & CAPACITY | | CROSS & LONGITUDINAL, "A" CAR [70]; "B" CAR [76] |
| | #2 END | "A" & "B" CAR: DRESSER, W/O CENTERING DEVICE | SEAT MANUFACTURER | | AMERICAN SEATING |
| ELECTRIC PORTION | | WALTON | SEAT MATERIAL | | FIBERGLASS |
| HAND BAR | | ELLCON NATIONAL | SIGN | | TRANSIGN |
| FIRE EXTINGUISHER | | 2-3/4 LBS DRY POWDER | SIGNAL DEVICE | | PNEUPHONIC HORN |
| HAND BRAKE | | ELLCON NATIONAL | STANCHION | | ELLCON NATIONAL [8] |
| HEATER | | E.L. WEIGAND | WINDOW | | MORTON |
| HEAD & TAIL LIGHTS | | LUMINATOR | WINDSHIELD WIPER | | AMERICAN BOSCH, ELECTRIC |
| | | **EVEN-NUMBERED CARS ONLY (A CAR)** | | | **ODD-NUMBERED CARS ONLY (B CAR)** |
| BATTERY | | EXIDE, SM120 NICKEL CADMIUM 25 CELL | AIR COMPRESSOR | | WABCO, D4; NYAB, N9412 |
| STATIC CONVERTER | | GE, 17KG326 | SWITCH PANEL | #1 END | VAPOR |
| SWITCH PANEL | #1 END | VAPOR | | | |

## TRUCK

| | | | |
|---|---|---|---|
| MANUFACTURER | BUCKEYE STEEL | JOURNAL BEARING | SKF, 5x9 ROLLER |
| TYPE | C.S.F. EQUALIZER BAR MOTOR | SHOCK ABSORBER | DELCO |
| WEIGHT | #1 END: 17,953 LBS; #2 END: 17,520 LBS | TRACTION MOTOR | GE, 1257 (115 HP) [2] |
| BRAKE RIGGING | AMERICAN STEEL FOUNDRY, WABCO, UNIT CYLINDER BRAKE | TRIP COCK | WABCO, D1 |
| GEAR UNIT | GE, 6A - 64A (117:23) | | |

NOTE 1: CARS WERE CONSECUTIVELY NUMBERED 500 - 1227. CARS WERE EVEN NUMBERED 1228 - 1278.

CAR CONSIST: A-B-B-A.
ORIGINAL TRUCKS WERE ROCKWELL AIR SUSPENSION SYSTEM.
NO. 1 END A CAR HAS TRAIN OPERATOR/CONDUCTOR CAB.
CARS 680 & 681 DELIVERED 6/26/76 AS 1776 & 1976, RESPECTIVELY. HAD RED, WHITE & BLUE STRIPES WITH STARS TAPED OVER BLUE BAND.
CAR 816 SENT TO CARACAS, VENEZUELA IN 1976 AS A DEMONSTRATOR.
CARS 941 & 1054 SCRAPPED (11/1987).

CAR NUMBERS: 5482 - 6258, SEE NOTE 1
TOTAL: 752 CARS, SEE NOTE 2
BUILT BY: PULLMAN STANDARD
DATE: 1975 - 78
AVERAGE COST PER CAR: $275,381 (BID)

OVERHAULED BY: MK
DATE: 1990 - 91
OVERHAUL COST PER CAR: $464,000

# R46 (BMT/IND)

## GOH

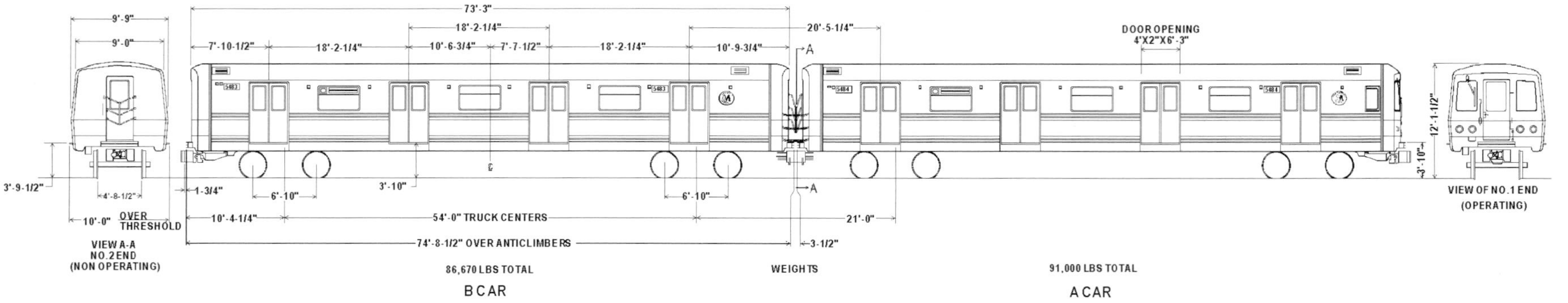

9'-9" 9'-0" 3'-9-1/2" 4'-8-1/2" 10'-0" OVER THRESHOLD
VIEW A-A
NO. 2 END
(NON OPERATING)

73'-3" 18'-2-1/4" 7'-10-1/2" 18'-2-1/4" 10'-6-3/4" 7'-7-1/2" 18'-2-1/4" 10'-9-3/4" 20'-5-1/4"
1-3/4" 6'-10" 3'-10" 6'-10" 10'-4-1/4" 54'-0" TRUCK CENTERS 21'-0" 74'-8-1/2" OVER ANTICLIMBERS 3-1/2"
DOOR OPENING 4'X2"X6'-3"
12'-1-1/2" 3'-10"
VIEW OF NO. 1 END
(OPERATING)

86,670 LBS TOTAL
B CAR

WEIGHTS

91,000 LBS TOTAL
A CAR

Copyright © 1997 by NYC Transit

## CAR BODY - EQUIPMENT

| | | | | |
|---|---|---|---|---|
| AIR CONDITIONING | | THERMO KING,10-TON UNITS [2] | HVAC CONTROL | THERMO-KING |
| AUXILIARY CIRCUIT PROTECTION | | CIRCUIT BREAKERS | LIGHTING | LUMINATOR, ENCLOSED PLASTIC LENS & BACK LIGHTED "AD" CARDS |
| BRAKE EQUIPMENT TYPE | | NYAB (DYNAMIC & FRICTION) | MASTER CONTROLLER | NYAB, NEWTRAN I, 763966 |
| COUPLER | | NYAB, RTC201P | PROPULSION CONTROL TYPE | GE, SCM, 17KG192AH1 GROUP |
| DOOR HANGER | | J.T. NELSON | PUBLIC ADDRESS SYSTEM | COMCO |
| DOOR PANEL | | MORTON | SASH (CAB) | J. T. NELSON |
| DOOR OPERATOR MOTOR | | WESTCODE | SEAT ARRANGEMENT & CAPACITY | CROSS & LONGITUDINAL, "A" CAR [70]; "B" CAR [76] |
| DRAFT GEAR | #1 END | "A" CAR: HADADY, RTD107 WITH CENTERING DEVICE | SEAT MANUFACTURER | AMERICAN SEATING (PASSENGER); EBONEX (TRAIN OPERATOR) |
| | | "B" CAR: MK LINKBAR | SEAT MATERIAL | FIBERGLASS |
| | #2 END | "A" & "B" CAR: MK LINKBAR | SIGN | LUMINATOR/TELEWELD |
| ELECTRIC PORTION | | NYAB | SIGNAL DEVICE | PNEUPHONIC HORN |
| HAND BAR | | CON TECH | STANCHION | CON TECH [8] |
| HAND BRAKE | | ELLCON NATIONAL/WESTCODE | WINDOW | J. T. NELSON |
| HEATER | | TELEWELD | WINDSHIELD WIPER | AMERICAN BOSCH, ELECTRIC |
| HEAD & TAIL LIGHTS | | LUMINATOR | | |
| EVEN-NUMBERED CARS ONLY (A CAR) | | | ODD-NUMBER CARS ONLY (B CAR) | |
| BATTERY | | ALCAD | AIR COMPRESSOR | WABCO, D4S |
| STATIC CONVERTER | | GE, 17KG417B1 | SWITCH PANEL | WESTCODE |
| SWITCH PANEL | #1 END | VAPOR | | |

## TRUCK

| | | | |
|---|---|---|---|
| MANUFACTURER | BUCKEYE STEEL | SHOCK ABSORBER, VERT. | MONROE, 80132 |
| TYPE | CAST STEEL FRAME EQUALIZER MOTOR | TRACTION MOTOR | GE, 1257E1 (115 HP) [2] |
| WEIGHT | #1 END: 18,790 LBS; #2 END: 18,460 LBS | TREAD BRAKE UNIT | D7587719 |
| GEAR UNIT | GE, 7GA64D2 (117:23) | TRIP COCK | NYAB, D1 |
| JOURNAL BEARING | 5X9 AP ROLLER | WHEEL | MAFERSA - SPECIAL |
| SHOCK ABSORBER, HORIZ. | MONROE, 70059 | | |

NOTE 1: CARS ARE CONSECUTIVELY NUMBERED FROM 5482 - 6207. CARS ARE EVEN NUMBERED FROM 6208-6258.
NOTE 2: CARS 941 & 1054 DAMAGED IN COLLISION 6/4/1987 AND SCRAPPED FROM ORIGINAL CONTRACT AMOUNT (11/1987).

CAR CONSISTS: 5482 - 6207: A-B-B-A. 6208-6258: AA.
NO. 1 END A CAR HAS TRAIN OPERATOR/CONDUCTOR CAB.
ORIGINAL TRUCKS WERE ROCKWELL AIR SUSPENSION SYSTEM.
ORIGINAL CAR NUMBERS WERE 500 - 1278 (SEE PAGE 497).

CARS EQUIPPED WITH LINK BAR BETWEEN CONSIST & COUPLERS AT "A" CAR NO. 1 ENDS.
ORIGINAL SPEEDOMETERS REMOVED. REINSTALLED GEAR BOX SPEEDOMETER BY BROCKSOPP (1992).
INSTALLED DOOR OBSTRUCTION SENSING DEVICE BY VAPOR (1995).
AMETEK ALTERNATING CURRENT BLOWER MOTORS/INVERTERS INSTALLED ON ALL CARS (BY 2005).

CAR NUMBERS: 1301 - 1625
TOTAL: 325 CARS
BUILT BY: KAWASAKI

DATE: 1983 - 85
AVERAGE COST PER CAR: $918,293

# R62 (IRT)

8'-10-7/16"
8'-7-3/16"
11'-10-5/8"
4'-8-1/2"
9'-8-3/4"
15'-9"
9'-9-3/4"
NO. 1 END
DOOR OPENING 4'-2"X6'-2-1/2"
3'-7-3/4"
2'-6-3/8"
6'-10"
6'-10"
1-3/4"
7'-6-1/4"
36'-0" TRUCK CENTERS
7'-6-1/4"
1-3/4"
51'-1/2" OVER ANTICLIMBERS

74,900 LBS TOTAL (ODD CAR)
74,540 LBS TOTAL (EVEN CAR WITH COMPRESSOR)
73,380 LBS TOTAL (EVEN CAR WITHOUT COMPRESSOR)

Copyright © 1997 by NYC Transit

## CAR BODY - EQUIPMENT

| | | | |
|---|---|---|---|
| AIR COMPRESSOR | WABCO, D4 | LIGHTING SYSTEM | GULTON, ENCLOSED PLASTIC LENS, BACK LIGHTED 'AD' CARDS |
| AIR CONDITIONING | STONE SAFETY, 12 TON | MASTER CONTROLLER | GE,17KG192AA1 |
| AUXILIARY CIRCUIT PROTECTION | CIRCUIT BREAKERS | PROPULSION CONTROL TYPE | GE, CAM, 17KG1924A1 GROUP |
| BATTERY | SAFT NIFE, PR80F; SAFT,SMT8 | PUBLIC ADDRESS SYSTEM | COMCO |
| BRAKE EQUIPMENT TYPE | WABCO, RT2 (DYNAMIC & FRICTION) | SASH (CAB) | ALNA |
| CONVERTER | GE, 17KG415A1 | SEAT ARRANGEMENT & CAPACITY | LONGITUDINAL [44] |
| COUPLER | WABCO: #1 END, H2C; #2 END, H2C | SEAT MATERIAL | FIBERGLASS |
| DOOR HANGER & PANEL | O.M. EDWARDS | SEAT TYPE | INDIVIDUALLY CONTOURED |
| DOOR HARDWARE | J.L. HOWARD | SEAT, PASSENGER | K.H.I. |
| DOOR OPERATOR MOTOR | VAPOR: R.H. , 58166495-10; L.H., 58166496-10 | SEAT, TRAIN OPERATOR | NIPPON, RECLINING SEAT |
| DRAFT GEAR | DRESSER, RTD 112 | SIGN | M. DENKI |
| ELECTRIC PORTION | WABCO, BL26B | SIGNAL DEVICE | PNEUPHONIC HORN |
| HAND BAR | K.H.I. | STANCHION | K.H.I. |
| HAND BRAKE | ELLCON NATIONAL, 840FL1 | TAIL LIGHT | GULTON |
| HEAD & TAIL LIGHTS | GULTON | WINDOW | ALNA |
| HEATER | WIEGAND | WINDOW HARDWARE | J.L.HOWARD |
| LIGHTING INVERTER | GULTON | WINDSHIELD WIPER | AMERICAN BOSCH, ELECTRIC |

## TRUCK

| | | | |
|---|---|---|---|
| MANUFACTURER | NIPPON SHARYO | GEAR UNIT | GE, 1GA64 (117.23) |
| TYPE & QUANTITY | C.S.F. EQUALIZER BAR MOTOR [650] (INCLUDES 40 SPARES) | JOURNAL BEARING | KOYO SEIKO, 5X9 AP ROLLER |
| TRUCK SERIAL NUMBERS | 20869-21518 | SHOCK ABSORBER | MONROE |
| WEIGHT | #1 END: 18,590 LBS; #2 END: 18,200 LBS | TRACTION MOTOR | GE 1257E1 (115 HP) [2] |
| BRAKE RIGGING | WABCO, TBU | TRIP COCK | WABCO, DIA |

MODIFIED SINGLE CARS TO FIVE CAR CONSISTS AND REPLACED COUPLERS WITH LINKBARS BETWEEN CONSISTS(1991).
INSTALLED DOPPLER SPEEDOMETER BY EDO (1994).
CAR 1369 OUT OF SERVICE AT CONCOURSE YARD.
CARS 1435, 1436 AND 1439 ARE DAMAGED AND OUT OF SERVICE.
CARS 1437 & 1440 WERE SCRAPPED.
INSTALLED DOOR OBSTRUCTION SENSING DEVICE BY VAPOR (4/1996 - 1/1997).
INSTALLED AMETEK ALTERNATING CURRENT BLOWER MOTORS/ INVERTERS (STARTED 11/1996) AND COMPLETED ON ALL CARS (1997).

CARS 1431, 1432, 1433, 1434 AND 1438 ARE UNITIZED.

| CAR NUMBERS: 1651 - 2475<br>TOTAL: 825 CARS<br>BUILT BY: BOMBARDIER | DATE: 1984 - 87<br>AVERAGE COST PER CAR: $798,770 (BID) | # R62A (IRT) |
|---|---|---|

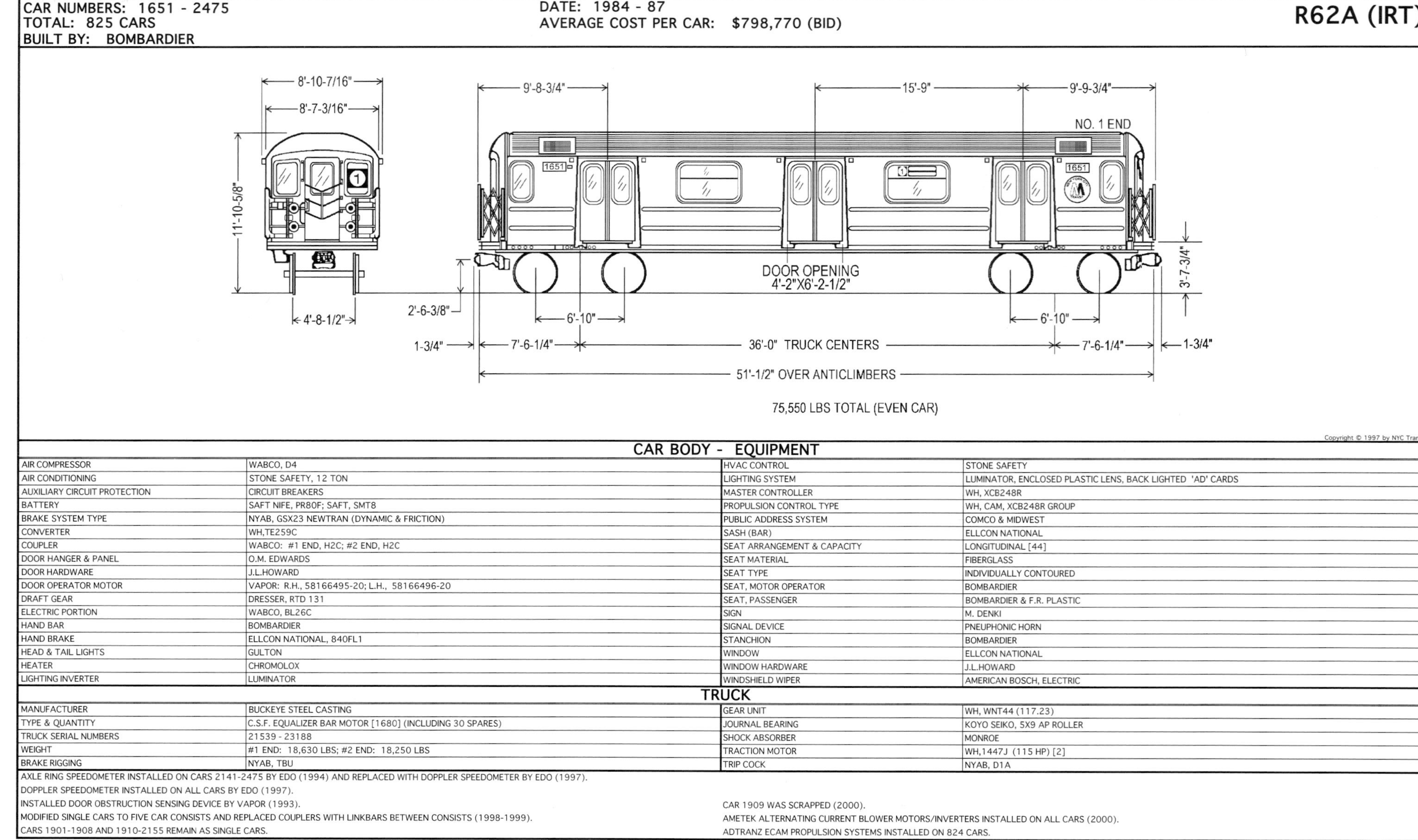


Copyright © 1997 by NYC Transit

## CAR BODY - EQUIPMENT

| | | | |
|---|---|---|---|
| AIR COMPRESSOR | WABCO, D4 | HVAC CONTROL | STONE SAFETY |
| AIR CONDITIONING | STONE SAFETY, 12 TON | LIGHTING SYSTEM | LUMINATOR, ENCLOSED PLASTIC LENS, BACK LIGHTED 'AD' CARDS |
| AUXILIARY CIRCUIT PROTECTION | CIRCUIT BREAKERS | MASTER CONTROLLER | WH, XCB248R |
| BATTERY | SAFT NIFE, PR80F; SAFT, SMT8 | PROPULSION CONTROL TYPE | WH, CAM, XCB248R GROUP |
| BRAKE SYSTEM TYPE | NYAB, GSX23 NEWTRAN (DYNAMIC & FRICTION) | PUBLIC ADDRESS SYSTEM | COMCO & MIDWEST |
| CONVERTER | WH,TE259C | SASH (BAR) | ELLCON NATIONAL |
| COUPLER | WABCO: #1 END, H2C; #2 END, H2C | SEAT ARRANGEMENT & CAPACITY | LONGITUDINAL [44] |
| DOOR HANGER & PANEL | O.M. EDWARDS | SEAT MATERIAL | FIBERGLASS |
| DOOR HARDWARE | J.L.HOWARD | SEAT TYPE | INDIVIDUALLY CONTOURED |
| DOOR OPERATOR MOTOR | VAPOR: R.H., 58166495-20; L.H., 58166496-20 | SEAT, MOTOR OPERATOR | BOMBARDIER |
| DRAFT GEAR | DRESSER, RTD 131 | SEAT, PASSENGER | BOMBARDIER & F.R. PLASTIC |
| ELECTRIC PORTION | WABCO, BL26C | SIGN | M. DENKI |
| HAND BAR | BOMBARDIER | SIGNAL DEVICE | PNEUPHONIC HORN |
| HAND BRAKE | ELLCON NATIONAL, 840FL1 | STANCHION | BOMBARDIER |
| HEAD & TAIL LIGHTS | GULTON | WINDOW | ELLCON NATIONAL |
| HEATER | CHROMOLOX | WINDOW HARDWARE | J.L.HOWARD |
| LIGHTING INVERTER | LUMINATOR | WINDSHIELD WIPER | AMERICAN BOSCH, ELECTRIC |

## TRUCK

| | | | |
|---|---|---|---|
| MANUFACTURER | BUCKEYE STEEL CASTING | GEAR UNIT | WH, WNT44 (117.23) |
| TYPE & QUANTITY | C.S.F. EQUALIZER BAR MOTOR [1680] (INCLUDING 30 SPARES) | JOURNAL BEARING | KOYO SEIKO, 5X9 AP ROLLER |
| TRUCK SERIAL NUMBERS | 21539 - 23188 | SHOCK ABSORBER | MONROE |
| WEIGHT | #1 END: 18,630 LBS; #2 END: 18,250 LBS | TRACTION MOTOR | WH,1447J (115 HP) [2] |
| BRAKE RIGGING | NYAB, TBU | TRIP COCK | NYAB, D1A |

AXLE RING SPEEDOMETER INSTALLED ON CARS 2141-2475 BY EDO (1994) AND REPLACED WITH DOPPLER SPEEDOMETER BY EDO (1997).
DOPPLER SPEEDOMETER INSTALLED ON ALL CARS BY EDO (1997).
INSTALLED DOOR OBSTRUCTION SENSING DEVICE BY VAPOR (1993).
MODIFIED SINGLE CARS TO FIVE CAR CONSISTS AND REPLACED COUPLERS WITH LINKBARS BETWEEN CONSISTS (1998-1999).
CARS 1901-1908 AND 1910-2155 REMAIN AS SINGLE CARS.

CAR 1909 WAS SCRAPPED (2000).
AMETEK ALTERNATING CURRENT BLOWER MOTORS/INVERTERS INSTALLED ON ALL CARS (2000).
ADTRANZ ECAM PROPULSION SYSTEMS INSTALLED ON 824 CARS.

CAR NUMBERS: 2500 - 2924
TOTAL: 425 CARS
BUILT BY: WEST-AMRAIL

DATE: 1986 - 88
AVERAGE COST PER CAR: $ 915,000 (BID)

# R68 (BMT/IND)

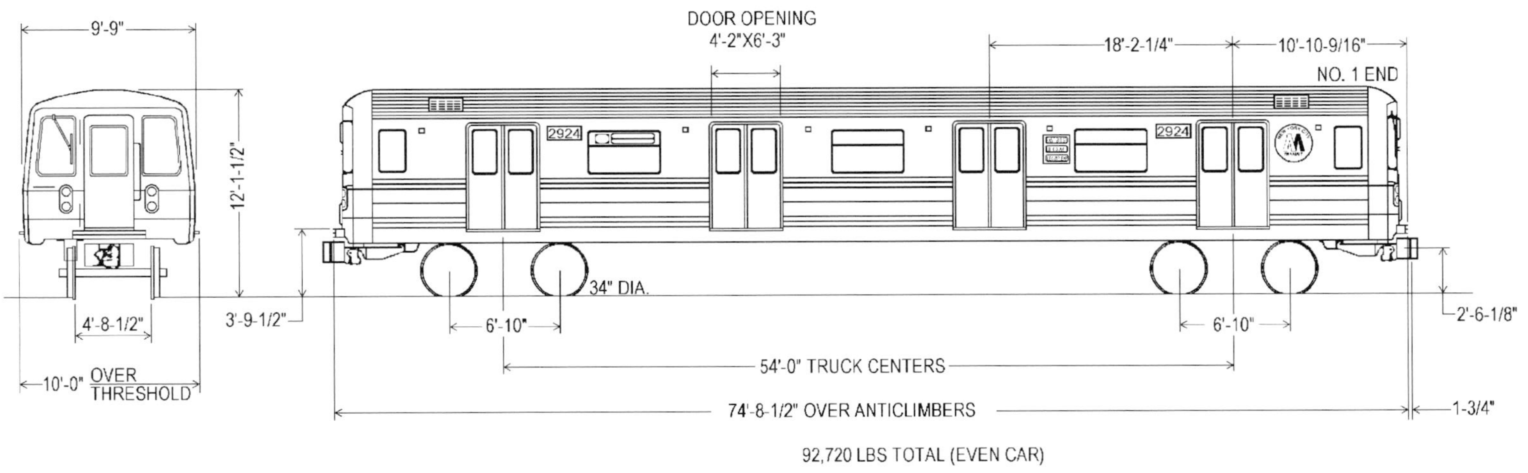

9'-9"
12'-1-1/2"
4'-8-1/2"
3'-9-1/2"
10'-0" OVER THRESHOLD
DOOR OPENING 4'-2"X6'-3"
18'-2-1/4"
10'-10-9/16"
NO. 1 END
34" DIA.
6'-10"
6'-10"
2'-6-1/8"
54'-0" TRUCK CENTERS
74'-8-1/2" OVER ANTICLIMBERS
1-3/4"
92,720 LBS TOTAL (EVEN CAR)

Copyright © 1997 by NYC Transit

## CAR BODY - EQUIPMENT

| | | | |
|---|---|---|---|
| AIR COMPRESSOR | WABCO, D4S; KNORR | HVAC CONTROL | SAFETY ELECTRIC |
| AIR CONDITIONING | STONE SAFETY, 10 TON [2] | LIGHTING INVERTER | LUMINATOR |
| AUXILIARY CIRCUIT PROTECTION | CIRCUIT BREAKERS | LIGHTING SYSTEM | LUMINATOR, ENCLOSED LIGHTED 'AD' CARD PLASTIC LENS, BACK |
| BATTERY | SAFT, S.M.T.8 | MASTER CONTROLLER | WH, XM880 |
| BRAKE EQUIPMENT TYPE | NYAB, GSX23 NEWTRAN (DYNAMIC & FRICTION) | PROPULSION CONTROL TYPE | WH, CAM, XCB248S GROUP |
| CONVERTER | WH, TE259D | PUBLIC ADDRESS SYSTEM | COMCO |
| COUPLER | WABCO: #1 END, H2C; #2 END, H2C | SASH (CAB) | ELLCON NATIONAL |
| DOOR HANGER | FAIVELEY | SASH HARDWARE | J.L. HOWARD |
| DOOR HARDWARE | J.L.HOWARD | SEAT | ANF |
| DOOR OPERATOR MOTOR | VAPOR, FAIVELY: R.H., 58166679; L.H., 58166680 | SEAT ARRANGEMENT & CAPACITY | CROSS & LONGITUDINAL [70] |
| DOOR PANEL | O.M. EDWARDS/MORTON | SEAT MATERIAL | FIBERGLASS |
| DRAFT GEAR | HADADY (DRESSER) | SEAT TYPE | INDIVIDUAL & CONTOURED |
| ELECTRIC PORTION | WABCO, BL33E | SIGN | CNA |
| FIRE EXTINGUISHER | AMEREX, 441 | SIGNAL DEVICE | PNEUPHONIC HORN |
| HAND BAR | ANF | STANCHION | ANF |
| HAND BRAKE | ELLCON-NATIONAL, 840FL1 | WINDOW | ELLCON NATIONAL |
| HEAD & TAIL LIGHTS | GABRIEL | WINDSHIELD WIPER | AMERICAN BOSCH, ELECTRICAL |
| HEATER | T.E.E.N. | | |

## TRUCK

| | | | | |
|---|---|---|---|---|
| MANUFACTURER | | CASTING -SANBRE ET MEUSE | TRACTION MOTOR | WH, 1447J (115 HP) [2] |
| ASSEMBLER | | ALSTHOM CREUSOT | GEAR UNIT | WH, WNT44-1 (117:23) |
| TYPE | | C.S.F. EQUALIZER BAR MOTOR | BRAKE RIGGING | NYAB, TBU190 |
| TRUCK SERIAL NUMBERS | | 23239 - 24088 | JOURNAL BEARING | KOYO SEIKO, 5X9 AP ROLLER |
| WEIGHT | #1 END | 18,420 LBS | SHOCK ABSORBER | VERTICAL- SJR, 16-98-150; HORIZONTAL.- SJR, 16-97-137 |
| | #2 END | 18,150 LBS | TRIP COCK | NYAB, D1 |

REPLACED BATTERY WITH SAFT NIFE SRM/80 (1997).
REPLACED TRACTION MOTOR WITH JEUMONT-SCHNEIDER, J1447 (1986 - 1988).
INSTALLED GEAR BOX SPEEDOMETER BY BACH-SIMPSON.
INSTALLED DOOR OBSTRUCTION SENSING DEVICE BY WESTCODE (1997).

CARS 2916-2924 REMAIN AS SINGLE CARS ASSIGNED TO THE FRANKLIN SHUTTLE.
CARS 2500-2915 ARE NOW ARRANGED IN 4 CAR UNITS (1998-99).
ADTRANZ ECAM PROPULSION SYSTEMS INSTALLED ON ALL CARS (1999).
AMETEK ALTERNATING CURRENT BLOWER MOTORS/INVERTERS INSTALLED ON ALL CARS (1999).

CAR NUMBERS: 5001 - 5200
TOTAL: 200 CARS
BUILT BY: KAWASAKI

DATE: 1988 - 89
AVERAGE COST PER CAR: $915,000 (BID)

# R68A (BMT/IND)

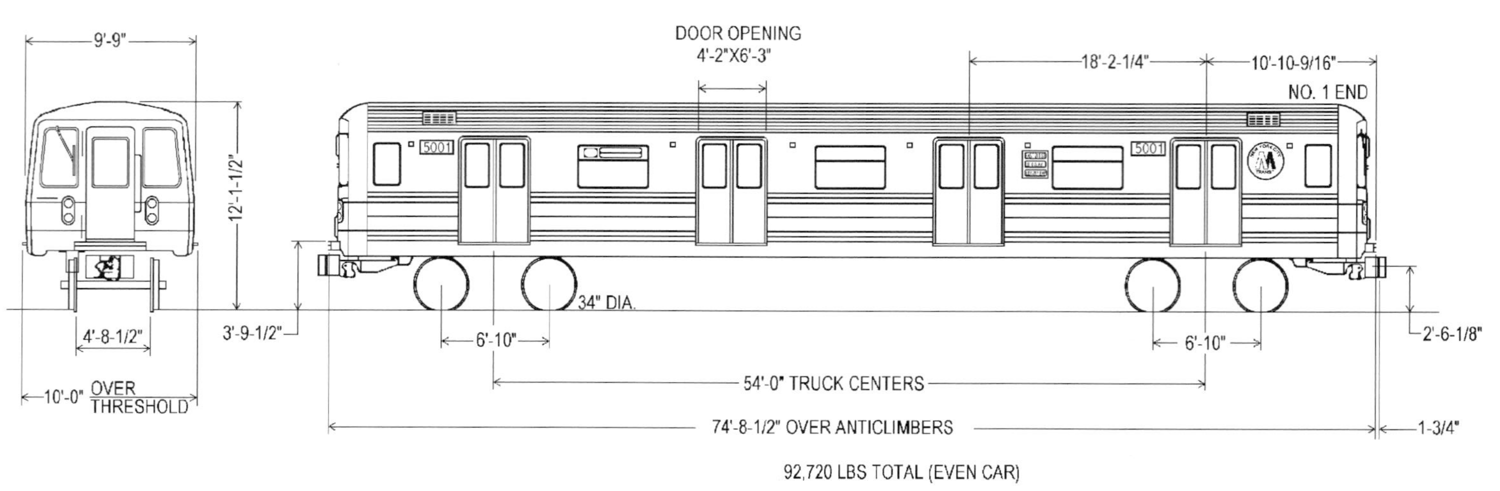

9'-9"
12'-1-1/2"
4'-8-1/2"
3'-9-1/2"
10'-0" OVER THRESHOLD
DOOR OPENING 4'-2"X6'-3"
18'-2-1/4"
10'-10-9/16"
NO. 1 END
34" DIA.
6'-10"
6'-10"
2'-6-1/8"
54'-0" TRUCK CENTERS
74'-8-1/2" OVER ANTICLIMBERS
1-3/4"
92,720 LBS TOTAL (EVEN CAR)

Copyright © 1997 by NYC Transit

## CAR BODY EQUIPMENT

| | | | |
|---|---|---|---|
| AIR COMPRESSOR | WABCO, D4S | HVAC CONTROL | SAFETY ELECTRIC |
| AIR CONDITIONING | STONE SAFETY, 20 TON | INVERTER BALLAST | LUMINATOR |
| AUX. CIRCUIT PROTECTION | CIRCUIT BREAKERS | LIGHTING SYSTEM | M. DENKI, ENCLOSED PLASTIC LENS BACK LIGHTED 'AD' CARD |
| BATTERY | YASHIRO, KA8E3 | MASTER CONTROLLER | WH, XM880 |
| BRAKE EQUIPMENT TYPE | WABCO (DYNAMIC & FRICTION) | PROPULSION CONTROL TYPE | WH, CAM, XCB248S GROUP |
| CONVERTER | WH,TE259D | PUBLIC ADDRESS SYSTEM | COMCO |
| COUPLER | WABCO, H2C | SASH (CAB) | ALNA-KOKI/MARITA SEISAKUSHO |
| DOOR HANGER & PANEL | O.M. EDWARDS/MORTON | SEAT ARRANGEMENT & CAPACITY | CROSS & LONGITUDINAL [70] |
| DOOR HARDWARE | J.L. HOWARD | SEAT MATERIAL | FIBERGLASS |
| DOOR OPERATOR MOTOR | VAPOR: R.H., 58166878; L.H., 58166879 | SEAT TYPE | INDIVIDUALLY CONTOURED |
| DRAFT GEAR | HADADY | SEAT, PASSENGER | YASHIRO KAKO |
| ELECTRIC PORTION | WABCO, BL33G | SEAT, TRAIN OPERATOR | KAWASAKI |
| FIRE EXTINGUISHER | AMEREX, 441 | SIGN | M. DENKI, CLASS ABC |
| HAND BAR | KAWASAKI | SIGNAL DEVICE | PNEUPHONIC HORN |
| HAND BRAKE | ELLCON NATIONAL, 840FL1 | STANCHION | KAWASAKI |
| HEAD & TAIL LIGHTS | MORIO DENKI | WINDOW | KAWASAKI |
| HEATER | TEEN | WINDSHIELD WIPER | AMERICAN BOSCH, ELECTRIC |

## TRUCK

| | | | |
|---|---|---|---|
| MANUFACTURER | CASTING-FUKUSHIMA STEEL | GEAR UNIT | WH, WN445 (117.23) |
| ASSEMBLER | KAWASAKI | JOURNAL BEARING | TIMKEN, 5X9 CLASS D AP ROLLER |
| TYPE | C.S.F. EQUALIZER BAR MOTOR | SHOCK ABSORBER | VERTICAL; HORIZONTAL.- ALSTHOM CREUSOT |
| TRUCK SERIAL NUMBERS | 24089 - 24488 | TRACTION MOTOR | WH, 1447J (115 HP) [2] |
| WEIGHT | #1 END: 18,370 LBS; #2 END: 18,350 LBS | TRIP COCK | WABCO, D1 |
| BRAKE RIGGING | WABCO, TBU GR90 | | |

REPLACED BATTERY WITH SAFT NIFE SRM/100 (1997).
INSTALLED DOPPLER SPEEDOMETER BY EDO ON 110 CARS (10/1996 - END/1997).
ALL CARS ARE NOW UNITIZED INTO 4 CAR UNITS (1995-97).
ADTRANZ ECAM PROPULSION INSTALLED ON ALL CARS (2000).
AMETEK ALTERNATING CURRENT BLOWER MOTORS/INVERTERS INSTALLED ON ALL CARS (2001).
INSTALLED DOOR OBSTRUCTION SENSING DEVICE BY WESTCODE (1996).

CAR NUMBERS: 8001, 8005, 8006, 8010 (A CAR); 8002, 8003, 8004, 8007, 8008, 8009 (B CAR)
TOTAL: 10 CARS
DATE: 1992
BUILT BY: KAWASAKI
AVERAGE COST PER CAR: $2,209,000

# R110A (A DIVISION)

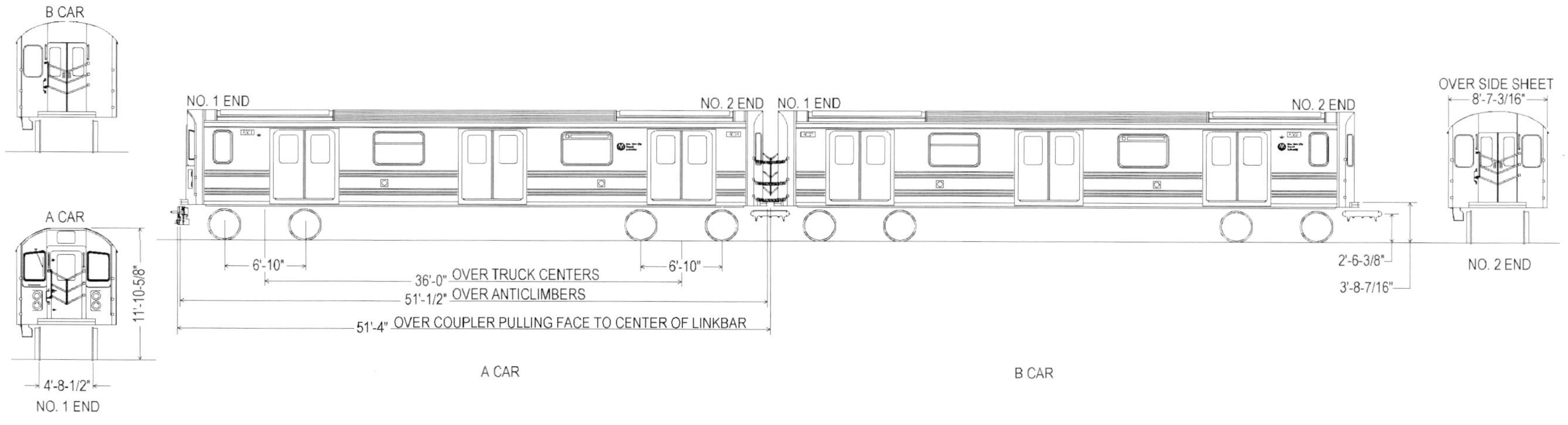

B CAR
A CAR
11'-10-5/8"
4'-8-1/2"
NO. 1 END
NO. 1 END NO. 2 END NO. 1 END NO. 2 END
6'-10" 6'-10"
36'-0" OVER TRUCK CENTERS
51'-1/2" OVER ANTICLIMBERS
51'-4" OVER COUPLER PULLING FACE TO CENTER OF LINKBAR
2'-6-3/8"
3'-8-7/16"
OVER SIDE SHEET 8'-7-3/16"
NO. 2 END
A CAR
B CAR

Copyright © 1997 by NYC Transit

## DIMENSIONS

| DIMENSIONS | |
|---|---|
| DOORWAY HEIGHT (END) | 6' - 3" |
| DOORWAY HEIGHT (SIDE) | 6' - 3" |
| DOORWAY WIDTH (END-CLEAR OPENING) | 2' - 9" |
| DOORWAY WIDTH (SIDE-CLEAR OPENING) | 5' - 4" |
| FLOOR TO CEILING HT. (HIGH CEILING) | 7' - 2" |
| FLOOR TO CEILING HT. (LOW CEILING) | 6' - 6 3/4" |
| HEIGHT (TOP OF RAIL TO FLOOR) | 3' - 8 1/2" |
| TRACK GAUGE | 4' - 8 1/2" |
| WIDTH (OVER THRESHOLD) | 8' - 9 1/2" |
| **PERFORMANCE** | |
| ACCELERATION RATE (SERVICE) | 2.0 MPHPS |
| BRAKING RATE (EMERGENCY) | 3.2 MPHPS |
| NOISE LEVEL - 40 MPH | |
| STATION-INTERIOR | 79 dBA |
| STATION-EXTERIOR | 88 dBA |
| ELEVATED TRACK STATIONARY | 60 dBA |
| RIDE QUALITY @ AW3; 40 MPH | 6.5 HRS |
| SERVICE SPEED (MAXIMUM) | 55 MPH |
| NOMINAL LINE VOLTAGE | 600 VDC |

## WEIGHT & CAPACITY

| WEIGHT & CAPACITY | |
|---|---|
| **CAR WEIGHT (EMPTY)** | |
| A CAR | 72,000 LBS |
| B CAR | 67,400 LBS |
| **CAR WEIGHT (LOADED)** | |
| A CAR (182 CUSTOMERS) | 100,000 LBS |
| B CAR (190 CUSTOMERS) | 95,400 LBS |
| **BUFF LOAD (AT ANTICLIMBER)** | |
| A&B CAR | 200,000 LBS |
| **SEATING CAPACITY** | |
| A CAR | 24 |
| B CAR | 28 |
| **STANDING CAPACITY** | |
| A CAR | 182 |
| B CAR | 190 |
| **TRUCK WEIGHT** | |
| MOTOR CAR W/PARKING BRAKE | 15,748 LBS |
| TRAILER CAR W/PARKING BRAKE | 9,800 LBS |

## CAR EQUIPMENT

| CAR EQUIPMENT | |
|---|---|
| **AIR SUPPLY SYSTEM & FRICTION BRAKING** | |
| AIR SUPPLY UNIT | WABCO, MODEL D-4-A |
| BRAKE SYSTEM TYPE | WABCO RT7 |
| PARKING BRAKE | SAB WABCO |
| **AUXILIARY POWER SYSTEM** | |
| BATTERY | SAFT, 300AH (B CAR) |
| INTERMEDIATE VOLTAGE POWER SUPPLY | AEG, 350 VDC (60 KW) |
| LOW VOLT. POWER SUPPLY | AEG, 37.5 VDC (17 KW) |
| **CAR BODY** | |
| BODY | STAINLESS STEEL |
| BONNET | FIBERGLASS |
| FLOORING | PLYWOOD SANDWICHED BETWEEN STEEL SHEETS |
| ROOF | STAINLESS STEEL |
| CAB PARTITION | STAINLESS STEEL |
| **COUPLER ASSEMBLY** | |
| NO. 1 END | WABCO N-2-C (A CAR);LINKBAR (B CAR). |
| NO. 2 END | LINKBAR (A&B CAR) |
| ELECTRICAL PORTION | 104 PINS |
| **CAB & CREW OPERATION CONTROL** | |
| MASTER CONTROLLER | WABCO, SINGLE HANDLE |
| MASTER DOOR CONTROLLER | VAPOR |
| MONITORING SYS. - TLM SCREEN & CAR MONITORING UNIT | AEG, ON BOARD COMPUTER CONTROLLED |
| RADIO | AEROTRON |
| SIGNAL DEVICE | PNEUPHONIC HORN |
| SPEEDOMETER | KAWASAKI |
| **CUSTOMER ENVIRONMENT** | |
| INFORMATION SIGNS | MORIO DENKI |
| ALARM SYSTEM | TAPE SWITCH |
| COMMUNICATION SYSTEMS | TELEPHONICS/AEROTRON |
| CUSTOMER EMERGENCY BRAKE | WABCO |
| LIGHTING | M. DENKI |
| **DOOR SYSTEMS** | |
| DOOR OPERATOR | VAPOR |
| DOOR PANEL | KAWASAKI |
| SIDE DOOR | 6 DOUBLE SLIDING |
| END DOOR | 2 DOUBLE SLIDING B CAR<br>1 DOUBLE SLIDING A CAR |
| **HEATING, VENTILATING & AIR CONDITIONING (HVAC)** | |
| A/C CAPACITY | MITSUBISHI, TWO 6 TONS UNITS, O/H |
| HEATING | FORCED AIR O/H<br>CONVECTION FLOOR HEAT |
| VENTILATING | FORCED AIR |
| **PROPULSION SYSTEM & ELECTRIC BRAKING** | |
| INVERTER DRIVE | AEG POWERS AC TRACTION MOTORS |
| TRACTION MOTOR | AEG, MODEL 1501A, 150 HP., 3 PHASE, 4 POLE |
| **SIGNALING & TRAIN CONTROL SYSTEM** | |
| | FIXED BLOCK |
| **TRUCK** | |
| TRUCK TYPE | WELDED, FABRICATED STEEL FRAME |
| PRIMARY SUSPENSION | CONICAL RUBBER |
| SECONDARY SUSPENSION | AIR BAG |

NEW TECHNOLOGY TEST TRAIN STARTED NO. 2 LINE SERVICE ON 6/30/93. B CAR HAS MOTOR AT NO. 2 END ONLY. CONSIST: 5 CAR UNIT (ABBBA), 10 CAR TRAIN.
CARS EQUIPPED WITH SPEEDOMETER.

CAR NUMBERS: 3001, 3003, 3004, 3006, 3007, 3009, (MOTOR), 3002, 3005, 3008, (TRAILER)
TOTAL: 9 CARS
BUILT BY: BOMBARDIER
DATE: 1992
AVERAGE COST PER CAR: $2,167,000

# R110B (B DIVISION)

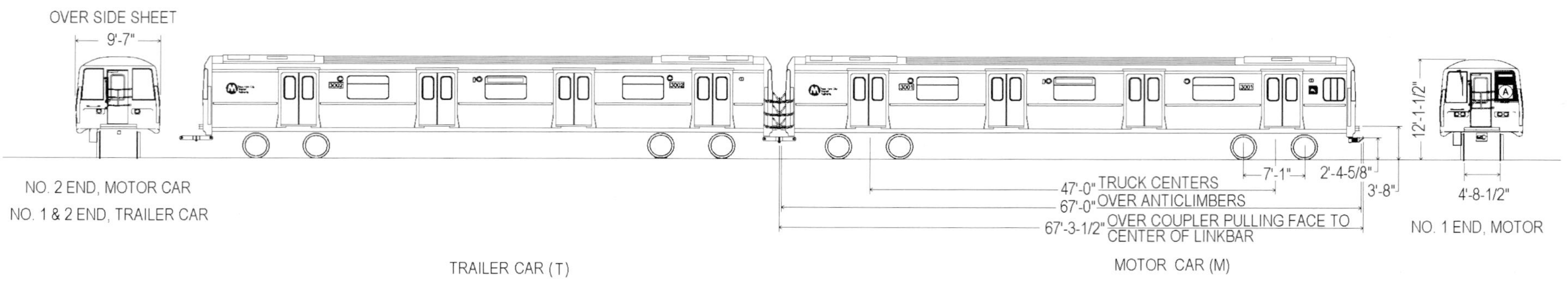

OVER SIDE SHEET 9'-7"

NO. 2 END, MOTOR CAR
NO. 1 & 2 END, TRAILER CAR

47'-0" TRUCK CENTERS
67'-0" OVER ANTICLIMBERS
67'-3-1/2" OVER COUPLER PULLING FACE TO CENTER OF LINKBAR
7'-1" 2'-4-5/8" 3'-8" 12'-1-1/2" 4'-8-1/2"

TRAILER CAR (T)
MOTOR CAR (M)
NO. 1 END, MOTOR

Copyright © 1997 by NYC Transit

## DIMENSIONS

| | |
|---|---|
| DOORWAY HEIGHT (END) | 6'-3" |
| DOORWAY HEIGHT (SIDE) | 6' - 3" |
| DOORWAY WIDTH (END OPENING) | 2' - 6" |
| DOORWAY WIDTH (SIDE OPENING) | 4' - 2" |
| FLOOR TO CEILING HT. (HIGH CEILING) | 7' - 1/2" |
| FLOOR TO CEILING HT. (LOW CEILING) | 6' - 7 1/2" |
| HEIGHT (TOP OF RAIL TO FLOOR) | 3' - 8 7/8" |
| TRACK GAUGE | 4' - 8 1/2" |
| WHEEL DIAMETER | 34" |
| WIDTH (OVER THRESHOLD) | 10'-0" |
| PERFORMANCE | |
| ACCELERATION RATE (SERVICE) | 2.0 MPHPS |
| BRAKING RATE (EMERGENCY) | 3.2 MPHPS |
| NOISE LEVEL - 40 MPH | |
| STATION-INTERIOR | 79 - 85 dBA |
| STATION-EXTERIOR | 85 dBA |
| ELEVATED TRACK STATIONARY | 58 dBA |
| RIDE QUALITY @ AW3; 40 MPH | 6.5 HRS |
| SERVICE SPEED (MAXIMUM) | 55 MPH |
| NOMINAL LINE VOLTAGE | 600 VDC |

## WEIGHT & CAPACITY

| | |
|---|---|
| **CAR WEIGHT (EMPTY)** | |
| MOTOR CAR | 83,348 LBS |
| TRAILER CAR | 68,759 LBS |
| **CAR WEIGHT (LOADED)** | |
| MOTOR CAR (259 CUSTOMERS) | 123,234 LBS |
| TRAILER CAR (260 CUSTOMERS) | 108,799 LS |
| **BUFF LOAD (AT ANTICLIMBER)** | |
| MOTOR & TRAILER CARS | 400,000 LBS |
| **SEATING CAPACITY** | |
| MOTOR & | 42 |
| TRAILER CARS | 46 |
| **STANDING CAPACITY** | |
| MOTOR CAR | 217 |
| TRAILER CAR | 214 |
| **TRUCK WEIGHT** | |
| MOTOR CAR W/PARKING BRAKE | 14,383 LBS |
| MOTOR CAR WO/PARKING BRAKE | 14,207 LBS |
| TRAILER CAR W/PARKING BRAKE | 11,044 LBS |
| TRAILER CAR WO/PARKING BRAKE | 10,593 LBS |

## CAR EQUIPMENT

| | |
|---|---|
| **AIR SUPPLY SYSTEM & FRICTION BRAKING** | |
| AIR SUPPLY UNIT | NYAB/KNORR, ES |
| BRAKE TYPE | NYAB, TREAD, MODEL 590 (M/T) |
| | KNORR, DISC, MODEL 660x110 (T) |
| BRAKE SYSTEM TYPE | NYCOTRON, TC-1 |
| PARKING BRAKE | KNORR/NYAB |
| **AUXILIARY POWER SYSTEM** | |
| BATTERY | SAFT-SRX, 300 AH (T) |
| INTERMEDIATE VOLTAGE POWER SUPPLY | GE, 370 VDC/75 KW |
| LOW VOLTAGE POWER SUPPLY | GE, 37.5 VDC/15 KW |
| **CAR BODY** | |
| BODY | STAINLESS STEEL |
| BONNET | FIBERGLASS |
| FLOORING | PLYWOOD SANDWICHED BETWEEN STAINLESS STEEL SHEETS |
| ROOF | STAINLESS STEEL |
| CAB PARTITION (M) | TRANSVERSE |
| PANELS/SEATS | BOMBARDIER |
| **COUPLER & DRAWBAR ASSEMBLY** | |
| NO. 1 END | NYAB, RTC-201P (M); LINKBAR (T) |
| NO. 2 END | LINKBAR (M&T) |
| ELECTRICAL PORTION | COAXIAL/RADIOLINK |
| **CAB & CREW OPERATIONAL CONTROLS** | |
| MASTER CONTROLLER | KNORR, SINGLE HANDLE |
| MASTER DOOR CONTROLLER | BOMBARDIER |
| MONITORING TERMINAL WITH CONSOLE CONTROLS | PRIMETCH |
| RADIO | AEROTRON |
| SIGNAL DEVICE | PNEUPHONIC HORN |
| SPEEDOMETER | BOMBARDIER |

| | |
|---|---|
| **CUSTOMER ENVIRONMENT** | |
| INFORMATION SIGNS | LUMINATOR/POCATEC |
| ALARM SYSTEM | BOMBARDIER, TAPE SWITCH |
| COMMUNICATION SYSTEMS | POCATEC |
| CUSTOMER EMERGENCY BRAKE | NYAB |
| LIGHTING | BOMBARDIER/LUMINATOR |
| **DOOR SYSTEMS** | |
| DOOR CONTROL | BOMBARDIER, SERIAL BUS |
| DOOR OPERATOR | BOMBARDIER, O/H, ONE PER PANEL |
| DOOR PANEL | BOMBARDIER |
| SIDE/END DOOR | 8 DBL. SLIDING/2 SINGLE SWINGING |
| **HEATING, VENTILATING & AIR CONDITIONING (HVAC)** | |
| A/C CAPACITY | BOMBARDIER, TWO 9 TONS, O/H |
| HEATING | FORCED CONVECTION O/H & FLOOR HEATERS |
| VENTILATING | FORCED AIR |
| **PROPULSION SYSTEM & ELECTRIC BRAKING** | |
| INVERTER DRIVE | GE, POWERS TRACTION MOTORS |
| TRACTION MOTOR | GE, GEB 7-B, 202 HP, 3 PHASE 4 POLE, 3.9% RATED SLIP |
| **SIGNALING & TRAIN CONTROL SYSTEM** | |
| | FIXED BLOCK |
| **TRUCK** | |
| TRUCK TYPE | WELDED, FABRICATED STEEL FRAME, INBOARD BEARING |
| PRIMARY SUSPENSION | CHEVRON SPRING |
| SECONDARY SUSPENSION | AIR BAG |
| POWER COLLECTION | FERRAZ, CURRENT COLLECTOR |

NEW TECHNOLOGY TEST TRAIN STARTED "A" LINE SERVICE ON 6/15/93. CONSIST: 3 CAR UNIT (MOTOR-TRAILER- MOTOR), 9 CAR TRAIN
CARS EQUIPPED WITH SPEEDOMETER AND DOOR OBSTRUCTION SENSING DEVICE.

CAR NUMBERS: 6301 - 7180
TOTAL: 1030 CARS
BUILT BY: BOMBARDIER

OPTION CARS: 1101 - 1250
DATE: 1999-2003
AVERAGE COST PER CAR: $1,217,312.26

# R142 (A DIVISION)

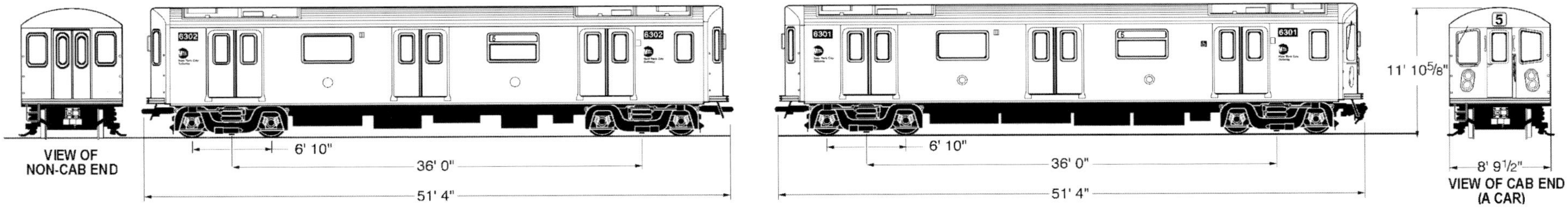

B CAR

A CAR

VIEW OF NON-CAB END

6' 10"

36' 0"

51' 4"

11' 10 5/8"

8' 9 1/2"

VIEW OF CAB END (A CAR)

Copyright © 2003 by NYC Transit

## DIMENSIONS

| DIMENSIONS | |
|---|---|
| DOORWAY HEIGHT (END) | 6'- 3" |
| DOORWAY HEIGHT (SIDE) | 6' - 3" |
| DOORWAY WIDTH (END-CLEAR OPENING) | 2' - 6" |
| DOORWAY WIDTH (SIDE-CLEAR OPENING) | 4' - 6" |
| FLOOR TO CEILING HT. (HIGH CEILING) | 7' - 1/2" |
| FLOOR TO CEILING HT. (LOW CEILING) | 6' - 7 1/2" |
| HEIGHT (TOP OF RAIL TO FLOOR) | 3' - 8 3/4" |
| TRACK GAUGE | 4' - 8 1/2" |
| WIDTH (OVER THRESHOLD) | 8' - 9 1/2" |
| **PERFORMANCE** | |
| ACCELERATION RATE | 2.5 MPHPS |
| BRAKING RATE (EMERGENCY) | 3.2 MPHPS |
| SERVICE SPEED (MAXIMUM) | 55 MPH |
| NOMINAL LINE VOLTAGE | 600 VDC |

## WEIGHT & CAPACITY

| WEIGHT & CAPACITY | |
|---|---|
| **CAR WEIGHT (EMPTY)** | |
| A CAR | 72,000 LBS |
| B CAR | 66,300 LBS |
| **CAR WEIGHT (LOADED)** | |
| A CAR (AW3) | 100,000 LBS |
| B CAR (AW3) | 94,300 LBS |
| **BUFF LOAD (AT ANTICLIMBER)** | |
| A&B CAR | 240,000 LBS |
| **SEATING CAPACITY** | |
| A CAR | 34 |
| B CAR | 40 |
| **STANDING CAPACITY** | |
| A CAR | 142 |
| B CAR | 148 |
| **TRUCK WEIGHT** | |
| MOTOR CAR W/PARKING BRAKE | 14,794 LBS |
| MOTOR CAR W/O PARKING BRAKE | 14,640 LBS |
| TRAILER CAR W/PARKING BRAKE | 10,490 LBS |

## CAR EQUIPMENT

| CAR EQUIPMENT | |
|---|---|
| **AIR SUPPLY SYSTEM & FRICTION BRAKING** | |
| AIR SUPPLY UNIT | WABCO, MODEL D-4-A |
| BRAKE SYSTEM TYPE | WABCO RT-5 |
| PARKING BRAKE | SAB WABCO |
| **AUXILIARY POWER SYSTEM** | |
| BATTERY | SAFT, 195AH (B CAR) |
| INTERMEDIATE VOLTAGE POWER SUPPLY | SEPSA, 320 VDC (25.2 KW) |
| LOW VOLT. POWER SUPPLY | SEPSA, 37.5 VDC (15 KW) |
| **CAR BODY** | |
| BODY | STAINLESS STEEL |
| BONNET | FIBERGLASS |
| FLOORING | PLYWOOD SANDWICHED BETWEEN STEEL SHEETS |
| ROOF | STAINLESS STEEL |
| CAB PARTITION | STAINLESS STEEL |
| **COUPLER ASSEMBLY** | |
| NO. 1 END | WABCO; HOOK, FLAT FACE |
| NO. 2 END | LINKBAR (A&B CAR) |
| ELECTRICAL PORTION | 46 PINS |
| **CAB & CREW OPERATION CONTROL** | |
| MASTER CONTROLLER | WABCO, SINGLE HANDLE |
| MASTER DOOR CONTROLLER | VAPOR |
| MONITORING SYS. - MDL/TOD SCREEN & CAR MONITORING UNIT | VAPOR, ON BOARD COMPUTER CONTROLLED |
| RADIO | BENDIX-KING |
| SIGNAL DEVICE | PNEUPHONIC HORN |
| SPEEDOMETER | BACH SIMPSON |
| EVENT RECORDER | BACH SIMPSON |

| CAR EQUIPMENT | |
|---|---|
| **CUSTOMER ENVIRONMENT** | |
| INFORMATION SIGNS | POCATEC |
| PASSENGER EMERGENCY INTERCOM | TELEPHONICS |
| COMMUNICATION SYSTEMS | TELEPHONICS |
| PASSENGER EMERGENCY HANDLE UNIT | WABCO |
| LIGHTING | LUMINATOR |
| **DOOR SYSTEMS** | |
| DOOR OPERATOR | VAPOR |
| DOOR PANEL | BOMBARDIER/MILUFAB |
| SIDE DOOR | 6 DOUBLE SLIDING |
| END DOOR | A-CAR #1 END: STORM, SWING<br>B-CAR AND A-CAR #2 END: BI-PARTING SLIDING |
| **HEATING, VENTILATING & AIR CONDITIONING (HVAC)** | |
| A/C CAPACITY | 2 BOMBARDIER ROOF-MOUNTED 6.5 TONS UNITS |
| HEATING | FORCED AIR O/H<br>CONVECTION FLOOR HEAT |
| VENTILATING | FORCED AIR |
| **PROPULSION SYSTEM & ELECTRIC BRAKING** | |
| INVERTER DRIVE | GEC ALSTOM AC TRACTION MOTORS |
| TRACTION MOTOR | GEC ALSTOM, MODEL 4LCA1640A, 147.5 HP./230V/ 400V, 3 PHASE, 4 POLE, 1750 RPM |
| **SIGNALING & TRAIN CONTROL SYSTEM** | |
| | FIXED BLOCK |
| **TRUCK** | |
| TRUCK TYPE | WELDED, FABRICATED STEEL FRAME |
| PRIMARY SUSPENSION | CHEVRON RUBBER |
| SECONDARY SUSPENSION | AIR BAG |
| CURRENT COLLECTOR | WABCO |

THE R142 NEW MILLENNIUM TRAINS STARTED NO. 2 LINE SERVICE ON 10/20/2000. B CAR HAS MOTOR AT NO. 2 END ONLY. CONSIST: 5 CAR UNIT (ABBBA), 10 CAR TRAIN

CAR NUMBERS: 7211 - 7610
TOTAL: R142A: 600
BUILT BY: KAWASAKI

OPTION CARS: 7611 - 7730; 7731 - 7810
DATE: 1999-2005
AVERAGE COST PER CAR: $1,215,466.66

# R142A (A DIVISION)

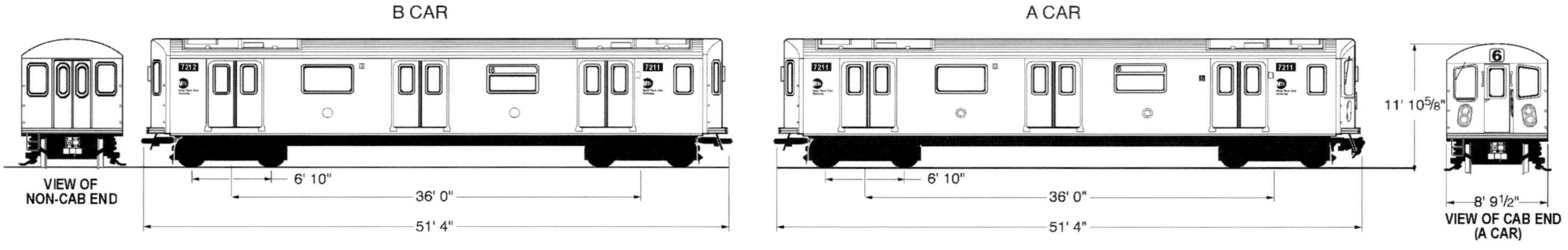

B CAR

A CAR

VIEW OF NON-CAB END

6' 10"

36' 0"

51' 4"

11' 10 5/8"

8' 9 1/2"

VIEW OF CAB END (A CAR)

Copyright © 2003 by NYC Transit

## DIMENSIONS

| | |
|---|---|
| DOORWAY HEIGHT (END) | 6'-3" |
| DOORWAY HEIGHT (SIDE) | 6'- 2.94" |
| DOORWAY WIDTH (CAB END-CLEAR OPENING) | 2' - 6 1/2" |
| DOORWAY WIDTH (SIDE-CLEAR OPENING) | 4' - 6" |
| FLOOR TO CEILING HT. (HIGH CEILING) | 7' - 2 1/2" |
| FLOOR TO CEILING HT. (LOW CEILING) | 6' - 7 1/2" |
| HEIGHT (TOP OF RAIL TO FLOOR) | 3' - 8 3/4" |
| TRACK GAUGE | 4' - 8 1/2" |
| WIDTH (OVER THRESHOLD) | 8' - 9.39" |
| **PERFORMANCE** | |
| ACCELERATION RATE | 2.5 MPHPS |
| BRAKING RATE (EMERGENCY) | 3.2 MPHPS |
| NOISE LEVEL - 40 MPH | |
| STATION-INTERIOR | 79 dBA |
| STATION-EXTERIOR | 88 dBA |
| ELEVATED TRACK STATIONARY | 60 dBA |
| RIDE QUALITY @ AW3; 40 MPH | VERTICAL 4.0 HR |
| | HORIZONTAL 2.5 HR |
| SERVICE SPEED (MAXIMUM) | 55 MPH |
| NOMINAL LINE VOLTAGE | 600 VDC |

## WEIGHT & CAPACITY

| | |
|---|---|
| **CAR WEIGHT (EMPTY)** | |
| A CAR | 73,300 LBS |
| B CAR | 67,800 LBS |
| **CAR WEIGHT (LOADED)** | |
| A CAR (182 CUSTOMERS) | 101,328 LBS |
| B CAR (182 CUSTOMERS) | 95,828 LBS |
| **BUFF LOAD (AT ANTICLIMBER)** | |
| A&B CAR | 200,000 LBS |
| **SEATING CAPACITY** | |
| A CAR | 34 |
| B CAR | 40 |
| **STANDING CAPACITY** | |
| A CAR | 142 |
| B CAR | 148 |
| **TRUCK WEIGHT** | |
| MOTOR CAR W/PARKING BRAKE | 14,900 LBS |
| TRAILER CAR W/PARKING BRAKE | 10,350 LBS |

## CAR EQUIPMENT

| | |
|---|---|
| **AIR SUPPLY SYSTEM & FRICTION BRAKING** | |
| AIR SUPPLY UNIT | WABCO, MODEL D-4-AS |
| BRAKE SYSTEM TYPE | WABCO RT-96 |
| PARKING BRAKE | WABCO BFCF |
| **AUXILIARY POWER SYSTEM** | |
| BATTERY | SAFT, 195 AH (B CAR) |
| AUXILIARY INVERTER POWER SUPPLY | TOSHIBA, 16KVA, 120 VAC, 60 HZ |
| LOW VOLT. POWER SUPPLY | TOSHIBA, 37.5 VDC (15 KW) |
| **CAR BODY** | |
| BODY | STAINLESS STEEL |
| BONNET | FIBERGLASS |
| FLOORING | PLYWOOD SANDWICHED BETWEEN STEEL SHEETS |
| ROOF | STAINLESS STEEL |
| CAB PARTITION | TRANSVERSE |
| **COUPLER ASSEMBLY** | |
| NO. 1 END | WABCO; HOOK, FLAT FACE |
| NO. 2 END | LINKBAR (A&B CAR) |
| ELECTRICAL PORTION | 46 PINS |
| **CAB & CREW OPERATION CONTROL** | |
| MASTER CONTROLLER | WABCO, SINGLE HANDLE |
| MASTER DOOR CONTROLLER | VAPOR |
| MONITORING SYS. - TOD SCREEN & CAR MONITORING SYSTEM | KOITO |
| RADIO | BENDIX-KING |
| SIGNAL DEVICE | PNEUPHONIC HORN |
| SPEEDOMETER | BACH SIMPSON |
| EVENT RECORDER | WABTEC |

| | |
|---|---|
| **CUSTOMER ENVIRONMENT** | |
| INFORMATION SIGNS | LUMINATOR |
| COMMUNICATION SYSTEMS | TELEPHONICS |
| PASSENGER EMERGENCY HANDLE UNIT | WABCO |
| LIGHTING | LUMINATOR |
| **DOOR SYSTEMS** | |
| DOOR OPERATOR | VAPOR |
| DOOR PANEL | KAWASAKI |
| SIDE DOOR | 6 DOUBLE SLIDING |
| END DOOR | A-CAR #1 END: STORM, SWING<br>B-CAR AND A-CAR #2 END: BI-PARTING SLIDING |
| **HEATING, VENTILATING & AIR CONDITIONING (HVAC)** | |
| A/C CAPACITY | 2 MELCO ROOF-MOUNTED 6.5 TONS UNITS |
| HEATING | FORCED AIR O/H<br>CONVECTION FLOOR HEAT |
| VENTILATING | FORCED AIR |
| **PROPULSION SYSTEM & ELECTRIC BRAKING** | |
| INVERTER DRIVE | BTPC AC TRACTION MOTOR INVERTERS |
| TRACTION MOTOR | BTPC, MODEL 1508C, 150 HP., 3 PHASE, 4 POLE |
| **SIGNALING & TRAIN CONTROL SYSTEM** | |
| | FIXED BLOCK |
| **TRUCK** | |
| TRUCK TYPE | WELDED, FABRICATED STEEL FRAME |
| PRIMARY SUSPENSION | RADIUS ARM |
| SECONDARY SUSPENSION | AIR BAG |
| CURRENT COLLECTOR | WABCO |

THE R142A NEW MILLENNIUM TRAINS STARTED NO. 6 LINE SERVICE ON 11/02/2000. B CAR HAS MOTOR AT NO. 1 END ONLY. CONSIST: 5 CAR UNIT (ABBBA), 10 CAR TRAIN

CAR NUMBERS: 8101 - 8312
TOTAL: 212 CARS
BUILT BY: KAWASAKI

DATE: 2001-2003
AVERAGE BID COST PER CAR: $1,576,475

# R143 (B DIVISION)

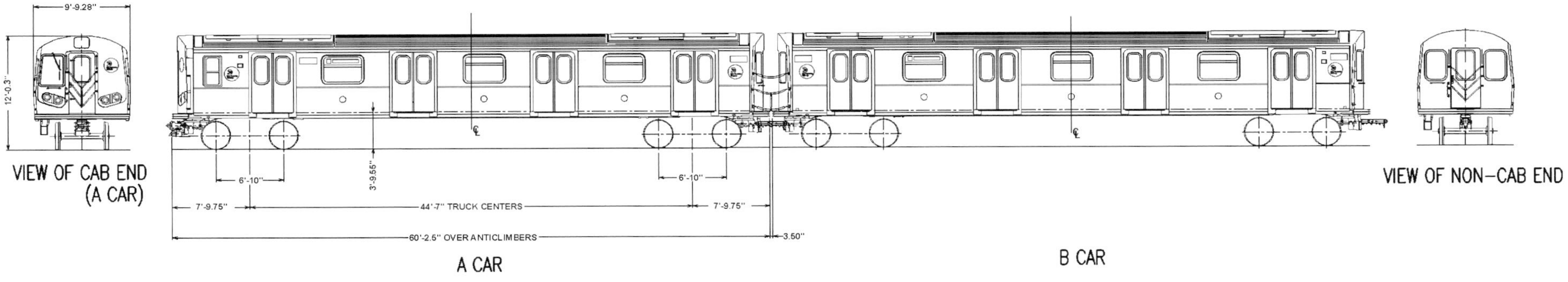

9'-9.28"
12'-0.3"
VIEW OF CAB END (A CAR)
6'-10"
3'-9.55"
6'-10"
7'-9.75"
44'-7" TRUCK CENTERS
7'-9.75"
60'-2.5" OVER ANTICLIMBERS
3.50"
A CAR
B CAR
VIEW OF NON-CAB END

Copyright © 2003 by NYC Transit

## DIMENSIONS

| | |
|---|---|
| DOORWAY HEIGHT (END) | 6'- 3" |
| DOORWAY HEIGHT (SIDE) | 6' - 3" |
| DOORWAY WIDTH (END-CLEAR OPENING) | 2' - 6" |
| DOORWAY WIDTH (SIDE-CLEAR OPENING) | 4' - 2" |
| FLOOR TO CEILING HT. (HIGH CEILING) | 7' - 2 1/2" |
| FLOOR TO CEILING HT. (LOW CEILING) | 6' - 7 1/2" |
| HEIGHT (TOP OF RAIL TO FLOOR) | 3' - 9.55" |
| TRACK GAUGE | 4' - 8 1/2" |
| WIDTH (OVER THRESHOLD) | 10' - 0" |
| **PERFORMANCE** | |
| ACCELERATION RATE | 2.5 MPHPS |
| BRAKING RATE (EMERGENCY) | 3.2 MPHPS |
| NOISE LEVEL - 40 MPH | |
| STATION-INTERIOR | 80 dBA |
| STATION-EXTERIOR | 80 dBA |
| ELEVATED TRACK STATIONARY | 65 dBA |
| RIDE QUALITY @ AW3; 40 MPH | VERTICAL 4.0 HR |
| | HORIZONTAL 2.5 HR |
| SERVICE SPEED (MAXIMUM) | 55 MPH |
| NOMINAL LINE VOLTAGE | 600 VDC |

## WEIGHT & CAPACITY

| | |
|---|---|
| **CAR WEIGHT (EMPTY)** | |
| A CAR | 83,700 LBS |
| B CAR | 81,900 LBS |
| **CAR WEIGHT (LOADED)** | |
| A CAR (240 CUSTOMERS) | 120,700 LBS |
| B CAR (240 CUSTOMERS) | 118,900 LBS |
| **BUFF LOAD (AT ANTI-CLIMBER)** | |
| A&B CAR | 200,000 LBS |
| **SEATING CAPACITY** | |
| A CAR | 42 |
| B CAR | 44 |
| **STANDING CAPACITY** | |
| A CAR | 198 |
| B CAR | 202 |
| **TRUCK WEIGHT** | |
| WEIGHT (AVG.) | 15,711 LBS |

## CAR EQUIPMENT

| | |
|---|---|
| **AIR SUPPLY SYSTEM AND FRICTION BRAKING** | |
| AIR SUPPLY UNIT | WABCO, MODEL D-4-AS |
| BRAKE SYSTEM TYPE | WABCO RT96 |
| PARKING BRAKE | WABCO |
| **AUXILIARY POWER SYSTEM** | |
| BATTERY | SAFT, 250AH (B CAR) |
| AUXILIARY INVERTER | TOSHIBA, 13 KVA, 120V 60Hz |
| LOW VOLT. POWER SUPPLY | TOSHIBA, 37.5 VDC (18 KW) |
| **CAR BODY** | |
| BODY | STAINLESS STEEL |
| BONNET | FIBERGLASS |
| FLOORING | PLYWOOD SANDWICHED BETWEEN STEEL SHEETS |
| ROOF | STAINLESS STEEL |
| CAB PARTITION | TRANSVERSE |
| **COUPLER ASSEMBLY** | |
| NO. 1 END | WABCO; HOOK, FLAT FACE |
| NO. 2 END | LINKBAR (A&B CAR) |
| ELECTRICAL PORTION | 70 PINS |
| **CAB AND CREW OPERATION CONTROL** | |
| MASTER CONTROLLER | WABCO, SINGLE HANDLE |
| MASTER DOOR CONTROLLER | VAPOR |
| MONITORING SYS. - TLM | KOITO |
| SCREEN & CAR MONITORING UNIT | |
| RADIO | BENDIX-KING |
| SIGNAL DEVICE | PNEUPHONIC HORN |
| SPEEDOMETER | BACH-SIMPSON |
| EVENT RECORDER | WABTEC |

| | |
|---|---|
| **CUSTOMER ENVIRONMENT** | |
| INFORMATION SIGNS | SIDE INTERIOR INFORMATION SIGN, TELECITE |
| | CEILING INTERIOR INFORMATION SIGN, TELECITE |
| | STRIP MAP & SIDE DESTINATION, LUMINATOR |
| | END ROUTE SIGN, LUMINATOR |
| PASSENGER EMERGENCY INTERCOM | TELEPHONICS |
| COMMUNICATION SYSTEMS | AUTOMATIC ANNOUNCEMENT SYSTEM, TELEPHONICS |
| PASSENGER EMERGENCY HANDLE UNIT | WABCO |
| LIGHTING | LUMINATOR |
| **DOOR SYSTEMS** | |
| DOOR OPERATOR | VAPOR |
| DOOR PANEL | KAWASAKI |
| SIDE DOOR | 8 DOUBLE SLIDING |
| END DOOR | A-CAR #1 END: STORM, SWING |
| | B-CAR AND A-CAR #2 END: SINGLE SLIDING |
| **HEATING, VENTILATING & AIR CONDITIONING (HVAC)** | |
| A/C CAPACITY | 2 MELCO ROOF-MOUNTED 7.5 TONS UNITS |
| HEATING | FORCED AIR |
| | CONVECTION FLOOR HEAT |
| VENTILATING | FORCED AIR |
| INVERTER DRIVE | BOMBARDIER POWERS AC TRACTION MOTORS |
| TRACTION MOTOR | BOMBARDIER, MODEL 1508C, 150 HP., 3 PHASE, 4 POLE |
| **SIGNALING & TRAIN CONTROL SYSTEM** | |
| | FIXED BLOCK - CBTC READY |
| **TRUCK** | |
| TRUCK TYPE | WELDED, FABRICATED STEEL FRAME |
| PRIMARY SUSPENSION | RADIUS ARM |
| SECONDARY SUSPENSION | AIR BAG |
| CURRENT COLLECTOR | WABCO |

THE R143 NEW MILLENNIUM TRAINS STARTED SERVICE ON 2/12/2002. ALL AXLES ARE MOTORIZED. CONSIST: 4 CAR UNIT (ABBA), 8 CAR TRAIN

CAR NUMBERS: 8313-8972
TOTAL: 660 CARS
BUILT BY: ALSTOM TRANSPORTATION, INC. & KAWASAKI RAIL CAR INC.

DATE: 2005 - 2008
AVERAGE COST FOR CAR IN 4 CAR UNIT: $1,319,589
AVERAGE COST FOR CAR IN 5 CAR UNIT: $1,308,342

# R160 (B DIVISION)

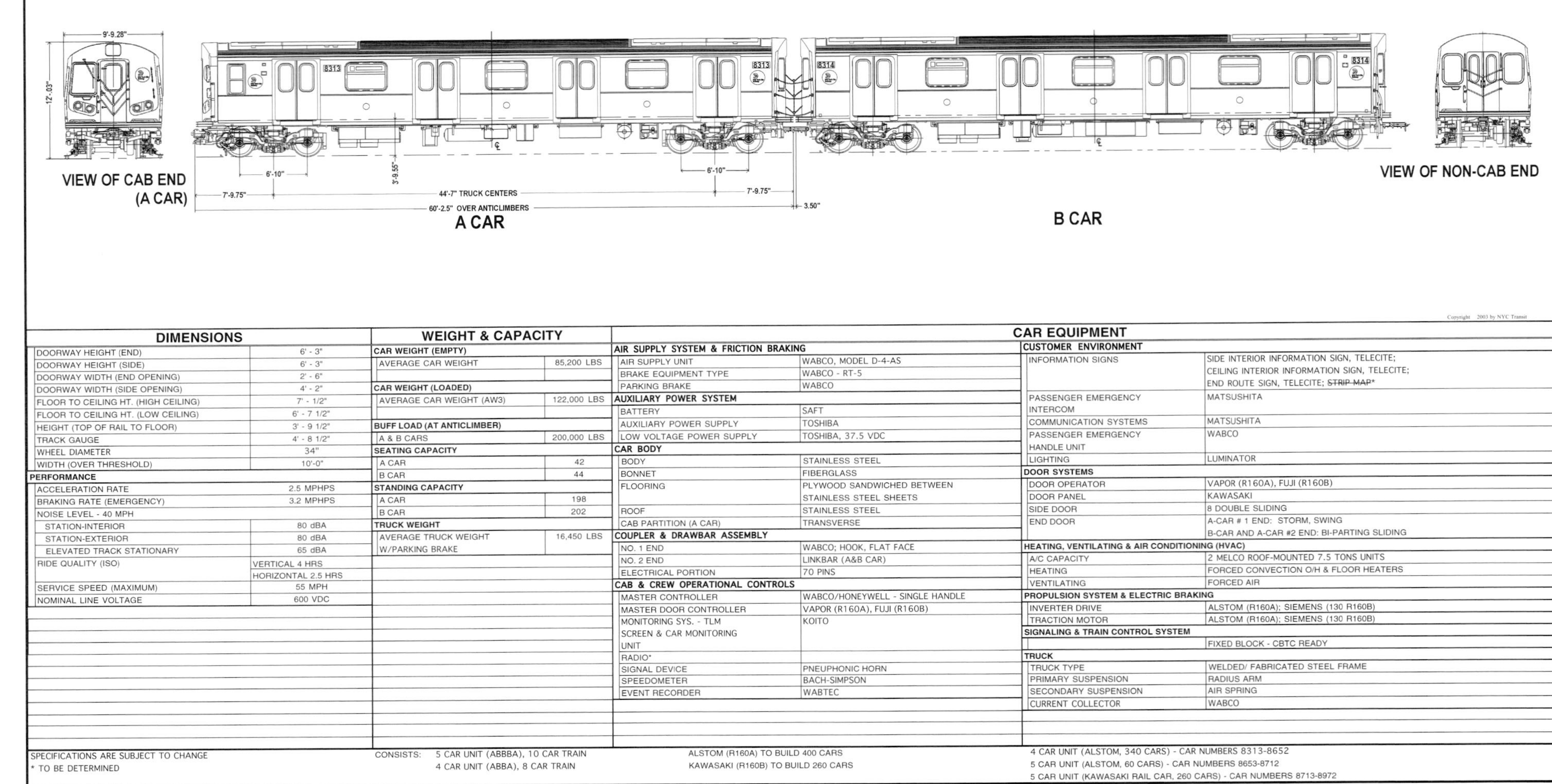


Copyright 2003 by NYC Transit

## DIMENSIONS

| | |
|---|---|
| DOORWAY HEIGHT (END) | 6' - 3" |
| DOORWAY HEIGHT (SIDE) | 6' - 3" |
| DOORWAY WIDTH (END OPENING) | 2' - 6" |
| DOORWAY WIDTH (SIDE OPENING) | 4' - 2" |
| FLOOR TO CEILING HT. (HIGH CEILING) | 7' - 1/2" |
| FLOOR TO CEILING HT. (LOW CEILING) | 6' - 7 1/2" |
| HEIGHT (TOP OF RAIL TO FLOOR) | 3' - 9 1/2" |
| TRACK GAUGE | 4' - 8 1/2" |
| WHEEL DIAMETER | 34" |
| WIDTH (OVER THRESHOLD) | 10'-0" |
| **PERFORMANCE** | |
| ACCELERATION RATE | 2.5 MPHPS |
| BRAKING RATE (EMERGENCY) | 3.2 MPHPS |
| NOISE LEVEL - 40 MPH | |
| STATION-INTERIOR | 80 dBA |
| STATION-EXTERIOR | 80 dBA |
| ELEVATED TRACK STATIONARY | 65 dBA |
| RIDE QUALITY (ISO) | VERTICAL 4 HRS<br>HORIZONTAL 2.5 HRS |
| SERVICE SPEED (MAXIMUM) | 55 MPH |
| NOMINAL LINE VOLTAGE | 600 VDC |

## WEIGHT & CAPACITY

| | |
|---|---|
| **CAR WEIGHT (EMPTY)** | |
| AVERAGE CAR WEIGHT | 85,200 LBS |
| **CAR WEIGHT (LOADED)** | |
| AVERAGE CAR WEIGHT (AW3) | 122,000 LBS |
| **BUFF LOAD (AT ANTICLIMBER)** | |
| A & B CARS | 200,000 LBS |
| **SEATING CAPACITY** | |
| A CAR | 42 |
| B CAR | 44 |
| **STANDING CAPACITY** | |
| A CAR | 198 |
| B CAR | 202 |
| **TRUCK WEIGHT** | |
| AVERAGE TRUCK WEIGHT W/PARKING BRAKE | 16,450 LBS |

## CAR EQUIPMENT

| | |
|---|---|
| **AIR SUPPLY SYSTEM & FRICTION BRAKING** | |
| AIR SUPPLY UNIT | WABCO, MODEL D-4-AS |
| BRAKE EQUIPMENT TYPE | WABCO - RT-5 |
| PARKING BRAKE | WABCO |
| **AUXILIARY POWER SYSTEM** | |
| BATTERY | SAFT |
| AUXILIARY POWER SUPPLY | TOSHIBA |
| LOW VOLTAGE POWER SUPPLY | TOSHIBA, 37.5 VDC |
| **CAR BODY** | |
| BODY | STAINLESS STEEL |
| BONNET | FIBERGLASS |
| FLOORING | PLYWOOD SANDWICHED BETWEEN STAINLESS STEEL SHEETS |
| ROOF | STAINLESS STEEL |
| CAB PARTITION (A CAR) | TRANSVERSE |
| **COUPLER & DRAWBAR ASSEMBLY** | |
| NO. 1 END | WABCO; HOOK, FLAT FACE |
| NO. 2 END | LINKBAR (A&B CAR) |
| ELECTRICAL PORTION | 70 PINS |
| **CAB & CREW OPERATIONAL CONTROLS** | |
| MASTER CONTROLLER | WABCO/HONEYWELL - SINGLE HANDLE |
| MASTER DOOR CONTROLLER | VAPOR (R160A), FUJI (R160B) |
| MONITORING SYS. - TLM SCREEN & CAR MONITORING UNIT | KOITO |
| RADIO* | |
| SIGNAL DEVICE | PNEUPHONIC HORN |
| SPEEDOMETER | BACH-SIMPSON |
| EVENT RECORDER | WABTEC |

| | |
|---|---|
| **CUSTOMER ENVIRONMENT** | |
| INFORMATION SIGNS | SIDE INTERIOR INFORMATION SIGN, TELECITE; CEILING INTERIOR INFORMATION SIGN, TELECITE; END ROUTE SIGN, TELECITE; ~~STRIP MAP~~* |
| PASSENGER EMERGENCY INTERCOM | MATSUSHITA |
| COMMUNICATION SYSTEMS | MATSUSHITA |
| PASSENGER EMERGENCY HANDLE UNIT | WABCO |
| LIGHTING | LUMINATOR |
| **DOOR SYSTEMS** | |
| DOOR OPERATOR | VAPOR (R160A), FUJI (R160B) |
| DOOR PANEL | KAWASAKI |
| SIDE DOOR | 8 DOUBLE SLIDING |
| END DOOR | A-CAR # 1 END: STORM, SWING<br>B-CAR AND A-CAR #2 END: BI-PARTING SLIDING |
| **HEATING, VENTILATING & AIR CONDITIONING (HVAC)** | |
| A/C CAPACITY | 2 MELCO ROOF-MOUNTED 7.5 TONS UNITS |
| HEATING | FORCED CONVECTION O/H & FLOOR HEATERS |
| VENTILATING | FORCED AIR |
| **PROPULSION SYSTEM & ELECTRIC BRAKING** | |
| INVERTER DRIVE | ALSTOM (R160A); SIEMENS (130 R160B) |
| TRACTION MOTOR | ALSTOM (R160A); SIEMENS (130 R160B) |
| **SIGNALING & TRAIN CONTROL SYSTEM** | |
| | FIXED BLOCK - CBTC READY |
| **TRUCK** | |
| TRUCK TYPE | WELDED/ FABRICATED STEEL FRAME |
| PRIMARY SUSPENSION | RADIUS ARM |
| SECONDARY SUSPENSION | AIR SPRING |
| CURRENT COLLECTOR | WABCO |

SPECIFICATIONS ARE SUBJECT TO CHANGE
* TO BE DETERMINED

CONSISTS: 5 CAR UNIT (ABBBA), 10 CAR TRAIN
4 CAR UNIT (ABBA), 8 CAR TRAIN

ALSTOM (R160A) TO BUILD 400 CARS
KAWASAKI (R160B) TO BUILD 260 CARS

4 CAR UNIT (ALSTOM, 340 CARS) - CAR NUMBERS 8313-8652
5 CAR UNIT (ALSTOM, 60 CARS) - CAR NUMBERS 8653-8712
5 CAR UNIT (KAWASAKI RAIL CAR, 260 CARS) - CAR NUMBERS 8713-8972